Stochastic Systems: The Mathematics of Filtering and Identification and Applications

NATO ADVANCED STUDY INSTITUTES SERIES

Proceedings of the Advanced Study Institute Programme, which aims at the dissemination of advanced knowledge and the formation of contacts among scientists from different countries

The series is published by an international board of publishers in conjunction with NATO Scientific Affairs Division

A	Life Sciences	Plenum Publishing Corporation
B	Physics	London and New York
C	Mathematical and Physical Sciences	D. Reidel Publishing Company Dordrecht, Boston and London
D	Behavioural and Social Sciences	Sijthoff & Noordhoff International Publishers
E	Applied Sciences	Alphen aan den Rijn and Germantown U.S.A.

Series C – Mathematical and Physical Sciences

Volume 78 – Stochastic Systems: The Mathematics of Filtering and Identification and Applications

Stochastic Systems: The Mathematics of Filtering and Identification and Applications

Proceedings of the NATO Advanced Study Institute held at Les Arcs, Savoie, France, June 22 - July 5, 1980

edited by

MICHIEL HAZEWINKEL

Erasmus Universiteit, Rotterdam, The Netherlands

and

JAN C. WILLEMS

Rijksuniversiteit, Groningen, The Netherlands

D. Reidel Publishing Company

Dordrecht : Holland / Boston : U.S.A. / London : England

Published in cooperation with NATO Scientific Affairs Division

Library of Congress Cataloging in Publication Data

CIP

NATO Advanced Study Institute (1980 : Savoie, France)
Stochastic systems: The mathematics of filtering and identification and applications.

(NATO advanced study institutes series. Series C, Mathematical and physical sciences ; v. 78)
Includes index.
1. Stochastic systems–Congresses. 2. Filters (Mathematics)–Congresses. 3. System identification–Congresses. I. Hazewinkel, Michiel. II. Willems, Jan C. III. North Atlantic Treaty Organization. Division of Scientific Affairs. IV. Title. V. Series.
QA402.N33 1980 003 81-11944
ISBN 90-277-1330-8 AACR2

Published by D. Reidel Publishing Company
P.O. Box 17, 3300 AA Dordrecht, Holland

Sold and distributed in the U.S.A. and Canada
by Kluwer Boston Inc.,
190 Old Derby Street, Hingham, MA 02043, U.S.A.

In all other countries, sold and distributed
by Kluwer Academic Publishers Group,
P.O. Box 322, 3300 AH Dordrecht, Holland

D. Reidel Publishing Company is a member of the Kluwer Group

Printed in The Netherlands

TABLE OF CONTENTS

**) Invited main series of lectures

*) Invited lectures

PREFACE

In the last five years or so there has been an important renaissance in the area of (mathematical) modeling, identification and (stochastic) control. It was the purpose of the Advanced Study Institute of which the present volume constitutes the proceedings to review recent developments in this area with particular emphasis on <u>identification</u> and <u>filtering</u> and to do so in such a manner that the material is accessible to a wide variety of both embryo scientists and the various breeds of established researchers to whom identification, filtering, etc. are important (such as control engineers, time series analysts, econometricians, probabilists, mathematical geologists, and various kinds of pure and applied mathematicians; all of these were represented at the ASI).

For these proceedings we have taken particular care to see to it that the material presented will be understandable for a quite diverse audience. To that end we have added a fifth tutorial section (besides the four presented at the meeting) and have also included an extensive introduction which explains in detail the main problem areas and themes of these proceedings and which outlines how the various contributions fit together to form a coherent, integrated whole. The prerequisites needed to understand the material in this volume are modest and most graduate students in e.g. mathematical systems theory, applied mathematics, econometrics or control engineering will qualify. And if one finds that one does not have the prerequisites necessary to understand the prerequisites (to quote Paul Halmos) there is no need to get discouraged; much can be done (particularly in this field) by means of osmosis or diffusion coupled with the ability to believe (results of others).

The mathematical tools and results used in the area of mathematical systems theory centering around filtering and identification have evolved considerably in the recent past. They do not remain restricted to the obviously relevant parts of statistics, probability and stochastic differential equations but also involve techniques, concepts and results from such fields as topology and geometry of manifolds, Lie algebras (or differential operators), Lie groups, functional analysis (evolution semigroups),

M. Hazewinkel and J. C. Willems (eds.), Stochastic Systems: The Mathematics of Filtering and Identification and Applications, ix–xi.

functional integration (Feynman-Kac formulae, path integrals), differential geometry (curvature), quantum field theory (Heisenberg commutation relations, Stone-v. Neumann theorem, stochastic quantization, stochastic mechanics) and in the not too distant future probably such topics as (harmonic) analysis on nil- and solv-manifolds, theta functions, ...

As indicated above there are diverse groups of scientists to whom it is of importance to become acquainted woth the various mathematical tools available and above all to see how they work in more or less applied situations.

These groups do not always communicate perfectly with each other and we hope that these proceedings will serve as a forum for a dialogue and facilitate interspecialist appreciation and understanding. To this end we aimed to present and discuss the new trends in filtering and identification in an integrated manner together with the underlying mathematics and its applications (both actual and potential) and in relation with the mathematical problems thrown up or suggested by the various engineering and time series approaches to filtering, identification, adaptive control and modeling.

We leave to the reader to judge whether these proceedings fulfill these noble intentions.

In any case the NATO division of scientific affairs agreed with us that it was definitely worth trying and supplied a substantial grant from their ASI program which put the meeting on a sound financial basis. It is a pleasure to record here our indebtedness to Dr T. Kester of International Transfer of Science and Technology and Dr M. diLullo of the NATO Scientific Affairs division for their help, advice, and encouragement also in the earlier planning stages of this ASI.

In addition we gave a rather large number of big industrial corporations and other organizations the opportunity to contribute something financial towards the success of the conference. Most resolutely refused to help in any way (a list is available on request) but it is pleasure to mention the favourable exceptions: ESSO Nederland BV which supplied support which enabled us to assist a number of participants from Non-NATO countries; NSF which supplied the air ticket for one junior participant fron the USA and Shell Nederland BV which donated full expenses for three junior participants from The Netherlands. In addition North Holland Publ. Co., Prentice Hall and especially D. Reidel Publ. Co. and Pitman Publ. Co. helped us with a modest book display, thus making the ASI more attractive.

Finally it remains for us to thank the many people in our respective institutes and the staff at Les Arcs for their manifold efforts towards the success of the meeting.

In addition to the tutorial and invited papers (marked with * or ** in the table of contents) these proceedings contain a number of contributed papers. Quite apart from the intrinsic merits of these papers their selection mainly reflects the idiosyncratic tastes of the editors, though we also believe that they fit the main themes of the conference well and together with the invited papers give a very fair picture of the emerging new trends in modeling, identification and adaptive control and linear and nonlinear filtering. This last topic, practically a new emerging field, which seems to have an unusually bright immediate future, has especially received much attention both because of the wealth of recent fundamental results and its glittering prospects.

LEITFADEN

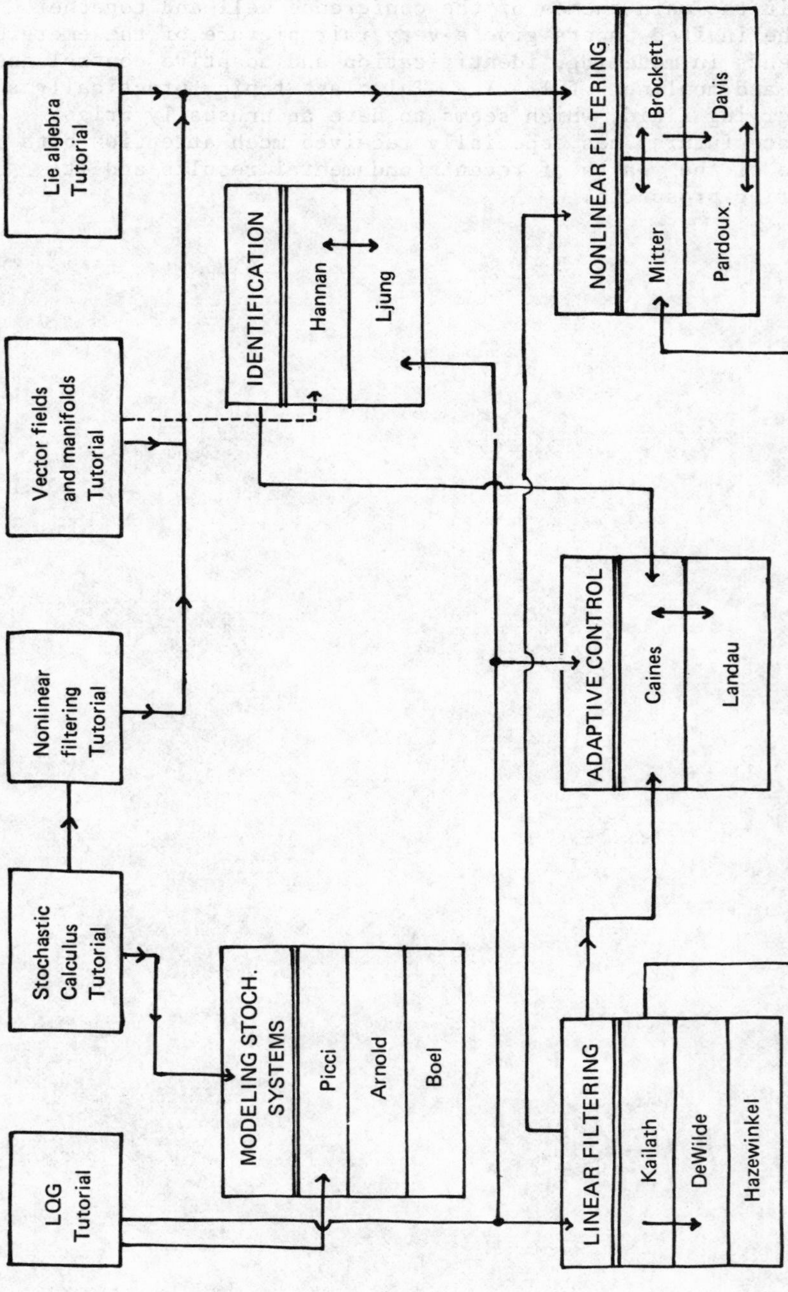

Lie algebra Tutorial
Vector fields and manifolds Tutorial
Nonlinear filtering Tutorial
Stochastic Calculus Tutorial
LQG Tutorial
IDENTIFICATION
Hannan
Ljung
MODELING STOCH. SYSTEMS
Picci
Arnold
Boel
NONLINEAR FILTERING
Brockett
Davis
Mitter
Pardoux
ADAPTIVE CONTROL
Caines
Landau
LINEAR FILTERING
Kailath
DeWilde
Hazewinkel

The purpose of the present
course is the deepening and
development of difficulties
underlying contemporary theory
.....

A.A. Blasov

Part 1

THE MAIN THEMES OF THE ASI "STOCHASTIC SYSTEMS: THE MATHEMATICS OF FILTERING AND IDENTIFICATION AND APPLICATIONS"

INTRODUCTION: AN ANNOTATED NAVIGATION CHART AND SKETCH OF THE MAIN THEMES IN THIS VOLUME AND THEIR INTERRELATIONS.

Michiel Hazewinkel
Dept. Math., Erasmus Univ.
Rotterdam,
P.O. Box 1738,
3000 DR Rotterdam
The Netherlands

Jan C. Willems
Dept. Math., Univ. of Groningen
P.O. Box 800,
9700 AV Groningen
The Netherlands

This volume is devoted to modeling and analysis of uncertain dynamical systems in an uncertain environment and the synthesis of filters, identifiers and adaptive controllers in such a setting. All this with particular emphasis on recursive (and/or on-line) techniques.

This is a large and varied field of inquiry.It was the intention of the conference,of which this volume constitutes the proceedings,to review the most important themes and new developments in a coherent manner without making too many demands on the audience in the matter of prerequisites.

As a result this volume contains tutorial material, reviews and surveys, as well as research papers on the topics of modeling, adaptive control, identification and filtering and applications. The present introduction is intended to provide an informal outline of the main themes of the volume: identification and filtering and recursiveness, and to indicate how the various contributions fit together. That is, it is essentially an (annotated) navigation chart. We have concentrated mostly on the

M. Hazewinkel and J. C. Willems (eds.), Stochastic Systems: The Mathematics of Filtering and Identification and Applications, 3–26.

tutorial and the invited survey-and-state-of-the-art papers (marked with a *) or **) in the table of contents).

1. THE SETTING AND THE BASIC THEMES.

An uncertain dynamical system may be defined as a map F from an input space $\mathcal{U}$ (which is a family of maps from the time axis $T \subset \mathbb{R}$ to the space of input values U) and an uncertainty space N to an output space $\mathcal{Y}$ (which is a family of maps from T to the space of output values Y) which is nonanticipating, that is to say that for all values of the uncertainty parameter n the output y is independent of future values of the input u. The uncertain system under consideration is often called the plant and is depicted by the following signal flow diagram (fig. 1):

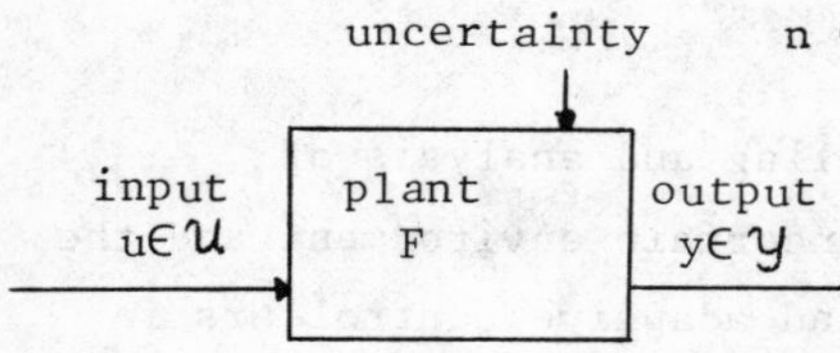

figure 1

We think of the inputs as variables which can be manipulated (controls) or, more generally, through which the environment can influence the system; we think of the outputs as variables which can be measured (observations) or, more generally, through which the system can influence the environment. The uncertainty reflects the fact that the dynamic behaviour is unknown (for example because the numerical value of a parameter is not known) or that it may depend on a stochastic phenomenon. We think of the uncertainty as a parameter n being choosen by 'nature'.

For the purpose of the contributions in this volume it is in fact insightful to assume that the uncertainty space N is a product space, $N = P \times R$ with P a set of unknown parameters and R the outcome space of a general random variable. Formally, there is a probability space $\{\Omega, A, P\}$ and a map $\tau : \Omega \to R$ which selects the value in R in a random fashion. Finding out from measurements the actual value (or 'best' approximation) of the unknown parameter $p \in P$ is the basic problem in system identification while finding the actual value (or 'best' estimate) of the random parameter (vector) $r \in R$ is (indirectly) the basic problem in filtering. Preferably one wants to do this in a recursive manner that is, roughly, by a technique which updates a 'state-type' parameter vector η_t by means of the new information gathered at time t while the desired unknown parameters $p \in P$ or $r \in R$ are calculated as (known) functions of η_t. The vector η_t, so to speak, embodies or codifies all the useful information gathered up to time t.

2. MODELING ISSUES.

The study of stochastic dynamic systems brings with it the problem of modeling, particularly if one wants to use differential equation models. The reason why one wants to use such models is, as in the deterministic case, connected with the fact that one much prefers, for good (computational) reasons and also from a basic mathematical point of view to use <u>recursive models</u>, that is, models which display the <u>state</u> of the system explicitely. In a stochastic framework the idea of state leads to modeling in terms of a Markov process (since in general there is also an input we should really think of a controlled Markov process). Writing down the evolution of a Markov process leads to differential equations with a white noise term on the right hand side and the rigorous interpretation of such equations leads to Itô calculus.

An <u>Itô equation</u> is a differential equation of the form

$$(1) \qquad dx = f(x)dt + g(x)dw \qquad x(t_o) = x_o$$

with $x \in \mathbb{R}^n$, $f: \mathbb{R}^n \to \mathbb{R}^n$, $g: \mathbb{R}^n \to \mathbb{R}^{nxm}$ (the nxm-matrices), w an $\mathbb{R}^m$-\ .ued Wiener stochastic process, and $x_o \in \mathbb{R}^n$ a random vector. Assume that w and x_o are defined on the probability space (Ω,A,P). The above equation can be thought of intuitively as the equation

$$\dot{x} = f(x) + g(x)\frac{dw}{dt} \qquad x(t_o) = x_o$$

(at least as long as w_1 is scalar valued) with $\frac{dw}{dt}$ 'white noise'. This, however, is not a process defined in the convential way. The rigorous interpretation of equation (1) is made in terms Itô calculus and is the subject of CURTAIN's tutorial [section 2.2 in this volume]. Under suitable assumptions, explained in the tutorial, (1) yields a well-defined Markov process x. We may add inputs and outputs to (1) which leads to the usual form of a stochastic differential system given by:

$$(2) \quad \begin{aligned} dx &= f(x,u(t))dt + g(x,u(t))dw & x(t_o) &= x_o \\ dy &= g(x)dt + dv & y(t_o) &= 0 \end{aligned}$$

where v is a stochastic Wiener process assumed to be independent of w. The noises w and v are respectively called the system noise and the output noise. (Problems where the system noise w and the output noise v are dependent are of interest but are usually not given much attention ; cf., however, e.g. section 7.3 in this volume; this introduces fundamental extra difficulties). Model (2) leads then to an uncertain dynamical system of the type informally discussed in section 1, with uncertainty random variables.

Two 'case studies' of modeling of stochastic systems are contained in part 3 of this volume, both taken from areas where

there is a great deal of recent activity in applied mathematics. The first one of these contributions is by BOEL [section 3.3] and describes how one may set up stochastic models of computer networks. The models proposed are in terms of queues and contrary to (1) involve mainly Poisson noise. An interesting feature in the analysis of these models is the important role played by 'quasi-(time) reversibility'.
The second paper about modeling is by ARNOLD [section 3.1] and treats chemical reactions. Such reactions show irregularities in space and one can consider the local behaviour versus the global behaviour where one expects to be able to derive some type of space average behaviour. Chemical reactions also have a stochastic feature due to the fact that particles react when they 'meet' which is modeled as a random phenomenon. The purpose of ARNOLD's paper is to demonstrate how global deterministic models may be viewed as suitable limits of global stochastic models or of local deterministic models both of which may in turn be viewed as a limit of a local stochastic model.

One of the important issues in mathematical control theory is the realization theory problem. This means essentially the realization (or modeling) of a given input/output operator by means of a 'machine' of type (2). It also means the construction of a stochastic process of a certain type with a pregiven covariance function.

As we have already mentioned many applications (in fact most of those discussed in this volume as Kalman filtering and nonlinear filtering) need, in order to carry out the required calculations, a model in state space form. Often, one starts with a model in input/output form - some model of the type introduced in section 1 - and the question then arises how to construct an equivalent state space model. In the context of random processes, this problem becomes the following: Let $y(t)$, $t \in T \subset \mathbb{R}$ be a given stochastic process with outcome space Y. The problem then is to construct a space X, a Markov process $x(t)$, $t \in T \subset \mathbb{R}$ with outcome space X, and a map $h: X \to Y$ such that $h(x(t))$ is in some sense equivalent to the original process $y(t)$. In the paper by LINDQUIST and PICCI the realization theory for multivariate stationary gaussian stochastic processes is presented.

In addition part 3 of this volume contains two papers on the more qualitative properties of stochastic differential equations

$$\dot{x}(t) = f(x(t),\xi(t)),\ x(0) = x_o$$

$$\dot{x}(t) = f(x(t),\ \xi(t)),\ x(0) = x_o$$

with random initial condition x_o and $\xi(t)$ a random process. Here solutions are to be interpreted pathwise, i.e. this equation is really a collection of equations, one for each noise trajectory (and initial condition).

The paper by Arnold [section 3.2] is a survey in extended abstract form of problems, concepts and results of the qualitative theory of such equations. Qualitative concepts include such things as stationary solutions, attractors, stability and ergodicity. This last topic is the subject of the paper by Wihstutz. Obviously something like ergodicity for instance is of relevance when discussing the compatibility between local (micro) stochastic models and global (average) deterministic models. Think of statistical mechanics.

3. NONLINEAR FILTERING.

The filtering problem takes up by far the largest part of this volume. In abstract terms the filtering problem is a stochastic version of an obtimal observer design problem. Take an uncertain plant as introduced in section 1, and make, to simplify the discussion, the (inessential) restriction that there are no inputs. Assume furthermore that there are two types of outputs: one output, which we will denote by y, which is a signal which can be measured - <u>the observations</u> - and another output, which we will denote by z, which is a signal which we would like to know - the <u>to-be-estimated output</u>. These outputs take on their values in a space Z; often z = x, the state of the plant processor, which accepts as inputs the observations y and produces as outputs estimates $\hat{z}$ of z. Formally we have a plant (F_y, F_z): $N \to Y \times Z$ and we wish to construct a nonanticipating map K: $Y \to Z$ such that, in some sense, $\hat{z} = Ky = K_y^F(n)$ is close to $z = F_z(n)$ (see Figure 2). Expressing

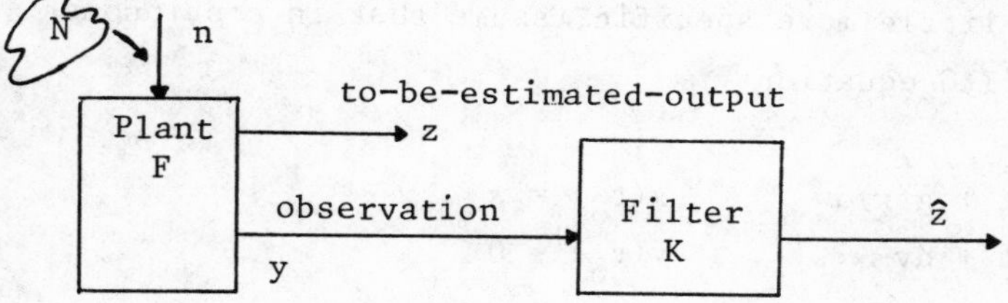

figure 2

'being close to' in terms of a loss functional and assuming the uncertainty to be a random variable it is natural to express this problem in terms of the minimization of the average loss $E\{d(z,\hat{z})\}$. It is furthermore clear that one can formulate this minimization for all times $t \in T$ which leads to the problem of finding, for all t, a K_t: $\underline{Y} \to Z$ which minimizes $E\{d(z(t))\}$, where d is an appropriate distance function. Now, since one wants to obtain this estimate $\hat{z}(t)$ for all t, it is very natural and advantageous to attempt to do this computation recursively. This is done by trying to find a 'state' of the observer s such that the computation of $\hat{z}$ may be carried out according to the diagram:

$$\frac{y(\tau)}{t_o < t < t'} \longrightarrow s(t') \qquad \frac{y(\tau)}{t' < \tau < t''} \longrightarrow s(t'')$$

$$\downarrow \qquad\qquad\qquad\qquad\qquad \downarrow$$

$$\hat{z}(t') \qquad\qquad\qquad\qquad\qquad \hat{z}(t'')$$

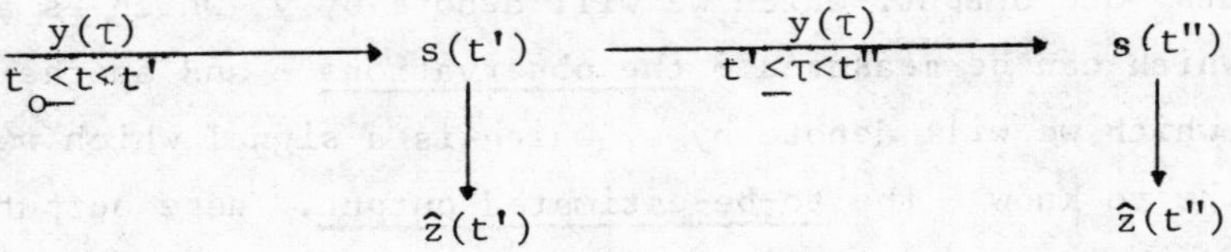

Storing $s(t')$ instead of $y(\tau)$ for $t_o \leq \tau < t'$ will hopefully also lead to an automatic data reduction which could be very advantageous from the point of view of computational complexity and memory storage requirements.

Let us be a little more specific. Assume that in continuous time we have the Itô equation

$$\begin{array}{ll} dx = f(x)dt + g(x)dw & x(t_o) = x_o \\ (3) \quad dy = h(x)dt + dv & y(t_o) = 0 \\ z = k(x) & \end{array}$$

with $x \in X := \mathbb{R}^n$, $y \in \mathbb{R}^p$, $z \in \mathbb{R}^q$, and v and w mutually independent Wiener processes and independent of the initial randomness $x_o \in \mathbb{R}^n$. Assume that we want to obtain the best estimate in the quadratic sense of $z(t)$ based on observations $y(\tau)$ for $t_o \leq \tau \leq t$. This is the filtering problem. The prediction problem asks for the best estimate of $z(s)$ given $y_\tau, t_o \leq \tau \leq t$, $t < s$ and the smoothing problem asks for the best estimate of $z(s)$ given $t_o \leq \tau \leq t$, $s < t$, i.e. given also future observations. It is wellknown that the conditional expectation $z^*(t) := E\{z(t)|y(\tau), t_o \leq \tau \leq t\}$ is the best estimate in the least squares sense, i.e. it minimizes every quadratic loss of the form $E\{||z(t)-\hat{z}(t)||^2\}$. The filtering problem is then to give a (recursive) algorithm for computing this conditional expectation.

Because of the special structure of the system (3), in particular, because of the Markov property of x, it follows that the conditional distribution $\pi_t := p(x(t)\,|\,y(\tau),\ t_o \le \tau \le t)$ can act as a state for the filter. That is to say that there exists an update equation of the type

$$(4)\qquad \begin{aligned} d\pi &= A_1(\pi)dt + B_1(\pi)dy(t) \\ z^*(t) &= \int_X k(x)\pi(t)dx \end{aligned}$$

with $\pi(t_o)$ = the distribution of x_o. Since $x \in \mathbb{R}^n$, π is a function on $\mathbb{R}^n$ and hence one may expect that (4) will be a type of partial differential equation. In fact A_1 and B_1 are integro-differential operators on X.

In the tutorial article of DAVIS & MARCUS [section 2.3] this equation and the rigorous derivation of it is discussed together with the role of the so-called Duncan-Mortensen-Zakai-equation which is an unnormalized version of (4). That is, instead of having an update equation for $\pi(t)$, the D-M-Z equation computes a function $\rho(t)$ with the property that $\pi(t)$ is related to it by a simple formula of the type

$$\pi(t) = \frac{\rho(t)}{\int_X \rho(t)dx}$$

Working with $\rho(t)$ has certain advantages: ρ satisfies a much simpler looking equation than π. The equation of ρ is a stochastic partial differential equation:

$$(5)\qquad d\rho = A_2\rho\, dt + h\rho\, dy(t)$$

from which $z^*(t)$ is calculated by means of the (output) map

$$(6)\qquad z^*(t) = \left(\int_X \rho(t)dx\right)^{-1} \int_X k(x)\rho(t)dx$$

Here A_2 is a suitable linear differential operator defined in DAVIS-MARCUS [section 2.3]. This is a <u>bilinear</u> equation in the sense that ρ satisfies a linear equation in which the driving term is a linear function of the 'input' y.

This bilinear structure of the Zakai-equation is very much exploited by BROCKETT [section 7.1] in his expository article in which he explains the geometric structure of the Zakai-equation, with an eye towards finding conditions for the existence of finite-dimensional filters.

The issue of the finite-dimensionality of the filter receives a great deal of attention in this volume. Let us explain in an informal way what this fuss is all about. Consider equation (4) or (5). This defines (the filtering problem was precisely set-up this way) a non-anticipating map from the observation y which acts as inputs to the filter to produce estimates z^* which are the outputs of the filter. Now (4) and (5) are realizations of this map, but they are infinite dimensional realizations because the state $\pi(t)$ or $\rho(t)$ is a map from $X = \mathbb{R}^n$ to $\mathbb{R}$, i.e. it is an infinite dimensional object (a function space). Now, it may be the case that this filter (input/output map) admits a finite dimensional realization.

This means that there would be a finite dimensional manifold M and a differential equation with output map

$$\dot{m} = v(m,\dot{y},t), \quad z^* = w(m) \tag{7}$$

on it such that (7) defines the same input/output map as (5) and (6). Obviously finite dimensionality of a filter is a very desirable (if not necessary) feature if one actually wants to implement it.

Thus assuming that a finite dimensional machine for calculating z^* (a filter) exists we would have two equivalent ways for processing the data y_s, $0 \le s \le t$ to produce $z^*(t)$).

The finite dimensional machine can be assumed to be of minimal dimension and assuming this one expects that there exists a map from (the from ρ_o or π_o accessible part of the function space) to the manifold M which takes the evolution equation for ρ_t (or π_t) to the equation for m. (This infinite dimensional extension of a result of Sussmann still has to be proved; it seems now very likely to be true in one sense or another). In the case of ρ_t there would result a filter of the form

(8) $$\dot{m} = \alpha(m) + \beta(m)\dot{y}_t, \quad z^* = \gamma(m)$$

where $\alpha(m)$ and $\beta(m)$ are vectorfields on the manifold M.

It is also definitely not unreasonable to look for a filter of the form (8) because (for linear systems) the Kalman-Bucy filter of considerable fame and enormous applicability is precisely a machine of the form (8). And so is for that matter the extended Kalman filter.

A main tool in this analysis is the Lie algebra of operators generated by the two operators A_2 and 'multiplication by h' which occur in the equation (5). This Lie algebra is called the estimation Lie algebra. The necessary differential topology and Lie-algebra background material for all this can be found in the tutorials of Hazewinkel [sections 2.4 and 2.5].

One particular most interesting feature of the estimation Lie algebra of a system (5) is that it is intrinsic. That is, it is (up to isomorphism) invariant under (nonlinear) changes of coordinates (cf. Brockett's lectures [section 7.1]). As such it could help e.g. in recognizing that a certain highly non-linear looking system is in fact a linear system to which a nonlinear change of coordinates has been applied. This Lie algebraic criterion will not be a sufficient, though, e.g. because the estimation Lie algebra is also invariant under socalled Gauge transformations, which do not correspond to coordinate changes.

One consequence of the existence of a map as discussed just above equation (8) above is the existence of a homomorphism of Lie algebras from the estimation Lie algebra to the Lie algebra of vectorfields on M generated by the vectorfields α and β in (8). In the particular case of linear systems and the Kalman-Bucy filter this can be checked by hand (Brockett [section 7.1]). Thus finite dimensional exact filters give rise to certain homomorphisms of Lie algebras and as a matter of fact there is evidence for a reverse statement as well. One collection of results which we shall need for this are uniqueness existence and regularity results for stochastic partial differential equations of the type (5), which is the subject of the contributions by Michel (section. 7.12) and Sussmann (section 7.14) cf. also Pardoux (section 7.4) (Michel uses the socalled Malliavin Stochastic variational calculus (currently a hot topic which was the subject of a conference in Durham later in 1980); additional or similar results on existence, regularly, uniqueness will probably result from the variational path integral formulation of Fleming and Mitter discussed in [section 7.2]). Given these one can exploit certain theorems concerning Lie algebras discussed in Hazewinkel - Marcus [section 7.9] to conclude e.g. that there exist no finite dimensional exact filters for any nonconstant statistic of the socalled cubic sensor. Though some of the things mentioned above are still conjectural this is now a firm theorem. Indeed it seems likely that we shall be able to prove that as a rule finite dimensional exact filters will not exist, which brings us to approximate calculation devices, a topic to which we shall return below.

Meanwhile there is obvious interest in analysing the estimation algebra in various cases. Finite dimensionality of this algebra would be nice to have and this is the topic of Ocone [section 7.13], though of course a Lie algebra of vectorfields on a finite dimensional manifold need not be finite dimensional. Low dimensionality of the estimation algebra and ease of computation ought to be related, cf. Baillieul [section 7.5] and the question whether similar estimation algebras correspond to filtering problems of equal computational complexity is addressed by Baras [section 7.6]. It is perhaps too early in the game to say just how useful the estimation Lie algebra and its concomitant geometrical considerations will be in the actual construction of (approximate) filters. (Its intrinsic nature exerts of course a powerful appeal and the writers of the present words are quite optimistic in this regard). Meanwhile, however, these geometrical ideas have certainly helped our theoretical understanding and have also helped in the actual construction of unexpectedly low dimensional filters (for finite state Markov chains, cf. Brockett [section 7.1]).

In our informal exposition of the nonlinear filtering problem we have up to now skipped over an important point or rather several much related points. Equations (4) and (5) are stochastic differential equations. This implies that abstractly they define a map from the probability space Ω to the observations $\dot{y}$ and then via the non-anticipating filter map to the optimal estimates z*. However, from the construction of stochastic integrals it follows that in principle these maps depend on the probability measure on Ω. This is, of course, an unpleasant situation since it says that we cannot just consider the filter map as simply acting on realizations of the observation process, in other words the filter map does not act (necessarily) 'sample pathwise'. In DAVIS' contribution [section 7.3] it is shown that in a large class of filtering problems one can in fact prove that the filter acts indeed sample-pathwise.

There is a second point, much related, as it turns out, to the first. The conditional expectation

$z^* = E[k(x)|y^t] = \int k(x)\pi_t(x)dx$ is a functional of y_t. I.e. given by some function ϕ which is only determined up to sets of measure zero (with respect to the measure on the function space $C([0,T])$ induced by the y and this measure has the same sets of measure zero as Wiener measure. Since the set of functions of bounded variations has measure zero ϕ is so to speak undefined on these. However, <u>physical</u> observation paths will be of bounded variation and so this approach to filtering would seem to be inapplicable unless there exists a version which is e.g. continuous w.r.t. the supremum norm on $C([0,T])$, giving us a 'robust' form of the filter (Robustness is, roughly, the property of a statistical procedure, or observer, or model, or ... to perform well even when the assumptions underlying its construction are not fully met). This fortunately turns out to be the case if the observation noises are independent of the system noises and also more generally provided the output y_t is scalar. The issue is much related to the pathwise issue discussed above; cf. Davis [section 7.3].

This robust-pathwise approach goes via a Feynman-Kac formula and thus suggests links with the path-integral approach to Quantum mechanics (à la Nelson). Another interesting and stimulating observation in this respect is that the estimation Lie algebra of the simplest (nonzero) linear system $dx = dw_t$, $dy_t = x_t dt + dv_t$ is the four dimensional oscillator Lie algebra (of some fame), whose derived Lie algebra is the even more famous Heisenberg Lie algebra of the canonical quantum mechanical commutation relations. And indeed the Kalman-filter for this system turns out to be gauge equivalent to a forced (euclidean) harmonic oscillator. The deep and fundamental relations of (nonlinear) filtering with quantum theory of which the two observations above are symptomatic are the subject of Mitter [section 7.2].

As the quantum constant h goes to zero quantum mechanics

goes to deterministic mechanics and one may ask to what deterministic limit nonlinear filtering converges if the noise intensity goes to zero. This matter is discussed in Hijab [section 7.10].

Both the estimation algebra approach and the robust-pathwise approach offer approximation possibilities. For the former some speculations are offered in Hazewinkel-Marcus [section 7.9]. Approximation by continuous time Markov chains is the subject of Di Masi-Runggaldier [section 7.8]. As soon as one starts approximating the question of a priori lower and upper bounds on the errors arises and whether these bounds are perhaps attained asymptotically. This is discussed by Bobrovsky-Zakai [section 7.7]. Finally Le-Gland uses the nonlinear filtering equations (and robustness) as an approach to maximum likelihood estimation for an astronomical observation problem.

So far, in this section we have concentrated on the filtering problem, neglecting the closely related and equally interesting prediction and smoothing problems. Pardoux [section 7.4] discusses the matter of finding DE's driven by the observations for optimal smoothers and predictors by means of a novel method involving both backward and forward equations. (The latter is the Duncan-Mortensen-Zakai-equation). This also yields new results for the smoothing problem extending the known results for finite state Markov processes.

4. LINEAR FILTERING.

Of course, there is one case in which the filtering problem may be solved explicitely, namely where the maps f, g, h, and k of (3) are linear, which leads to the model

$$\begin{aligned} dx &= F(t)x\,dt + G(t)dw \qquad & x(t_o) &= x_o \text{ (gaussian)} \\ dy &= H(t)x\,dt + I(t)dw \qquad & y(t_o) &= 0 \\ z &= K(t)x \end{aligned}$$

with F,G,H and I matrices of suitable dimension. The solution of the filtering problem in this case is given by the celebrated Kalman-Bucy filter. These filtering equations are very wellknown and play an important role in some of the other papers of these volume. The tutorial by WILLEMS [section 2.1] gives a brief introduction to the Kalman filter in the context of the general LQG (linear-quadratic-gaussian) stochastic control problem.

If one drops the assumption that there is a state-space model the filtering, smoothing and prediction problems take the following form. We have the following model for the observed process y_t

$$y_t = z_t + v_t$$

where z_t is the (stationary) signal and v_t is white random noise. The smoothing problem, filtering problem and prediction problem now take the form: find the best estimate $\hat{y}_{t|\tau}$ given observations up to and including time τ where respectively $\tau > t$ (smoothing), $\tau = t$ (filtering), $\tau < t$ (prediction). This is the problem studied and solved by Wiener and Kolmogorov in the early forties. The techniques involved in this solution, their extension to the case of finite time interval observations and associated problems of (efficient) computation are discussed in Kailath [section 5.1] (Wiener-Hopf technique, Ambartzumian-Chandrasekhar equations, Krein-Levinson equations). Kailath then goes on to discuss an extension to nonstationary models and a scattering theory framework for linear estimation.

Now scattering theory compares the asymptotic behaviour of an evolving system as $t \to -\infty$ with its asymptotic behaviour as $t \to \infty$. It is especially relevant when comparing the behaviour of a reference system (no scattering object) with that of a perturbed system (a scattering object is present) when the perturbations are negligible for large $|t|$. Think e.g. of a wave packet traveling from left to right being scattered by some object at the origin. Let U^t and U_o^t denote the evolution operators giving the state of the system at time t in terms of the state at time 0 for the perturbed and unperturbed system respectively. Then there are two states x_+ and x_- of the unperturbed system such that $U^t x$ behaves as $U_o^t x_-$ for $t \to -\infty$ and $U_o^t x_+$ for $t \to +\infty$. The scattering operator is the mapping $S: x_- \to x_+$ and the inverse scattering problem is the reconstruction of the scatterer from the scattering operator.

The relation of inverse scattering with linear prediction is the main theme of DeWilde, Fokkema en Widya [section 5.2]. Here, as in Kailath, the 'scatterer' is a transmission line with incident and reflected (light) waves from both sides. DeWilde e.a. first discuss (Redheffer) scattering, then the main theoretical result which says that the predictor filter may be obtained by solving a (very special) inverse scattering problem and then proceed how this fact can be used to produce concrete algorithms.

As was mentioned above (in the section on nonlinear filtering) there are links between quantum theory and the Duncan-Mortensen-Zakai-equation-approach to state-space-model filtering. This is not the first time that links between filtering problems and quantum theory have appeared. In fact, in a Seminaire Bourbaki exposé in 1961 Cartier discusses how a certain number of results of the spectral theory of Wiener and Kolmogorov filtering can be grouped around the ideas related to the Stone-von Neumann uniqueness theorem on representations of the Heisenberg Lie algebra (canonical commutation relations), and how the Wiener-Kolmogorov theory can be deduced from this point of view. This was the subject of the lectures by Hazewinkel [section 5.3].

In this connection it is interesting to observe that Wiener-Kolmogorov filtering can be viewed as a limit of Kalman-Bucy filtering and that on the other hand a main result of scattering theory (the translation representation theorem) is in fact equivalent to (the Weyl form of) the Stone-von Neumann theorem. There seems to be room for future work here.

5. IDENTIFICATION.

In the context of Section 1, the identification problem typically arises in a context where an uncertain system has, in addition to a stochastic component, also an uncertain non-stochastic 'parameter'. The basic problem is then to find out from measurements of the input and the output variables what the value of this unknown parameter is. There are, of course, more general situations where one may use identification ideas. For example one could try to fit a linear model to a nonlinear plant or one could try to fit a low dimensional linear model to a (very) high dimensional linear plant. In these cases it is not really fair to say that one tries to determine the unknown parameters of the plant. However, for the purposes of the present discussion, it suffices to think of the identification problem in this simple minded context.

Let us denote the unknown parameter(s) by Θ
If the input used is u then we will observe $y = F(u,\Theta,\omega)$ which, of course, will in general also depend on the parameters Θ and the random element $\omega \in \Omega$. In a dynamic situation it is natural to introduce also the time $t \in T$. At each instant one will then have available the past of u and y and an identification scheme will give us an estimate $\hat{\Theta}(t)$ of Θ(see figure 3).

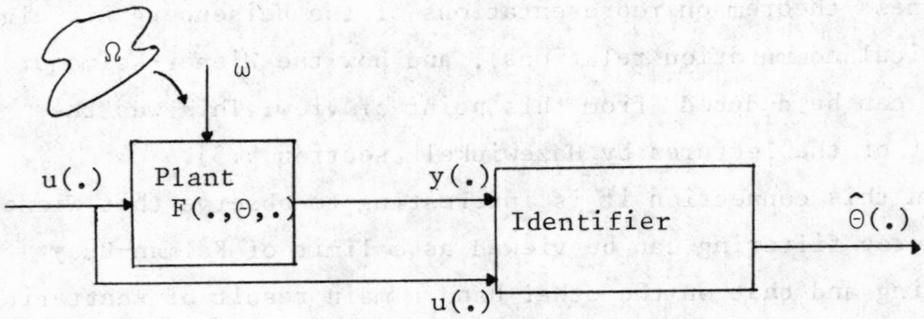

figure 3

There are two basic issues which are discussed in this volume regarding identification:

1. conditions for convergence of $\Theta^*(t)$ to the true parameter

2. recursive implementation of identification algorithms.

The article by HANNAN [section 4.1] contains a general convergence result for a class of identification problems. The model considered is a discrete time multivariable ARMAX (= autoregressive moving average with exogenous components) model which relates the input and output by

$$(10) \quad \sum_{j=o}^{p} A(j)y(t-j) = \sum_{j=1}^{m} D(j)u(t-j) + \sum_{j=o}^{q} B(j)\varepsilon(t-j)$$

where one can think of ε as white random noise (the precise assumptions are given in the paper) and $A(0)$, ..., $A(p)$, $D(1)$, ..., $D(m)$, and $B(0)$, ..., $B(9)$ as matrices with unknown coefficients. Let us denote this string of matrices by Θ. In this case 4 is thus a high dimensional Euclidean space. The identification principle used is that of maximum likelihood. The principle behind this idea is wellknown: at each instance of time there is a certain probability density $p(u(0), ..., u(t-1), y(0), ..., y(t); \Theta)$ which expenses the likelihood that the string $y(0)$, ..., $y(t)$ would have been observed with the input $u(.)$ and the parameter value Θ. At each instant of time one then chooses the parameter $\hat{\Theta}(t)$ such that it maximizes this likelihood function over all possible Θ. The convergence question is whether or not $\lim_{t\to\infty} \hat{\Theta}(t) = \Theta^*$, where Θ^* equals the true value of the parameter matrices $A(0)$, ..., $B(q)$ which generate the data y from the input u. HANNAN proves a nice and very general result in this direction.

Of course to state and prove such a result one needs a

topology on the space of all possible models and if one wants to go beyond this result and discuss also how fast the convergence is, one needs more, namely a metric or a Riemannian structure on the space of all possible systems of a given type. It is here that the geometry and topology of the space of linear systems enter the picture and the discrete invariants called Kronecker indices turn out to have an important role to play. As such the results presented in this paper are a primeur in giving hard evidence of the relevance of this geometric structure issue in system identification.

Basically the same questions as in Hannan's paper are addressed by Deistler [section 4.3] for the case that some initial structural information on the to be identified system is already given.

Statistical tests to decide whether ARMA models will be adequate are considered by Guegan in [section 4.4].

Both from a conceptual and from a practical point of view it is important to implement an identification scheme in a recursive algorithm. The idea behind this is basically the same as explained in the context of nonlinear filtering. However, since one in general does not like to treat the unknowns as random variables, the procedure for obtaining recursive algorithms goes differently. In addition there are many different ways of approaching an identification problem (contrary to the situation in nonlinear filtering where there are many reasons for considering in the first place the conditional mean of the to be estimated variables). LJUNG's contribution [section 4.2] provides a very readable account of various aspects of recursive system identification basically all in the context of <u>scalar</u> ARMAX models as (10). He describes a number of identification routines and discusses their convergence properties. He then gives some practical guidelines for the implementation of these algorithms and closes by giving some results on the application of identification algorithms in adaptive control (see section 6 of this article).

It is possible, of course, to treat an identification problem from a so-called Bayesian point of view. In the context of the model introduced in the beginning of this section, one then puts a probability measure on Θ, the space of unknown parameters. By considering now the product measure on $\Theta \times \Omega$, the total uncertainty space, this problem becomes a purely stochastic one and it is possible, for instance, to use nonlinear filtering ideas in system identification. This approach applied to ARMAX models (written, however, in state space form) is the subject of the article by KRISHNAPRASAD & MARCUS [section 4.5]. The estimation Lie algebras of these problems have a particularly pleasing structure with interesting possibilities for the existence of explicit recursive (approximate) filters.

6. ADAPTIVE CONTROL.

The last main topic discussed in this book is that of adaptive control. This is really one of the very early motivations of control theory: the design of control algorithms which will automatically learn the value of the (changing) plant parameters and self-adjust their control strategy accordingly.

Most of the adaptive control strategies proposed in the literature work according to a separation principle of identification and control. This is easily explained in the context of the general set-up discussed in Section 1. Assume that we have given an uncertain plant F with observed output $y = F(x,\Theta,\omega)$, with control input $u \in U$, unknown parameter $\Theta \in \Theta$, and stochastic uncertainty $\omega \in \Omega$. The problem is to design a feedback compensator, i.e., a nonanticipating map $G: Y \to U$, such that the closed loop system has some desirable properties. This control design purpose may be expressible in terms of closed loop stability, an optimal stochastic control criterion, or some of the design formulations of multivariable control as, for example, model matching, pole placement, disturbance decoupling, etc. The difficulty, however, is that the unknown parameter Θ is indeed unknown.

If one uses a recursive identification scheme as explained in section 5 one will have at each instant of time an estimate $\hat{\Theta}(t)$ of the unknown parameter. Assume now that if Θ were known one would use the feedback control law G which, since it will depend on Θ, we denote by G_Θ. If G_Θ is implemented recursively, this will lead to a set of update equations with coefficients depending on Θ. The idea of using separation is to use for these parameters the estimate $\hat{\Theta}(t)$ at time t. This is illustrated in figure 4

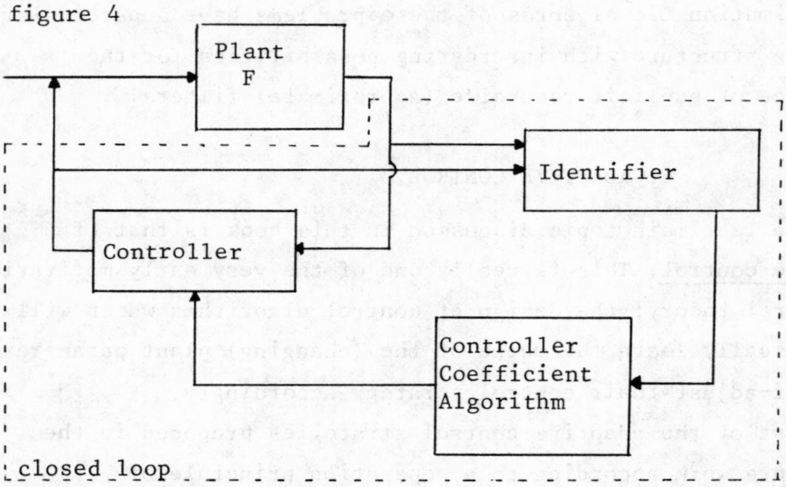

figure 4

The ensuing closed loop system will be very nonlinear and its properties are difficult to analyze. Moreover, one cannot simply conclude that a convergent identification routine will remain convergent when used in this closed loop framework. Indeed, assumptions like u is bounded, deterministic, and if it is stochastic, independent of the stochastic disturbance of the plant, which one may have to make in order to prove the convergence of the identification scheme, need not be satisfied.

The adaptive separation scheme induces a map $\tilde{G}: Y \to U$. The adaptive control scheme is said to be self-tuning if, as $t \to \infty$, the map $\tilde{G}$ converges to G_{Θ^*} in some sense. Here Θ^* denotes the true value of the parameter Θ and G_{Θ^*} denotes the controller which achieves the control objective (optimal performance, model matching, etc.) at the true value Θ^*. The surprising part of the results obtained so far is that self-tuning may occur even when inside the controller the convergence $\hat{\Theta}(t) \underset{t\to\infty}{\to} \Theta^*$ does not hold.

In this volume we have three papers on self-tuning control. The first one is by LANDAU [section 6.1] and treats self-tuning results for model reference adaptive control algorithms for scalar systems of the ARMAX type (10). The second paper is by CAINES & DORER [section 6.2]. It discusses a stabilization property for a class of (TV) ARMAX models, that is models of the type (10) but with time-varying coefficients. These coefficients are assumed to be stochastically time varying and the purpose of the paper is to prove asymptotic stability of the closed loop system.

The third paper in this chapter by FUCHS [section 6.3] discusses the stability of the overall system in terms of properties of the separate control subsystem and the identification subsystem.

7. CONCLUSIONS.

It is perhaps safer to leave the conclusions and statements of future prospects to the reader (after he has carefully read and digested the papers in this volume). For ourselves let us say that the future seems very bright, strong new impulses seem present everywhere in this field of filtering and identification and there seems to be a most promising gathering of forces in the sense that more and more new mathematical subjects are brought in to bear upon the subject, which, when knowledgeably used, seem likely to enhance our understanding and improve our techniques.

To quote Joseph Louis Lagrange:

'As long as algebra and geometry proceeded along separate paths their advance was slow and their applications limited. But when these sciences joined company, they drew from each other fresh vitality and whence-forward marched on at a rapid pace toward perfection'.

It may well be (in our opinion) that in the field of enquiry of these proceedings we are witnessing today the beginnings of a similar joining of forces.

"The time has come", the Walrus said "To talk of many things: of shoes - and ships - and sealing wax - of cabbages and kings - of why the sea is boiling hot - and whether pigs have wings,"

L. Carroll

Part 2

INTRODUCTORY TUTORIAL MATERIAL FOR IDENTIFICATION AND FILTERING PROBLEMS

In order to make the material presented at the summer institute accessible to all participants - apart from qualifications a random selection of control and electrical engineers, probabilists, statisticians, time series and identification specialists, and other pure and applied mathematicians of diverse kinds - we included four tutorials, respectively on: linear quadratic gaussian control, stochastic calculus, manifolds and vectorfields and the basic equations of nonlinear filtering theory. The first two are here reproduced in condensed form with ample references to the standard literature, the second two are reproduced more or less in extenso. We have also added a fifth tutorial section on Lie algebras which concentrates on those aspects of the theory which, so far, have proved to be of importance in control and filtering and identification theory.

THE LQG-PROBLEM: A BRIEF TUTORIAL EXPOSITION

Jan C. WILLEMS

Mathematics Institute
P.O. Box 800, 9700 AV GRONINGEN, The Netherlands

ABSTRACT

The purpose of this article is to provide a brief tutorial exposition of the formal setting, the main ideas, and the formulas for the linear quadratic gaussian stochastic optimal control problem (the so-called *LQG-problem*).

INTRODUCTION

The LQG-Problem occupies in all its facets the central position in the contemporary teaching of control theory and, eventhough it is not a particularly active source of theoretical research problems, it provides an important background for ideas when more general situations are being considered. Therefore it appeared useful to include, for completeness sake, such an exposition in this volume.

We will not give proofs. In more complete treatments, we refer the reader to Kalman's original papers [1, 2, 3], our recent exposition in *Statistica Neerlandica* [4], or the textbook [5]. Extensive bibliographies may be found in [6] or [7].

A few words about notation and nomenclature: $\mathbb{R}$, $\mathbb{R}^e$, and $\mathbb{Z}$ denote respectively the real line, the extended real line $(:= \mathbb{R} \cup \{-\infty\} \cup \{+\infty\})$, and the integers; X^Y denotes the set of all maps from Y into X, and $L_p([t_o,t_1]; \mathbb{R}^q)$ denotes all $\mathbb{R}^q$-valued maps defined on the interval $[t_o,t_1]$ with components in L_p; a.e. stands for 'almost everywhere', ^{T} denotes transposition, and ≥ 0 denotes that a symmetric matrix is nonnegative definite;

M. Hazewinkel and J. C. Willems (eds.), Stochastic Systems: The Mathematics of Filtering and Identification and Applications, 29–44.

$\{\Omega, A, P\}$ denotes a probability triple, $L(x_o)$ denotes the 'law' of a random vector, i.e., its distribution or its density, for example $L(x_o) = N(\bar{x}_o, \Pi_o)$ signifies that x_o has a normal (or gaussian) distribution with mean $\bar{x}_o$ and covariance Π_o, a process on $[t_o, t_1]$ is said to be a *Wiener process* if it is a vector valued random process and each component consists of an independent Brownian motion on $[t_o, t_1]$. Let $T \subseteq \mathbb{R}$ and f a map from $X \subset X^T$ into Y^T; then f is said to be *(strictly) nonanticipating* if $x_1, x_2 \in X$, $t \in T$, and $x_1(t') = x_2(t')$ for $t' \leq t$ $(t' < t)$ imply $f(x_1)(t') = f(x_2)(t')$ for $t' \leq t$.

1. STOCHASTIC OPTIMAL CONTROL: FORMAL SET-UP

An abstract stochastic optimal control problem is defined by:

(i) a *plant;*

(ii) a *cost criterion;*

(iii) a class of *admissible control laws;*

(iv) a *probability space.*

The plant is an uncertain dynamical system defined over a time axis $T \subseteq \mathbb{R}$, a control input space $U \subset U^T$ (with U the control input alphabet), an uncertainty space Ω, an observation output space $Y \subset Y^T$ (with Y the observed output alphabet), by the *system function* G: $U \times \Omega \rightarrow Y$ which models the observed output $y(\cdot) = G(u(\cdot), \omega)$ which results from the control input $u(\cdot)$ and the uncertainty parameter ω. The system function $G(\cdot, \omega)$ is assumed to be nonanticipating in u for all $\omega \in \Omega$. This formalizes the natural constraint that, although what we observe may depend on the control used, the present observation does not depend on the future control.

We think of u as to be chosen by the decision maker and of ω as being chosen by nature. In selecting which $u \in U$ will be used, the decision maker may exploit the information which he has about the outcome of the choice of ω by nature. This information is obtained through the observation y. However, since this observation itself depends on u, we obtain an implicit set of ('feedback') equations and some care is be taken in formalizing this set-up. It is important to realize that in stochastic control we cannot a priori assume that our information about ω is limited but fixed, because by choosing u cleverly we are actually also in a position to influence to some extent which information we will obtain about ω. Ideas as *learning* and *probing* stem form this fact.

The most logical approach appears to be by proceeding to define a family of admissible control laws $F \subset U^Y$. Each admissible control law is thus a map $F : Y \to U$. We assume that each element $F \in F$ satisfies the following conditions:

(A.1) F is *nonanticipating;*

(A.2) for all $\omega \in \Omega$ *there exist a unique solution* $u \in U$ which solves the equation $u = FG(u,\omega)$. We denote this solution by $F_u(\omega)$.

(A.3) *The closed loop information flow* from ω to u *is nonanticipating* with respect to G, which is formalized as follows: if $\omega_1, \omega_2 \in \Omega$, $t \in T$ are such that $G(u,\omega_1)(t') = G(u,\omega_2)(t')$ for all $u \in U$ and $t' \leq t$, then we must also have $F_u(\omega_1)(t') = F_u(\omega_2)(t')$, $t' \leq t$.

There is an important class of problems where the conditions (A.2) and (A.3) follow from (A.1), namely discrete time problems with $T = \{t \in \mathbb{Z} \mid -\infty < t_0 \leq t \leq t_1 \leq \infty\}$, $U = U^T$ and with the system function strictly nonanticipating in u.

The cost is a map $J: U \times \Omega \to \mathbb{R}^e$.

Now, to every $F \in F$ and $\omega \in \Omega$ there corresponds the cost $J(F_u(\omega), \omega)$. We denote by $\hat{J}$ the resulting map $\hat{J} : F \times \Omega \to \mathbb{R}^e$ defined by $\hat{J}(F,\omega) := J(F_u(\omega),\omega)$. The purpose of the decision maker is to choose this F such that $\hat{J}(F,\omega)$ is as small as possible. Of course, it will only be in exceptional circumstances that there exists a uniformly best decision, i.e. a superior control law $F_{sup} \in F$ such that $\hat{J}(F_{sup},\omega) \leq \hat{J}(F,\omega)$ for all $\omega \in \Omega$ and $F \in F$. If such a uniform best decision does not exist, it is logical to minimize the average performance obtained by averaging the cost over $\omega \in \Omega$. This is formalized by assuming that the uncertainty space Ω is a probability space $\{\Omega, A, P\}$ and defining the map $J_{av} : F \to \mathbb{R}^e$ by $J_{av}(F) := E\{\hat{J}(F,\cdot)\} = \int_\Omega \hat{J}(F,\omega)dP(\omega)$, where it is assumed that the required integral is well-defined. Of course, the idea of modelling the uncertainty ω as random is often also very much justified by the experimental background of the problem.

The problem in stochastic optimal control is to find, for a given plant G, a given cost J, a given family of admissible control laws F, and a given probability space $\{\Omega, A, P\}$, the optimal control law $F^* \in F$ defined by the fact that it must satify $J_{av}(F^*) \leq J_{av}(F)$ for all $F \in F$. The information flow in this set-up is illustrated in Figure 1.

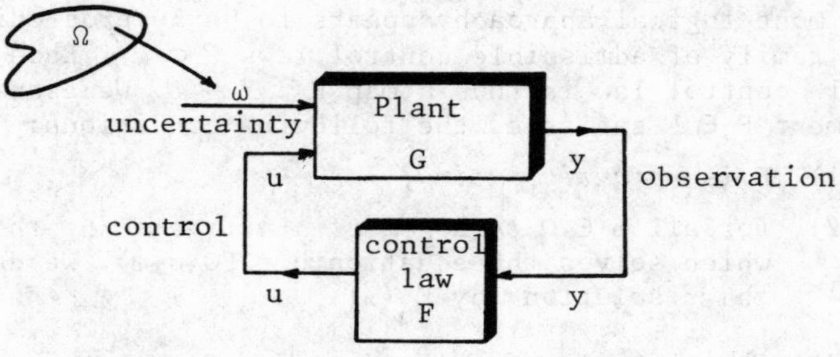

Figure 1

2. THE LQG-PROBLEM: FORMAL SET-UP

In this section we will set up the finite time horizon LQG-problem in the framework of Section 1. We will consider the discrete time case first, because it is technically easier to explain.

Discrete Time Case: The time axis is given by $T = \{t \in \mathbf{Z} \mid -\infty < t_o \leq t \leq t_1 < \infty\}$. The uncertainty is modelled on a given probability space $\{\Omega, A, P\}$ by a family of random vectors $x_o \in \mathbb{R}^n$ and $w(t_o), w(t_o + 1), \ldots, w(t_1) \in \mathbb{R}^q$. We assume these be a independent gaussian random vectors with $L(x_o) = N(\bar{x}_o, \Pi_o)$ and $L(w(t)) = N(0, \Gamma(t))$. These random vectors define respectively the initial uncertainty x_o and the 'disturbance input', the stochastic process $\{w(t), t \in T\}$.

The plant is defined over $U = \mathbb{R}^m$, $U = (\mathbb{R}^m)^T$, $Y = \mathbb{R}^p$, $Y = (\mathbb{R}^p)^T$ and with the system function defined by

$$(1) \quad \begin{cases} x(t+1) = A(t)x(t) + B(t)u(t) + G(t)w(t) \\ y(t) = C(t)x(t) + J(t)w(t);\ x(t_o) = x_o \end{cases}$$

The matrices $A(t)$, $B(t)$, $G(t)$, $C(t)$, and $J(t)$, $t \in T$, are given matrices of appropriate dimension which together with the statistical data, $\bar{x}_o$, Π_o, and Γ, completely specify the model of the plant. The generation of the observed output y from the control input u and the uncertainty ω hence proceeeds as follows: 'nature' chooses the basic random variable ω which generates the initial condition $x(t_o) = x_o(\omega)$ out the disturbance input $w_\omega(\cdot)$. Together with u, this $x(t_o)$ and w then define x and y. The fact that this model involves the intermediate variable x is not accidental. Indeed, this variable has the property of *'state'* and we will comment on the implication of this later.

The cost criterion is given by

$$(2) \quad x^T(t_1+1)Mx(t_1+1) + \sum_{t=t_0}^{t_1}[u^T(t)R(t)u(t) + 2\,u^T(t)S(t)x(t) + x^T(t)Q(t)x(t)]$$

where $M = M^T$, $R(t) = R^T(t)$, $S(t)$, and $Q(t) = Q^T(t)$, $t \in T$, are given matrices of appropriate dimensions which specify the cost as a combination of the *running cost* (the terms in the $\sum_t$) and the *terminal cost*. More general expressions are possible and useful but the critical thing is that the cost is a - not necessarily homogeneous - quadratic form in u and x. It is clear that every control sequence u and every value of x_o and the disturbance input w yield a unique real number for the cost. The map J is hence defined by the composition:

$$(u,\omega) \mapsto (u,(x_o,w)) \overset{(1)}{\longmapsto} (u,x) \overset{(2)}{\longmapsto} J(u,\omega).$$

The resulting signal flow graph, which also holds for the continuous time case, is shown is Figure 2.

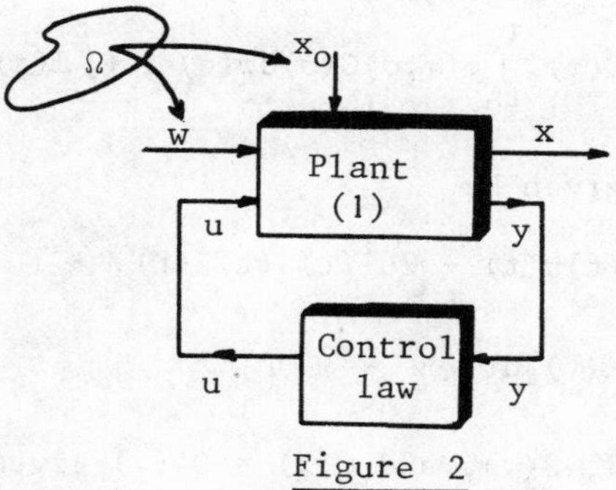

Figure 2

Continuous Time Case: In the continuous time case the time axis is given by $T = \{t \in \mathbb{R} \mid -\infty < t_o \leq t \leq t_1 < \infty\}$. The uncertainty is modelled on a given probability space $\{\Omega, A, P\}$ by a gaussian random vector $x_o \in \mathbb{R}^n$, $L(x_o) = N(\bar{x}_o, \Pi_o)$ and an independent q-dimensional Wiener process w on $[t_o, t_1]$.

The plant is defined over $U = \mathbb{R}^m$, $\mathcal{U} = L_2([t_o,t_1];\mathbb{R}^m)$, $Y = \mathbb{R}^p$, $\mathcal{Y} = L_2([t_o,t_1];\mathbb{R}^p)$ by the system function defined by the stochastic differential equation*

* Readers not familiar with this type of differential equation may consult the tutorial by R.F. Curtain on Stochastic Calculus in this volume.

$$(3)\quad \begin{cases} dx(t) = A(t)x(t)dt + B(t)u(t)dt + G(t)dw(t) \\ dy(t) = C(t)x(t)dt + J(t)dw(t);\ x(t_o) = x_o,\ y(t_o) = 0 \end{cases}$$

with $A(\cdot)$, $B(\cdot)$, $C(\cdot)$, $C(\cdot)$, and $J(\cdot)$ given matrix-valued functions of appropriate dimensions defined on $[t_o,t_1]$and with the elements of A in $L_1([t_o,t_1];\mathbb{R})$ and those of B,G,C, and J in $L_2([t_o,t_1];\mathbb{R})$.

Eventhough this model passes, as in the discrete time, via the intermediate variable x, the state, in the end we obtain a system function G of the type

$$(4)\quad (u,\omega) \mapsto (u,(x_o,w)) \xmapsto{(3)} y = Lu + r$$

with L a linear map from $L_2([t_o,t_1];\mathbb{R}^m)$ into $L_2([t_o,t_1];\mathbb{R}^p)$ given by

$$(Lu)(t) = \int_{t_o}^{t} C(\tau) \int_{t_o}^{\tau} \Phi(\tau,\sigma)B(\sigma)u(\sigma)d\sigma d\tau$$

with Φ the transition matrix of $\dot{x}(t) = A(t)x(t)$ defined by $\dot{\Phi}(t,\tau) = A(t)\Phi(t,\tau)$; $\Phi(\tau,\tau) = I$, and with r the gaussian $\mathbb{R}^p$-valued stochastic process on $[t_o,t_1]$ given by

$$r(t) = \int_{t_o}^{t} C(\tau)\Phi(\tau,t_o)x_o d\tau + \int_{t_o}^{t} C(\tau) \int_{t_o}^{\tau} \Phi(\tau,\sigma)G(\sigma)dw(\sigma)d\tau + \int_{t_o}^{t} J(\tau)dw(\tau)$$

The cost criterion is given by

$$(5)\quad x^T(t_1)Mx(t_1) + \int_{t_o}^{t_1} [u^T(t)R(t)u(t) + 2u^T(t)S(t)x(t) + x^T(t)Q(t)x(t)]dt$$

with $M = M^T$, and $R(\cdot) = R^T(\cdot)$, $S(\cdot)$, and $Q(\cdot) = Q^T(\cdot)$ given matrix-valued functions of appropriate dimensions defined $[t_o,t_1]$ and with elements in $L_\infty([t_o,t_1];\mathbb{R})$. Similarly as in the discrete time case the expression (4) yields a well defined real number for all $u \in U$ and (almost all) $\omega \in \Omega$.

It is natural in this situation to define the class of admissible control laws to consist of all maps $F: L_2([t_o,t_1];\mathbb{R}^p) \to L_2([t_o,t_1];\mathbb{R}^m)$ such that:

(i) F is nonanticipating;

(ii) I-FL has a nonanticipating inverse on $L_2([t_o,t_1];\mathbb{R}^p)$.

It is easily seen that (ii) is precisely the implementation of

(A.1 - A.3) to the case at hand. Note that the control input corresponding to the control law F is given by $(I-FL)^{-1}r$ and is thus a nonanticipating function of the stochastic process r(the measurement which would have been observed if the control u = 0 were used). Actually, in the case at hand one could define as admissible control inputs all the $L_2([t_o,t_1]; \mathbb{R}^m)$ stochastic processes u(.) such that for all t, u(t) is measurable w.r.t. the σ-field induced by the random vectors $r(\tau)$, $t_o \leq \tau \leq t$. Other than the fact that this formulation cannot easily be generalized beyond the linear case, it is, in our opinion, a very unnatural way of setting up this stochastic control problem and it yields only a slighty more general class of admissible control input processes. Note, finally, that the admissible control laws include as special cases all maps $F: L_2([t_o,t_1]; \mathbb{R}^p) \rightarrow L_2([t_o,t_1]; \mathbb{R}^m)$ which are locally Lipschitz continuous as defined for example on [8] and used by Wonham [9] for the LQG-problem.

We close this section be remarking that the formulation of the discrete time and the continuous time LQG-problem are not parallel as they stand. They are if instead of considering y as observation, one considers its derivative as observation and uses as system model the *white noise* formulation

$$\begin{cases} \dot{x}(t) = A(t)x(t) + B(t)u(t) + G(t)\frac{dw}{dt}(t) \\ y(t) = C(t)x(t) + J(t)\frac{dw}{dt}(t) \;;\; x(t_o) = x_o \end{cases}$$

Eventhough it remains possible to define what one means with a solution of this differential equation there is the difficulty that in general it is not possible to make sense of nonlinear functions of distributions, and since one wants to allow nonlinear control laws, this is very inconvenient indeed.

The solution of the LQG-problem may be explained by dividing it into three different stages:

1. First one solves the problem with *exact state measurements*, i.e., when in (1) or (3) one has the measurement y(t) = x(t). It turns out that problem is equivalent to a *deterministic linear quadratic optimal control problem*.

2. The second stage involves the recursive optimal estimation of the state from observations. This stage involves the celebrated *Kalman filter*.

3. The third stage involves the combination of (1) and (2) where the optimal estimate of the state constructed in 2 is substituted in the optimal state feedback control

law obtained in 1. This procedure is called the *separation principle* and the fact that it yields the optimal control law is one of the very appealing aspects of the LQG-theory.

3. THE DETERMINISTIC LQ-PROBLEM

Consider now the following (deterministic) optimal control problem derived from the stochastic optimal control problem posed in Section 2:

<u>discrete time version</u>:

model: $x(t+1) = A(t)x(t) + B(t)u(t);\ x(t_o) = x_o$

cost : $x^T(t_1+1)Mx(t_1+1) + \sum_{t=t_o}^{t_1} [u^T(t)R(t)u(t)$

$+ 2u^T(t)S(t)x(t) + x^T(t)Q(t)x(t)]$

<u>continuous time version</u>:

model: $\dot{x}(t) = A(t)x(t) + B(t)u(t);\quad x(t_o) = x_o$

cost : $x^T(t_1)Mx(t_1) + \int_{t_o}^{t_1} [u^T(t)R(t)u(t)$

$+ 2u^T(t)S(t)x(t) + x^T(t)Q(t)x(t)]dt$

Let $J(u;x_o)$ denote the cost resulting form the initial condition x_o by using the control input $u \in U$.

In deterministic optimal control it is, on the level of problem definition, easiest to think of the problem as an *open loop* control problem, that is to say that we seek, for a fixed $x_o \in \mathbb{R}^n$, an element $u^* \in U$ such that $J(u^*; x_o) \leq J(u,x_o)$. Let us tacitly assume for the time being that his minimum exists for all x_o.

It is also possible to think of this problem in *closed loop* form and consider feedback control laws, in the fashion explained in Section 1, which use the state x as measurement and where x_o parametrizes the uncertainty. Now, since the state x_o is assumed to be measured, we will have the uncertainty $x(t_o) = x_o$ available as information for all $t \in T$ and consequently there will exist a uniformly best control law (use, when the uncertainty is x_o, the control u^* which minimizes $J(u;x_o)$), and thus, as the performance is concerned, there is no advantage in using closed loop versus open loop control.

The next point brings in the property of state. As a consequence of what is sometimes called *Bellman's principle* of *optimality* and because of the fact that x has the property of state (some abstract ideas on this may be found in [10, 11] and for the stochastic case in the article by Picci and its references in this volume) one will always find an optimal state feedback control law in the class of memoryless feedback control laws. That is to say that there will exist a function $f^*:\mathbb{R}^n \times T \to \mathbb{R}^m$ such that the solution $x^*(t)$ (which we assume to exist uniquely) of

$$x(t+1) = A(t)x(t) + B(t)f(x^*(t),t) \quad \text{(discrete time)}$$

$$\dot{x}(t) = A(t)x(t) + B(t)f(x^*(t),t) \quad \text{(continuous time)}$$

with $x(t_o) = x_o$ will generate via $u^*(t) = f^*(x^*(t),t)$ the optimal open loop control and will hence yield for all x_o the minimum of $J(u;x_o)$ as the performance. Of course, this fact that it suffices to look into the class of memoryless state feedback control laws is one of the basic principles in operations research in general and dynamic programming in particular.

In the previous paragraph we have completely ignored the question of existence of the optimal control. It should be clear that this requires $R(t) = R^T(t) \geq 0$ $(t \in T)$. In fact, we will assume that $R(t) + R^T(t) \geq \varepsilon I > 0$. This condition is called the nonsingularity condition. Of course, this condition alone does not ensure the existence of the required minimum and further conditions on M, $R(\cdot)$, $S(\cdot)$, and $Q(\cdot)$ need to be imposed:

<u>Proposition 1</u>: *Assume that the LQ-problem is nonsingular, i.e. that there exist* $\varepsilon > 0$ *such that* $R(t) = R^T(t) \geq \varepsilon I > 0$ *for* $t \in T$.

(i) *(<u>discrete time</u>): Consider the Riccati difference equation:*

(6) $$K(t) = A^T(t)K(t+1)A(t) - [S(t)+B^T(t)K(t+1)A(t)]^T [R(t) + B^T(t)K(t+1)B(t)]^{-1}[S(t)+B^T(t)K(t+1)A(t)] - Q(t).$$

$K(t_1 + 1) = M$

in the $(n \times n)$ *matrix* $K(\cdot)$

(ii) (<u>continuous time</u>): *Consider the Riccati differential equation:*

$$(7)\quad \dot{K}(t) = -A^T(t)K(t) - K(t)A(t) + [S(t) + B^T(t)k(t)]^T R^{-1}(t)\,[S(t) + B^T(t)K(t)] - Q(t)$$

$$K(t_1) = M$$

in the $(n \times n)$ *matrix* $K(\cdot)$.

Then the LQ *problem has a minimum for all* x_o *iff the above Riccati equation, solved backwards, has a solution up to* t_o.

Although the condition of the above proposition is sharp, it is not as explicit ar one may like to have it. Sufficient conditions for the existence of a solution to the Riccati equations are:

1. $t_1 - t_o$ is 'sufficiently small' , or

2. $M \geq 0$ and $\begin{bmatrix} R(t) & S(t) \\ S^T(t) & Q(t) \end{bmatrix} \geq 0 \quad (t \in T)$

The positivity conditions in 2. are the classical assumptions which one finds in the expositions of this topic. It should be realized however that these conditions which are very natural from many points of view, are nevertheless much more restrictive from the potential applications point of view (e.g. second variation problems) than is often advertized.

The optimal control for the LQ-problem is given in the following

<u>Theorem 2:</u> *Assume that the Riccati equation (6) or (7) has a unique solution* K(t) *on* T. *Then the optimal control (in*

memoryless state feedback form) is given by $u^*:(x,t) \mapsto N(t)\,x$ *with* N *given by*

(8) *(discrete time)* $$N(t) = -\,[R(t) + B^T(t)K(t+1)B(t)]^{-1}\,[S(t) + B^T(t)K(t+1)A(t)]$$

(9) *(continuous time)* $$N(t) = -\,R^{-1}(t)\,[B^T(t)K(t) + S(t)]$$

4. THE KALMAN FILTER

A general formulation of the filtering problem is as follows: Let $T \subset \mathbb{R}$ be the time axis and $\{z(t),\ t \in T\}$ and $\{y(t),\ t \in T\}$ be two stochastic processes defined on some probability space $\{\Omega, A, P\}$ and with outcome spaces Z and Y respectively. The process z is the *to-be-estimated* process while the process y is the *observed* process. The filtering problem is to estimate for all $t \in T$ the value of $z(t)$ on the basis of the observations $y_t^- := \{y(\tau),\ \tau < t\}$. Let $\hat{z}(t)$ denote this estimate.

Let Y denote the space of all realizations of y, i.e., $Y = \{y' \in Y^T \mid y' = y_\omega \text{ some } \omega \in \Omega\}$. In effect in a filtering problem we are thus asking to construct a nonanticipating map $H : Y \to Z^T$ defined by $(Hy)(t) := \hat{z}(t)$. The signal flow graph is shown in Figure 3.

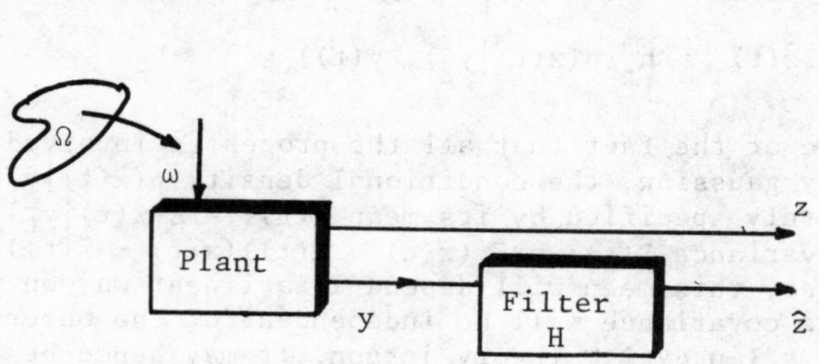

Figure 3

It is natural to quantify the performance of the filter map H in terms of an optimality criterion involving the expected estimation error $E\{d(z(t), \hat{z}(t))\}$ with d a measure of the error between z(t) and $\hat{z}(t)$. It is well known that under very broad conditions the optimal estimate will then be the conditional mean of z(t) given y_t^-. In particular, if $Z = \mathbb{R}^k$, $Y = \mathbb{R}^m$, and (z,y) is a gaussian process then $\hat{z}(t) = E\{z(t)|y_t^-\}$ minimizes $E\{||z(t) - z^*(t)||^2\}$ (or any other nonnegative homogeneous quadratic functional in $z(t) - z^*(t)$). The optimal least squares filter in this gaussian context is hence given by the map $y \to H_y^*$ with $(H_y^*)(t) := E\{z(t)|y_t^-\}$. The problem now is to give a convenient expression for this map. An important feature of the Kalman filter is that it is recursive. This again is connected with the crucial role played by the state.

Consider the following filtering problem given by (1) in discrete time and by (3) in continuous time with y the observed and $z = H(t)x(t)$ the to-be-estimated process. In here u is a process such that u(t) is measurable w.r.t. the σ-field induced by $y(\tau)$, $\tau < t$. The problem is to construct a recursive implementation of the map $y_t^- \mapsto E\{z(t)|y_t^-\}$. The solution of this problem is given in Theorem 3. The following considerations serve to carify the structure of the filter given in Theorem 3:

(i) because of the property of state which the process x enjoys its conditional density (or, more generally, its conditional measure) $p(x(t)|y_t^-)$ will be a state of the filter map that is to say that for any $t' \le t''$, $p(x(t'')|y_{t''})$ may be expressed as

$$p(x(t'')|y_{t''}^-) = F_{t',t''}(p(x(t')|y_{t'}^-), y(\tau)\ t' \le \tau < t'')$$

and $\hat{z}(t) := E\{z(t)|y_t^-\}$ may be expressed as

$$\hat{z}(t) = h_t(p(x(t)|y_t^-), y(t))$$

(ii) because of the fact that all the processes involved are jointly gaussian, the conditional density $p(x(t)|y_t^-)$ is completely specified by its mean $\hat{x}(t) := E\{x(t)|y_t^-\}$ and its covariance $\Sigma(t) := E\{(x(t) - \hat{x}(t))(x(t) - \hat{x}(t))^T\}$. Moreover, this mean will depend in a linear way on y_t^-, but the covariance will be independent of the observation y (in system theory jargon, it may hence be calculated 'off-line'). The state of the filter is completely specified by $\hat{x}(t)$ (since it is independent of y, there is formally no need to include $\Sigma(t)$ in the specification of the filter state). The expression of the Kalman filter provides the precise explicit formula by which $\hat{x}(t)$ is updated by y.

(iii) Since $z(t) = H(t)x(t)$, we obtain $\hat{z}(t) = H(t)\hat{x}(t)$.

Theorem 3: *Assume that the filtering problem is nonsingular, i.e., that there exist* $\varepsilon > 0$ *such that* $J(t)J^T(t) \geq \varepsilon I$ *for* $t \in T$.

(i) *(discrete time)* $\hat{z}(t) := E\{z(t)|y_t^-\}$ *is given by*

$$\hat{x}(t+1) = A(t)\hat{x}(t) + B(t)u(t) + L(t)(y(t) - C(t)\hat{x}(t)) \tag{9}$$

$$\hat{z}(t) = H(t)\hat{x}(t); \quad \hat{x}(t_o) = \bar{x}_o$$

where L *is given by*

$$L(t) = [A(t)\Sigma(t)C^T(t) + G(t)\Gamma(t)J^T(t)][J(t)\Gamma(t)J^T(t) + C(t)\Gamma(t)C^T(t)]^{-1}$$

with Σ *defined by the Riccati difference equation*

$$\Sigma(t+1) = [A(t) - L(t)C(t)]\Sigma(t)A^T(t) + G(t)\Gamma(t)G^T(t) - L(t)J(t)\Gamma(t)G^T(t)$$

$$\Sigma(t_o) = \Pi_o$$

(ii) *(continuous time)* $\hat{z}(t) = E\{z(t)|y_t^-\}$ *is given by*

$$d\hat{x}(t) = A(t)\hat{x}(t)dt + B(t)u(t)dt + L(t)(dy(t) - C(t)\hat{x}(t)dt) \tag{10}$$

$$\hat{z}(t) = H(t)\hat{x}(t) \; ; \quad \hat{x}(t_o) = \bar{x}_o$$

where L *is given by*

$$L(t) = (\Sigma(t)C^T(t) + G(t)J^T(t))(J(t)J^T(t))^{-1}$$

where Σ *is defined by the Riccati differential equation*

$$\dot{\Sigma}(t) = A(t)\Sigma(t) + \Sigma(t)A^T(t) - (\Sigma(t)C^T(t) + G(t)J^T(t))(J(t)J^T(t))^{-1}(\Sigma(t)C^T(t) + G(t)J^T(t))^T + G(t)G^T(t)$$

$$\Sigma(t_o) = \Pi_o$$

In fact, $\hat{x}(t) = E\{x(t)|y_t^-\}$ *and* $\Sigma(t) = E\{(x(t) - \hat{x}(t))(x(t) - \hat{x}(t))^T\}$

We conclude this section with two remarks concerning points which play a role in other contributions in these proceedings:

1. An important feature of the Kalman filter is its *innovations property*, i.e. the fact that, in the discrete time case, the random variables $y(t) - C(t)\hat{x}(t) + y(t) - E\{y(t)|y^-_{t-1}\}$ are independent and that, in the continuous time case, the process $y(t) - \int_{t_0}^{t} C(t)\hat{x}(t)dt$ is a gaussian independent increment process on $[t_0,t_1]$ i.e., a non-standard Wiener process. In the continuous time case the analogous property is valid under much more general circumstances, but the precise extent of its validity remains a matter of conjecture.

2. Finally, a word about the *smoothness* of the Kalman filter. As it stands in Theorem 3 it is unclear in the continuous time case of the map y into $\hat{x}$ is smooth, this because $\hat{x}$ in (10) is updated by dy. However be defining $v(t): = \hat{x}(t) - L(t)y(t)$ one obtained the equivalent representation

$$\dot{v}(t) = (A(t) - L(t)C(t)v(t) + B(t)u(t) +$$

$$[(A(t) - L(t)C(t)L(t) - \dot{L}(t)]y(t)$$

$$\hat{x}(t) = v(t) + L(t)y(t) \;\; ; \; v(0) = \bar{x}_o$$

which shows (assuming that $\dot{L}$ exists) the desired continuity, It is in to call the above representation the *robust form* of the Kalman Filter.

5. THE SOLUTION OF THE LQG-PROBLEM

From Sections 4 and 5 we have learned the following facts:

1. If there is no uncertainty (or, which is hardly different, if the initial state is uncertain but measured), then the optimal least squares memoryless state feedback control law is given by $N(t)x$ with $N(t)$ given by (8) in Theorem 2.

2. The optimal least squares estimate of $N(t)x(t)$ given y^-_t, $E\{N(t)x(t)|y^-_t\}$, may be computed as $N(t)\hat{x}(t)$ as shown in Theorem 3. Common sense suggests that using the feedback control law $u(t) = N(t)\hat{x}(t)$ should provide a reasonable admissible feedback control law. This is called the *separation principle* or the *certainty equivalence principle* (in non LQG situations these two principles will, in general, not coincide). The LQG-problem is one where this procedure actually provides the optimal feedback control law. Indeed:

Theorem 4: *Consider the LQG stochastic optimal control problem introduced in Section 2. Assume that the nonsingularity conditions of both Theorem 2 and 3 are satisfies, and that the Riccati equation of Theorem 2 has a solution up to* t_0. *Then*

(discrete time):

$$\hat{x}(t+1) = A(t)\hat{x}(t) + B(t)N(t)\hat{x}(t) + L(t)(y(t) - C(t)\hat{x}(t))$$

$$u(t) = N(t)\hat{x}(t) \; ; \quad \hat{x}(t_0) = \bar{x}_0$$

(continuous time):

$$d\hat{x}(t) = A(t)\hat{x}(t)dt + B(t)N(t)\hat{x}(t)dt + L(t)(dy(t) - C(t)\hat{x}(t)dt)$$

$$u(t) = N(t)\hat{x}(t) \; ; \; \hat{x}(t_0) = \bar{x}_0$$

with N *and* L *as given in Theorem 2 and 3 respectively, defines the optimal feedback control law.*

Signal flow graphs illustrating the structure of the Kalman filter and the optimal LQG controller may be found in [4] and [5].

We conclude with a few closing comments:

1. In the LQ-problem the nonsingularity condition is needed to ensure that the optimal control would be smooth and contain no impulses (and higher order singularities). Similarly in the Kalman filter, the nonsingularity assumption ensures that the filter has no impulses in its weighting pattern. However, whereas the condition is necessary for the LQ-controller, there are many examples of singular filtering and LQG-problems which have nevertheless a smooth solution. For example, if exact measurements in the state are available then it is not difficult to see that the optimal control law is given by $u(t) = N(t)x(t)$. Other singular type control problems are used in self-tuning regulators used in adaptive control algorithms.

2. An important extension of the LQG-problem treated in Theorem 4 is the time-invariant case which may be obtained by taking the matrices in the model and the cost to be independent of t and letting $t_0 \to -\infty$ and $t_1 \to \infty$ (see [1] - [5]).

3. As we have seen in Section 2, the LQG-problem treats models where the observation is given in the form of $y = Lu + r$ with L a linear map and r a gaussian process. In order to use Theorem 4, one needs to have this model in the form (1) or (2). Algorithms and the theory which perform this step are the so-called realization algorithms. Both the deterministic version (for L) and the stochastic version (for r) are relevant here. It would be of interest to extend the existing theory such that on the one hand it enables one to treat the deterministic and

the stochastic problems simultaneously and on the other hand that it enables one to consider also the cost functional into the realization problem.

REFERENCES

[1] Kalman, R.E. (1960), Contributions to the theory of optimal control, Bol. Soc. Mat. Mexicana, Vol. 5, pp. 102-119.

[2] Kalman, R.E. (1960), A new approach to linear filtering and prediction problems, J. Basic Eng. (Trans, ASME Ser. D), Vol. 82, pp. 34-35.

[3] Kalman, R.E. and Bucy, R.S. (1961), New results in linear filtering and prediction theory, J. Basic Eng. (Trans. ASME Ser. D), Vol. 83, pp. 95-107.

[4] Willems, J.C. (1978), Recursive filtering, Statistica Neerlandica, Vol. 32, pp. 1-39.

[5] Kwakernaak, H, and Sivan R. (1972), Linear Optimal Control Systems, Wiley.

[6] Kailath, T. (1974), A view of three decades of linear filtering, IEEE Trans. on Information Theory, Vol. IT-20, pp. 146-180.

[7] Special Issue on the Linear-Quadratic-Gaussian Problem, IEEE Trans. on Automatic Control, Vol. AC-16, 1971.

[8] Willems, J.C. (1971), The Analysis of Feedback Systems, MIT Press.

[9] Wonham, W.M. (1970), Random differential equations in control theory, in Probabilistic Methods in Applied Mathematics, A.T. Barucha - Reid, Ed., pp. 131-212, Academic Press.

[10] Kalman, R.E., Falb, P.L. and Arbib, M.A. (1969), Topics in Mathematical Systems Theory, McGraw Hill.

[11] Willems, J.C., Systems theory models for the analysis of physical systems, Ricerche di Automatica, Vol. 10, to appear.

A TUTORIAL ARTICLE ON THE ITÔ INTEGRAL AND THE STOCHASTIC CALCULUS

Ruth F. Curtain

Rijksuniversiteit Groningen
Mathematisch Instituut
Postbus 800,
9700 AV Groningen, The Netherlands

ABSTRACT

This article briefly reviews the prerequisite material on Itô stochastic integrals and stochastic differential equations assumed for the study institute. For details the reader is referred to the recommended text "Stochastic Differential Equations" by L. Arnold, John Wiley 1974.

1. INTRODUCTION - STOCHASTIC PROCESSES

In this tutorial lecture, for simplicity, we shall only consider real-valued stochastic processes, that a family of random variables $x(t)$ parametrized by $t \in R$. All the concepts generalize to the case of vector-valued stochastic processes (see [1]). We shall be concerned with two types of stochastic processes, martingales and Markov processes.

Definition 1 - Markov Process

A Markov process is a stochastic process $\{x_t \; ; \; t \in T \subset R \text{ or } Z\}$ if for any $s, t \in T$, $s < t$ and any Borel set B of R, we have that

$$P(x_t \in B | x_S) = P(x_t \in B \mid x_s)$$

where $P(x_t \in B | x_s)$ denotes the conditional probability that the event $x_t \in B$ shall occur given that x_s has occurred.

M. Hazewinkel and J. C. Willems (eds.), Stochastic Systems: The Mathematics of Filtering and Identification and Applications, 45–51.

Intuitively this means that the future probability distribution of x_t depends only on the present and not on the whole past and so it is a natural mathematical model for a stochastic dynamical system. The martingale concept is harder to motivate from an intuitive point of view, so perhaps it is better to say that the mathematical structure of martingales lends itself particularly to an elegant theory of stochastic integration.

Definition 2 - Martingale

Let x_t $t \in T$ be a stochastic process such that it is F_t measurable for all $t \in T$, where F_t is a family of increasing sigma algebras ($F_s \subset F_t \subset F$ for $s > t$ and (Ω, F, p) the basic probability space). Then $\{x_t, F_t, t \in T\}$ is a martingale if $E\{x_t | F_s\} = x_s$ with probability one.
Some elementary examples of stochastic processes which occur frequently in the stochastic systems literature are

Example 1 - Discrete-time white noise

This is the process $\{e_k, k = 0, 1, 2, \ldots \infty\}$ where e_k are a zero mean mutually independent random variables with variance 1. This is clearly Markov and if we let F_k be the sigma algebra generated by $e_0, e_1 \ldots e_k$, then $\{y_k, F_k, k = 0, 1, \ldots\}$ is a martingale, where $y_k = \sum_{j=0}^{k} e_j$.

Example 2 - Scalar linear system driven by white noise.

Consider the process for $k = 0, 1, 2, \ldots$ defined by $x_{k+1} = ax_k + be_k$, x_0 given. Where a and b are scalars and e_k is white noise. One can regard this as the scalar linear system (a,b), driven by a white noise input $\{e_k\}$. x_k is a Markov process.

Example 3 - Wiener process

The Wiener Process $\{w(t), t \geq 0\}$ is a scalar stochastic process with independent stationary increments and such that $w(0) = 0$ and $w(t)$ is normal with zero expectation and variance t.

It has very interesting properties, being a Markov process, a martingale, a gaussian process and having continuous but non-differentiable sample paths.

Just as in example 2 we obtain a Markov Process from a discrete time linear system with white noise input, one could think of continuous time white noise as the input to the continuous-time dynamical system $\dot{x} = ax + bu$. While this notion is very popular with engineers, there is in fact no second order-continuous time stochastic process which fits the white-noise concept, due to the fact that requiring

$E\{x(t)x(s)\} = 0$ for $t \neq s$ leads to an infinite variance. While it is possible to use a generalized stochastic process definition (see §3 of [1]), the more usual approach is to consider the stochastic differential equation

$$dx(t) = a\, x(t)dt + b\, dw(t) \quad ; \; x(0) = x_o$$

by which we mean the integral equation

$$x(t) = x_o + \int_o^t a\, x(s)ds + \int_o^t b\, dw(s)$$

A more general model for a stochastic dynamical system would be

$$dx(t) = f(t, x(t))dt + g(t,x(t))dw(t) \quad ; \; x(0) = x_o .$$

Since the Wiener process is not of bounded variation, the Stieltjes definitions of integration are not applicable here and so one needs to define the Itô integral

$$\int_o^T g(t)dw(t).$$

2. THE ITÔ INTEGRAL

We proceed to define the Itô integral $\int_o^T f(s,\omega)dw(s)$ for a special class of random functions $f(s,\omega)$, such that for each $t \in [0,T]$, $f(t,\cdot)$ is measurable with respect to the sigma algebra $\mathcal{F}_t$ generated by $w(s)$; $0 \leq s \leq t$ and furthermore

$$E\{\int_o^T f^2(s)ds\} < \infty .$$

If in addition, f happens to be a step function in t, then we define

$$\int_o^T f(s,\omega)dw(s) = \sum_{i=o}^{n} f(t_i)\,[w(t_{i+1}) - w(t_i)]$$

where $\quad 0 = t_o \leq t_1 < \dots < t_n = T$

Then using it is easy to prove that

$$E\{\int_o^T f(s,\omega)dw(s)\} = 0 \tag{2.1}$$

and

$$E\{\int_o^T f(s,\omega)dw(s)\}^2 = \int_o^T E\{f^2(s)\}ds \tag{2.2}$$

For general f we approximate it by a sequence of step functions $\{f_n(s,\omega)\}$, such that

$$E\{\int_0^T (f_n(t) - f(t))^2\, dt\} \to 0 \quad \text{as} \quad n \to \infty .$$

Then one can show that the quadratic mean limit of $\int_0^T f_n(t)dw(t)$ exists and is independent of the approximating sequence and so we choose this as our definition of $\int_0^T f(s,\omega)dw(s)$. This Itô integral also has the properties (2.1) and (2.2) and it is a gaussian random variable. The indefinite integral $\int_0^t f(s,\omega)dw(s)$ has a version with continuous sample paths and it is a martingale with respect to F_t.

If the integrand, f, is non-random then we call it the Wiener integral and this is sufficient to generate a large class of linear stochastic dynamical systems.

<u>Example 4</u> - <u>Linear stochastic differential equation</u>

$$dx(t) = a\, x(t)dt + b\, dw(t)\ ;\ x(0) = x_o .$$

This has the unique solution $x(t) = e^{at}x_o + \int_0^t e^{a(t-s)}b\,dw(s)$ which is a gaussian Markov process (if x_o is gaussian).

The Wiener integral has properties similar to those for ordinary integration, but if the integrand is stochastic the situation becomes rather different.

<u>Example 5</u>

It is readily calculated by using discrete approximations that for the Itô integral

$$\int_0^t w(s)\, dw(s) = \frac{w(t)^2}{2} - \frac{t}{2}$$

which is rather surprizing.

This arises from our choice of $f(t_i)$ as the approximation to f on $[t_i, t_{i+1}]$. Other choices are also possible and they all lead to different answers. If we choose $f(\frac{t_i + t_{i+1}}{2})$, then we obtain $w(t)^2$ and if we choose $f((1-a)t_i + at_{i+1})$, we obtain $\frac{w(t)^2}{2} + (a - \frac{1}{2})t$.

They all produce a Markov process, but only the Itô integral with $a = 0$ ensures that $\int_0^t w(s)dw(s)$ is a martingale.

For most theoretical uses, the martingale property is desirable; however, the Stratonovich integral where one chooses $f(\frac{t_i + t_{i+1}}{2})$

is also attractive, because of its simpler calculus.

3. THE ITÔ CALCULUS

As illustrated by $\int_0^t w(s)dw(s)$, if we wish to use the Itô integral we need to learn a new calculus. Suppose that x(t) has the Itô differential

$$dx(t) = f(t)dt + g(t)dw(t) \tag{3.1}$$

or equivalently

$$x(t) = x_0 + \int_0^t f(s)ds + \int_0^t g(s)dw(s) \tag{3.2}$$

where it is assumed that f and g are such that the integrals are well-defined. Then if $\Phi : R^2 \to R$ is C^2 in the first argument and C^1 in the second, the stochastic process $y(t) = \ (x(t),t)$ has the Itô differential

$$dy(t) = [\frac{\partial\Phi}{\partial t}(x(t),t) + \frac{\partial\Phi}{\partial x}(x(t),t)f(t) + \frac{1}{2}\frac{\partial^2\Phi}{\partial x^2}g^2(t)]dt$$

$$+ \frac{\partial\Phi}{\partial x}(x(t),t)\ g(t)dw(t).$$

If one use the Stratonovich integral, the extra term $\frac{1}{2}\frac{\partial^2\Phi}{\partial x^2}g^2(t)dt$ would not appear.

4. STOCHASTIC DIFFERENTIAL EQUATIONS

In seeking a more general class of stochastic dynamical systems one is led to the study of the following stochastic differential equation

$$\begin{cases} dx(t) = a(t)x(t)dt + b(t)x(t)dw(t) \\ x(t_0) = x_0 \end{cases} \tag{4.1}$$

where $a(\cdot,\cdot)$ and $b(\cdot,\cdot)$ are deterministic maps from R^2 to R which are measurable and satisfy

$$|a(t,x) - a(t,y)| \leq L_1 |x - y| \tag{4.2}$$

$$|b(t,x) - b(t,y)| < L_2 |x - y| \tag{4.3}$$

$$|a(t,x)|^2 + |b(t,x)|^2 \leq K^2 (1 + |x|^2) \tag{4.4}$$

If x_0 is independent of the future increments $w(t)-w(t_0)$ of the Wiener process, then (4.1) has a unique solution which is a Markov process.

We have already considered the linear case in Example 4 and one of the few nonlinear equations with an explicit solution is the following

Example 5

$$dx(t) = a\,x\,dt + b\,x\,dw(t)$$

$$x(0) = x_0$$

This has the unique solution $x(t) = x_0 \exp\{(a-\frac{b^2}{2})t + bw(t)\}$.

The theorem quoted here has generalizations to the vector case, to the case where a and b are random and where w(t) is replaced by a martingale. ([4]) Another interesting aspect is that (4.1) models a large class of continuous Markov Processes, called diffusion processes ([5]).

5. APPROXIMATION OF STOCHASTIC DIFFERENTIAL EQUATIONS AND THE MODELLING QUESTION

Suppose that we have an essentially deterministic dynamical system system (4.1) $\dot{x} = f(x) + g(x)u(t)$, and we wish to choose a stochastic input, $u(t) = \xi(t)$, where $\xi(t)$ approximates "white noise", that is, it has very irregular sample paths. Real noise is continuous and a possible model is

$$\dot{x} = f(x) + g(x)\xi(t,\omega) \tag{4.2}$$

where one solves a deterministic equation for each sample path ω. The disadvantage of this model is that the solution is not a Markov process, contrary to one's intuition. Alternatively we could use the Itô model

$$dx(t) = f(x)dt + g(x)dw(t) \tag{4.3}$$

or the Stratonovich model with the different interpretation of integration. What one would want to know is the relationship between the solution of (4.3) and solutions of (4.2) with $\xi(t,\omega)=w_n(t,\omega)$, a smooth approximation of the Wiener process, which converges to w(t) as $n \to \infty$ in some suitable manner. If we denote the solutions of (4.2) with $\xi(t,\omega) = w_n(t,\omega)$ by $x_n(t)$ and those of (4.3) by x(t), then for smooth f and g we find that as $n \to \infty$, $x_n(t)$ converges to the solution of the Stratonovich version and not to the solution of the Itô version. $x_n(t)$ converges to the Itô solution of

(4.4) $$d\tilde{x}(t) = (f(\tilde{x}) + \tfrac{1}{2}(g'(\tilde{x}))^2\, dt + g(\tilde{x})dw(t)$$

In other words, when modelling physical phenomena with white noise disturbances (4.2), one should choose the Stratonovich version (4.3), or the Itô equation (4.4). This problem does not arise if g is independent of x, of course. Interesting discussions of the modelling question can be found in [1] and [5].

REFERENCES

1. L. Arnold, "Stochastic Differential Equations". John Wiley, 1974.

2. K. Åström, "Introduction to Stochastic Control Theory". Academic Press, 1970.

3. M.H.A. Davis, "Linear Estimation and Stochastic Control". Champman & Hall, 1977.

4. R.S. Liptser and A.N. Shiryayev, "Statistics of Random Processes I." General Theory, Springer Verlag, 1977.

5. E. Wong, "Stochastic Processes in Information and Dynamical Systems". McGraw Hill, New York, 1971.

AN INTRODUCTION TO NONLINEAR FILTERING

M.H.A. Davis

Department of Electrical Engineering
Imperial College, London SW7 2BT, England.

Steven I. Marcus

Department of Electrical Engineering
The University of Texas at Austin
Austin, Texas 78712, U.S.A.

ABSTRACT

In this paper we provide an introduction to nonlinear filtering from two points of view: the innovations approach and the approach based upon an unnormalized conditional density. The filtering problem concerns the estimation of an unobserved stochastic process $\{x_t\}$ given observations of a related process $\{y_t\}$; the classic problem is to calculate, for each t, the conditional distribution of x_t given $\{y_s, 0 \leq s \leq t\}$. First, a brief review of key results on martingales and Markov and diffusion processes is presented. Using the innovations approach, stochastic differential equations for the evolution of conditional statistics and of the conditional measure of x_t given $\{y_s, 0 \leq s \leq t\}$ are given; these equations are the analogs for the filtering problem of the Kolmogorov forward equations. Several examples are discussed. Finally, a less complicated evolution equation is derived by considering an "unnormalized" conditional measure.

M. Hazewinkel and J. C. Willems (eds.), Stochastic Systems: The Mathematics of Filtering and Identification and Applications, 53–75.

I. INTRODUCTION

Filtering problems concern "estimating" something about an unobserved stochastic process $\{x_t\}$ given observations of a related process $\{y_t\}$; the classic problem is to calculate, for each t, the conditional distribution of x_t given $\{y_s, 0 \leq s \leq t\}$. This was solved in the context of linear system theory by Kalman and Bucy [1],[2] in 1960, 1961, and the resulting "Kalman filter" has of course enjoyed immense success in a wide variety of applications. Attempts were soon made to generalize the results to systems with nonlinear dynamics. This is an essentially more difficult problem, being in general infinite-dimensional, but nevertheless equations describing the evolution of conditional distributions were obtained by several authors in the mid-sixties; for example, Bucy [3], Kushner [4], Shiryaev [5], Stratonovich [6], Wonham [7]. In 1969 Zakai [8] obtained these equations in substantially simpler form using the so-called "reference probability" method (see Wong [9]).

In 1968 Kailath [10] introduced the "innovations approach" to linear filtering, and the significance for nonlinear filtering was immediately appreciated [11], namely that the filtering problem ought to be formulated in the context of martingale theory. The definitive treatment from this point of view was given in 1972 by Fujisaki, Kallianpur and Kunita [12]. Textbook accounts including all the mathematical background can be found in Liptser and Shiryaev [13] and Kallianpur [14].

More recent work on nonlinear filtering has concentrated on the following areas (this list and the references are not intended to be exhaustive):

(i) Rigorous formulation of the theory of stochastic partial differential equations (Pardoux [15], Krylov and Rozovskii [16]);

(ii) Introduction of Lie algebraic and differential geometric methods (Brockett [17]);

(iii) Discovery of finite dimensional nonlinear filters (Benes [18]);

(iv) Development of "robust" or "pathwise" solutions of the filtering equations (Davis [19]);

(v) Functional integration and group representation methods (Mitter [30]).

All of these topics are dealt with in this volume and all of them use the basic equations of nonlinear filtering theory: the Fujisaki, et.al., equation [12] and/or the Zakai equation [8]. These equations can be derived in a quick and self-contained way, modulo some technical results, the statements of which are readily appreciated and the details of which can be found in the references [13],[14]. This is the purpose of the present article.

The general problem can be described as follows. The *signal* or *state* process $\{x_t\}$ is a stochastic process which cannot be observed directly. Information concerning $\{x_t\}$ is obtained from the *observation* process $\{y_t\}$, which we will assume is given by

$$y_t = \int_0^t z_s ds + w_t \tag{1}$$

where $\{z_t\}$ is a process "related" to $\{x_t\}$ (e.g., $z_t = h(x_t)$) and $\{w_t\}$ is a Brownian motion process. The process $\{y_t\}$ is to be thought of as noisy nonlinear observations of the signal $\{x_t\}$. The objective is to compute least squares estimates of functions of the signal x_t given the "past" observations $\{y_s, 0 \leq s \leq t\}$ -- i.e. to compute quantities of the form $E[\phi(x_t)|y_s, 0 \leq s \leq t]$. In addition, it is desired that this computation be done *recursively* in terms of a statistic $\{\pi_t\}$ which can be updated using only new observations:

$$\pi_{t+\tau} = \gamma(t,\tau,\pi_t,\{y_{t+u}, 0 \leq u \leq \tau\}), \tag{2}$$

and from which estimates can be calculated in a "pointwise" or "memoryless" fashion:

$$E[\phi(x_t)|y_s, 0 \leq s \leq t] = \delta(t, y_t, \pi_t). \tag{3}$$

In general, π_t will be closely related to the conditional distribution of x_t given $\{y_s, 0 \leq s \leq t\}$, but in certain special cases π_t will be computable with a finite set of stochastic differential equations driven by $\{y_t\}$ (see [20] for some examples).

In order to obtain specific results, additional structure will be assumed for the process $\{x_t\}$; we will assume throughout that $\{x_t\}$ is a semimartingale (see Section II), but more detailed results will be derived under the assumption that $\{x_t\}$ is a Markov process or in particular a vector diffusion process of the form

$$x_t = x_0 + \int_0^t f(x_s)ds + \int_0^t G(x_s)d\beta_s, \tag{4}$$

where $x_t \in \mathbb{R}^n$ and $\beta_t \in \mathbb{R}^m$ is a vector of independent Brownian motion processes. General terminology and precise assumptions will be presented in Section II. In Section III, Markov processes of the form (4) will be studied, and Kolmogorov's equations for the evolution of the unconditional distribution (i.e. without observations) of the process $\{x_t\}$ will be presented. The corresponding equations for the conditional distribution of x_t given $\{y_s, 0 \leq s \leq t\}$ will be derived in Section IV using the "innovations approach". Finally, in Section V we derive a less complex set of equations for an unnormalized conditional distribution of x_t, in the form given by Zakai [8].

II. TERMINOLOGY AND ASSUMPTIONS

In this section we review certain notions concerning stochastic processes and martingales; for further tutorial material on martingale integrals and stochastic calculus, the reader is referred to the tutorial of R. Curtain in this volume and the paper of Davis [21] (see also [9],[13],[22]-[24]). All stochastic processes will be defined on a fixed probability space (Ω, F, P) and a finite time interval $[0,T]$, on which there is defined an increasing family of σ-fields $\{F_t, 0 \leq t \leq T\}$. It is assumed that each process $\{x_t\}$ is *adapted* to F_t -- i.e. x_t is F_t-measurable for all t. The σ-field generated by $\{x_s, 0 \leq s \leq t\}$ is denoted by $X_t = \sigma\{x_s, 0 \leq s \leq t\}$. (x_t, F_t) is a *martingale* if x_t is adapted to F_t, $E|x_t| < \infty$, and $E[x_t | F_s] = x_s$ for $t \geq s$. (x_t, F_t) is a *supermartingale* if $E[x_t | F_s] \leq x_s$ and a *submartingale* if $E[x_t | F_s] \geq x_s$. The process (x_t, F_t) is a *semimartingale* if it has a decomposition $x_t = x_0 + a_t + m_t$, where (m_t, F_t) is a martingale and $\{a_t\}$ is a process of bounded variation. Given two square integrable martingales (m_t, F_t) and (n_t, F_t), one can define the *predictable quadratic covariation* $(\langle m,n\rangle_t, F_t)$ to be the unique "predictable process of integrable variation" such that $(m_t n_t - \langle m,n\rangle_t, F_t)$ is a martingale [29, p.34]. For the purposes of this paper, however, the only necessary facts concerning $\langle m,n\rangle$ are that (a) $\langle m,n\rangle_t = 0$ if $m_t n_t$ is a martingale; and (b) if β is a standard Brownian motion process, then

$$\langle \beta, \beta \rangle_t = t \quad \text{and} \quad \left\langle \int_0^t \eta_s^1 d\beta_s, \int_0^t \eta_s^2 d\beta_s \right\rangle_t = \int_0^t \eta_s^1 \eta_s^2 ds.$$

In this tutorial exposition, the following hypotheses will be assumed for all nonlinear estimation problems:

H1. $\{y_t\}$ is a real-valued process;

H2. $\{w_t\}$ is a standard Brownian motion process;

H3. $E[\int_0^T z_s^2 ds] < \infty$

H4. $\{z_t\}$ is independent of $\{w_t\}$.

Hypotheses (H1) and (H4) can be weakened, but the calculations become more involved [8],[12],[13, Chapter 8]. Similar results to those derived here can also be derived in the case that $\{w_t\}$ is replaced by the sum of a Brownian motion and a counting process [25]. Hypotheses on the process $\{x_t\}$ and the relationship between x_t and w_t will be imposed as they are needed in the sequel.

Finally, we will need two special cases of Ito's differential rule. Suppose that (ξ_t^i, F_t), i=1,2, are semimartingales of the form

$$\xi_t^i = \xi_0^i + a_t^i + m_t^i, \tag{5}$$

where $\{m_t^i\}$, i=1,2, are square integrable martingales with $\{m_t^1\}$ and $\{a_t^1\}$ sample continuous. Then

$$\xi_t^1 \xi_t^2 = \xi_0^1 \xi_0^2 + \int_0^t \xi_s^1 d\xi_s^2 + \int_0^t \xi_s^2 d\xi_s^1 + \langle m^1, m^2 \rangle_t. \tag{6a}$$

Also, if ϕ is a twice continuously differentiable function of a process x of the form (4), then

$$\psi(x_t) = \psi(x_0) + \sum_{i=1}^{n} \frac{\partial\psi}{\partial x^i}(x_s)dx_s^i + \frac{1}{2}\sum_{i,j=1}^{n}\int_0^t \frac{\partial^2\psi}{\partial x^i \partial x^j}(x_s)a^{ij}(x_s)ds \tag{6b}$$

where $A(x) = [a^{ij}(x)] := G(x)G'(x)$ and x^i denotes the ith component of x.

III. MARKOV AND DIFFUSION PROCESSES

A very clear account of the material in this section can be found in Wong's book [9]. A stochastic process $\{x_t, t \in [0,T]\}$ is a *Markov process* if for any $0 \le s \le t \le T$ and any Borel set B of the state space S,

$$P(x_t \in B | X_s) = P(x_t \in B | x_s).$$

For any Markov process $\{x_t\}$, we can define the *transition probability function*

$$P(s,x,t,B) := P(x_t \in B | x_s = x),$$

which can easily be shown to satisfy the *Chapman-Kolmogorov equation:* for any $0 \leq s \leq u \leq t \leq T$,

$$P(s,x,t,B) = \int_S P(u,y,t,B)\, P(s,x,u,dy). \tag{7}$$

In addition, all finite dimensional distributions of a Markov process are determined by its initial distribution and transition probability function. A Markov process $\{x_t\}$ is *homogeneous* if $P(s+u,x,t+u,B) = P(s,x,t,B)$ for all $0 \leq s \leq t \leq T$ and $0 \leq s+u \leq t+u \leq T$.

For a homogeneous Markov process $\{x_t\}$ and $f \in \mathcal{B}(S)$ (i.e. f is a bounded measurable real-valued function on S), define

$$T_t f(x) = E_x[f(x_t)] := \int_S f(y)P(0,x,t,dy).$$

The Chapman-Kolmogorov equation then implies that T_t is a semigroup of operators acting on $\mathcal{B}(s)$; i.e. $T_{t+s}f(x) = T_t(T_s f)(x)$ for $t,s \geq 0$. The *generator* L of T_t (or, of $\{x_t\}$) is the operator acting on a domain $D(L) \subset \mathcal{B}(S)$ given by

$$L\phi = \lim_{t \downarrow 0} \frac{1}{t}(T_t\phi - \phi),$$

the limit being uniform in $x \in S$ and $D(L)$ consisting of all functions such that this limit exists. It is immediate from this and the semigroup property that

$$\frac{d}{dt} T_t\phi = LT_t\phi \tag{8}$$

and (8) is, in abstract form, the *backward equation* for the process. Writing it out in integral form and recalling the definition of T_t gives the *Dynkin formula:*

$$E_x[\phi(x_t)] - \phi(x) = E_x \int_0^t L\phi(x_s)ds. \tag{9}$$

This implies, using the Markov property again, that the process M_t^ϕ defined for $\phi \in D(L)$ by

$$M_t^\phi = \phi(x_t) - \phi(x) - \int_0^t L\phi(x_s)ds \tag{10}$$

is a martingale [26, p.4]. This property can be used as a definition of L; this is the approach pioneered by Stroock and Varadhan [26]. Then L is known as the *extended generator* of $\{x_t\}$, since there may be functions ϕ for which M_t^ϕ is a martingale but which are not in D(L) as previously defined.

There is another semigroup of operators associated with $\{x_t\}$, namely the operators which transfer the initial distribution of the process into the distributions at later times t. More precisely, let M(S) be the set of probability measures on S and denote

$$\langle\phi,\mu\rangle = \int_S \phi(x)\mu(dx)$$

for $\phi \in \mathcal{B}(S)$, $\mu \in M(S)$. Suppose x_0 has distribution $\pi \in M(S)$; then the distribution of x_t is given by

$$U_t\pi(A) = P[x_t \in A] = E(I_A(x_t)) = \langle T_t I_A, \pi\rangle.$$

This shows that U_t is adjoint to T_t in that

$$\langle\phi, U_t\pi\rangle = \langle T_t\phi,\pi\rangle \; (=E\phi(x_t))$$

for $\phi \in \mathcal{B}(S)$, $\pi \in M(S)$. Thus the generator of U_t is L^*, the adjoint of L, and $\pi_t := U_t\pi$ satisfies

$$\frac{d}{dt}\pi_t = L^*\pi_t, \quad \pi_0 = \pi. \tag{11}$$

This is the *forward equation* of x_t in that it describes the evolution of the distribution π_t of x_t. The objective of filtering theory is to obtain a similar description of the *conditional* distribution of x_t given $\{y_s, s \leq t\}$.

In order to get these results in more explicit form we consider in the remainder of this section a process $\{x_t\}$ satisfying a stochastic differential equation of the form (4), where $\{\beta_t\}$ is an $\mathbb{R}^m$-valued standard Brownian motion process independent of x_0. For simplicity we assume that f and G do not depend explicitly on t (this is no loss of generality, since the "process" $\tau(t) = t$ can be accommodated by augmenting (4) with the equation $d\tau/dt = 1$, $\tau(0) = 0$). Under the usual Lipschitz and growth assumptions which guarantee existence and uniqueness of (strong) solutions of (4), the following results can be proved [9],[22]-[24].

Theorem 1: The solution of (4) is a homogeneous Markov process with infinitesimal generator

$$L = \sum_{i=1}^{n} f^i(x) \frac{\partial^i}{\partial x^i} + \frac{1}{2} \sum_{i,j=1}^{n} a^{ij}(x) \frac{\partial^2}{\partial x^i \partial x^j} \tag{12}$$

where $A(x) = [a^{ij}(x)] := G(x)G'(x)$, and f^i and x^i denote the ith components of f and x, respectively.

Hence Ito's rule (6b) in this case can be written as

$$\psi(x_t) = \psi(x_0) + \int_0^t L\psi(x_s)ds + \int_0^t \nabla\psi'(x_s)G(x_s)d\beta_s,$$

emphasizing again that M_t^ψ (see (10)) is a martingale (here $\nabla\psi$ is the gradient of ψ with respect to x, expressed as a column vector). It can also be shown [24] that the solution of (4) satisfies the Feller and strong Markov properties, and is a diffusion process with drift vector f and diffusion matrix A. If this process has a smooth density then the abstract equations (8) and (11) translate into Kolmogorov's backward and forward equations for the transition density.

Theorem 2 [24, p.104]: Assume that the solution $\{x_t\}$ of (4) has a transition density:

$$P(s,x,t,B) = \int_B p(s,x,t,y)dy$$

satisfying

a) for $t-s > \delta > 0$, $p(s,x,t,y)$ is continuous and bounded in s, t, and x;

b) the partial derivatives $\frac{\partial p}{\partial s}, \frac{\partial^i p}{\partial x^i}, \frac{\partial^2 p}{\partial x^i \partial x^j}$ exist.

Then for $0 < s < t$, p satisfies the *Kolmogorov backward equation*

$$\frac{\partial}{\partial s} p(s,x,t,y) + Lp(s,x,t,y) = 0 \tag{13}$$

with $\lim_{s \uparrow t} p(s,x,t,y) = \delta(x-y)$ and L given by (12); i.e. p is the fundamental solution of (13).

Outline of Proof: From (7), we have

$$p(s+h,x,t,y) - p(s,x,t,y)$$
$$= \int p(s,x,s+h,z)[p(s+h,x,t,y) - p(s+h,z,t,y)]dz.$$

Dividing both sides by h and letting $h \to 0$ yields (13) by using the definition of L.

More relevant to filtering problems is the Kolmogorov forward equation.

Theorem 3 [24, p.102]: Assume that $\{x_t\}$ satisfying (4) has a transition density $p(s,x,t,y)$, and that $\frac{\partial f}{\partial x^i}, \frac{\partial A}{\partial x^i}, \frac{\partial^2 A}{\partial x^i \partial x^j}, \frac{\partial p}{\partial t}$, $\frac{\partial p}{\partial y}$, and $\frac{\partial^2 p}{\partial y^2}$ exist. Then for $0 < s < t$, p satisfies the *Kolmogorov forward equation*

$$\frac{\partial p}{\partial t}(s,x,t,y) = -\sum_{i=1}^{n} \frac{\partial}{\partial y^i}(f^i(y)p(s,x,t,y)) + \frac{1}{2}\sum_{i,j=1}^{n} \frac{\partial^2}{\partial y^i \partial y^j}(a^{ij}(y)p(s,x,t,y)) \tag{14}$$
$$:= L^* p(s,x,t,y)$$

where L^* is the formal adjoint of L. Also, the initial condition is $\lim_{t \downarrow s} p(s,x,t,y) = \delta(y-x)$.

Outline of Proof: Assume, for simplicity of notation, that $\{x_t\}$ is a scalar diffusion (n=1). From (9), we have

$$\frac{\partial}{\partial t}\int p(s,x,t,z)\phi(z)dz = \int p(s,x,t,z)L\phi(z)dz \tag{15}$$

for some twice continuously differentiable function which vanishes outside some finite interval. The derivative and integral on the left-hand side of (15) can be interchanged, and an integration by parts then yields

$$\int p(s,x,t,z)f(z)\frac{\partial\phi}{\partial z}(z)dz = -\int\phi(z)\frac{\partial}{\partial z}(f(z)p(s,x,t,z))dz,$$

$$\int p(s,x,t,z)g^2(z)\frac{\partial^2\phi}{\partial z^2}(z)dz = \int\phi(z)\frac{\partial^2}{\partial z^2}(g^2(z)p(s,x,t,z))dz,$$

hence

$$\int\Big\{\frac{\partial p}{\partial s}(s,x,t,z) + \frac{\partial}{\partial z}[f(z)p(s,x,t,z)]$$
$$-\frac{1}{2}\frac{\partial^2}{\partial z^2}[g^2(z)p(s,x,t,z)]\Big\}\phi(z)dz = 0.$$

Since the expression in curly brackets is continuous and $\phi(z)$ is an arbitrary twice differentiable function vanishing outside a finite interval, (14) follows.

We note that if x_0 has distribution P_0, then the density of x_t is $p(t,y) = \int p(0,x,t,y)P_0(dx)$, and $p(t,y)$ also satisfies (14). Conditions for the existence of a density satisfying the differentiability hypotheses of Theorems 2 and 3 are given in [24, pp.96-99] (see also Pardoux [15]).

IV. THE INNOVATIONS APPROACH TO NONLINEAR FILTERING

In this section we derive stochastic differential equations for the evolution of conditional statistics and of the conditional density for nonlinear filtering problems of the types discussed in Sections I and II; the equations will be the analogs of (9) and the Kolmogorov forward equation for the filtering problem. We will follow the innovations approach, as presented in [12] and [13]; this approach was originally suggested by Kailath [10] (for linear filtering) and Frost and Kailath [11].

Assume that the observations have the form (1) and that (H1)-(H4) hold. Define $\mathcal{Y}_t := \sigma\{y_s, 0 \le s \le t\}$; for any process η_t we use the notation $\hat{\eta}_t := E[\eta_t | \mathcal{Y}_t]$. Now introduce the *innovations process:*

$$\nu_t := y_t - \int_0^t \hat{z}_s ds. \tag{16}$$

The incremental innovations $\nu_{t+h} - \nu_t$ represent the "new information" concerning the process $\{z_t\}$ available from the observations between t and t+h, in the sense that $\nu_{t+h} - \nu_t$ is independent of $\mathcal{Y}_t$. The following properties of the innovations are crucial.

Lemma 1: The process $(\nu_t, \mathcal{Y}_t)$ is a standard Brownian motion process. Furthermore, $\mathcal{Y}_s$ and $\sigma\{\nu_u - \nu_t, 0 \le s \le t < u \le T\}$ are independent.

Proof: From (16) we have for $s < t$,

$$E[\nu_t | \mathcal{Y}_s] = \nu_s + E[\int_s^t (z_u - \hat{z}_u)du + w_t - w_s | \mathcal{Y}_s]. \tag{17}$$

The second term on the right-hand side of (17) is zero; here we have used the fact that $w_t - w_s$ is independent of $\mathcal{Y}_s$. Hence $(\nu_t, \mathcal{Y}_t)$ is a martingale. Consider now the quadratic variation of $\{\nu_t\}$: for $t \in [0,T]$ fix an integer n and define

$$Q_t^n = \sum_{0 \le k < 2^n t} [\nu((k+1)/2^n) - \nu(k/2^n)]^2.$$

The almost sure limit (as $n \to \infty$) of $Q_t^n := Q_t$ is the quadratic variation of ν_t. It is easy to see that the quadratic variation of $\int_0^t (z_u - \hat{z}_u)du$ is zero, so that the quadratic variation of ν_t is the same as that of w_t, or $Q_t = t$. But by a theorem of Doob [12, Lemma 2.1], a square integrable martingale with continuous sample paths and quadratic variation t is a standard Brownian motion, and the lemma follows.

Notice that the very specific conclusions of Lemma 1 regarding the structure of the innovations process are valid without any restrictions on the distributions of z_t. The next lemma is related to Kailath's "innovations conjecture". By definition ν_t is $\mathcal{Y}_t$-measurable and $\sigma\{\nu_s, 0 \le s \le t\} \subset \mathcal{Y}_t$. The innovations conjecture is that $\mathcal{Y}_t \subset \sigma\{\nu_s, 0 \le s \le t\}$, and hence that the two σ-fields are equal; i.e. the observations and innovations processes contain the same information. At the time that [12] was written, the answer to this question was not known under very general conditions on $\{z_t\}$; recently, it has been shown in [27] that the conjecture is true under the conditions (H1)-(H4). It is a well-known fact [13, Theorem 5.6] that all martingales of Brownian motion are stochastic integrals, and the point of a positive answer to the innovations conjecture is that it enables any Y_t-martingale to be written as a stochastic integral with respect to the *innovations* process $\{\nu_t\}$. The essential contribution of Fujisaki, Kallianpur and Kunita [12] was to show that this representation holds whether or not the innovations conjecture is valid. Specifically, they showed:

Lemma 2: Every square integrable martingale (m_t, Y_t) with respect to the observation σ-fields Y_t is sample continuous and has the representation

$$m_t = E[m_0] + \int_0^t \eta_s d\nu_s \tag{18}$$

where $\int_0^T E[\eta_s^2]\,ds < \infty$ and $\{\eta_t\}$ is jointly measurable and adapted to Y_t. In other words, m_t can be written as a stochastic integral with respect to the innovations process. (But note that $\{\eta_t\}$ is adapted to Y_t and not necessarily to F_t^{ν}.)

In order to obtain a general filtering equation, let us consider a real-valued F_t-semimartingale ξ_t and derive an equation satisfied by $\hat{\xi}_t$. We have in mind semimartingales $\phi(x_t)$ where ϕ is some smooth real-valued function and $\{x_t\}$ is the signal process, but it is just as easy to consider a general semimartingale of the form

$$\xi_t = \xi_0 + \int_0^t \alpha_s ds + n_t \tag{19}$$

where (n_t, F_t) is a martingale.

Theorem 4: Assume that $\{\xi_t\}$ and $\{y_t\}$ are given by (19) and (1), respectively, and that $\langle n,w\rangle_t = 0$. Then $\{\hat{\xi}_t\}$ satisfies the stochastic differential equation

$$\hat{\xi}_t = \hat{\xi}_0 + \int_0^t \hat{\alpha}_s ds + \int_0^t [\widehat{\xi_s z_s} - \hat{\xi}_s \hat{z}_s]\, d\nu_s. \tag{20}$$

Proof: First we define

$$\mu_t := \hat{\xi}_t - \hat{\xi}_0 - \int_0^t \hat{\alpha}_s ds$$

and show that $(\mu_t, \mathcal{Y}_t)$ is a martingale. Now, for $s<t$,

$$\begin{aligned} E[\hat{\xi}_t-\hat{\xi}_s|\mathcal{Y}_s] &= E[\xi_t-\xi_s|\mathcal{Y}_s] \\ &= E[\int_s^t \alpha_u du|\mathcal{Y}_s] + E[n_t-n_s|\mathcal{Y}_s] \\ &= E[\int_s^t E[\alpha_u|\mathcal{Y}_u]du|\mathcal{Y}_s] + E[E[n_t-n_s|F_s]|\mathcal{Y}_s]. \end{aligned} \tag{21}$$

The last term in (21) is zero, since (n_t, F_t) is a martingale; thus (21) proves that $(\mu_t, \mathcal{Y}_t)$ is a martingale. Hence,

$$\begin{aligned} \hat{\xi}_t &= \hat{\xi}_0 + \int_0^t \hat{\alpha}_s ds + \mu_t \\ &= \hat{\xi}_0 + \int_0^t \hat{\alpha}_s ds + \int_0^t \eta_s d\nu_s \end{aligned} \tag{22}$$

where the last term in (22) follows from Lemma 2.

It remains only to identify the precise form of η_t, using Ito's differential rule (6a) and an idea introduced by Wong [28]. From (1) and (19), and since $\langle n,w\rangle_t = 0$,

$$\xi_t y_t = \xi_0 y_0 + \int_0^t \xi_s(z_s ds + dw_s) + \int_0^t y_s(\alpha_s ds + dn_s). \tag{23}$$

Also, from (16) and (22),

$$\hat{\xi}_t y_t = \hat{\xi}_0 y_0 + \int_0^t \hat{\xi}_s(\hat{z}_s ds + d\nu_s) + \int_0^t y_s(\hat{\alpha}_s ds + \eta_s d\nu_s) + \int_0^t \eta_s ds. \tag{24}$$

Now it follows immediately from properties of conditional expectations that for $t \geq s$,

$$E[\xi_t y_t - \hat{\xi}_t y_t|\mathcal{Y}_s] = 0.$$

Calculating this from (23),(24) we see that

$$\eta_t = \widehat{\xi_t z_t} - \hat{\xi}_t \hat{z}_t. \tag{25}$$

Inserting (25) into (22) gives the desired result (20).

Formula (20) is not very useful as it stands (it is not a recursive equation for $\hat{\xi}_t$), but we can use it to obtain more explicit results for filtering of Markov processes.

Theorem 5: Assume that $\{x_t\}$ is a homogeneous Markov process with infinitesimal generator L, that $\{y_t\}$ is given by (1) with $z_t = h(x_t)$, and that $\{x_t\}$ and $\{w_t\}$ are independent. Then for any $\phi \in D(L)$, $\pi_t(\phi) := E[\phi(x_t)|\mathcal{Y}_t]$ satisfies

$$\pi_t(\phi) = \pi_0(\phi) + \int_0^t \pi_s(L\phi)ds + \int_0^t [\pi_s(h\phi) - \pi_s(h)\pi_s(\phi)]d\nu_s. \quad (26)$$

Proof: Notice that (M_t^ϕ, F_t) (see (10)) is a martingale, so that $\xi_t := \phi(x_t)$ is of the form (19) with $\alpha_t := L\phi(x_t)$, $n_t := M_t^\phi$. Also, it is shown in [12, Lemma 4.2] that the independence of $\{x_t\}$ and $\{w_t\}$ implies $\langle M^\phi, w\rangle_t = 0$. The theorem then follows immediately from Theorem 4.

Remarks: (i) Since $\{\pi_t(\phi): \phi \in D(L)\}$ determines a measured valued stochastic process π_t, (26) can be regarded as a recursive (infinite-dimensional) stochastic differential equation for the conditional measure π_t of x_t given $\mathcal{Y}_t$, and $\pi_t(\phi)$ is a conditional statistic computed from π_t in a memoryless fashion (see (2)-(3)). In general, however, it is not possible to derive a *finite dimensional* recursive filter, even for the conditional mean $\hat{x}_t$; some special cases in which finite dimensional recursive filters exist are given in Examples 1 and 3 below.

(ii) If w_t in (1) were multiplied by $r^{\frac{1}{2}}$ with $r > 0$, one would suspect that as $r \to \infty$ the observations would become infinitely noisy, thus giving no information about the state; i.e. $\pi_t(\phi)$ would reduce to the unconditional expectation $E[\phi(x_t)]$. In fact, in this case the last term in (26) is multiplied by r^{-1}, so (26) reduces to (9) as $r \to \infty$.

Example 1 [7]: Let $\{x_t\}$ be a finite state Markov process taking values in $S = \{s_1, \ldots, s_N\}$. Let p_t^i be the probability that

$x_t = s_i$, and assume that $p_t := [p_t^1,\dots,p_t^N]'$ satisfies

$$\frac{d}{dt} p_t = Ap_t.$$

(This is the forward equation for $\{x_t\}$; cf.(11).) Given the observations (1), the conditional distribution of x_t given $\mathcal{Y}_t$ can be determined from (26) as follows. Let $\phi(x) = [\phi_1(x),\dots,\phi_N(x)]'$, where

$$\phi_i(x) = \begin{cases} 1, & x = s_i \\ 0, & x \neq s_i. \end{cases}$$

Then applying (26) to each ϕ_i yields the following: let $B = \mathrm{diag}(h(s_1),\dots,h(s_N))$ and let $b = [h(s_1),\dots,h(s_N)]'$. Then if $\tilde{p}_t^i = P[x_t = s_i | \mathcal{Y}_t]$ and $\tilde{p}_t = [\tilde{p}_t^1,\dots,\tilde{p}_t^N]'$, we have

$$\tilde{p}_t = \tilde{p}_0 + \int_0^t A\tilde{p}_s ds + \int_0^t [B-(b'\tilde{p}_s)I]\,\tilde{p}_s(dy_s-(b'\tilde{p}_s)ds).$$

In this case, the conditional distribution is determined recursively by N stochastic differential equations.

Example 2: Assume that $\{x_t\}$ is a diffusion process given by (4) with infinitesimal generator (12) and that the conditional distribution of x_t given $\mathcal{Y}_t$ has a density $\tilde{p}(t,x)$. Then under appropriate differentiability hypotheses [13, Theorem 8.6], one can do an integration by parts in (26) (precisely as in Theorem 3 above) to obtain the stochastic partial differential equation

$$d\tilde{p}(t,x) = L^*\tilde{p}(t,x)dt + \tilde{p}(t,x)\,[h(x)-\pi_t(h)]\,d\nu_t \tag{27}$$

where

$$\pi_t(h) = \int h(x)\tilde{p}(t,x)dx. \tag{28}$$

This is a recursive equation for the computation of $\tilde{p}(t,x)$; it is not only infinite dimensional but has a complicated structure due to the presence of the integral in (28). Equation (27) is the analog of the Kolmogorov forward equation; in fact, (27) reduces to (13) as the observation noise approaches ∞ (see Remark (ii)).

The conditional mean cannot in general be computed with a finite dimensional recursive filter, as is seen by letting $\phi(x) = x$ in (26):

$$\hat{x}_t = \hat{x}_0 + \int_0^t \pi_s(f)ds + \int_0^t [\pi_s(hx) - \pi_s(h)\hat{x}_s]\, d\nu_s. \tag{29}$$

Hence, $\pi_t(f)$, $\pi_t(hx)$, and $\pi_t(h)$ are all necessary for the computation of $\hat{x}_t$, etc. One case in which this calculation is possible is given in the next example.

Example 3 (Kalman-Bucy Filter): Suppose, for simplicity, that $\{x_t\}$ and $\{y_t\}$ are given by the following scalar "linear-Gaussian" equations:

$$x_t = x_0 + \int_0^t ax_s ds + bw_t$$

$$y_t = \int_0^t cx_s ds + v_t$$

where x_0 is Gaussian and independent of $\{w_t\}$ and $\{v_t\}$. Then (29) yields

$$\hat{x}_t = \hat{x}_0 + \int_0^t a\hat{x}_s ds + c\int_0^t [\pi_s(x^2) - \hat{x}_s^2][dy_s - c\hat{x}_s ds]$$

$$= \hat{x}_0 + \int_0^t a\hat{x}_s + c\int_0^t P_t(dy_s - c\hat{x}_s ds) \tag{30}$$

where $P_t := E[(x_t - \hat{x}_t)^2 | Y_t]$ is the conditional error covariance. However, since $\{x_t\}$ and $\{y_t\}$ are jointly Gaussian, P_t is nonrandom and constitutes a "gain" process which can be precomputed and stored. P_t satisfies the differential equation (derived from (26) by noticing that the third central moment of a Gaussian distribution is zero):

$$\frac{d}{dt} P_t = 2aP_t + b^2 - c^2P_t^2.$$

Since P_t is nonrandom and the differential equation for $\hat{x}_t$

involves no other conditional statistics, it constitutes a recursive one-dimensional filter (the Kalman-Bucy filter) for the computation of the conditional mean.

V. THE UNNORMALIZED EQUATIONS

Throughout this section it will be assumed that $\{x_t\}$ is a homogeneous Markov process with infinitesimal generator L, $\{y_t\}$ is given by (1) with $z_t = h(x_t)$, and $\{x_t\}$ and $\{w_t\}$ are independent. In this case, the conditional measure π_t satisfies the equation (26), but it is often more convenient to work with a less complicated equation which is obtained by considering an "unnormalized" version of π_t. The unnormalized equations are derived in [9, Chapter 6] and [8]; the use of measure transformations will follow these references, but we will use a shorter derivation of the unnormalized equations, via (26) and Ito's rule.

The first step is to define a new measure P_0 on the measurable space $(\Omega, \mathcal{F})$ by

$$P_0(A) = \int_A \frac{dP_0}{dP}(\omega)\, P(d\omega)$$

for all $A \in \mathcal{F}$, where

$$\frac{dP_0}{dP} = \exp\left(-\int_0^T h(x_s)dy_s + \frac{1}{2}\int_0^T h^2(x_s)ds\right)$$

is the Radon-Nikodym derivative of P_0 with respect to P.

<u>Lemma 3 [9, p.232]</u>: P_0 has the following properties:

(a) P_0 is a probability measure -- i.e. $P_0(\Omega) = 1$;

(b) Under P_0, $\{y_t\}$ is a standard Brownian motion;

(c) Under P_0, $\{x_t\}$ and $\{y_t\}$ are independent;

(d) $\{x_t\}$ has the same distributions under P_0 as under P;

(e) P is absolutely continuous with respect to P_0 with Radon-Nikodym derivative

$$\frac{dP}{dP_0} = \left(\frac{dP_0}{dP}\right)^{-1} = \exp\left(\int_0^T h(x_s)dy_s - \frac{1}{2}\int_0^T h^2(x_s)ds\right).$$

It can also be shown [13, Section 6.2] that

$$\Lambda_t := \exp\left(\int_0^t h(x_s)dy_s - \frac{1}{2}\int_0^t h^2(x_s)ds\right)$$

is a martingale with respect to F_t and P_0, so that

$$\Lambda_t = E_0\left[\frac{dP}{dP_0} \,|\, F_t\right],$$

where E_0 is the expectation with respect to P_0. It can be shown [9, p.234] that

$$\pi_t(\phi) := E[\phi(x_t)|Y_t] = \frac{E_0[\phi(x_t)\Lambda_t|Y_t]}{E_0[\Lambda_t|Y_t]} =: \frac{\sigma_t(\phi)}{\sigma_t(1)}. \quad (31)$$

Hence conditional statistics of x_t given Y_t, in terms of the original measure P, can be calculated in terms of conditional statistics under the measure P_0. We now proceed to derive a recursive equation for the measure σ_t; an approach to solving (31) by a path integration of the numerator and denominator is pursued in some other papers in this volume.

Since $\sigma_t(\phi) = \sigma_t(1)\cdot\pi_t(\phi)$ and we have the equation (26) for $\pi_t(\phi)$, an equation for $\sigma_t(\phi)$ is derived by finding a stochastic differential equation for $\sigma_t(1) := E_0[\Lambda_t|Y_t]$ and applying Ito's rule.

Lemma 4: $E_0[\Lambda_t|Y_t]$ is given by the formula

$$\hat{\Lambda}_t := E_0[\Lambda_t|Y_t] = \exp\left(\int_0^t \pi_s(h)dy_s - \frac{1}{2}\int_0^t \pi_s^2(h)ds\right). \quad (32)$$

Proof: By Ito's rule, Λ_t satisfies

$$\Lambda_t = 1 + \int_0^t \Lambda_s h(x_s)dy_s. \quad (33)$$

It follows as in the proof of Theorem 4 that $\hat{\Lambda}_t$ is a martingale with respect to $\mathcal{Y}_t$. Since $\{y_t\}$ is a Brownian motion under P_0, there must exist a $\mathcal{Y}_t$-adapted process $\{\eta_t\}$ such that [13, Theorem 5.6]

$$\hat{\Lambda}_t = 1 + \int_0^t \eta_s dy_s. \tag{34}$$

We identify η_t by the same technique as in Theorem 4: from (33) and Ito's rule,

$$\Lambda_t y_t = \int_0^t \Lambda_s dy_s + \int_0^t y_s \Lambda_s h(x_s) dy_s + \int_0^t \Lambda_s h(x_s) ds. \tag{35}$$

From (34) and Ito's rule,

$$\hat{\Lambda}_t y_t = \int_0^t \hat{\Lambda}_s dy_s + \int_0^t y_s \eta_s dy_s + \int_0^t \eta_s ds. \tag{36}$$

Now $E[\Lambda_t y_t - \hat{\Lambda}_t y_t | \mathcal{Y}_s] = 0$ for $t \geq s$, and calculating this from (35) and (36) yields

$$\eta_t = \widehat{\Lambda_t h(x_t)} := E_0[\Lambda_t h(x_t) | \mathcal{Y}_t]. \tag{37}$$

But from (31),

$$E_0[\Lambda_t h(x_t) | \mathcal{Y}_t] = \pi_t(h) \hat{\Lambda}_t,$$

so (34) becomes

$$\hat{\Lambda}_t = 1 + \int_0^t \hat{\Lambda}_s \pi_s(h) dy_s. \tag{38}$$

However, this has the unique solution

$$\hat{\Lambda}_t = \exp\left(\int_0^t \pi_s(h) dy_s - \frac{1}{2} \int_0^t \pi_s^2(h) ds \right)$$

and the lemma is proved.

Theorem 6: For any $\phi \in D(L)$, $\sigma_t(\phi)$ satisfies

$$\sigma_t(\phi) = \sigma_0(\phi) + \int_0^t \sigma_s(L\phi)ds + \int_0^t \sigma_s(h\phi)dy_s. \tag{39}$$

Proof: By Ito's rule, we have from (26) and (38):

$$\begin{aligned} d(\hat{\Lambda}_t\pi_t(\phi)) &= \hat{\Lambda}_t[\pi_t(L\phi)dt + (\pi_t(h\phi)-\pi_t(h)\pi_t(\phi))(dy_t-\pi_t(h)dt)] \\ &\quad + \pi_t(\phi)[\hat{\Lambda}_t\pi_t(h)dy_t] + [\pi_t(h\phi)-\pi_t(h)\pi_t(\phi)]\hat{\Lambda}_t\pi_t(h)dt \\ &= \hat{\Lambda}_t\pi_t(L\phi)dt + \hat{\Lambda}_t\pi_t(h\phi)dy_t, \end{aligned}$$

which gives (39) since $\sigma_t(\phi) = \hat{\Lambda}_t\pi_t(\phi)$.

The remarks following Theorem 5 are also applicable here. In addition, we note that the Stratonovich version of (39), which is utilized in a number of papers in this volume, is:

$$\sigma_t(\phi) = \sigma_0(\phi) + \int_0^t \sigma_s(\tilde{L}\phi)ds + \int_0^t \sigma_s(h\phi)\circ dy_s \tag{40}$$

where

$$\tilde{L}\phi(x) = L\phi(x) - \frac{1}{2}\, h^2(x)\phi(x)$$

and $\circ$ denotes a Stratonovich (symmetric) stochastic integral [9],[22].

Example 4: Under the assumptions of Example 2, we can derive a stochastic differential equation for $q(t,x) := \hat{\Lambda}_t\tilde{p}(t,x)$; this is interpreted as an *unnormalized conditional density*, since then

$$\tilde{p}(t,x) = \frac{q(t,x)}{\int q(t,x)dx}.$$

As in Example 2, an integration by parts in (39) yields the stochastic partial differential equation:

$$dq(t,x) = L^*q(t,x)dt + h(x)q(t,x)dy_t. \tag{41}$$

Notice that (41) has a much simpler structure than (27): it does not involve an integral such as $\pi_t(h)$, and it is a bilinear stochastic differential equation with $\{y_t\}$ as its input. This structure is utilized by a number of papers in this volume.

ACKNOWLEDGMENT

The work of S. I. Marcus was supported in part by the U.S. National Science Foundation under grant ENG-76-11106.

REFERENCES

1. R. E. Kalman, "A new approach to linear filtering and prediction problems," J. Basic Eng. ASME, 82, 1960, pp. 33-45.

2. R. E. Kalman and R. S. Bucy, "New results in linear filtering and prediction theory," J. Basic Engr. ASME Series D, 83, 1961, pp. 95-108.

3. R. S. Bucy, "Nonlinear filtering," IEEE Trans. Automatic Control, AC-10, 1965, p. 198.

4. H. J. Kushner, "On the differential equations satisfied by conditional probability densities of Markov processes," SIAM J. Control, 2, 1964, pp. 106-119.

5. A. N. Shiryaev, "Some new results in the theory of controlled stochastic processes [Russian]," *Trans. 4th Prague Conference on Information Theory*, Czech. Academy of Sciences, Prague, 1967.

6. R. L. Stratonovich, *Conditional Markov Processes and Their Application to the Theory of Optimal Control*. New York: Elsevier, 1968.

7. W. M. Wonham, "Some applications of stochastic differential equations to optimal nonlinear filtering," SIAM J. Control, 2, 1965, pp. 347-369.

8. M. Zakai, "On the optimal filtering of diffusion processes," Z. Wahr. Verw. Geb., 11, 1969, pp. 230-243.

9. E. Wong, *Stochastic Processes in Information and Dynamical Systems*. New York: McGraw-Hill, 1971.

10. T. Kailath, "An innovations approach to least-squares estimation -- Part I: Linear filtering in additive white noise," IEEE Trans. Automatic Control, AC-13, 1968, pp. 646-655.

11. P. A. Frost and T. Kailath, "An innovations approach to least-squares estimation III," IEEE Trans. Automatic Control, AC-16, 1971, pp. 217-226.

12. M. Fujisaki, G. Kallianpur, and H. Kunita, "Stochastic differential equations for the nonlinear filtering problem," Osaka J. Math., 1, 1972, pp. 19-40.

13. R. S. Liptser and A. N. Shiryaev, *Statistics of Random Processes I*. New York: Springer-Verlag, 1977.

14. G. Kallianpur, *Stochastic Filtering Theory*. Berlin-Heidelberg-New York: Springer-Verlag, 1980.

15. E. Pardoux, "Stochastic partial differential equations and filtering of diffusion processes," Stochastics, 2, 1979, pp. 127-168 [see also Pardoux' article in this volume].

16. N. V. Krylov and B. L. Rozovskii, "On the conditional distribution of diffusion processes [Russian]," Izvestia Akad. Nauk SSSR, Math Series 42, 1978, pp. 356-378.

17. R. W. Brockett, this volume.

18. V. E. Benes, "Exact finite dimensional filters for certain diffusions with nonlinear drift," Stochastics, to appear.

19. M. H. A. Davis, this volume.

20. J. H. Van Schuppen, "Stochastic filtering theory: A discussion of concepts, methods, and results," in *Stochastic Control Theory and Stochastic Differential Systems*, M. Kohlmann and W. Vogel, eds. New York: Springer-Verlag, 1979.

21. M. H. A. Davis, "Martingale integrals and stochastic calculus," in *Communication Systems and Random Process Theory*, J. K. Skwirzynski, ed. Leiden: Noordhoff, 1978.

22. L. Arnold, *Stochastic Differential Equations*. New York: Wiley, 1974.

23. A. Friedman, *Stochastic Differential Equations and Applications, Vol. 1*. New York: Academic Press, 1975.

24. I. I. Gihman and A. V. Skorohod, *Stochastic Differential Equations*. New York: Springer-Verlag, 1972.

25. I. Gertner, "An alternative approach to nonlinear filtering," Stochastic Processes and their Applications, 7, 1978, pp. 231-246.

26. D. W. Stroock and S. R. S. Varadhan, *Multidimensional Diffusion Processes*. New York: Springer-Verlag, 1979.

27. D. Allinger and S. K. Mitter, "New results on the innovations problem for nonlinear filtering," Stochastics, to appear.

28. E. Wong, "Recent progress in stochastic process -- a survey," IEEE Trans. Inform. Theory, IT-19, 1973, pp. 262-275.

29. J. Jacod, *Calcul Stochastique et Problèmes de Martingales*. Berlin-Heidelberg-New York: Springer-Verlag, 1979.

30. S. K. Mitter, this volume.

A TUTORIAL INTRODUCTION TO DIFFERENTIABLE MANIFOLDS AND VECTOR FIELDS

Michiel Hazewinkel

Dept. Math., Erasmus Univ. Rotterdam

In this tutorial I try by means of several examples to illustrate the basic definitions and concepts of differentiable manifolds. There are few proofs (not that there are ever many at this level of the theory). This material should be sufficient to understand the use made of these concepts in the other contributions in this volume, or, at least, it should help in explaining the terminology employed.

1. INTRODUCTION AND A FEW MOTIVATIONAL REMARKS

Roughly an n-dimensional differentiable manifold is a gadget which locally looks like $\mathbb{R}^n$ but globally perhaps not; A precise definition is given below in section 2. Examples are the sphere and the torus, which are both locally like $\mathbb{R}^2$ but differ globally from $\mathbb{R}^2$ and from each other.

Such objects often arise naturally when discussing problems in analysis (e.g. differential equations) and elsewhere in mathematics and its applications. A few advantages which may come about by doing analysis on manifolds rather than just on $\mathbb{R}^n$ are briefly discussed below.

M. Hazewinkel and J. C. Willems (eds.), Stochastic Systems: The Mathematics of Filtering and Identification and Applications, 77–93.

1.1 Coordinate freeness ("Diffeomorphisms"). A differentiable manifold can be viewed as consisting of pieces of $\mathbb{R}^n$ which are glued together in a smooth (= differentiable) manner. And it is on the basis of such a picture that the analysis (e.g. the study of differential equations) often proceeds. This brings more than a mere extension of analysis on $\mathbb{R}^n$ to analysis on spheres, tori, projective spaces and the like; it stresses the "coordinate free approach", i.e. the formulation of problems and concepts in terms which are invariant under (nonlinear) smooth coordinate transformations and thus also helps to bring about a better understanding even of analysis on $\mathbb{R}^n$. The more important results, concepts and definitions tend to be "coordinate free".

1.2 Analytic continuation. A convergent power series in one complex variable is a rather simple object. It is considerably more difficult to obtain an understanding of the collection of all analytic continuations of a given power series, especially because analytic continuation along a full circle may yield a different function value than the initial one. The fact that the various continuations fit together to form a Riemann surface (a certain kind of 2-dimensional manifold usually different from $\mathbb{R}^2$) was a major and most enlightening discovery which contributes a great deal to our understanding.

1.3 Submanifolds. Consider an equation $\dot{x} = f(x)$ in $\mathbb{R}^n$. Then it often happens, especially in problems coming from mechanics, that the equation is such that it evolves in such a way that certain quantities (e.g. energy, angular momentum) are conserved. Thus the equation really evolves on a subset $\{x \in \mathbb{R}^n | E(x) = c\}$ which is often a differentiable submanifold. Thus it could happen that $\dot{x} = f(x)$, f smooth, is constrained to move on a 2-sphere which then immediately tells us that there is an equilibrium point.

Also one might meet 2 seemingly different equations, say, one in $\mathbb{R}^4$ and one in $\mathbb{R}^3$ (perhaps both intended as a description of the same process) of which the first has two conserved quantities and the second one. It will then be important to decide whether the surfaces on which the equations evolve are diffeomorphic, i.e. the same after a suitable invertible transformation and whether the equations on these submanifolds correspond under these transformations.

1.4 Behaviour at infinity. Consider a differential equation in the plane $\dot{x} = P(x,y)$, $\dot{y} = Q(x,y)$ where P and Q are relatively prime polynomials. To study the behavior of the paths far out in the plane and such things as solutions escaping to infinity and coming back, Poincaré already completed the plane to real projective 2-space (an example of a differential manifold). Also the projective plane is by no means the only smooth manifold compactifying $\mathbb{R}^2$ and it will be of some importance for the behaviour of the equation near infinity whether the "right" compactification to which the equation can be extended will be a projective 2-space, a sphere or a torus, or, ..., or, whether no such compactification exists at all. A good example of a set of equations which are practically impossible to analyse completely without bringing in manifolds are the matrix Riccati equations (which naturally live on Grassmann manifolds (which also gives in this case a very considerable saving in the number of dimensions needed)).

1.5 Avoiding confusion between different kinds of objects. Consider an ordinary differential equation $\dot{x} = f(x)$ on $\mathbb{R}^n$, where $f(x)$ is a function $\mathbb{R}^n \to \mathbb{R}^n$. When one now tries to generalize this idea of a differential equation to a differential equation on a manifold one discovers that $\dot{x}$ and hence $f(x)$ are a different kind of object; they are not functions, but as we shall see, they are vectorfields; in other words under a nonlinear change of coordinates they transform in a different way than functions do.

2. DIFFERENTIABLE MANIFOLDS

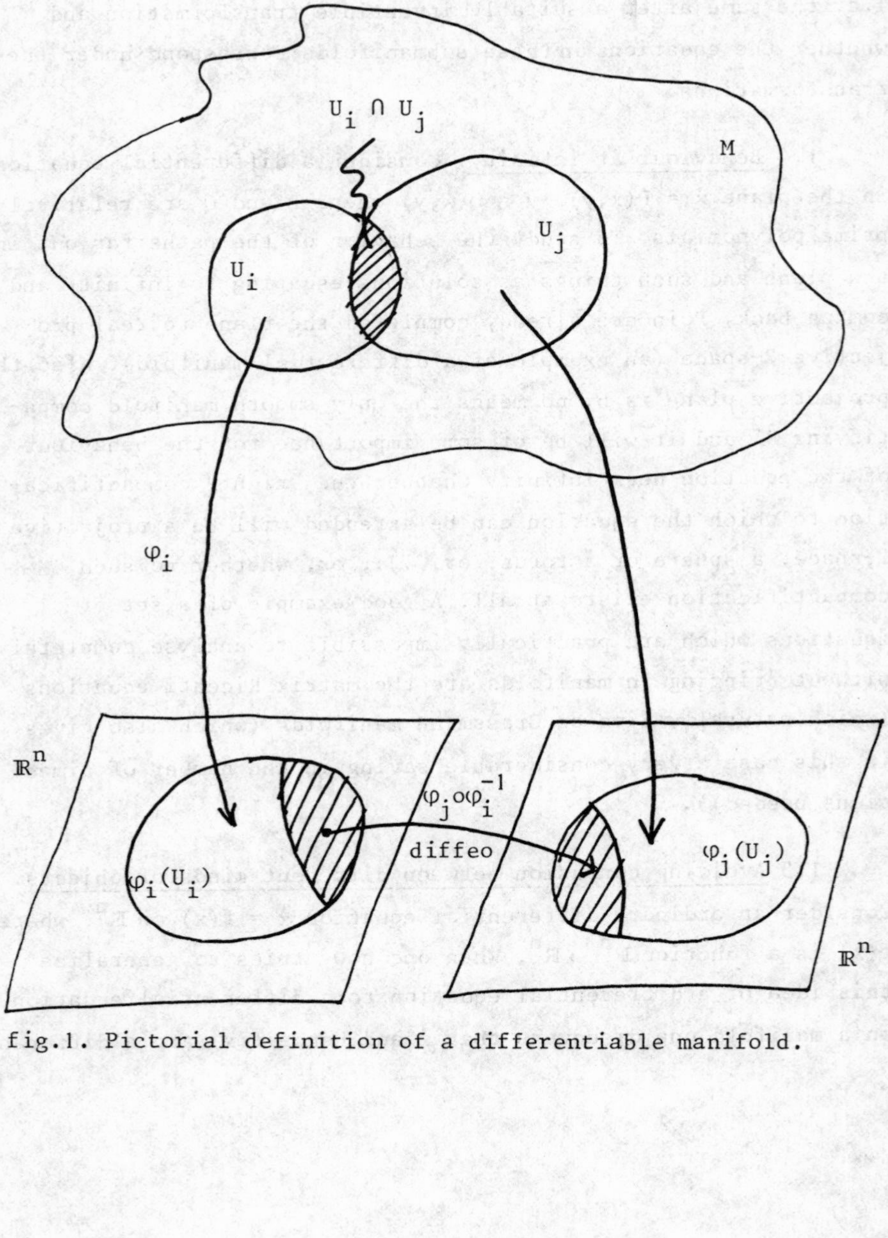

fig.1. Pictorial definition of a differentiable manifold.

Let U be an open subset of $\mathbb{R}^n$, e.g. an open ball. A function $f: U \to \mathbb{R}$ is said to be C^∞ or smooth if all partial derivatives (any order) exist at all $x \in U$. A mapping $\mathbb{R}^n \supset U \to \mathbb{R}^m$ is smooth if all components are smooth; $\varphi: U \to V$, $U \subset \mathbb{R}^n$, $V \subset \mathbb{R}^n$ is called a diffeomorphism if φ is 1 - 1, onto, and both φ and φ^{-1} are smooth.

As indicated above a smooth n-dimensional manifold is a gadget consisting of open pieces of $\mathbb{R}^n$ smoothly glued together. This gives the above pictorial definition of a smooth n-dimensional manifold M (fig.1).

2.1 Example. The circle $S^1 = \{(x_1,x_2) \mid x_1^2 + x_2^2 = 1\} \subset \mathbb{R}^2$

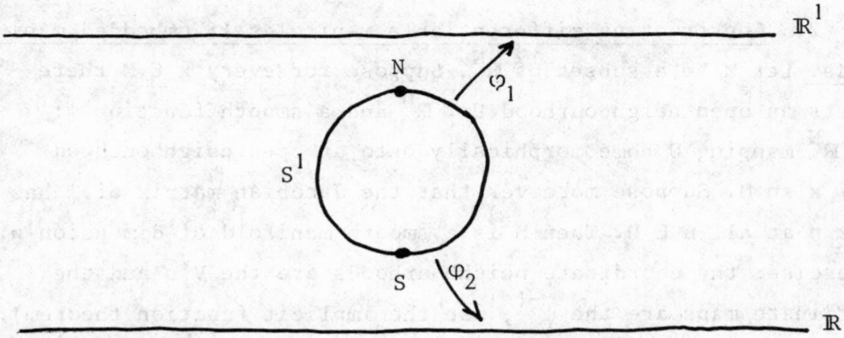

fig.2. Example: the circle

$U_1 = S^1 \setminus \{S\}$, $U_2 = S^1 \setminus \{N\}$ so $U_1 \cup U_2 = S^1$. The "coordinate charts" φ_1 and φ_2 are given by

$$\varphi_1(x_1,x_2) = \frac{x_1}{1+x_2}, \quad \varphi_2(x_1,x_2) = \frac{x_1}{1-x_2}$$

Thus $\varphi_1(U_1 \cap U_2) = \mathbb{R} \setminus \{0\}$, $\varphi_2(U_1 \cap U_2) = \mathbb{R} \setminus \{0\}$ and the map

$\varphi_2 \circ \varphi_1^{-1} : \mathbb{R}\setminus\{0\} \to \mathbb{R}\setminus\{0\}$ is given by $x \mapsto x^{-1}$ which is a diffeomorphism.

2.2 Formal definition of a differentiable manifold.
The data are
- M, a Hausdorff topological space
- A covering $\{U_i\}_{i\in I}$ by open subsets of M
- Coordinate maps $\varphi_i : U_i \to \varphi_i(U_i) \subset \mathbb{R}^n$, $\varphi_i(U_i)$ open in $\mathbb{R}^n$.

These data are subject to the following condition
- $\varphi_i \circ \varphi_j^{-1} : \varphi_j(U_i \cap U_j) \to \varphi_i(U_i \cap U_j)$ is a diffeomorphism

Often one also adds the requirement that M be paracompact. We shall however disregard these finer points; nor shall we need them in this volume.

2.3 Constructing differentiable manifolds 1: embedded manifolds. Let M be a subset of $\mathbb{R}^N$. Suppose for every $x \in M$ there exists an open neighbourhood $U \subset \mathbb{R}^n$ and a smooth function ψ: $U \to \mathbb{R}^N$ mapping U homeomorphically onto an open neighbourhood V of x in M. Suppose moreover that the Jacobian matrix of ψ has rank n at all $u \in U$. Then M is a smooth manifold of dimension n. (Exercise: the coordinate neighbourhoods are the V's and the coordinate maps are the ψ^{-1}; use the implicit function theorem). Virtually the same arguments show that if $\varphi: U \to \mathbb{R}^k$, $U \subset \mathbb{R}^{n+k}$, is a smooth map and the rank of the Jacobian matrix $J(f)(x)$ is k for all $x \in \varphi^{-1}(0)$, then $\varphi^{-1}(0)$ is a smooth n-dimensional manifold. We shall not pursue this approach but concentrate instead on:

2.4 Constructing differentiable manifolds 2: gluing. Here the data are as follows
- an index set I
- for every $i \in I$ an open subset $U_i \subset \mathbb{R}^n$
- for every ordered pair (i,j) an open subset $U_{ij} \subset U_i$

- diffeomorphisms $\varphi_{ij}: U_{ij} \to U_{ji}$ for all $i,j \in I$

These data are supposed to satisfy the following compatibility conditions

- $U_{ii} = U_i$, $\varphi_{ii} = id$
- $\varphi_{jk} \circ \varphi_{ij} = \varphi_{ik}$ (where appropriate)

(where the last identity is supposed to imply also that $\varphi_{ij}(U_{ij} \cap U_{ik}) \subset U_{jk}$ so that $\varphi_{ij}(U_{ij} \cap U_{ik}) = U_{jk} \cap U_{ji}$).

These are not all conditions but the present lecturer e.g. has often found it advantageous to stop right here so to speak, and to view a manifold simply as a collection of open subsets of $\mathbb{R}^n$ together with gluing data (coordinate transformation rules).

From the data given above one now defines an abstract topological space M by taking the disjoint union of the U_i and then identifying $x \in U_i$ and $y \in U_j$ iff $x \in U_{ij}$, $y \in U_{ji}$, $\varphi_{ij}(x) = y$. This gives a natural injection $U_i \to M$ with image U_i' say. Let $\varphi_i: U_i' \to U_i$ be the inverse map. Then this gives us a differentiable manifold M in the sense of definition 2.2 provided that M is Hausdorff and paracompact, and these are the conditions which must be added to the gluing compatibility conditions above.

2.5 <u>Functions on a "glued manifold".</u> Let M be a differentiable manifold obtained by the gluing process described in 2.4 above. Then a differentiable function $f: M \to \mathbb{R}$ consist simply of a collection of functions $f_i: U_i \to \mathbb{R}$ such that $f_j \circ \varphi_{ij} = f_i$ on U_{ij}, as illustrated in fig. 3.

Thus for example a function on the circle S^1, cf. figure 2, can be described either as a function of two variables restricted to S^1 or as two functions f_1, f_2 of one variable on U_1 and U_2 such that $f_1(x) = f_2(x^{-1})$. Obviously the latter approach can have considerable advantages.

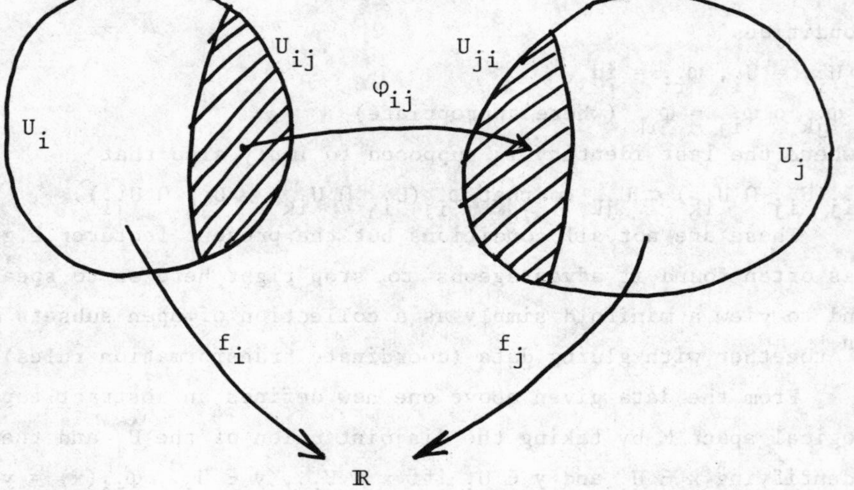

fig.3. Functions on a glued manifold

2.6 <u>Example of a 2 dimensional manifold: the Möbius band.</u> The (open) Möbius band is obtained by taking a strip in $\mathbb{R}^2$ as indicated below in fig. 4 without its upper and lower edges and identifying the left hand and right hand edges as indicated

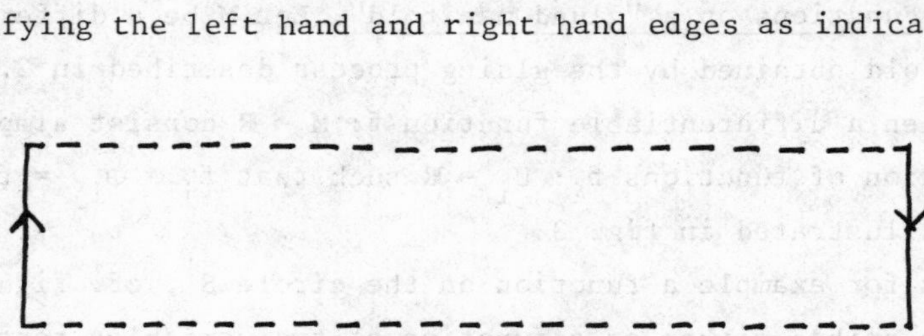

fig.4. Construction of the Möbius band

The resulting manifold (as a submanifold of $\mathbb{R}^3$) looks something like the following figure 5.

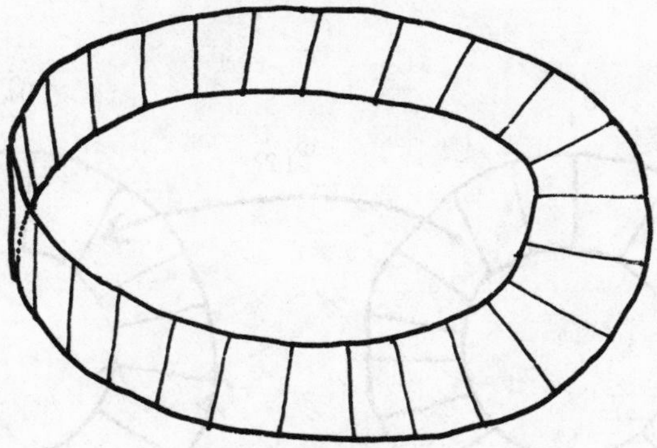

fig.5. The Möbius band

It is left as an exercise to the reader to cast this description in the form required by the gluing description of 2.4 above. The following pictorial description (fig. 6) will suffice.

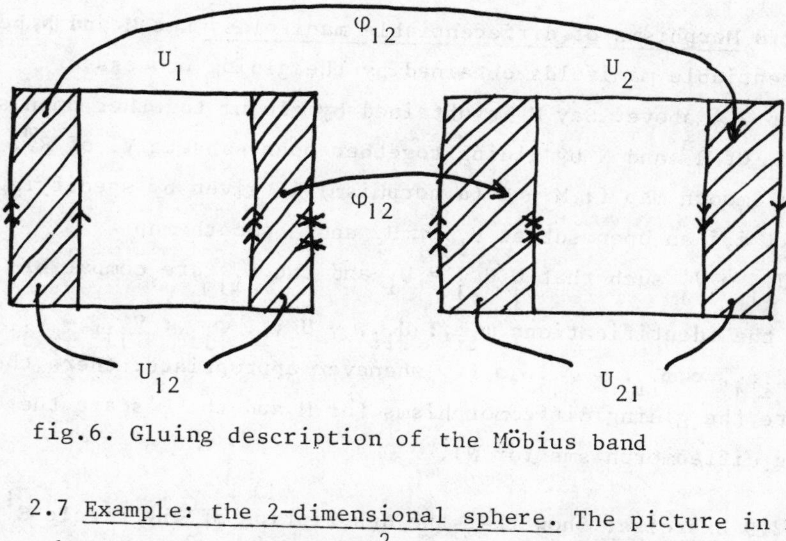

fig.6. Gluing description of the Möbius band

2.7 <u>Example: the 2-dimensional sphere.</u> The picture in fig. 7 below shows how the 2-sphere $S^2 = \{(x_1,x_2,x_3)\,|\,x_1^2 + x_2^2 + x_3^2 = 1\}$ can be obtained by gluing two disks together. If the surface of the earth is viewed as a model for S^2 the first disk covers everything north of Capricorn and the second everything south of Cancer.

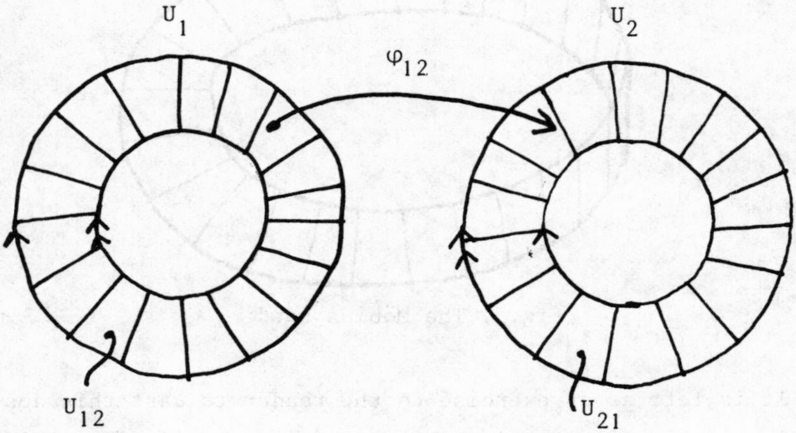

fig.7. Gluing description of the 2-sphere S^2

2.8 Morphisms of differentiable manifolds. Let M and N be differentiable manifolds obtained by the gluing process of section 2.4 above. Say M is obtained by gluing together open subsets U_i of $\mathbb{R}^n$ and N by gluing together open subsets V_j of $\mathbb{R}^m$. Then a smooth map $f: M \to N$ (a morphism) is given by specifying for all i,j an open subset $U_{ij} \subset U_i$ and a smooth map $f_{ij}: U_{ij} \to V_j$ such that $\bigcup_j U_{ij} = U_i$ and the f_{ij} are compatible under the identifications $\varphi_{ii'}: U_{ii'} \to U_{i'i}$, $\varphi_{jj'}: V_{jj'} \to V_{j'j}$, i.e. $f_{i'j'} \circ \varphi_{ii'} = \varphi_{jj'} \circ f_{ij}$ whenever appropriate. (Here the φ's are the gluing diffeomorphisms for M and the ψ's are the gluing diffeomorphisms for N).

2.9 Exercise: Show that the description of the circle S^1 as in 2.1 above gives an injective morphism $S^1 \to \mathbb{R}^2$.

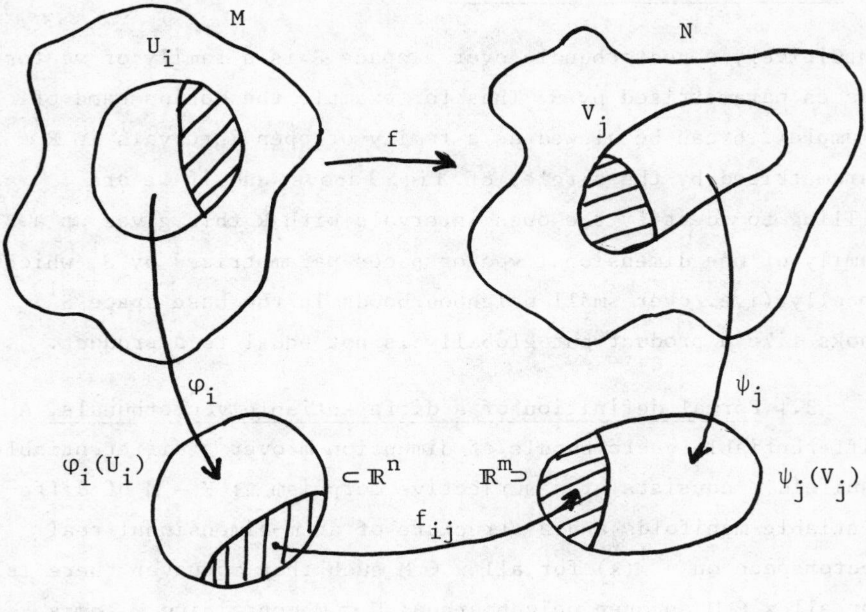

fig.8. Morphisms

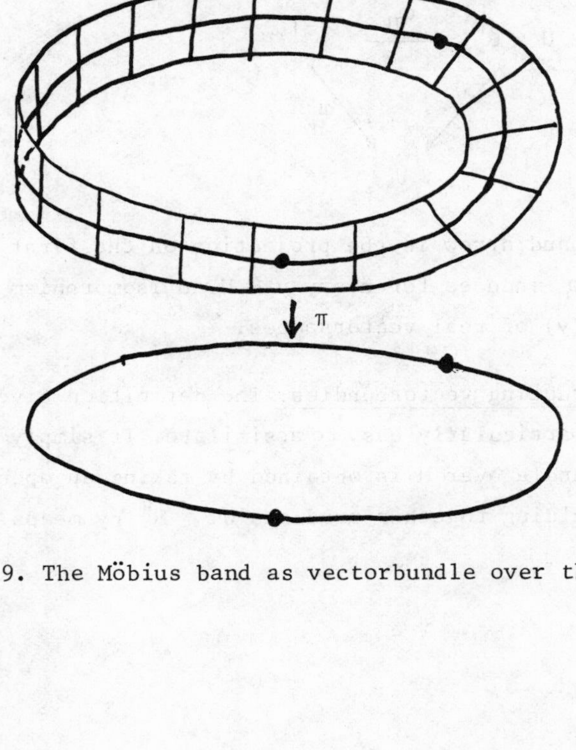

fig.9. The Möbius band as vectorbundle over the circle

3. DIFFERENTIABLE VECTORBUNDLES

Intuitively a vectorbundle over a space S is a family of vectorspaces parametrized by S. Thus for example the Möbius band of example 2.6 can be viewed as a family of open intervals in $\mathbb{R}$ parametrized by the circle, cf. fig. 9 above, and if we are willing to identify the open intervals with $\mathbb{R}$ this gives us a family of one dimensional vectorspaces parametrized by S^1 which locally (i.e. over small neighbourhoods in the base space S^1) looks like a product but globally is not equal to a product.

3.1 Formal definition of a differentiable vectorbundle. A differentiable vectorbundle of dimension m over a differentiable manifold M consists of a surjective morphism $\pi: E \to M$ of differentiable manifolds and a structure of an m-dimensional real vectorspace on $\pi^{-1}(x)$ for all $x \in M$ such that moreover there is for all $x \in M$ an open neighbourhood $U \subset M$ containing x and a diffeomorphism $\varphi_U: U \times \mathbb{R}^m \to \pi^{-1}(U)$ such that the following diagram commutes

$$\begin{array}{ccc} U \times \mathbb{R}^m & \xrightarrow{\varphi_U} & \pi^{-1}(U) \\ & \searrow \quad \swarrow \pi & \\ & U & \end{array}$$

where the lefthand arrow is the projection on the first factor, and such that φ_U induces for every $y \in U$ an isomorphism $\{y\} \times \mathbb{R}^m \to \pi^{-1}(y)$ of real vectorspaces.

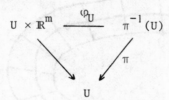

3.2 Constructing vectorbundles. The definition given above is not always particularly easy to assimilate. It simply means that a vectorbundle over M is obtained by taking an open covering $\{U_i\}$ of M and gluing together products $U_i \times \mathbb{R}^m$ by means of diffeomorphisms which are linear (i.e. vectorspace structure preserving) in the second coordinate. Thus an m-dimensional vectorbundle over M is given by the following data

- an open covering $\{U_i\}_{i\in I}$ of M.
- for every i,j a smooth map $\varphi_{ij}: U_i \cap U_j \to GL_m(\mathbb{R})$ where $GL_m(\mathbb{R})$ is the space of all invertible real m × m matrices considered as an open subset of $\mathbb{R}^{m^2}$. These data are subject to the following compatibility conditions
- $\varphi_{ii}(x) = I_m$, the identity matrix, for all $x \in U_i$
- $\varphi_{jk}(x)\,\varphi_{ij}(x) = \varphi_{ik}(x)$ for all $x \in U_i \cap U_j \cap U_k$

From these data E is constructed by taking the disjoint union of the $U_i \times \mathbb{R}^m$, $i \in I$ and identifying $(x,v) \in U_i \times \mathbb{R}^m$ with $(y,w) \in U_j \times \mathbb{R}^m$ if and only if $x = y$ and $\varphi_{ij}(x)v = w$. The morphism π is induced by the first coordinate projections $U_i \times \mathbb{R}^m \to U_i$.

3.3 Constructing vectorbundles 2. If the base manifold M is itself viewed as a smoothly glued together collection of open sets in $\mathbb{R}^n$ we can descripe the gluing for M and a vectorbundle E over M all at once. The combined data are than as follows

- open sets $U_i \times \mathbb{R}^m$, $U_i \subset \mathbb{R}^n$ for all $i \in I$
- open subsets $U_{ij} \subset U_i$ for all $i,j \in I$
- diffeomorphisms $\varphi_{ij}: U_{ij} \to U_{ji}$
- diffeomorphisms $\tilde{\varphi}_{ij}: U_{ij} \times \mathbb{R}^m \to U_{ji} \times \mathbb{R}^m$ of the form $(x,v) \mapsto (\varphi_{ij}(x), A_{ij}(x)v)$ where $A_{ij}(x)$ is an m × m invertible real matrix depending smoothly on x.

These data are then subject to the same compatibility conditions for the $\tilde{\varphi}_{ij}$'s (and hence the φ_{ij}) as described in 2.4 above.

3.4 Example: the tangent vectorbundle of a smooth manifold. Let the smooth manifold M be given by the data U_i, U_{ij}, φ_{ij} as in 2.4. Then the tangent bundle TM is given by the data

- $U_i \times \mathbb{R}^n$, $U_{ij} \times \mathbb{R}^n \subset U_i \times \mathbb{R}^n$
- $\tilde{\varphi}_{ij}: U_{ij} \times \mathbb{R}^n \to U_{ji} \times \mathbb{R}^n$, $\tilde{\varphi}_{ij}(x,v) = (\varphi_{ij}(x), J(\varphi_{ij})(x)v)$

where $J(\varphi_{ij})(x)$ is the Jacobian matrix of φ_{ij} at $x \in U_{ij}$.

Exercise: check that these gluing morphisms do indeed define a vectorbundle; i.e. check the compatibility (chain rule !)

3.5 Morphisms of vectorbundles. A morphism of vectorbundles from the vectorbundle $\pi: E \to M$ to the vectorbundle $\pi': E' \to M'$ is a pair of smooth maps $\tilde{f}: E \to E'$, $f: M \to M'$ such that $\pi' \circ \tilde{f} = f \circ \pi$ and such that the induced map $\tilde{f}_x: \pi^{-1}(x) \to \pi^{-1}(f(x))$ is a homomorphism of vectorspaces for all $x \in M$. We leave it to the reader to translate this into a local pieces and gluing data description.

As an example consider two manifolds M, N both described in terms of local pieces and gluing data. Let $f: M \to N$ be given in these terms by the $f_{ij}: U_{ij} \to V_j$ (cf. 2.8 above). Then the maps $\tilde{f}_{ij}: U_{ij} \times \mathbb{R}^n \to V_j \times \mathbb{R}^m$ defined by $\tilde{f}_{ij}(x,v) = (f_{ij}(x), J(f_{ij})(x)v)$ combine to define a morphism of vectorbundles $\tilde{f} = Tf: TM \to TN$.

4. VECTORFIELDS

A vectorfield on a manifold M assigns in a differentiable manner to every $x \in M$ a tangent vector at x, i.e. an element of the fibre $T_xM = \pi^{-1}(x)$ of the tangent bundle TM. Slightly more precisely this gives the

4.1 Definitions. Let $\pi: E \to M$ be a vectorbundle. Then a section of E is a smooth map $s: M \to E$ such that $\pi \circ s = id$. A section of the tangent vectorbundle $TM \to M$ is called a vectorfield.

Suppose that M is given by a local pieces and gluing data description as in 2.4 above. Then a vectorfield s is given by "local sections" $s_i^!: U_i \to U_i \times \mathbb{R}^n$ of the form $s_i^!(x) = (x, s_i(x))$, i.e. by a collection of functions $s_i: U_i \to \mathbb{R}^n$ such that $J(\varphi_{ij})(x)(s_i(x)) = s_j(x)$ for all $x \in U_{ij}$.

4.2 <u>Derivations.</u> Let A be an algebra over $\mathbb{R}$. Then a derivation is an $\mathbb{R}$-linear map $D: A \to A$ such that $D(fg) = (Df)g + f(Dg)$ for all $f,g \in A$.

4.3 <u>Derivations and vectorfields.</u> Now let M be a differentiable manifold and let S(M) be the $\mathbb{R}$-algebra of smooth functions $M \to \mathbb{R}$. Then every vectorfield s on M defines a derivation of S(M), (which assigns to a function f its derivative along s), which can be described as follows. Let M be given in terms of local pieces U_i and gluing data U_{ij}, φ_{ij}. Let $f: M \to \mathbb{R}$ and the section $s: M \to TM$ be given by the local functions $f_i: U_i \to \mathbb{R}$, $s_i: U_i \to \mathbb{R}^n$. Now define $g_i: U_i \to \mathbb{R}$ by the formula

$$g_i(x) = \sum_k s_i(x)_k \frac{\partial f_i}{\partial x_k}(x) \qquad (4.4)$$

where $s_i(x)_k$ is the k-th component of the n-vector $s_i(x)$. It is now an easy exercise to check that $g_j(\varphi_{ij}(x)) = g_i(x)$ for all $x \in U_{ij}$ (because $(\varphi_{ij})(x)s_i(x) = s_j(x)$ for these x) so that the $g_i(x)$ combine to define a function $g = D_s(f): M \to \mathbb{R}$. This defines a map $D: S(M) \to S(M)$ which is seen to be a derivation. Inversely every derivation of S(M) arises in this way.

4.5 <u>The Lie bracket of derivations and vectorfields.</u> Let D_1,D_2 be derivations of an $\mathbb{R}$-algebra A. Then, as is easily checked, so is

$$[D_1,D_2] = D_1D_2 - D_2D_1$$

So if s_1,s_2 are vectorfields on M, then there is a vectorfield $[s_1,s_2]$ on M corresponding to the derivation $[D_{s_1},D_{s_2}]$. This vectorfield is called the Lie bracket of s_1 and s_2 and $(s_1,s_2) \mapsto [s_1,s_2]$ defines a Lie algebra structure on the vectorspace V(M) of all vectorfields on M.

If M is given in terms of local pieces U_i and gluing data U_{ij}, φ_{ij} then the Lie bracket operation can be described as follows. Let the vectorfields s and t be given by the local functions $s_i, t_i: U_i \to \mathbb{R}^n$. Then [s,t] is given by the local functions

$$\sum_j s_j \frac{\partial t_i}{\partial x_j} - \sum_j t_j \frac{\partial s_i}{\partial x_j}$$

4.6 <u>The $\frac{\partial}{\partial x}$ notation.</u> Let the vectorfield s: M → TM be given by the functions $s_i: U_i \to \mathbb{R}^n$. Then using the symbols $\frac{\partial}{\partial x_k}$ in first instance simply as labels for the coordinates in $\mathbb{R}^n$ we can write

$$s_i = \sum s_i(x)_k \frac{\partial}{\partial x_k} \tag{4.7}$$

This is a most convenient notation because as can be seen from (4.4) this gives precisely the local description of the differential operator (derivation) D_s associated to s.

4.7 <u>Differential equations on a manifold.</u> A differential equation on a manifold M is given by an equation

$$\dot{x} = s(x) \tag{4.8}$$

where s: M → TM is a vectorfield, i.e. a section of the tangentbundle. At every moment t, equation (4.8) tells us in which direction and how fast x(t) will evolve by specifying a tangent vector s(x(t)) at x(t).

Again it is often useful to take a local pieces and gluing data point of view. Then the differential equation (4.8) is given by a collection of differential equations $\dot{x} = s_i(x)$ in the usual sense of the word on U_i where the functions $s_i(x)$ satisfy $J(\varphi_{ij})(x)s_i(x) = s_j(x)$ for all $x \in U_{ij}$.

In these terms a solution of the differential equation is simply a collection of solutions of the local equations, i.e. a collection of maps $f_i: V_i \to U_i$, $V_i \subset \mathbb{R}(\geq 0)$ such that $\cup V_i = \mathbb{R}(\geq 0)$, $\frac{d}{dt} f_i(t) = s_i(f_i(t))$ which fit together to define a morphism $\mathbb{R}(\geq 0) \to M$, i.e. such that $\varphi_{ij}(f_i(t)) = f_j(t)$ if $t \in V_i \cap V_j$.

In more global terms a solution of (4.8) which passes through x_0 at time 0 is a morphism of smooth manifolds $f: \mathbb{R} \to M$ such that $Tf: T\mathbb{R} \to TM$ satisfies $Tf(t,1) = s(f(t))$ for all $t \in \mathbb{R}$ (or a suitable subset of $\mathbb{R}$), i.e. Tf takes the vectorfield (section) $1: \mathbb{R} \to T\mathbb{R} = \mathbb{R} \times \mathbb{R}$, $t \mapsto (t,1)$ into the vectorfield (section) $s: M \to TM$.

4.9 <u>Conclusion.</u> Here, where it starts to get interesting, is, according to a developing tradition in textbook writing, a good place to stop.

A SHORT TUTORIAL ON LIE ALGEBRAS

Michiel Hazewinkel
Dept. Math., Erasmus Univ. Rotterdam
P.O. Box 1738,
3000 DR ROTTERDAM

This tutorial does not correspond to an actual oral lecture during the conference at Les Arcs in June, 1980. However, to improve accessibility and understandability of the material in this volume it seemed wise to include a small section on the basic facts and definitions concerning Lie algebras which play a role in control and nonlinear filtering theory. This is what these few pages attempt to do.

1. DEFINITION OF LIE ALGEBRAS. EXAMPLES. Let k be a field and V a vectorspace over k. (For the purpose of this volume it suffices to take $k = \mathbb{R}$ or (rarely) $k = \mathbb{C}$; the vectorspace V over k need not be finite dimensional). A Lie algebra structure on V is then a bilinear map (called brackett multiplication)

(1.1) $[\ ,\ ]: V \times V \to V$

such that the two following conditions hold

(1.2) $[u,u] = 0$ for all $u \in V$

M. Hazewinkel and J. C. Willems (eds.), Stochastic Systems: The Mathematics of Filtering and Identification and Applications, 95–108.

(1.3) $[u,[v,w]] + [v,[w,u]] + [w,[u,v] = 0$ for all $u,v,w, \in V$.

The last identity is called the Jacobi identity. Of course the bilinearity of (1.1) means that $[au+bv,w] = a[u,w] + b[v,w]$, $[u,bv+cw] = b[u,v] + c[u,w]$. From (1.2) it follows that

(1.4) $$[u,v] = -[v,u]$$

by considering $[u+v,u+v] = 0$ and using bilinearity.

1.5. Example. The Lie algebra associated to an associative algebra.

Let A be an associative algebra over k. Now define a new multiplication (brackett) on A by the formula

(1.6) $$[v,w] = vw - wv, \quad w,v \in A$$

Then A with this new multiplication is a Lie algebra. (Exercise: check the Jacobi identity (1.3)).

1.9. Remark. In a certain precise sense all Lie algebras arise in this way. That is for every Lie algebra L there is an associative algebra A containing L such that $[u,v] = uv - vu$. I.e. every Lie algebra arises as a subspace of an associative algebra A which happens to be closed under the operation $(u,v) \to uv - vu$. Though this "universal enveloping algebra" construction is quite important it will play no role in the following and the remark is intended to make Lie algebras easier to understand for the reader.

1.7. Example. Let $M_n(k)$ be the associative algebra of all n x n matrices with coefficients in k. The associated Lie algebra is written $g\ell_n(k)$; i.e. $g\ell_n(k)$ is the n^2-dimensional vectorspace of all n x n matrices with the brackett multiplication $[A,B] = AB - BA$.

1.8. Example. Let $s\ell_n(k)$ denote the subspace of all n x n matrices of trace zero. Because $Tr(AB-BA) = 0$ for all n x n matrices A,B, we see that $[A,B] \in s\ell_n(k)$ if $A,B \in s\ell_n(k)$ giving us an (n^2-1)-dimensional sub Lie algebra of $g\ell_n(k)$.

1.10. Example. The Lie algebra of first order differential operators with C^∞-coefficients.

Let V_n be the space of all differential operators (on the space $F(\mathbb{R}^n)$ of C^∞-functions (i.e. arbitrarily often differentiable functions in $x_1,\ldots,x_n$)) of the form

$$X = \sum_{i=1}^{n} f_i(x_1,\ldots,x_n)\frac{\partial}{\partial x_i} \tag{1.11}$$

where the f_i, $i = 1, \ldots, n$ are C^∞-functions. Thus $X: F(\mathbb{R}^n) \to F(\mathbb{R}^n)$ is the operator $X(\phi) = \sum_{i=1}^{n} f_i \frac{\partial\phi}{\partial x_i}$. Now define a brackett operation on V_n by the formula

$$[X,Y] = \sum_{i,j} \left(f_i \frac{\partial g_j}{\partial x_i}\frac{\partial}{\partial x_j} - g_j \frac{\partial f_i}{\partial x_j}\frac{\partial}{\partial x_i}\right) \tag{1.12}$$

if $X = \sum f_i \frac{\partial}{\partial x_i}$, $Y = \sum g_j \frac{\partial}{\partial x_j}$. This makes V_n a Lie algebra.

Check that $[X,Y](\phi) = X(Y(\phi)) - Y(X(\phi))$ for all $\phi \in F(\mathbb{R}^n)$.

1.13. Example. Derivations. Let A be any algebra (i.e. A is a vectorspace together with any bilinear map (multiplication) $A \times A \to A$: in particular A need not be associative). A derivation on A is a linear map $D: A \to A$ such that

(1.14) $D(uv) = (Du)v + u(Dv)$

For example let $A = \mathbb{R}[x]$ and D the operator $\frac{d}{dx}$. The D is a derivation. The operators (1.11) of the example above are derivations on $F(\mathbb{R}^n)$.

Let Der(A) be the vectorspace of all derivations. Define $[D_1,D_2] = D_1D_2 - D_2D_1$. Then $[D_1,D_2]$ is again a derivation and this brackett multiplication makes Der(A) a Lie algebra over k.

1.15. Example. The Weyl algebra W_1. Let W_1 be the vectorspace of all (any order) differential operators in one variable with polynomial coefficients. I.e. W_1 is the vectorspace with basis $x^i \frac{d^j}{dx^j}$, $i,j \in \mathbb{N} \cup \{0\}$. ($x^i$ is considered as the operator $f(x) \to x^i f(x)$). Consider W_1 as a space of operators acting, say, on k[x]. Composition of operators makes W_1 an associative algebra and hence gives W_1 also the structure of a Lie algebra. For example one has

$$[x \frac{d^2}{dx^2}, x^2 \frac{d}{dx}] = 3x^2 \frac{d^2}{dx^2} + 2x \frac{d}{dx}, \quad [x \frac{d}{dx}, x^i \frac{d^j}{dx^j}] = (i-j)x^i \frac{d^j}{dx^j}$$

1.16. Example. The oscillator algebra. Consider the four dimensional subspace of W_1 spanned by the four operators $\frac{1}{2}\frac{d^2}{dx^2} - \frac{1}{2}x^2$, x, $\frac{d}{dx}$, 1. One easily checks that (under the brackett multiplication of W_1)

$$[\frac{1}{2}\frac{d^2}{dx^2} - \frac{1}{2}x^2, x] = \frac{d}{dx}, \quad [\frac{1}{2}\frac{d^2}{dx^2} - \frac{1}{2}x^2, \frac{d}{dx}] = x, \quad [\frac{d}{dx}, x] = 1$$

(1.17)

$$[\frac{1}{2}\frac{d^2}{dx^2} - \frac{1}{2}x^2, 1] = [x,1] = [\frac{d}{dx}, 1] = 0$$

Thus this four dimensional subspace is a sub-Lie-algebra of W_1. It is called the oscillator Lie algebra (being intimately associated to the harmonic oscillator).

2. HOMOMORPHISMS, ISOMORPHISMS, SUBALGEBRAS AND IDEALS.

2.1. Sub-Lie-algebras. Let L be a Lie algebra over k and V a subvectorspace of L. If $[u,v] \in V$ for all $u,v \in L$. Then V is a sub-Lie-algebra of L. We have already seen a number of examples of this, e.g. the oscillator algebra of example 1.16 as a sub-Lie-algebra of the Weyl algebra W_1 and the Lie-algebra $s\ell_n(k)$ as a sub-Lie-algebra of $g\ell_n(k)$. Some more examples follow.

2.2. The Lie-algebra $so_n(k)$. Let $so_n(k)$ be the subspace of $g\ell_n(k)$ consisting of all matrices A such that $A + A^T = 0$ (where the upper ^{T} denotes transposes). Then if $A,B \in so_n(k)$

$[A,B] + [A,B]^T = AB - BA + (AB-BA)^T = A(B+B^T) - B(A+A^T) +$

$+ (B^T+B)A^T - (A^T+A)B^T = 0$ so that $[A,B] \in so_n(k)$. Thus $so_n(k)$ is a sub-Lie-algebra of $g\ell_n(k)$.

2.3. The Lie-algebra $t_n(k)$. Let $t_n(k)$ be the subspace of $g\ell_n(k)$ consisting of all upper triangular matrices. Because product and sum of upper triangular matrices are again upper triangular $t_n(k)$ is a sub-Lie-algebra of $g\ell_n(k)$.

2.4. The Lie-algebra $sp_n(k)$. Let Q be the 2n x 2n matrix

$Q = \begin{pmatrix} 0 & I_n \\ -I_n & 0 \end{pmatrix}$. Now let $sp_n(k)$ be the subspace of all 2n x 2n matrices A such that $AQ + QA^T = 0$. Then as above in example 2.2 one sees that $A,B \in sp_n(k) \Rightarrow [A,B] \in sp_n(k)$ so that $sp_n(k)$ is a sub-Lie-algebra of $g\ell_{2n}(k)$.

2.5. Ideals. Let L be a Lie-algebra over k. A subvectorspace $I \subset L$ with the property that for all $u \in I$ and all $v \in L$ we have $[u,v] \in I$ is called an ideal of L. An example is $s\ell_n(k) \subset g\ell_n(k)$, cf. example 1.8 above. Another example follows.

2.6. Example. The Heisenberg Lie-algebra. Consider the 3-dimensional subspace of W_1 spanned by the operators $x, \frac{d}{dx}, 1$. The formulas (1.17) show that this subspace is an ideal in the oscillator algebra.

2.7. Example. The centre of a Lie algebra. Let L be a Lie algebra. The centre of L is defined as the subset $Z(L) = \{z \in L | [u,z] = 0$ for all $u \in L\}$. Then $Z(L)$ is a subvector space of L and in fact an ideal of L. As an example it is easy to check that the centre of $g\ell_n(k)$ consists of scalar multiples of the unit matrix I_n.

2.8. Homomorphisms and isomorphisms. Let L_1 and L_2 be two Lie algebras over k. A morphism of $\alpha: L_1 \to L_2$ vectorspaces (i.e. a k-linear map) is a homomorphism of Lie algebras if $\alpha[u,v] = \alpha(u),\alpha(v)$ for all $u,v \in L_1$. The homomorphism α is called an isomorphism if it is also an isomorphism of vectorspaces.

2.9. Example. Consider the following three first-order differential operators in two variables x,P

$$a = (1-p^2)\frac{\partial}{\partial p} - Px\frac{\partial}{\partial x}, \quad b = P\frac{\partial}{\partial x}, \quad c = \frac{\partial}{\partial x}$$

Then one easily calculates (cf. (1.9)) $[a,b] = c$, $[a,c] = b$, $[b,c] = 0$. Now define α from the oscillator algebra of example 1.16 to this 3-dimensional Lie algebra as the linear map $\frac{1}{2}\frac{d^2}{dx^2} - \frac{1}{2}x^2 \to a$, $x \to b$, $\frac{d}{dx} \to c$, $1 \to 0$. Then the formulas above and (1.17) show that α is a homomorphism of Lie algebras.

2.10. Kernel of a homomorphism. Let $\alpha: L_1 \to L_2$ be a homomorphism of Lie algebras. Let $\mathrm{Ker}(\alpha) = \{u \in L_1 | \alpha(u) = 0\}$. Then $\mathrm{Ker}(\alpha)$ is an ideal in L_1.

2.11. Quotient Lie algebras. Let L be a Lie algebra and I an ideal in L. Consider the quotient vector space L/I and the quotient morphisms of vector spaces $L \overset{\alpha}{\to} L/I$. For all $\bar{u},\bar{v} \in L/I$ choose $u,v \in L$ such that $\alpha(u) = \bar{u}$, $\alpha(v) = \bar{v}$. Now define $[\bar{u},\bar{v}] = \alpha[u,v]$. Check that this does not depend on the choice of u,v.

This then defines a Lie-algebra structure on L/I and $\alpha: L \to L/I$ becomes a homomorphism of Lie-algebras.

2.12. Image of a homomorphism. Let $\alpha: L_1 \to L_2$ be a homomorphism of Lie algebras. Let $Im(\alpha) = \alpha(L_1) = \{u \in L_2 | \exists v \in L_1, \alpha(v) = u\}$. Then $Im\alpha$ is a sub-Lie-algebra of L_2 and α induces an isomorphism $L_1/Ker(\alpha) \simeq Im(\alpha)$.

2.13. Exercise. Consider the 3-dimensional vector space of all real upper triangular 3 x 3 matrices with zero's on the diagonal. Show that this a sub-Lie-algebra of $g\ell_3(\mathbb{R})$, and show that it is isomorphic to the 3-dimensional Heisenberg-Lie-algebra of example 2.6 but that it is not isomorphic to the 3-dimensional Lie-algebra $s\ell_2(\mathbb{R})$ of example 1.8.

2.14. Exercise. Show that the four operators x^2, $\frac{d^2}{dx^2}$, $x\frac{d}{dx}$, 1 span a 4-dimensional subalgebra of W_1, and show that this 4-dimensional Lie algebra contains a three dimensional Lie algebra which is isomorphic to $s\ell_2(\mathbb{R})$.

2.15. Exercise. Show that the six operators x^2, $\frac{d^2}{dx^2}$, x, $\frac{d}{dx}$, $x\frac{d}{dx}$, 1 space a six dimensional sub-Lie-algebra of W_1. Show that x, $\frac{d}{dx}$, 1 space a 3-dimensional ideal in this Lie-algebra and show that the corresponding quotient algebra is $s\ell_2(\mathbb{R})$.

3. LIE ALGEBRAS OF VECTORFIELDS.

Let M be a C^∞-manifold (cf. the tutorial on manifolds and vectorfields in this volume). Intuitively a vectorfield on M specifies a tangent vector t(m) at every point $m \in M$. Then given a C^∞-function f on M we can for each $m \in M$ take the derivation of f at m in the direction t(m), giving us a new function g on

M. This can be made precise in varying ways; e.g. as follows.

3.1. The Lie algebra of vectorfields on a manifold M. Let M be a C^∞-manifold, and let F(M) be the $\mathbb{R}$-algebra (pointwise addition and multiplications) of all smooth (= C^∞) functions f: $M \to \mathbb{R}$. By definition a C^∞-vectorfield on M is a derivation X: F(M) → F(M). The Lie algebra of derivations of F(M) cf. example 1.13, i.e. the Lie-algebra of smooth vectorfields on M, is denoted V(M).

3.2. Derivations and vectorfields. Now let $M = \mathbb{R}^n$ so that F(M) is simply the $\mathbb{R}$-algebra of C^∞-functions in $x_1, \ldots, x_n$. Then it is not difficult to show that every derivation X: $F(\mathbb{R}^n) \to F(\mathbb{R}^n)$ is necessarily of the form

$$X = \sum_{i=1}^{n} g_i \frac{\partial}{\partial x_i} \tag{3.3}$$

with $g_i \in F(\mathbb{R}^n)$. For a proof cf. [4, Ch.I,§2]. The corresponding vectorfield on $\mathbb{R}^n$ now assigns to $x \in \mathbb{R}^n$ the tangent vector $(g_1(x), \ldots, g_n(x))^T$.

On an arbitrary manifold we have representations (3.3) locally around every point and these expressions turn out to be compatible in precisely the x way needed to define a vectorfield as described in the tutorial on manifolds and vectorfields in this volume [3].

3.4. Homomorphisms of Lie algebras of vectorfields. Let M and N be C^∞-manifolds and let α: L → V(N) be a homomorphism of Lie algebras where L is a sub-Lie-algebra of V(M). Let ϕ: M → N be a smooth map. Then α and ϕ are said to be compatible if

$$\phi^*(\alpha(X)f) = X(\phi^*(f)) \quad \text{for all } f \in F(N) \tag{3.5}$$

where ϕ^* is the homomorphism of algebras F(N) → F(M), $f \to \phi^*(f) = f \circ \phi$.

In terms of the Jacobian of ϕ (cf.[3]), this means that

$$(3.6) \qquad J(\phi)(X_m) = \alpha(X)_{\phi(m)}$$

where X_m is the tangent vector at m of the vectorfield X.

If $\phi : M \to N$ is an isomorphism of C^∞-manifolds there is always precisely one homomorphism of Lie-algebras $\alpha : V(M) \to V(N)$ compatible with ϕ (which is then an isomorphism). It is defined (via formula (3.5)) by

$$(3.7) \qquad \alpha(X)(f) = (\phi^*)^{-1}X(\phi^*f), \quad f \in F(N)$$

3.8. <u>Isotropy subalgebras</u>. Let L be a sub-Lie-algebra of V(M) and let m M. The isotropy subalgebra L_m of L at m consists of all vectorfields in L whose tangent vector in m is zero, or, equivalently

$$(3.9) \qquad L_m = \{X \in L | Xf(m) = 0 \quad \text{all } f \in F(M)\}$$

Now suppose that $\alpha: L \to V(N)$ and $\phi: M \to N$ are compatible in the sense of 3.4 above. Then it follows easily from (3.5) that

$$(3.10) \qquad \alpha(L_m) \subset V(N)_{\phi(m)}$$

i.e. α isotropy subalgebras into isotropy subalgebras. Inversely if we restrict our attention to analytic vectorfields then condition (3.10) on α at m implies that locally there exists a ϕ which is compatible with α [7].

4. SIMPLE, NILPOTENT AND SOLVABLE LIE ALGEBRAS.

4.1. Nilpotent Lie algebras let L be a Lie-algebra over k.

The descending central seems of L is defined inductively by

$$(4.2) \qquad C^1L = L,\ C^{i+1}L = [L, C^iL],\ i \geq 1$$

It is easy to check that the C^iL are ideals. The Lie algebra L is called nilpotent if $C^nL = \{0\}$ for n big enough.

For each $x \in L$ we have the endomorphism adx: $L \to L$ defined by $y \to [x,j]$. It is now a theorem that if L is finite dimensional then L is nilpotent iff the endomorphisms adx are nilpotent for all $x \in L$. Whence the terminology.

4.3. Solvable Lie algebras. The derived seems of Lie algebras of a Lie algebra L is defined inductively by

$$(4.4) \qquad D^1L = L,\ D^{i+1}L = [D^iL, D^iL],\ i \geq 1$$

It is again easy to check that the D^iL are ideals. The Lie algebra L is called solvable if $D^nL = \{0\}$ for n large enough.

4.5. Examples. The Heisenberg Lie algebra of example 2.6 is nilpotent. The Oscillator algebra of example 1.16 is solvable but not nilpotent. The sub-Lie-algebra of W_1 with vector-space basis x^2, $\frac{d^2}{dx^2}$, x, $\frac{d}{dx}$, 1, $x\frac{d}{dx}$ is neither nilpotent, nor solvable. The Lie-algebra $t_n(k)$ of example 2.3 is solvable and in a way is typical of finite dimensional solvable Lie algebras in the sense that if k is algebraically closed (e.g. $k = C$), then every finite dimensional solvable Lie algebra over k is isomorphic to a sub-Lie-algebra of some $t_n(k)$.

4.6. Exercise. Show that sub-Lie-algebras and quotient-Lie-algebras of solvable Lie algebras (resp. nilpotent Lie algebras) are solvable (resp. nilpotent).

4.7. Abelian Lie-algebras. A lie algebra L is called abelian if $[L,L] = \{0\}$, i.e. if every brackett product is zero.

4.8. Simple Lie-algebras. A Lie algebra L is called simple if it is not abelian and if it has no other ideals than 0 and L. (Given the second condition the first one only rules out the zero- and one-dimensional Lie algebras). These simple-Lie-algebras and the abelian ones are in a very precise sense the basic building blocks of all Lie algebras.

The finite dimensional simple Lie algebras over $\mathbb{C}$ have been classified. They are the Lie algebras $sl_n(\mathbb{C})$, $sp_n(\mathbb{C})$, $so_n(\mathbb{C})$ of examples 1.8, 2.4 and 2.2 above and five additional exceptional Lie algebras. For infinite dimensional Lie algebras things are more complicated. The socalled filtered, primitive, transitive simple Lie algebras have also been classified (cf. e.g. [2]). One of these is the Lie-algebra $\hat{V}_n$ of all formal vector fields $\Sigma f_i(x_1,\ldots,x_n)\frac{\partial}{\partial x_i}$, where the $f_i(x)$ are (possibly non converging) formal power series in $x_1,\ldots, x_n$. This class of infinite dimensional simple Lie algebras by no means exhausts all possibilities. E.g. the quotient-Lie-algebras $W_n/\mathbb{R}.1$ are simple and non-isomorphic to any of those just mentioned.

4.9. Exercise. Let $V_{alg}(\mathbb{R}^n)$ be the Lie algebra of all differential operators (vector fields) of the form $\Sigma f_i(x_1,\ldots,x_n)\frac{\partial}{\partial x_i}$ with $f_i(x_1,\ldots,x_n)$ polynomial. Prove that $V_{alg}(\mathbb{R}^n)$ is simple.

5. REPRESENTATIONS.

Let L be a Lie algebra over k and M a vectorspace over k. A representation of L in M is a homomorphism of Lie algebras.

$$(5.1) \qquad \rho : L \to \mathrm{End}_k(M)$$

where $\mathrm{End}_k(M)$ is the vectorspace of all k-linear maps $M \to M$ which is of course given the Lie algebra structure

$[A,B] = AB - BA$. Equivalently a representation of L in M consists of a k-bilinear map

(5.2) $\sigma : L \times M \to M$

such that, writing xm for $\sigma(x,m)$, we have $[x,y]m = x(ym) - y(xm)$ for all $x,y \in L$, $m \in M$. The relation between the two definitions is of course $\sigma(x,m) = \rho(x)(m)$.

Instead of speaking of a representation of L in M we also speak (equivalently) of the L-module M.

5.3. Example. The Lie algebra $gl_n(k)$ of all $n \times n$ matrices naturally acts on k^n by $(A,v) \to Av \in k^n$ and this defines a representation $gl_n(k) \times k^n \to k^n$. The Lie algebra V(M) of vectorfields on a manifold M acts (by its definition) on F(M) and this is a representation of V(M). A quite important theorem concerning the existence of representations is

5.4. Ado's theorem. Cf. e.g. [1,§7]. If k is a field of characteristic zero, e.g. $k = \mathbb{R}$ or C and L is finite dimensional then there is a faithful representation $\rho: L \to End(k^n)$ for some n. (Here faithful means that ρ is injective).

Thus every finite dimensional Lie algebra L over k (of characteristic zero) can be viewed as a subalgebra of some $gl_n(k)$, and this subalgebra can then be viewed as a more concrete matrix "representation" of the "abstract" Lie algebra L.

5.5. Realizing Lie-algebras in V(M). A question of some importance for filtering theory is when a LIe algebra L can be realized as a sub-Lie-algebra of V(M), i.e. when L can be represented in F(M) by means of derivations of several papers in this volume for a discussion of the relevance of this problem. For finite dimensional Lie algebras Ado's theorem gives the answer because $(a_{ij}) \to \Sigma a_{ij} x_i \frac{\partial}{\partial x_j}$ defines an injective homomorphism of Lie-algebras $gl_n(\mathbb{R}) \to V(\mathbb{R}^n)$ (Exercise: check this)

6. LIE ALGEBRAS AND LIE GROUPS.

6.1. Lie groups. A (finite dimensional) Lie group is a finite dimensional smooth manifold G together with smooth maps $G \times G \to G$, $(x,y) \to xy$, $G \to G$, $x \to x^{-1}$ and a distinguished element $e \in G$ which make G a group. An example is the open subset of $\mathbb{R}^{n^2}$ consisting of all invertible $n \times n$ matrices with the usual matrix multiplication.

6.2. Left invariant vectorfields and the Lie algebra of a Lie group.

Let G be a Lie group. Let for all $g \in G$, $L_g: G \to G$ be the smooth map $x \to gx$. A vectorfield $X \in V(G)$ is called left invariant if $X(L_g^* f) = XL_g^*(Xf)$ for all functions f on G. Or, equivalently, if $J(L_g)X_x = X_{gx}$ for all $x \in G$, cf. section 3.4 above. Especially from the last condition it is easy to see that $X \to X_e$ defines an isomorphism between the vectorspace of left invariant vectorfields on G and the tangent space of G at e. Now the brackett product of two left invariant vectorfields is easily seen to be left invariant again so the tangent space of G at e (which is $\mathbb{R}^n$ if G is n-dimensional) inherits a Lie algebra structure. This is the Lie algebra Lie(G) of the Lie group G. It reflects so to speak the infinitesimal structure of G. A main reason for the importance of Lie algebras in many parts of mathematics and its applications is that this construction is reversible to a great extent making it possible to study Lie groups by means of their Lie algebras.

6.3. Exercise. Show that the Lie algebra of the Lie group $GL_n(\mathbb{R})$ of invertible real $n \times n$ matrices is the Lie algebra $gl_n(\mathbb{R})$.

7. POSTSCRIPT.

The above is a very rudimentary introduction to Lie algebras. Especially the topic "Lie algebras and Lie groups" also called "Lie theory" has been given very little space, in spite of the fact that it is likely to become of some importance in filtering (integration of a representation of a Lie algebra to a representation of a LIe (semi)group). The books [1, 4, 5, 6, 8] are all recommended for further material. My personal favourite (but by no means the easiest) is [4]; [6] is a classic and in its present incarnation very good value indeed.

REFERENCES.

1. A. Bourbaki, Groupes et algèbres de Lie, Chap. 1: Algèbres de Lie, Hermann, 1960.
2. M. Demazure, Classification des algèbres de Lie filtres, Sém. Bourbaki 1966/1967, Exp. 326, Benjamin, 1967.
3. M. Hazewinkel, Tutorial on Manifolds and Vectorfields. This volume.
4. S. Helgason, Differential Geometry, Lie Groups and Symmetric Spaces, Acad. Pr., 1978.
5. J.E. Humphreys, Introduction to Lie Algebras and Representation Theory, Springer 1972.
6. N. Jacobson, Lie Algebras, Dover reprint, 1980.
7. A.J. Krener, On the Equivalence of Control Systems and the Linearization of nonlinear Systems SIAM J. Control 11(1973), 670-676.
8. J.P. Serre, Lie Algebras and Lie Groups, Benjamin 1965.

It isn't that they can't see the
solution. It is that they can't
see the problem.

G.K. Chesterton

Part 3

FOUNDATIONS OF STOCHASTIC SYSTEMS AND MODELING ISSUES AND APPLICATIONS.

MATHEMATICAL MODELS OF CHEMICAL REACTIONS

Ludwig Arnold

Fachbereich Mathematik, Universität Bremen

1. Introduction

There are two main principles according to which chemical reactions in a spatial domain are modeled:

(i) *global* description (i.e. without diffusion, spatially homogeneous or 'well-stirred' case) versus *local* description (i.e. including diffusion, spatially inhomogeneous case),

(ii) *deterministic* description (macroscopic, phenomenological, in terms of concentrations) versus *stochastic* description (on the level of numbers of particles, taking into account internal fluctuations).

The combination of these two principles gives rise to essentially four mathematical models, see e.g. HAKEN [9] and NICOLIS and PRIGOGINE [18].

This paper briefly describes these models and deals with the question of whether and in what mathematical sense those models are consistent.

These are questions of approximation of smooth functions by jump processes and of Markov processes by non-Markov processes. In the proofs semigroups and martingales play an important role.

M. Hazewinkel and J. C. Willems (eds.), Stochastic Systems: The Mathematics of Filtering and Identification and Applications, 111–134.

As far as the mathematical tools are concerned, we have the following:

- global deterministic model: ordinary differential equations
- local deterministic model: partial differential equations
- global stochastic model: jump Markov processes/stochastic ordinary differential equations
- local stochastic model: space-time jump Markov processes/stochastic partial differential equations

Since we want to be conceptual rather than aim at highest generality, we restrict ourselves to the case of one single reactant X whose concentration is not kept constant, to one-dimensional volume $\Omega \subset \mathbb{R}^1$ and to a reaction scheme of the following type:

$$\sum_{j=1}^{r} \alpha_{ij} A_j + \beta_i X \underset{k_i'}{\overset{k_i}{\rightleftarrows}} \sum_{j=1}^{r} \alpha_{ij}' A_j + (\beta_i + 1) X,$$

$$\alpha_{ij}, \alpha_{ij}', \beta_i \in \mathbb{N} = \{0,1,2,\ldots\}; k_i, k_i' \geq 0, \; i=1,\ldots,s. \tag{1}$$

This survey paper is based on my papers [2] and [3].

2. Mathematical Models of Chemical Reactions

2.1 Global deterministic model

One can read-off from the reaction scheme (1) the gain term

$$\lambda(x) = \sum_{i=1}^{s} \alpha_i k_i x^{\beta_i} = \sum_{i=0}^{p} a_i x^i, a_i \geq 0,$$

and the loss term

$$\mu(x) = \sum_{i=1}^{s} \alpha_i' k_i' x^{\beta_i + 1} = \sum_{i=1}^{q} b_i x^i, \; b_i \geq 0.$$

Thus, the evolution of the concentration $\varphi = \varphi(t)$, $t \in [o,\infty)$, of the reactant X is described by the nonlinear ODE (kinetic equation)

$$\frac{d\varphi(t)}{dt} = f(\varphi(t)), \ \varphi(o) = \varphi_o = o,$$
$$f(x) = \lambda(x) - \mu(x) = \sum_{i=o}^{d} c_i x^i, \ c_o \geq o, c_d \neq o. \tag{2}$$

If $c_d < o$ then (2) has a non-negative solution existing for all times t which we assume from now on.

The right-hand side of (2) contains parameters. If they are allowed to fluctuate in a random manner, we are dealing with so-called *external noise* or *systems in a random environment* (see e.g. the forthcoming book by HORSTHEMKE and LEFEVER [10]).

2.2 Local deterministic model

Now besides reaction transport of matter via diffusion is taken into account. If $\varphi = \varphi(r,t)$, $r \in \Omega \subset \mathbb{R}^1$, $t \in [o,\infty)$, is the concentration, $D > o$ the diffusion coefficient,

$\Delta = \frac{\partial^2}{\partial r^2}$ the Laplacian, then

$$\frac{\partial\varphi}{\partial t}(r,t) = D\,\Delta\,\varphi(r,t) + f(\varphi(r,t)) \tag{3}$$

(equation of reaction and diffusion, see e.g. FIFE [6]). The domain Ω can be bounded or unbounded, and φ is subjected to boundary and initial conditions. For the problem of existence and uniqueness of solutions and invariant sets we refer to KUIPER [12] and AMANN [1], for asymptotic behavior and steady states see FIFE [7] and [8].

For later use we state the following existence and uniqueness result which can be obtained from KUIPER [12]: Take $\Omega = (o,1)$, assume $c_d < o$, and pick a $\rho > o$ to the right of the biggest zero of $f(x) = o$. Let the initial condition $\varphi(r,o) = \varphi_o(r)$ satisfy $O \leqq \varphi_o(r) \leqq \rho$, $r \in [o,1]$, $\varphi_o \in \overset{o}{H}{}^2_{bc}$, the closure in $H^2(o,1)$ of those functions in $C^2[o,1]$ which satisfy the boundary conditions. Choose as boundary conditions $\varphi(o,t) = \varphi(1,t) = o$ or

$\partial\varphi/\partial r(o,t) - \alpha\varphi(o,t) = o$, $\partial\varphi/\partial r(1,t) + \beta\varphi(1,t) = o$ $(\alpha,\beta \geqq o$
Then there exists a unique global solution φ of (3) satisfying the initial and boundary conditions such that $\varphi \in C^1([o,\infty), L_2(o,1)) \cap C([o,\infty), H^2(o,1))$
and
$o \leqq \varphi(r,t) \leqq \rho$ for all $r \in [o,1]$ and all $t \in [o,\infty)$.

2.3 Global stochastic model

We now look at the total number of particles $X(t)$ of reactant X in a volume of length L at time t. This function is modeled as a Markov jump process, in particular as a birth and death process with state space $\mathbb{N}$ and transition probabilities

$$P(X(t+s) = k+1 \mid X(s) = k) = p_{k,k+1}(t) = \lambda_k t + o(t), k \in \mathbb{N}$$

$$P(X(t+s) = k-1 \mid X(s) = k) = p_{k,k-1}(t) = \mu_k t + o(t), k-1 \in \mathbb{N}$$

where the birth rates λ_k and the death rates μ_k are given by

$$\lambda_k = L(a_o + a_1\frac{k}{L} + a_2\frac{k(k-1)}{L^2} + \ldots + a_p\frac{k(k-1)\ldots(k-p+1)}{L^p})$$

and

$$\mu_k = L(b_1\frac{k}{L} + b_2\frac{k(k-1)}{L^2} + \ldots + b_q\frac{k(k-1)\ldots(k-q+1)}{L^q}. \tag{4}$$

Introducing the functions

$$\lambda_L(x) = \sum_{i=o}^{p} a_i\, x(x-\frac{1}{L})\ldots(x-\frac{i-1}{L}),$$

$$\mu_L(x) = \sum_{i=1}^{q} b_i\, x(x-\frac{1}{L})\ldots(x-\frac{i-1}{L}), \tag{5}$$

we can write

$$\lambda_k = L\,\lambda_L(k/L),\ \mu_k = L\mu_L(k/L). \tag{6}$$

Observe that for $L \to \infty$ and each x

$$\lambda_L(x) = \lambda(x) + O(1/L),\ \mu_L(x) = \mu(x) + O(1/L).$$

The transition rates λ_k and μ_k together with an initial distribution uniquely determine the stochastic process $X(t)$ provided that either the reaction is linear or $c_d < o$. In particular, the probabilities $p_k(t) = P(X(t) = k)$ are the unique solution of the so-called *Master equation* (or KOLMOGOROV's second equation)

$$\dot{p}_k(t) = \lambda_{k-1}p_{k-1}(t) + \mu_{k+1}p_{k+1}(t) - (\lambda_k+\mu_k)p_k(t),\ k\in\mathbb{N}$$

for initial distribution $p_k(o) = p_k^o,\ \sum_{k\in\mathbb{N}} p_k^o = 1$.

2.4 Local stochastic model

Now the total volume of length L is divided into N cells of equal size $\ell = L/N$, where adjacent cells are connected by diffusion while in each cell reaction goes on. If $X_j(t)$ denotes the number of particles of reactant X in the j-th cell at time t , then

$$X(t) = (X_1(t),\ldots,X_N(t))$$

is modeled as a Markov jump process with state space $\mathbb{N}^N$ and the following transition intensities:

$$q_{k,k+e_j} = \lambda_{k_j},\ q_{k,k-e_j} = \mu_{k_j},\ j=1,\ldots,N,\ \text{(reaction)}$$

$$q_{k,k+e_{j+1}-e_j} = (D^*/2)k_j,\ j=1,\ldots,N-1, \qquad (7)$$

(diffusion)

$$q_{k,k+e_{j-1}-e_j} = (D^*/2)k_j,\ j=2,\ldots,N,$$

$q_{k,k+m} = o$, otherwise. Here
$k = (k_1,\ldots,k_j,\ldots,k_N)$, $k\pm e_j$, $k+e_{j\pm 1}-e_j$, $k+m \in \mathbb{N}^N$, e_j is the j-th unit vector in $\mathbb{R}^N$, and

$$\lambda_{k_j} = \ell\lambda_\ell(k_j/\ell),\ \mu_{k_j} = \ell\mu_\ell(k_j/\ell), \qquad (8)$$

and $D^* > o$ are the birth rate, death rate, and diffusion parameter, resp., and $\lambda_\ell(x)$ and $\mu_\ell(x)$ are the functions defined by (5).

We also have to fix boundary conditions. Saying nothing additional amounts to reflection at the boundary (zero flux boundary conditions). The coupling of the system to reservoirs at its boundaries with prescribed fixed (e.g. zero) particle concentrations can be modeled by adding a cell at each boundary with a fixed number of particles and by coupling the new cell to its neighbor by diffusion with intensities as given above.

The transition intensities as fixed above together with an initial distribution on $\mathbb{N}^N$ uniquely determine the stochastic process $X(t)$ provided that reaction is either linear of $c_d < o$. In particular, the probabilities $p_k(t) = P(X(t)=k)$, $k \in \mathbb{N}^N$, are the unique solution of the *multivariate Master equation* (written down e.g. for zero flux boundary conditions)

$$\dot{p}_k(t) = \sum_{j=1}^{N} (\lambda_{k_j-1}\, p_{k-e_j}(t) + \mu_{k_j+1}\, p_{k+e_j}(t) - (\lambda_{k_j} + \mu_{k_j}) p_k(t)$$

$$+ \frac{D^*}{2} \sum_{j=2}^{N} ((k_j+1) p_{k+e_j-e_{j-1}}(t) - k_j p_k(t))$$

$$+ \frac{D^*}{2} \sum_{j=1}^{N-1} ((k_j+1) p_{k+e_j-e_{j+1}}(t) - k_j p_k(t))$$

with initial distribution $(p_k^o)_{k \in \mathbb{N}^N}$.

3. Relations between the models

Since the stochastic (global or local) model is considered more detailed than the deterministic one, it should be possible to recover from the master equations the deterministic equations if the volume L tends to infinity (thermodynamic limit). On the other hand, one should be able to recover the global (stochastic or deterministic) description from the local one by letting the diffusion become dominant compared to reaction.

The subject of this section is to make the above statements precise and show in what sense the models are consistent.

3.1 Relation between the global stochastic and deterministic models

This question has been completely settled by KURTZ ([13], [14], [16]) by proving the following results:

Theorem 1. Given the set-up of sections 2.1 and 2.3. Then we have:

(i) *Law of large numbers*: Suppose $\lim_{L\to\infty} X(o)/L = \varphi_o$ in probability, then for each finite T and $\delta > o$

$$\lim_{L\to\infty} P\left(\sup_{o\leq t\leq T} \left| \frac{X(t)}{L} - \varphi(t) \right| > \delta \right) = o,$$

(ii) *Central limit theorem*: Suppose that in addition

$$\lim_{L\to\infty} \sqrt{L}\left(\frac{X(o)}{L} - \varphi_o\right) = y_o \quad \text{in probability, then}$$

$$\lim_{L\to\infty} \sqrt{L}\left(\frac{X(t)}{L} - \varphi(t)\right) = y(t) \quad \text{weakly}$$

(i.e. on the level of the corresponding probability measures in function space), where $y(t)$ is a Gaussian diffusion process which is the solution of the linear stochastic differential equation

$$dy(t) = \frac{d(\lambda-\mu)}{dx}(\varphi(t))y(t)dt + (\lambda+\mu)(\varphi(t))^{1/2}dW_t,$$

$$y(o) = y_o.$$

(iii) *Diffusion approximation*: There is a probability space on which

$$\sup_{o\leq t\leq T} \left| \frac{X(t)}{L} - u(t) \right| \leq K_T \frac{\log L}{L},$$

where K_T is a random variable depending on T and $u(t)$ is a diffusion process idential in distribution to the solution of the nonlinear stochastic differential equation

$$du(t) = (\lambda-\mu)(u(t))dt + \frac{1}{\sqrt{L}}((\lambda+\mu)(u(t)))^{1/2}dW_t,$$

$$u(o) = X(o)/L.$$

Remark: The approximation in Theorem 1 (i) and (ii) can also be made sample-wise (at least at the expense of restriction to linear reaction, see KURTZ [16]) with the result

$$X(t)/L = \varphi(t) + O(1/\sqrt{L}),$$
$$= \varphi(t) + (1/\sqrt{L})y(t) + O(\log L/L),$$
$$= u(t) + O(\log L/L)$$

so that it turns out that the error of the Gaussian and of the diffusion approximation have the same order of magnitude.

3.2 Relation between the local stochastic and deterministic models

As it turns out, the presence of diffusion forces us to rescale the lenght so that on a new r-axis the system is defined on the interval [O,R] with new cell size h given by

$$h = R\,\ell/L = R/N$$

where necessarily $h \to 0$. Depending on whether equation (3) is considered on a finite or infinite interval, $R = \text{const}$ or $R \to \infty$. For simplicity we assume $R = 1$, so that $h = \ell/L = 1/N$.

The N-vector process $X(t)/\ell$ corresponds to a step function on [O,1] defined by

$$x(r,t) = X_j(t)/\ell \quad \text{on} \quad ((j-1)/N, j/N],\ j=1,\ldots,N.$$

We will compare $x(\cdot,t) = x(t)$ with the solution $\varphi(t)$ of (3) as elements of $L_2(0,1)$, where the parameters L,ℓ and D^* in the stochastic model are moving such that $L \to \infty$ and $\ell/L = 1/N \to 0$. The parameter D in (3) is considered fixed.

Theorem 2. (ARNOLD and THEODOSOPULU [3]).
Given the set-up of sections 2.2 and 2.4. Take fixed or zero flux boundary conditions in both models.
Suppose $L \to \infty$, $N \to \infty$ such that

(i) $\lim \| x(o)-\varphi_o \|_{L_2} = 0$ in probability,

(ii) $D^*/2 = DN^2$ (entailing $D^* \to \infty$),

(iii) $N^2/\ell = L^2/\ell^3 \to 0$ (entailing $\ell \to \infty$).

Then we have the *law of large numbers*: For each finite T and $\delta > o$

$$\lim \sup_{o \leq t \leq T} P(\|x(t)-\varphi(t)\|_{L_2} > \delta) = o.$$

Remark: Condition (ii) is needed to end up with the Laplacian. Condition (iii) requires that the cell size has to increase to infinity quite fast, e.g. like $\ell = L^{2/3 + \epsilon}$, $o < \epsilon < 1/3$. On the other hand, if $L^2/\ell^3 \to \infty$ then the law of large numbers breaks down, so (iii) cannot be relaxed if one aimes at convergence in

L_2-norm. However, (iii) can be dismissed if one is content with a weaker kind of convergence, e.g.

$$\int_0^1 \psi(r)x(r,t)dr \to \int_0^1 \psi(r)\varphi(r,t)dr \quad \text{in probability}$$

for all $\psi \in C^1[o,1]$ which amounts to convergence of $x(t)$ to $\varphi(t)$ in the Sobolev space H^{-1} in the case of zero flux boundary conditions. The moral is that in any case we have to "turn on" diffusion to get the Laplacian. This makes the step function $x(\cdot,t)$ look very irregular so that it is only close to the smooth $\varphi(t)$ in H^{-1} norm. If we want to have $x(\cdot,t)$ close to $\varphi(t)$ in L_2 norm we have to make $x(\cdot,t)$ smoother by "turning on" reaction, i.e. by increasing the cell size so that we have more particles in each cell.

Proof of Theorem 2.

We assume zero flux boundary conditions.

Step 1: Limit behavior of an accompanying martingale:

According to a result of KURTZ [14] the process

$$z(t) = x(t) - x(O) - \int_0^t F(x(s))ds,$$

where

$$F(x) = \sum_m mf(x,m) = F^{(1)}(x)+F^{(2)}(x), \quad F^{(1)}(x)=f(x)=\lambda(x)-\mu(x)$$

$$F^{(2)}(x) = \frac{D^*}{2}\Delta_h^2 x, \quad \Delta_h^2 x = x(\cdot+h) - 2x(\cdot) + x(\cdot-h),$$

where $x((N+1)h) = O$ due to our boundary conditions, is a (step-function-valued) martingale provided

$$\sup_x \tau(x)<\infty \text{ and } \sup_x \tau(x)\sum_m \frac{|m|}{\ell}\sigma(x,m/\ell) <\infty.$$

Here the waiting time parameter $\tau(x)$ and the jump distribution function $\sigma(x,m/\ell)$ are given by

$$\tau(x)=\ell\sum_{j=1}^N (\lambda(x_j)+\mu(x_j))+\ell D^*\sum_{j=1}^N x_j$$

and

$$\sigma(x,m/\ell) = \ell f(x,m)/\tau(x), \quad q_{k,k+m} = \ell f(k/\ell,m).$$

Since our original process does not satisfy these conditions we shift to the stopped process

$$\tilde{x}(t) = x(t\wedge\tau),$$

where τ is the first-exit time of $x(t)$ from the set of step functions

$$S_\rho = \left\{ x\geq 0: \sup_{0\leq r\leq 1} x(r)\leq\rho \right\},$$

$\rho>O$ arbitrary. Then $\tilde{x}(t)$ is a jump Markov process with $\tilde{\tau}(x) = \tau(x)\chi_{S_\rho}(x)$,

$$\tilde{\sigma}(x,m/\ell) = \begin{cases} \sigma(x,m/\ell), & x \in S_\rho, \\ \text{arbitrary}, & x \notin S_\rho, \end{cases}$$

$$\tilde{f}(x,m)=f(x,m)\chi_{S_\rho}(x), \quad \tilde{F}(x)=F(x)\chi_{S_\rho}(x).$$

Thus for $\tilde{x}(t)$

$$\sup_x \tilde{\tau}(x) = \sup_{x\in S_\rho} \tau(x)\leq\ell N(\lambda_\rho+\mu_\rho+D^*\rho)<\infty$$

with

$$\lambda_\rho = \max_{[O,\rho]} \lambda(x), \quad \mu_\rho = \max_{[O,\rho]} \mu(x),$$

and

$$\sup_x \tilde{\tau}(x) \sum_m \frac{|m|}{\ell} \tilde{\sigma}(x,m/\ell) = \sup_{x\in S_\rho} \sum_m |m| f(x,m)$$

$$\leq N(\lambda_\rho+\mu_\rho+\sqrt{2}D^*\rho)<\infty,$$

so that $\tilde{z}(t)$ is a martingale with respect to $P_x, x\geq 0$. Note that for arbitrary initial distribution

$$E\tilde{z}(t) = O \quad \text{for all} \quad t\geq O.$$

Doob's inequalities read for the second moments

$$E_x \sup_{0\leq s\leq t} \|\tilde{z}(s)\|^2\leq 4E_x \|\tilde{z}(t)\|^2$$

and

$$P_x\left(\sup_{0\leq s\leq t} \|\tilde{z}(s)\|>\delta\right) \leq\delta^{-2}E_x\|\tilde{z}(t)\|^2.$$

We will estimate $E_x\|\tilde{z}(t)\|^2$. From a result of Kurtz [14] we derive

$$E_x\|\tilde{z}(t)\|^2 = \int_0^t E_x\tilde{\tau}(\tilde{x}(s))\Big(\sum_m \|m/\ell\|^2\tilde{\sigma}(\tilde{x}(s),m/\ell)\Big)\,ds$$

$$= (h/\ell)\int_0^t E_x(\sum_m |m|^2\tilde{f}(\tilde{x}(s),m)\,ds$$

$$= (1/\ell)\int_{s=0}^t E_x\int_{r=0}^1 (\lambda(\tilde{x}(r,s))+\mu(\tilde{x}(r,s)))\,dr\,ds$$

$$+\ (2D^*/\ell)\int_{s=0}^t E_x\int_{r=0}^1 \tilde{x}(r,s)\,dr\,ds$$

for $x \in S_\rho$, and $E_x\|\tilde{z}(t)\|^2 = 0$ for $x \notin S_\rho$. From this we obtain

$$\frac{2D^*}{\ell}\int_0^t E_x\bar{x}(s)\,ds \leq E_x\|\tilde{z}\ (t)\|^2 \leq \frac{t}{\ell}(\lambda_\rho + \mu_\rho + 2D^*\rho), \qquad (9)$$

where

$$\bar{x}(s) = \int_0^1 \tilde{x}(r,s)\,dr$$

is the global concentration (= total number of particles divided by total length).
From (9) we immediately have the following result:

(a) If $D^*/\ell \to 0$ then for each $t>0$ $\sup_{[0,t]}\|\tilde{z}(s)\| \to 0$ in probability and in mean square for any sequence of initial distributions for $x(0)$.

(b) If $\|x-x_0\| \to 0$, $x_0 \neq 0$, $x_0 \in L_2[0,1]$, and $D^*/\ell \to \infty$, then for each $t>0$

$$E_x\|\tilde{z}(t)\|^2 \to \infty.$$

Part (a) follows immediately from (9) and Doob's inequalities. Part (b) would follow from (9) once we known that $\psi(x,t) = E_x\bar{x}(t) \geq c > 0$ for all x and t small enough.
Take without restriction of generality $\sup_r x_0(r) \leq \rho$. From

$$\tilde{x}(t) = \tilde{z}(t) + \tilde{x}(0) + \int_0^t F(\tilde{x}(s))\,ds$$

we obtain by integrating over $r \in [0,1]$

$$\overline{x}(t) = \overline{z}(t) + \overline{x}(0) + \int_0^t \overline{F}^{(1)}(\tilde{x}(s))ds$$

and by taking the expectation

$$\psi(x,t) = \overline{x} + \int_0^t E_x \overline{F}^{(1)}(\tilde{x}(s))ds.$$

Thus $\psi(x,0) = \overline{x} \to \overline{x}_0 > 0$ and

$$\dot{\psi}(x,0) = E_x \overline{F}^{(1)}(\tilde{x}(0)) = \int_0^1 F^{(1)}(\tilde{x}(r))dr \to \int_0^1 F^{(1)}(x_0(r))dr$$

entailing that $\psi(x,t) \geqq c$ for all x and t small enough.

<u>Step 2:</u> <u>A Gronwall-Bellman lemma:</u>

In this step, we give an estimate for the random variable

$$\|\tilde{x}(t)-\varphi(t)\| = \left(\int_0^1 |\tilde{x}(r,t)-\varphi(r,t)|^2 dr\right)^{\frac{1}{2}}.$$

Since we deal only with the stopped process, we omit the tilde from now on.

<u>Lemma.</u> Let $\varphi(t)$ be the unique global solution of problem (3) with $0 \leqq \varphi_0(r) \leqq \rho_0$. Let $x(t)$ be the jump Markov process stopped at first-exit time from a sphere with arbitrary radius $\rho \leqq \rho_0$, take $0 \leqq x(r,0) \leqq \rho$.

Assume

$$D^*/2 = DN^2.$$

Then

$$v\|x(t)-\varphi(t)\| = v(t) + L_\rho \int_0^t \exp(L_\rho(t-s))\, v(s)ds, \qquad (10)$$

where

$$v(t) = \|x(0)-\varphi_0\| + 3\|z(t)\| + \int_0^1 \frac{\|z(t)-z(s)\|}{t-s} ds + \varepsilon(t),$$

$\varepsilon(t)$ is a deterministic non-negative function independent of $x(t)$ satisfying

$$\sup_{[0,t]} \varepsilon(s) \to 0 \quad (N \to \infty),$$

L_ρ is the local Lipschitz constant of the polynomial $f = \lambda - \mu$ in the interval $[0,\rho]$.

Proof of the Lemma.

1. We have

$$x(t)-\varphi(t)=z(t)+y(t)-\varphi(t),$$

where

$$y(t)=x(0)+\int_0^t F(x(s))ds$$

$$= x(0) + \int_0^1 f(x(s))ds+\frac{D^*}{2}\int_0^1 \Delta_h^2 x(s)ds.$$

The semigroup generated by

$$\frac{D^*}{2}\Delta_h^2 = A_N \in L(\mathfrak{X}_N),$$

$\mathfrak{X}_N$ = step functions on $[0,1]$ partitioned into intervals of length $1/N$ as subspace of $L_2[0,1]$, is denoted by

$$T_N(t) = \exp(A_N t) \in L(\mathfrak{X}_N), \quad t \geq 0.$$

The solution of the equation

$$u(t)=x(0)+\int_0^1 A_N u(s)ds+\int_0^t f(x(s))ds$$

for known $x(t)$ is

$$u(t)=T_N(t)x(0)+\int_0^t T_N(t-s)f(x(s))ds.$$

For $y(t)-u(t)=w(t)$ we have

$$w(t)=\int_0^1 A_N w(s)ds+\int_0^1 A_N z(s)ds$$

with solution

$$w(t)=\int_0^1 T_N(t-s)A_N z(s)ds.$$

Writing the solution of (3) in its (mild) semigroup form with $A = D\partial^2/\partial r^2$, $T(t)$ the corresponding semigroup $\in L(L_2[0,1])$,

$$\varphi(t) = T(t)\varphi_0 + \int_0^t T(t-s)f(\varphi(s))ds,$$

we obtain

$$\begin{aligned} x(t)-\varphi(t) &= z(t)+w(t)+u(t)-\varphi(t) \\ &= z(t)+\int_0^t T_N(t-s)A_N z(s)\,ds \\ &\quad +T_N(t)x(0)-T(t)\varphi_0 \\ &\quad +\int_0^t T_N(t-s)f(x(s))\,ds-\int_0^t T(t-s)f(\varphi(s))\,ds. \end{aligned}$$

2. Next we treat $w(t)$. The operator A_N corresponds to the symmetric NxN matrix

$$\tilde{A}_N = \frac{D^*}{2}\begin{pmatrix} -2 & 1 & 0 & 0 & \dots & 0 & 0 \\ 1 & -2 & 1 & 0 & & & \\ 0 & 1 & -2 & 1 & & \vdots & \vdots \\ \vdots & & & 1 & -2 & 1 & 0 \\ 0 & \dots & & 0 & 1 & -2 & 1 \\ 0 & \dots & & 0 & 0 & 1 & -2 \end{pmatrix}$$

with eigenvalues

$$\lambda_j = D^*(\cos\frac{\pi j}{N+1}-1), \quad j=1,\dots,N,$$

$-D^*<\lambda_N<\dots<\lambda_1<0$. $T_N(t)$ is thus an analytic contraction semigroup with

$$\| T_N(t)\| = e^{\lambda_1 t} \leqq 1$$

and

$$\| T_N(t)A_N\| \leqq 1/t \quad \text{for} \quad t>0$$

(CURTAIN and PRITCHARD [6], p. 33).

We rewrite $w(t)$ as

$$w(t)=\int_0^t T_N(t-s)A_N(z(s)-z(t))\,ds+\int_0^t T_N(t-s)A_N z(t)\,ds.$$

For the second term

$$\int_0^t T_N(t-s)A_n z(t)\,ds = (T_N(t)-I)z(t),$$

therefore

$$\|w(t)\| \leqq \int_0^t \frac{\|z(t)-z(s)\|}{t-s} ds + 2\|z(t)\|.$$

3. Now we estimate

$$T_N(t)x(O)-T(t)\varphi_0 = T_N(t)x(O)-T_N(t)P_N\varphi_0 + T_N(t)P_N\varphi_0 - T(t)\varphi_0,$$

where the projection $P_N: L_2[O,1] \to \mathfrak{X}_N$ is given by

$$P_N\varphi(r) = \frac{1}{h}\int_{(j-1)h}^{jh} \varphi(q)dq, r \in ((j-1)h, jh].$$

If $D^*/2 = D/h^2 = DN^2$, then

$$A_N P_N u \to Au \quad \text{for} \quad u \in C^2$$

which entails

$$T_N(t)P_N u \to T(t)u \quad \text{for each} \quad u \in L_2$$

(KATO [11], pp. 512-513), in particular

$$\| T_N(t)P_N\varphi_0 - T(t)\varphi_0\| = \varepsilon_1(t) \to O$$

uniformly in [O,T]. Since $\|\varphi_0 - P_N\varphi_0\| = \varepsilon_2 \to O$, we have

$$\| T_N(t)x(O) - T(t)\varphi_0\| \leqq \varepsilon_1(t) + \varepsilon_2 + \|x(O)-\varphi_0\|.$$

4. Finally

$$\int_0^t T_N(t-s)f(x(s))ds - \int_0^t T(t-s)f(\varphi(s))ds$$

$$= \int_0^t T_N(t-s)(f(x(s))-P_N f(\varphi(s)))ds.$$

$$+ \int_0^1 (T_N(t-s)P_N - T(t-s))f(\varphi(s))ds.$$

For the second term we conclude that by the dominated convergence theorem

$$\| \int_0^t (T_N(t-s)P_N - T(t-s))f(\varphi(s))ds\| = \varepsilon_3(t) \to O$$

uniformly in [O,T], while for the first term

$$\| \int_0^t T_N(t-s)(f(x(s))-P_N f(\varphi(s)))ds\| \leqq L_\rho \int_0^t \|x(s)-\varphi(s)\| ds + \varepsilon_4(t),$$

where

$$\varepsilon_4(t) = \int_0^t \| P_N f(\varphi(s)) - f(\varphi(s))\| ds \to 0$$

uniformly in [0,T].

5. Collecting all terms we obtain (putting $\varepsilon_1+\varepsilon_2+\varepsilon_3+\varepsilon_4=\varepsilon(t)$)

$$\| x(t) - \varphi(t)\| \leqq v(t) + L_\rho \int_0^t \| x(s) - \varphi(s)\| ds,$$

from which the result (10) follows after applying an appropriate version of the Gronwall-Bellman lemma.

Step 3: Conditions (ii) and (iii) of Theorem 2 make sure that result (a) in step 1 and the lemma in step 2 are true. We show that the right-hand side of (10) tends to 0 in mean and in probability. The only critical term is

$$I(t) = \int_0^t \frac{\|z(t)-z(s)\|}{t-s} ds.$$

Since $z(t)$ is a martingale, so is $z(t)-z(s)$, $t \leqq s$, and inequality (9) yields for arbitrary initial distributions

$$E\|z(t)-z(s)\|^2 \leqq (t-s)(\lambda_\rho+\mu_\rho+2D^*\rho)/\ell = (t-s)a,$$

thus

$$EI(t) = \int_0^t \frac{E\|z(t)-z(s)\|}{t-s} ds \leqq \sqrt{a} \int_0^t \frac{\sqrt{t-s}}{t-s} ds = 2\sqrt{at} \to 0$$

uniformly in each interval [0,T] if conditions (ii) and (iii) are satisfied. Therefore, $I(t) \to 0$ in mean and thus in probability.

Altogether, we have $v(t) \to 0$ in mean and in probability uniformly in [0,T]. Consequently, the right-hand side of (10) tends to 0 in mean and in probability uniformly in [0,T] from which the result follows.

Remarks. 1. For the d-dimensional cube ($d \geqq 1$) of volume $V = L^d$ divided into N^d cells of size $v = \ell^d$, $\ell = L/N$, and zero boundary conditions we obtain the following result: If we transform our system onto the unit cube $[0,1]^d$ with new cell size $h^d = (\ell/L)^d$, use again the step function approach on $[0,1]^d$, we have

$$\sup_{[0,t]} \|\tilde{z}(s)\| \to 0 \text{ in probality}$$

provided $D^*/\ell^d \to 0$. The analogue of the Lemma in step 2 requires $D^*/2d = DN^2$ in order to arrive at the d-dimensional Laplacian, so that the condition for convergence analogous to condition (iii) of Theorem 2 is

$$\frac{N^2}{\ell^d} = \frac{L^2}{\ell^{d+2}} \to 0,$$

which is fulfilled e.g. for $\ell = L^{2/(d+2)+\varepsilon}$, $0<\varepsilon<d/(d+2)$.

2. The error $x(t)-\varphi(t)$ (central limit theorem in the local case) will be treated in a subsequent paper, see ARNOLD, KOTELENEZ and CURTAIN [4].

3.3 Relation between the local and global deterministic models

We now look at the solution $\varphi(r,t)$ of (3) on $[0,1]$ with the boundary conditions given in 2.2 for $D \to \infty$. We expect that immediately after start a strong diffusion straightens out any spatial inhomogeneity so that $\varphi(r,t)$ is approximately equal to a constant $\psi(t)$ with respect to r, and $\psi(t)$ solves the global kinetic equation (2).

Note first that the only boundary conditions in the class considered fulfilled for a non-zero straight line are the zero flux conditions. Therefore, only under zero flux conditions there is a chance of recovering the global law by "turning on" diffusion.

Theorem 3. Given the solution $\varphi(r,t)$ of (3) on $[0,1]$ with zero flux boundary conditions

$$\frac{\partial\varphi(0,t)}{\partial r} = 0, \quad \frac{\partial\varphi(1,t)}{\partial r} = 0$$

and initial condition φ_0. Let $\psi(t)$ be the solution of (2) with initial condition

$$\psi_0 = \int_0^1 \varphi_0(r)\,dr.$$

Then for all $\delta > 0$ and finite $T > 0$

$$\lim_{D\to\infty} \sup_{\delta\le t\le T} \|\varphi(r,t)-\psi(t)\|_{L_2} = 0.$$

Proof. The operator D Δ generates the analytic contractio semigroup $T(t)$ on L_2 given by

$$T(t)x(r) = c_o + \frac{1}{\sqrt{2}} \sum_{n=1}^{\infty} c_n e^{-n^2\pi^2 Dt} \cos n\pi r$$

where $c_o = \int_0^1 x(r)dr$, $c_n = \frac{1}{\sqrt{2}} \int_0^1 x(r) \cos n\pi r \, dr$, $x \in L_2$

(see CURTAIN and PRITCHARD [6], p. 46). We have $T(t)c =$ for any spatial constant and $\| T(t) \| = 1$. We have $\lim_{D\to\infty} T(t) = \Pi$ uniformly for $\delta \leq t \leq T$, where Π is the projection onto the one-dimensional subspace generated by 1 defined by

$$\Pi x(r) \equiv \int_0^1 x(r)dr = c_o .$$

Writing both equations in their semigroup form taking into account $\Pi\varphi_o = \psi_o$, we obtain

$$\varphi(t) - \psi(t) = (T(t) - \Pi)\varphi_o + \int_0^t T(t-s)(f(\varphi(s))-f(\psi(s)))$$

For any interval $[o,T]$ and given φ_o there is a sup-topology sphere S_ρ of radius ρ such that $\varphi(r,t) \in S_\rho$ and $\psi(t) \in S_\rho$ for all $r \in [o,1]$, $t \in [o,T]$. In S_ρ f is Lipschitz with constant L_ρ. Therefore

$$\| \varphi(t)-\psi(t)\| \leqq \| T(t)-\Pi\| \|\varphi_o\| + L_\rho \int_0^t \|\varphi(s)-\psi(s)\| \, ds.$$

Putting $\| T(t) - \Pi\| \|\varphi_o\| = \epsilon(t)$, the Bellman-Gronwall lemma yields

$$\|\varphi(t)-\psi(t)\| \leqq \quad T(t)-\Pi \quad \varphi_o \quad + L_\rho \quad {}_0^t \quad \varphi(s)-\psi(s) \quad ds.$$

Putting $T(t) - \Pi \quad \varphi_o \quad = \epsilon(t)$, the Bellman-Gronwall lemma yields

$$\| \varphi(t)-\psi(t)\| \leqq \epsilon(t) + L_\rho \int_0^t \exp(L_\rho(t-s))\epsilon(s)ds$$

from which the result follows.
Results similar to Theorem 3 can be found in CONWAY, HOFF and SMOLLER [5].

3.4 Relation between the local and global stochastic models

We again assume zero flux boundary conditions for the local stochastic model (i.e.reflection at the boundary) and consider the case $D^* \to \infty$, L and ℓ fixed. The total number of particles

$$\overline{X}(t) = \sum_{j=1}^{N} X_j(t)$$

is not a Markov process anymore, but we expect that for big D^* $\overline{X}(t)$ is close in a certain sense to the process pertaining to the global stochastic description. Attempts have been made in this direction e.g. by MALEK-MANSOUR [17], but to our knowledge no rigorous proof has been given.

Theorem 4. Let $X(t)$ be the stochastic process belonging to the local stochastic model with transition intensities given by (7) and (8), let $\overline{X}(t)$ be the total number of particles in the local model. Then

$$\lim_{D^*\to\infty} \overline{X}(t) = Y(t) \quad \text{weakly} \qquad (11)$$

(i.e. on the level of the corresponding probability measures in function space), where $Y(t)$ is the stochastic process belonging to the global stochastic model with transition intensities given by (6) startet with the distribution of $\overline{X}(o)$.

Proof. We use a semigroup approximation method developed by KURTZ [15]. Without restricting the generality we can assume that our process is being stopped at first exit time from a set

$$\mathbb{N}_\rho^N = \{k \in \mathbb{N}^N : \overline{k} = \sum_1^N k_j \leqq \rho\}$$

with arbitrarily big ρ. Thus, we can restrict our whole investigation to this set.

The infinitesimal generator A_{D^*} of the Markov process $X(t)$ is defined on the Banach space $B(\mathbb{N}_\rho^N)$ of all bounded functions on $\mathbb{N}_\rho^N$ by

$$A_{D^*} = D^* L_1 + L_2$$

where

$$L_1 f(k) = \frac{1}{2}\sum_{j=1}^{N-1} (f(k+e_{j+1}-e_j)-f(k))k_j + \frac{1}{2}\sum_{j=2}^{N} (f(k+e_{j-1}-e_j)-f(k))k_j$$

and

$$L_2f(k) = \sum_{j=1}^{N} ((f(k+e_j)-f(k))\lambda_{k_j} + (f(k-e_j)-f(k))\mu_{k_j})$$

with λ_{k_j}, μ_{k_j} given by (8), and for $k \in \mathbb{N}_\rho^N$.

The infinitesimal generator B of the Markov process $Y(t)$ is defined on the Banach space $B(\mathbb{N}_\rho)$ of all bounded functions on $\mathbb{N}_\rho$ by

$$Bf(j) = (f(j+1) - f(j))\lambda_j + (f(j-1) - f(j))\mu_j, \; j \in \mathbb{N}_\rho,$$

with λ_j, μ_j given by (6).

Let M be the set of all $f \in B(\mathbb{N}_\rho)$ such that there is a sequence $f_{D*} \in B(\mathbb{N}_\rho^N)$ with

$$\lim_{D*\to\infty} \sup_k | f_{D*}(k) - f(\overline{k}) | = o \tag{12}$$

and

$$\lim_{D*\to\infty} \sup_k | A_{D*} f_{D*}(k) - Bf(\overline{k}) | = o . \tag{13}$$

We will prove that $M = B(\mathbb{N}_\rho)$. Since $\lambda - B$ is invertible for some $\lambda > o$, a theorem of KURTZ ([15], S. 630-631) assures that under these conditions (11) holds.

We proceed as in PAPANICOLAOU [19] and put for any $f \in B(\mathbb{N}_\rho)$

$$f_{D*}(k) = f(\overline{k}) + \frac{1}{D*}g(k), \; k \in \mathbb{N}_\rho^N,$$

with g to be determined. Certainly (12) is true for any g.

Furthermore,

$$A_{D*} f_{D*}(k) = D*L_1f(\overline{k}) + L_1g(k) + L_2f(\overline{k}) + \frac{1}{D*} L_2g(k).$$

Because $\overline{k+e_{j\pm 1}-e_j} = \overline{k}$ we have $L_1f(\overline{k}) = o$.

Thus (13) will be satisfied if g can be chosen such that

$$L_1g(k) = - L_2f(\overline{k}) + Bf(\overline{k}).$$

Observe that L_1 is the infinitesimal generator of the Markov process on $\mathbb{N}_\rho^N$ modeling pure diffusion of intensity 1 with reflection at the boundary. This process is ergodic with stationary distribution P_o being the multinomial distribution

$$p_k = \frac{m!}{k_1!k_2!\ \dots\ k_N!}\, N^{-m}\ ,\ \text{if } \bar{k}=m,\ k \in \mathbb{N}_\rho^N\ ,$$

where $m \leqq \rho$ is the initial number of particles in the system (which is conserved), and $p_k = o$ otherwise (see VAN DEN BROECK, HORSTHEMKE and MALEK-MANSOUR [20]).

Therefore $L_1 g = u$ has a solution provided the solvability condition

$$\int u(x)P_o(dx) = \sum_k u(k)p_k = o$$

holds for our particular u. This is the only thing that remains to be checked.

We have

$$\sum_k (-L_2 f(\bar{k}) + Bf(\bar{k}))p_k =$$

$$=(f(m+1)-f(m))(\lambda_m - \sum_k(\sum_{j=1}^N \lambda_{k_j})p_k) + (f(m-1)-f(m))(\mu_m - \sum_k(\sum_{j=1}^N \mu_{k_j})p_k$$

For the multinomial distribution of m particles into $N = L/\ell$ cells

$$\sum_k(\sum_{j=1}^N \lambda_{k_j})p_k = \sum_k(\sum_{j=1}^N \ell\lambda_\ell(k_j/\ell))p_k$$

$$= N\ell \sum_{j=o}^m \lambda_\ell(k_j/\ell)(\tfrac{1}{N})^j(1-\tfrac{1}{N})^{m-j}\binom{m}{j}$$

$$= L\lambda_L(m/L) = \lambda_m,$$

similarly for the μ-terms, so that indeed the solvability condition is satisfied.

4. Summary

We summarize the relations between the four models in the following scheme:

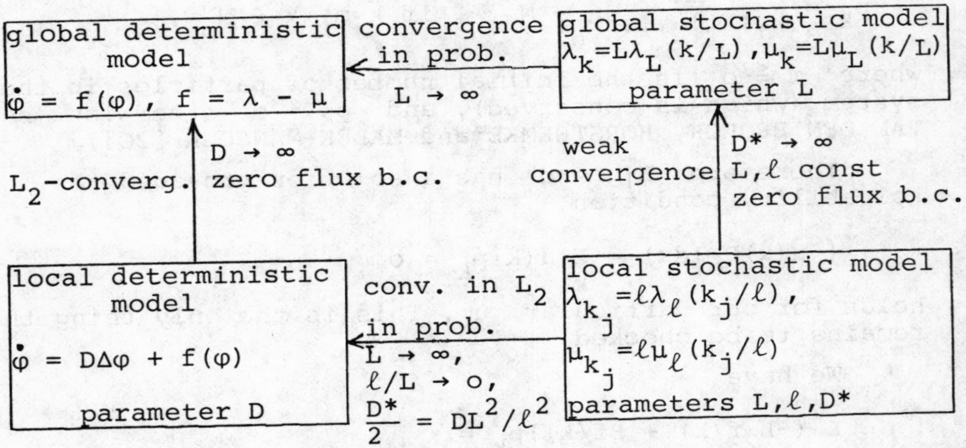
global deterministic model
$\dot{\varphi} = f(\varphi)$, $f = \lambda - \mu$
convergence in prob.
$L \to \infty$
global stochastic model
$\lambda_k = L\lambda_L(k/L)$, $\mu_k = L\mu_L(k/L)$
parameter L
$D \to \infty$
L_2-converg. zero flux b.c.
weak convergence
$D^* \to \infty$
L, ℓ const
zero flux b.c.
local deterministic model
$\dot{\varphi} = D\Delta\varphi + f(\varphi)$
parameter D
conv. in L_2
in prob.
$L \to \infty$,
$\ell/L \to o$,
$\frac{D^*}{2} = DL^2/\ell^2$
local stochastic model
$\lambda_{k_j} = \ell\lambda_\ell(k_j/\ell)$,
$\mu_{k_j} = \ell\mu_\ell(k_j/\ell)$
parameters L, ℓ, D^*

REFERENCES

1. Amann, H.: Invariant Sets and Existence Theorems for Semilinear Parabolic and Elliptic Systems. J. Math. Anal. Appl. 65 (1978), 432-467.
2. Arnold, L.: On the consistency of the mathematical models of chemical reactions. In: Haken, H. (ed.): Dynamics of Synergetic Systems, Springer-Verlag, Berlin-Heidelberg-New York 1980, p. 107-118
3. Arnold, L. and Theodosopulu, M.: Determistic Limit of the Stochastic Model of Chemical Reactions with Diffusion. Adv. Appl. Prob. 12 (1980), 367-379.
4. Arnold, L., Curtain, R. and Kotelenez, P.: Linear stochastic evolution equation models for chemical reactions. Report Forschungsschwerpunkt Dynamische Systeme, Universität Bremen 1980
5. Conway, E., Hoff. D. and Smoller, J.: Large time behavior of solutions of systems of nonlinear reaction-diffusion equations. SIAM J. Appl. Math. 35 (1978), 1-16.
6. Curtain, R.F. and Pritchard, A.J.: Infinite Dimensional Linear Systems Theory. Lecture Notes in Control and Information Sciences 8. Springer, Berlin-Heidelberg-New York 1978.
7. Fife, P.C.: Asymptotic States for Equations of Reaction and Diffusion. Bull. Amer. Math. Soc. 84 (1978), 693-726.
8. Fife, P.C.: Mathematical Aspects of Reacting and Diffusing Systems. Lecture Notes in Biomathematics 28. Springer, Berlin-Heidelberg-New York 1979.
9. Haken, H.: Synergetics. Springer, Berlin-Heidelberg-New York 1978 (second edition).
10. Horsthemke, W. and Lefever, R.: Forthcoming book on external noise, to appear in Springer Series in Synergetics.
11. Kato, T.: Perturbation theory for linear operators. Springer, Berlin-Heidelberg-New York 1966.
12. Kuiper, H.J.: Existence and Comparison Theorems for Nonlinear Diffusion Systems. J. Math. Anal. Appl. 60 (1977), 166-181.

13. Kurtz, T.: Solutions of Ordinary Differential Equations as Limits of Pure Jump Markov Processes. J. Appl. Prob. 7 (1970), 49-58.

14. Kurtz, T.: Limit Theorem for Sequences of Jump Markov Processes Approximating Ordinary Differential Processes. J. Appl. Prob. 8 (1971), 344-356.

15. Kurtz, T.: Semigroups of Conditioned Shifts and Approximation of Markov Processes. Annals of Prob. 3 (1975), 618-642.

16. Kurtz, T.: Strong Approximation Theorems for Density Dependent Markov Chains. Stoch. Proc. and their Appl. 6 (1978), 223-240.

17. Malek-Mansour, M.: Fluctuation et Transition de Phase de Non-equilibre dans les Systems Chimiques. These Université Libre de Bruxelles 1979.

18. Nicolis, G. and Prigogine, I.: Selforganization in Nonequilibrium Systems. Wiley, New York 1977.

19. Papanicolaou, G. C.: Asymptotic Analysis of Stochastic Equations. In: Rosenblatt, M. (ed.): Studies in Probability Theory. Studies in Mathematics Vol.18 The Mathematical Association of America 1978.

20. Van den Broeck, C., Horsthemke, W. and Malek-Mansour, M.: On the Diffusion Operator of the Multivariate Master Equation. Physica 89 A (1977), 339-352.

QUALITATIVE THEORY OF REAL NOISE EXCITED SYSTEMS

Ludwig Arnold

Universität Bremen

Abstract: Some of the problems and results of the qualitative theory of stochastic dynamical systems (SDS) under the influence of real noise are sketched. Reference is given to papers where details can be found. General references are ARNOLD [5] and ARNOLD and KLIEMANN [6].

1. Scope of the theory

1.1 Notion of an SDS

An SDS $\Sigma = (f, x_o, \xi)$ is an ordinary differential equation in $\mathbb{R}^d$,

$$\dot{x}(t) = f(x(t), \xi(t)),\ x(o) = x_o,\ t \in \mathbb{R}^+ \text{ or } \mathbb{R},$$

with random initial condition x_o and a stochastic noise process ξ with values in $\mathbb{R}^m$ in the (deterministic) right-hand side f. The differential equation is to be interpreted sample-wise as ordinary differential equation for each noise trajectory. Note that x_o is allowed to anticipate ξ.

Classification of Σ according to f:

- f nonlinear,
- f linear (in state): $f(x,\xi) = A(\xi)x + b(\xi)$,
  A = multiplicative noise, b = additive noise,
- f linear in noise: $f(x,\xi) = f(x) + G(x)\xi$,
- f linear in state and noise: $f(x,\xi) = Ax + B\xi$.

M. Hazewinkel and J. C. Willems (eds.), Stochastic Systems: The Mathematics of Filtering and Identification and Applications, 135–140.

Classification of Σ according to ξ :

Qualitative theory has mainly been developed for white ξ . In this case, x is a Markov process. However, the whiteness assumption is often not realistic (f has to be linear in ξ , ξ cannot be constrained to subsets of $\mathbb{R}^m$). Real (as opposed to 'white') noise is any ordinary stochastic process. Important cases are:

- ξ Markov, in particular diffusion: $d\xi=a(\xi)dt+b(\xi)dW$, "colored noise" = output of a system with white input. Advantage: (x,ξ) Markov. Drawback: excludes smooth (e.g. differentiable) processes
- ξ (strictly) stationary: models non - evolutionary environment.

1.2 Qualitative theory

Qualitative theory of SDS was founded by KUSHNER [17] and KHASMINSKI [12] in the 1960ies. It studies the general nature of a solution on the entire time interval (asymptotic or long-term behavior) without solving the equation. Due to the various topologies for random variables, there are usually various randomized versions of each of the following concepts of the deterministic theory founded by H. Poincaré and A. M. Lyapunov: invariant sets, critial configurations, limit sets, recursive concepts (recurrent/transient points, Poisson and Lagrange stability), Lyapunov stability, attractors, Lyapunov numbers, Floquet theory, invariant measures, ergodic theory etc. For example, the trivial solution $x \equiv o$ is called strongly stable with probability 1, in probability or in p-th mean if

$$\lim_{x_o \to o} \sup_{t \geq o} |x(t,x_o)| = o$$

with probability 1, in probability or in p-th mean, resp. The stochastic analogue of 'critical point' is a stationary solution.

2. Nonlinear f

2.1 Stationary ξ

There is a powerful criterion for the existence of a stationary solution of $\dot{x} = f(x,\xi)$ given by KHASMINSKII [12].

2.2 Markovian ξ

Write the original equation as $dx = f(x,\xi)dt$ and add

$$d\xi = a(\xi)dt + b(\xi)dW, \ \xi \in Y \subset \mathbb{R}^m,$$

then (x,ξ) is an everywhere degenerate Markov process. There is a complete recurrence theory developed by KLIEMANN [14] yielding a disjoint decomposition of the state space of x with respect to various recurrence properties, where these properties correspond to control properties of an accompanying deterministic control system $\dot{x} = f(x,u)$.

3. Linear f

3.1 Stationary ξ

Additive noise: For a given A, the class of processes ξ allowing a stationary solution of $\dot{x} = Ax + \xi$ can be exactly determined, see BUNKE [8], ARNOLD, HORSTHEMKE and STUCKI [2], OREY [19] and ARNOLD and WIHSTUTZ [7].

Multiplicative noise: For the growth

$$\lambda(x_o) = \limsup_{t\to\infty} \frac{1}{t} \log|x(t,x_o)|$$

of a solution of $\dot{x} = A(t)x$, $A(t)$ a matrix-valued stationary process, we have the multiplicative ergodic theorem of OSELEDEC [20] (for a more recent account see RUELLE [21]). The angular behavior of the solutions has been clarified by WIHSTUTZ [23] by proving a Floquet type representation of a fundamental matrix.

3.2 Markovian ξ

Multiplicative noise: For Markovian $A(t)$ stability and growth of $\dot{x} = A(t)x$ can be treated by KLIEMANN's recurrence analysis via control theory ([13], [14]) resulting in exact stability diagrams (KLIEMANN and RÜMELIN [15]). On this base the destabilizing or stabilizing effect of noise can be discussed (ARNOLD [3]).

4. Linearization

If $\dot{x} = f(x,\xi)$ is linearized around the trivial solution (say), we obtain a multiplicative noise linear system $\dot{y} = A(\xi(t))y$. For a stationary ξ, the stable manifold theorem of RUELLE [21] tells us that the multiplicative ergodic theorem for the linear system carries over to the nonlinear one.

5. Generalizations

One can consider SDS as stochastic flows on manifolds (ELWORTHY [10], RUELLE [21]), SDS in infinite dimensions (CHOW [9], RUELLE [22]), SDS with more general noise processes, e.g. semimartingales (METIVIER and PELLAUMAIL [18]), etc.

6. Applications

Qualitative theory of SDS applies whenever a system is subject to the action of something which can be modeled as a random process. This is particularly the case if one includes fluctuations (internal noise) or considers changes of the environment or of parameters (external noise), see e.g. ARNOLD and LEFEVER [4].

Qualitative theory is the mathematical base for the calculation of long-term objects like invariant measures, growth rates etc. by numerical or Monte-Carlo procedures. It tells us whether and where those objects (uniquely) exist and how they depend on parameters. Then one can go on and investigate noise-induced phase transitions or bifurcations (ARNOLD, HORSTHEMKE and LEFEVER [1], HORSTHEMKE [11]), stabilization by noise (ARNOLD [3]) and other noise-induced changes of qualitative behavior (KLIEMANN [16]).

References

[1] Arnold, L.; Horsthemke, W.and Lefever, R.: White and coloured external noise and transition phenomena in non-linear systems. Z. Physik B 29, 367-373 (1978)

[2] Arnold, L.; Horsthemke, W. and Stucki, J.: The influence of external real and white noise on the Lotka-Volterra model. Biometrical Journal 21, 451-471 (1979)

[3] Arnold, L.: A new example of an unstable system being stabilized by random parameter noise. Inform. Communication of Math. Chemistry 7, 133-140 (1979)

[4] Arnold, L.; Lefever, R. (eds.): Stochastic nonlinear systems in Physics, Chemistry and

Biology. Springer Series in Synergetics, Berlin-Heidelberg-New York 1981

[5] Arnold, L.: Qualitative theory of stochastic nonlinear systems. In [4].

[6] Arnold, L.; Kliemann, W.: Qualitative theory of stochastic systems. In: Bharucha-Reid, A. (ed.): Probabilistic Analysis and Related Topics, Vol. 3. Academic Press, New York 1981

[7] Arnold, L.; Wihstutz, V.: Stationary solutions of linear systems with stationary additive noise. Preprint Universität Bremen 1981

[8] Bunke, H.: Gewöhnliche Differentialgleichungen mit zufälligen Parametern. Akademie-Verlag, Berlin 1972

[9] Chow, P.-L.: Stochastic partial differential equations in turbulence related problems. In: Bharucha-Reid, A. (ed.): Probabilistic Analysis and Related Topics, Vol. 1, 1-43. Academic Press, New York 1978

[10] Elworthy, D.: Stochastic differential equations on manifolds. Lecture notes, Dept. of Mathematics, University of Warwick, 1978

[11] Horsthemke, W.: Noise induced non-equilibrium phase transitions. In [4]

[12] Khasminskii, R. Z.: Stability of systems of differential equations with random disturbances of their parameters (in Russian). Nauka, Moscow 1969

[13] Kliemann, W.: Some exact results on stability and growth of linear parameter excited stochastic systems. In: Kohlmann, M.; Vogel, W. (eds.): Stochastic Control and Stochastic Differential Systems. Lecture Notes in Control and Information Sciences No. 16, Springer-Verlag, Berlin-Heidelberg-New York 1979, 456-471

[14] Kliemann, W.: Qualitative Theorie nichtlinearer stochastischer Systeme. PhD Thesis Universität Bremen 1980

[15] Kliemann, W.; Rümelin, W.: Stability of linear

systems with random parameters: Theory and Simulations. Report Forschungsschwerpunkt Dynamische Systeme, Universität Bremen 1980

[16] Kliemann, W.: Qualitative theory of stochastic dynamical systems-applications to life sciences. In: Ianelli, M.; Koch, G. (eds.): Stochastic methods in life sciences. LN in Biomathematics, Springer-Verlag, Berlin-Heidelberg-New York 1981

[17] Kushner, H.: Stochastic stability and control. Academic Press, New York 1967

[18] Metivier, M.; Pellaumail, J.: Stochastic integration. Academic Press, New York 1980

[19] Orey, S.: Stationary solutions for linear systems with additive noise. Preprint University of Minnesota 1980

[20] Oseledec, V. I.: A multiplicative ergodic theorem. Lyapunov characteristic numbers for dynamical sytems. Trans. Moscow Math. Soc. 19, 197-231 (1968)

[21] Ruelle, D.: Ergodic theory of differentiable dynamical systems. Publ. Math. IHES no. 50, 275-306 (1979).

[22] Ruelle, D.: Characteristic exponents and invariant manifolds in Hilbert space. Preprint IHES/P/80/11, March 1980

[23] Wihstutz, V.: Ergodic theory of linear parameter-excited systems. Preprint Universität Bremen 1980

STOCHASTIC MODELS OF COMPUTER NETWORKS

René Boel

Electrical Engineering Department,
Imperial College, London, SW7 2BZ, U.K.

ABSTRACT

This paper reviews some recent advances in the use of queueing network models for computers and communications networks. It emphasizes the relation between the martingale approach and quasi-reversibility. A brief discussion of optimal design and control is included.

I. INTRODUCTION

This paper reviews recent progress in the use of queueing models to study the behaviour of computers and communications networks. The stochastic dynamical equations which describe this behaviour are highly non-linear. Hence one can in most cases only analyze the equilibrium distribution. This is similar to the stability analysis of nonlinear systems theory.

The paper attempts to show the relation between two approaches which have been developed in the last few years. One method studies directly the special structure of the transition matrix required for the equilibrium distribution to have an easy form. This work has been described in a recent book by Kelly [12] on quasi-reversible queueing systems. It will be summarized in §3. On the other hand Brémaud [7], Walrand and Varaiya [19] have used martingale models and nonlinear filtering theory to obtain similar structural properties, which simplify the analysis. The emphasis here is on proving independence of certain random variables. This is treated in §4 . A common underlying model for both the above approaches is developed in §2, together with

M. Hazewinkel and J. C. Willems (eds.), Stochastic Systems: The Mathematics of Filtering and Identification and Applications, 141–167.

examples of how to use these queueing models for modelling delays in computers and communications networks. Finally §5 and §6 deal with some tentative applications for optimal design and control.

Readers who like an extensive study of classical and modern results in queueing, and applications to computers, should read both volumes of Kleinrock [14]. A good survey, with an extensive bibliography, is by Kobayashi and Konheim [15].

2. QUEUEING NETWORK MODELS

Messages in a communications network and programs executed on a time shared computer all experience delays due to the limited capacity of the resources (transmission line, CPU, input-output equipment, etc.). This causes the different messages or programs to interact as customers in a queue, competing for a limited resource. Moreover each message or program will use sequentially several of these resources or use the same resource repeatedly. A queueing network will be used as a mathematical model for this behaviour, each resource forming a node of the network.

A single queue is modeled as follows. An integer valued random process A_t counts the number of arriving customers in $[0,t]$, i.e. $A_o = 0$ and A_t increases by +1 at the random times T_n^A when an arrival occurs. Arriving customers enter the waiting room or start service if the server is free. Each customers occupies the server for a random time, S_n. It is assumed that service requirements of different customers are independent. Customers leave the server at random times T_n^D and

$$D_t = \sum_{n=1}^{\infty} I_{T_n^D \leq t} \quad \text{counts the number of departures in } [0,t].$$

The queue length $Q_t = Q_o + A_t - D_t$ denotes the number of customers present in the waiting room or in service.

All the above stochastic processes are defined on a probability space (Ω, F, P) with an increasing family of σ-algebras $F_t \supset \sigma(A_s, D_s, s \leq t; Q_o)$. This means that A_t, D_t, Q_t are F_t-adapted. A complete probabilistic descritpion is then obtained by specifying the arrival rate λ_t and departure rate μ_t (assumed to exist) :

$$P(dA_t = 1|F_t) = E(dA_t|F_t) = \lambda_t \, . \, dt$$

$$P(dD_t = 1|F_t) = E(dD_t|F_t) = \mu_t \, . \, dt$$

Notice that d (as in $dA_t = A_{t+dt} - A_t$) always denotes forward increments. The above definitions of rates are equivalent to requiring that

$$M_t^A = A_t - \int_0^t \lambda_s \,.\, ds \text{ and } M_t^D = D_t - \int_0^t \mu_s \,.\, ds$$

are (P,F_t)- martingales. M_t is a (P,F_t)-martingale if for all $s \geq t$: $E(M_s|F_t) = M_t$ or equivalently $E(dM_t|F_t) = 0$.

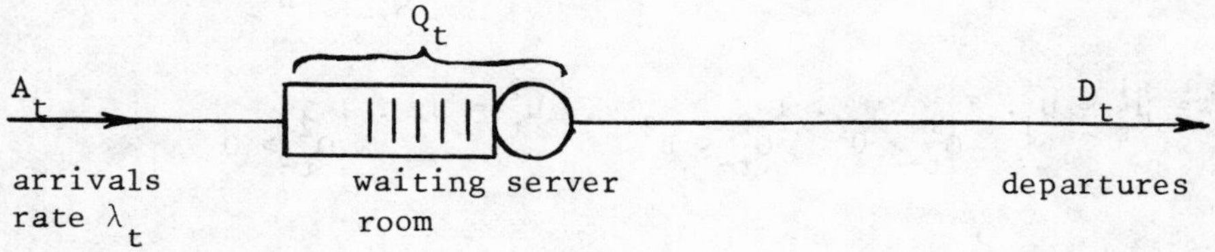

Fig. 1

Note that if no customers are present ($Q_{t-} = 0$) then $\mu_t = 0$ necessarily.

The simplest queueing model, denoted an M/M/1 queue, has A_t a Poisson process, rate λ, and the service times S_n are independent, exponentially distributed, mean $\frac{1}{\mu}$. Then

$\lambda_t = \lambda$ and $\mu_t = \mu . I_{Q_{t-}>0}$. An M/M/1/B has the same assumptions as above but a finite buffer with B-1 waiting spaces only. Then $\lambda_t = \lambda . I_{Q_{t-} < B}$ and $\mu_t = \mu . I_{Q_{t-} > 0}$. Whenever the rates $\lambda_t(Q_{t-})$ and $\mu_t(Q_{t-})$ depend on the queue length only, Q_t is a Markov process with the positive integers N_+ or $\{0,1,\ldots,B\}$ as state space. This would not be the case if service times are not exponentially distributed, if there are different classes of customers with different service requirements or if the arrival rate depends on the state of other queues as in the queueing networks below.

In most computer or communications applications each customer has to pass through several queues. The simplest case is a tandem queue, two queues in series, where the

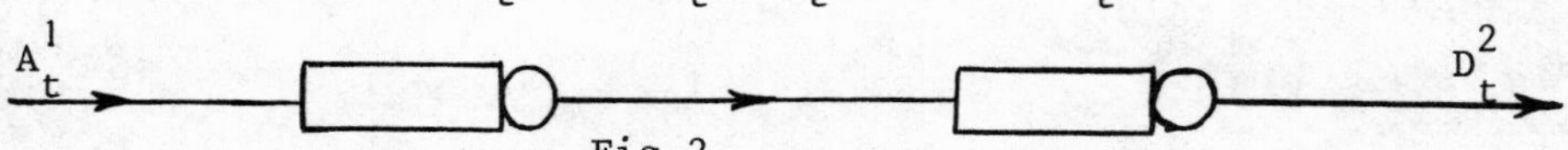

Fig.2

arrivals of the second queue are the departing customers of the first queue (without any extra delay). Then :

$$Q_t^1 = Q_o^1 + A_t^1 - D_t^1 ; \quad Q_t^2 = Q_o^2 + D_t^1 - D_t^2$$

If the arrival process to the first queue is Poisson and all service times are exponentially distributed (mean $\frac{1}{\mu_1}$, $\frac{1}{\mu_2}$ respectively for queue 1 and 2), then $\lambda_t^1 = \lambda \cdot I_{Q_{t-}^1 < B^1}$,

$$\lambda_t^2 = \mu_t^1 = \mu_1 \cdot I_{Q_{t-}^1 > 0} \cdot I_{Q_{t-}^2 < B^2} , \quad \mu_t^2 = \mu_2 \cdot I_{Q_{t-}^2 > 0}$$

where $B^1 - 1$ and $B^2 - 1$ are the number of waiting spaces available at the corresponding queues.

For a more general queueing network consider the following example of a data network connecting cities a, b, c and d, with transmission lines as indicated in fig. 3. Let λ_{ij} be the rate at which messages are generated in city i, of (exponentially distributed) random length, destined for city j.

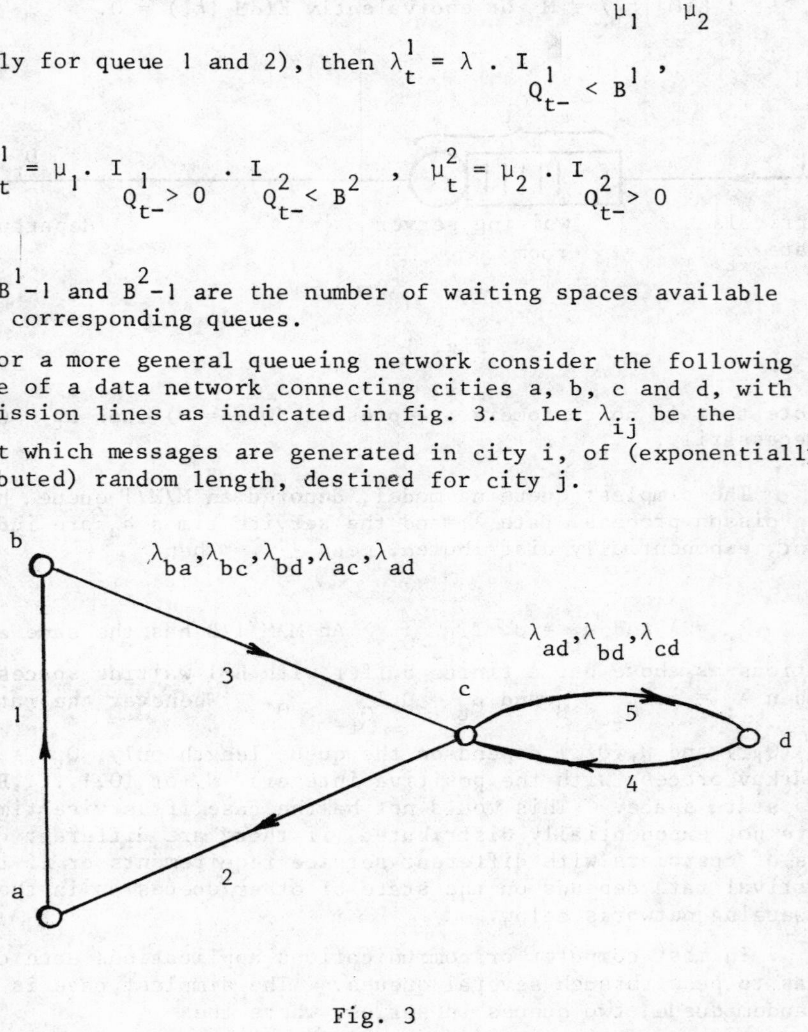

Fig. 3

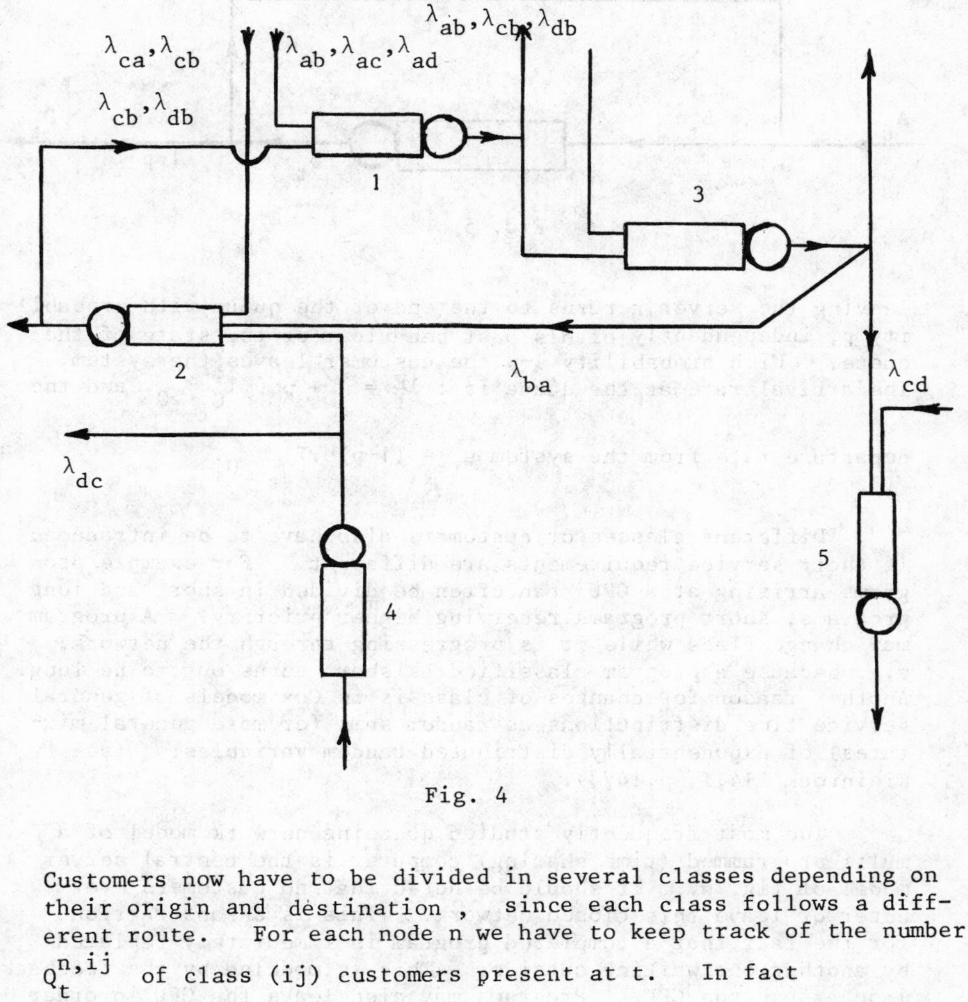

Fig. 4

Customers now have to be divided in several classes depending on their origin and destination , since each class follows a different route. For each node n we have to keep track of the number $Q_t^{n,ij}$ of class (ij) customers present at t. In fact $(Q_t^{n,ij}$; $n = 1,\ldots,5$; $i,j = a,b,c,d)$ will not be a Markov process unless the customer to be served next is chosen randomly or according to a fixed priority rule.

The above network was very simple because for each class of customers there is only one possible route. In general a customer on leaving node n will choose to go to node k randomly, according to a probability distribution $r^c(n,k)$, which depends on the class c to which the customer belongs. The simplest example of random routing is the M/M/1 queue with feedback of fig.5. A customer

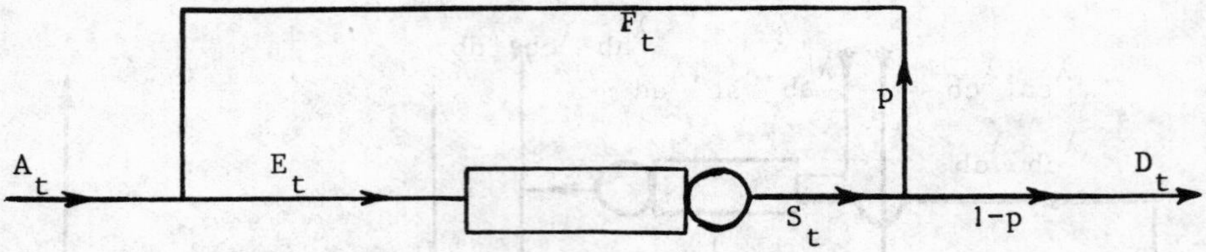

Fig. 5

leaving the server returns to the end of the queue with probability p, independently of his past behaviour or the state of the queue. With probability 1-p the customer leaves the system. The arrival rate at the queue is : $\lambda_t^E = \lambda + p.\mu.I_{Q_{t-}>0}$, and the departure rate from the system $\mu_t = (1-p)\mu.I_{Q_{t-}>0}$.

Different classes of customers also have to be introduced if their service requirements are different. For example programs arriving at a CPU can often be divided in short and long programs, short programs receiving higher priority. A program may change class while it is progressing through the network, e.g. because a program classified as short turns out to be long. Another reason for changes of class is in Cox models of general service time distributions as random sums (or more general mixtures) of exponentially distributed random variables. (see [Kleinrock 14,I, p.147]).

The most frequently studied queueing network model of a multi-programmed (time sharing) computer is the central server model of fig. 6. It should be noted that no customers ever enter or leave this closed network. This is an abstraction for the fact that a completed program is immediately replaced by another one waiting outside. This is modeled by the feedback loop around the CPU. Programs may also leave the CPU in order to get information stored on disk or tape memory, to print out results, etc. This is modeled by branching to one of the N-1 peripheral devices. Extensions of this model with different classes of customers, or several queues in series in the feedback lines, have been studied (Kleinrock [14, II §4]).

In modeling the behaviour of a process it may be necessary to keep track of other underlying stochastic processes. For example the rate at which a process leaves the CPU for use of memory devices (the"page fault" rate) depends on the program and on the information stored in the random access memory. Courtois [9, chpt. VIII] has modeled this as a finite state Markov process

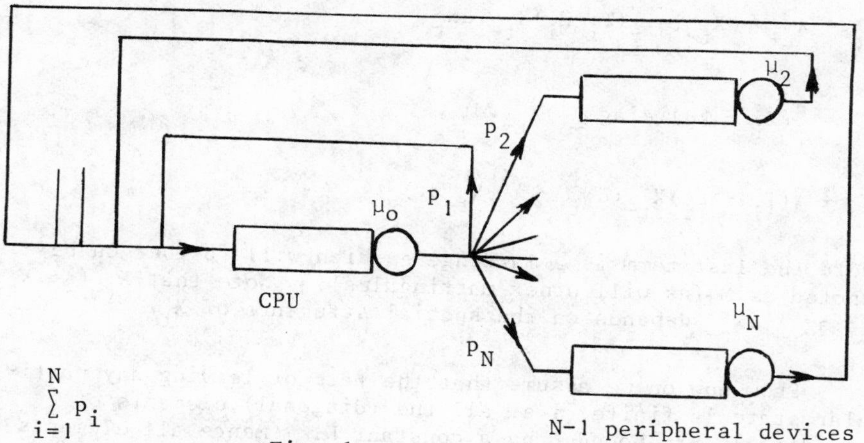

Fig. 6

X_t^p associated to the program (describing what computer scientists call the locality of the program). If a process with state X_t^p is being served by the CPU, the departure rate is $\mu(X_t^p)$. A complete state description of this requires the state and the exact position of each program in the system.

All the networks described above can be modeled as countable state Markov processes. For a closed network, or if all buffers are finite, the state space is finite. It will be helpful for notational purposes, to use as state the (finite or countably infinite) vector X_t, which has all elements 0 except for a 1 in the place corresponding to the state of the process at time t. A transition from one state to another (arrival, departure, internal change from one node to another, change of class) can be represented by multiplying X_t with a matrix which moves the single one. In all the above examples one can define a countable number of transition types, $i \in N_+$, such that N_t^i, a process which counts the number of type i transitions, has rate $\lambda^i(X_{t-}) = \underline{\lambda}^i . X_{t-}$ where $\underline{\lambda}^i$ is a row vector of rates (since the column vector has all zeros except for a one at a place corresponding to the present state, it picks up the proper $\lambda(X_t)$). The type i transition only influences the system if $X_{t-} \in E_i$, the enabling subset. It is represented by the matrix T_i ($X_t = T_i X_{t-}$ if $dN_t^i = 1$) which has all zeros except for a single one in each column corresponding to E_i moving the "one" in X_{t-} to its new position in X_t. By abuse of notation let E_i be the diagonal matrix having ones in places corresponding to elements of the set E_i, zero otherwise. This formalism (a slight modification of the notation used by Walrand and Varaya [18]) allows us to write stochastic dynamical system equations :

$$dX_t = X_{t+} - X_{t-} = \sum_i (T_i - E_i) X_{t-} . dN_t^i$$

$$dX_t = \sum_i (T_i - E_i)(\text{diag } \underline{\lambda}^i) X_{t-} . dt$$
$$+ \sum_i (T_i - E_i) X_{t-} (dN_t^i - \underline{\lambda}^i . X_{t-} dt) \tag{1}$$

where the last term is a martingale which will further on be denoted as M_t (as will other martingales). Note that $X_{t-} (\underline{\lambda}^i . X_{t-}) = \text{diag } \underline{\lambda}^i . X_{t-}$ depends on the special structure of X_t.

From now on we assume that the rate of leaving any particular state is finite, i.e. all the (diagonal) elements of $\sum_i E_i \text{ diag } \underline{\lambda}^i$ are bounded by a constant K. Hence all elements of $\sum_i T_i \text{ diag } \underline{\lambda}^i$ are bounded by the same constant. This allows the interchange of the limiting operations $\frac{d}{dt}$ and expectation E, necessary to obtain the Kolmogorov backward equation ($EX_t = \pi_t$ = ($\mathcal{P}$(system in state x at time t)) is differentiable) :

$$\frac{d}{dt} \pi_t = \sum_i (T_i - E_i)(\text{diag } \underline{\lambda}^i) . \pi_t \tag{2}$$

The bounded rate assumption also guarantees that (2) has a unique solution, with $\|\pi_t\|_1 = 1$ if $\|\pi_o\|_1 = 1$, where $\|.\|_1$ denotes the ℓ_1 - norm (see Feller [10, II, XIV-7].

To illustrate the above model consider an M/M/K/B queue wit arrival rate λ and K servers average service time $\frac{1}{\mu}$, and B-1 waiting spaces. Then $X_t = (I_{Q_t = n})$, a (B+1)-column vector has 2 types of transitions :

$A_t = N_t^1$ = arrivals, $\underline{\lambda}^1 = (\lambda, \lambda, \ldots, \lambda, \lambda)$, $E_1 = \{X_t : Q_t < B\}$,

$$T_1 = \begin{pmatrix} 0 & 0 & & & & \\ 1 & 0 & \cdot & & & \\ 0 & 1 & \cdot & \cdot & & \\ 0 & 0 & \cdot & \cdot & \cdot & \\ & & \cdot & 1 & 0 & 0 \\ & & & 0 & 1 & 0 \end{pmatrix} , \quad E_1 = \begin{pmatrix} 1 & 0 & & & \\ 0 & 1 & \cdot & & \\ \cdot & \cdot & \cdot & \cdot & \\ & \cdot & \cdot & \cdot & \cdot \\ & & \cdot & 1 & 0 \\ & & & 0 & 0 \end{pmatrix}$$

$D_t = N_t^2$ = departure, $\underline{\lambda}^2 = (0, \mu, 2\mu, \ldots, K\mu, K\mu, \ldots)$

$E_2 = \{X_t : Q_t > 0\}$

$$T_2 = \begin{pmatrix} 0 & 1 & 0 & \cdot & & & \\ 0 & 0 & 1 & \cdot & & & \\ \cdot & \cdot & \cdot & \cdot & \cdot & & \\ & \cdot & \cdot & \cdot & \cdot & \cdot & \\ & & \cdot & \cdot & \cdot & 1 & 0 \\ & & & \cdot & \cdot & 0 & 1 \\ & & & & \cdot & 0 & 0 \end{pmatrix}, \quad E_2 = \begin{pmatrix} 0 & 0 & & & & \\ 0 & 1 & \cdot & & & \\ & \cdot & \cdot & \cdot & & \\ & & \cdot & \cdot & \cdot & \\ & & & \cdot & \cdot & \\ & & & & 1 & 0 \\ & & & & 0 & 1 \end{pmatrix}$$

Then

$$dX_t = [(T_1-E_1)\operatorname{diag} \underline{\lambda}^1 + (T_2-E_2)\operatorname{diag} \underline{\lambda}^2]X_{t-}\,dt + dM_t$$

gives the more familiar looking equation :

$$dX_t = \begin{pmatrix} -\lambda & \mu & 0 & & & \\ \lambda & -\lambda-\mu & 2\mu & \cdot & & \\ 0 & \lambda & -\lambda-2\mu & \cdot & \cdot & \\ & \cdot & \cdot & \cdot & \cdot & \\ & & \cdot & \cdot & \cdot & K\mu \\ & & & \lambda & & -K\mu \end{pmatrix} X_{t-}\,dt + dM_t$$

Clearly $\lambda(T_1-E_1) + (T_2-E_2)\operatorname{diag} \underline{\lambda}^2$ is the transition matrix of the Markov process Q_t.

3. REVERSIBILITY AND QUASI-REVERSIBLE QUEUES

In this section a special class of queueing networks will be studied for which the equilibrium solution π of (2) $\sum_i [(T_i-E_i)\operatorname{diag} \underline{\lambda}^i]\pi = 0$ exists and takes a particularly simple form as product of functions of the local "sub-state" of component queues. This method of analysis was started by early work of Jackson [11] and has recently been expanded by Baskett et al. [2]. Here the very elegant approach of Kelly [12], using reversibility of Markov processes, will be followed.

A processes (X_t) is reversible if for all $t_1, t_2, \ldots, t_n$, $(X_{t_1}, \ldots, X_{t_n})$ and $(X_{T-t_1}, \ldots, X_{T-t_n})$ have the same distribution. Then (X_t) is necessarily stationary.

If (X_t) is a stationary, countable state Markov process with equilibrium distribution $\pi(j)$ and transition rates $P(j,k)$

(this is (j,k)-th element of $\sum_i (T_i - E_i)(\text{diag } \underline{\lambda}^i)$ in the notation of §2, for $j \neq k$) then Kolmogorov has shown that (X_t) is reversibile if and only if for all $j \neq k$: $\pi(j)P(j,k) = \pi(k)P(k,j)$. This is a local balance requirement, stating that the average rate of transitions from j to k equals the average transition rate from k to j. In the notation of §2 this local balance condition can be rewritten as

$$\pi(j)\sum_i (T_i)_{jk} \cdot \lambda^i(k) = \pi(k)\sum_i (T_i)_{kj} \cdot \lambda^i(j)$$

If for each ordered pair of state (j,k) there is only one transition type i(j,k) transforming j into k, then

$$\pi(j)\lambda^{i(j,k)}(k) = \pi(k)\lambda^{i(k,j)}(j) \qquad (3)$$

is the necessary and sufficient condition. This indicates that the equilibrium distribution of a reversible queueing network will in general be found by considering a state x_o from which all others can be reached, finding a chain $i(x_o,x_1)$, $i(x_1,x_2)\ldots$, $i(x_{n-1},x_n)$ and writing

$$\pi(x_n) = \prod_{\ell-1}^{n} \frac{\lambda^{i(x_{\ell-1},x_\ell)}(x_\ell)}{\lambda^{i(x_\ell,x_{\ell-1})}(x_{\ell-1})} \cdot \pi(x_o) \qquad (4)$$

and then finding $\pi(x_o)$ via the normalisation requirement $\|\pi\|_1 = 1$. A simple example of a reversible process is any Markov process when graph is a tree (link any nodes j, k such that "(j,k) > 0).

The graph of an M/M/1/B queue is a tree (fig.7), and (3) takes a very simple form : $\lambda.\pi(j) = \mu.\pi(j+1)$

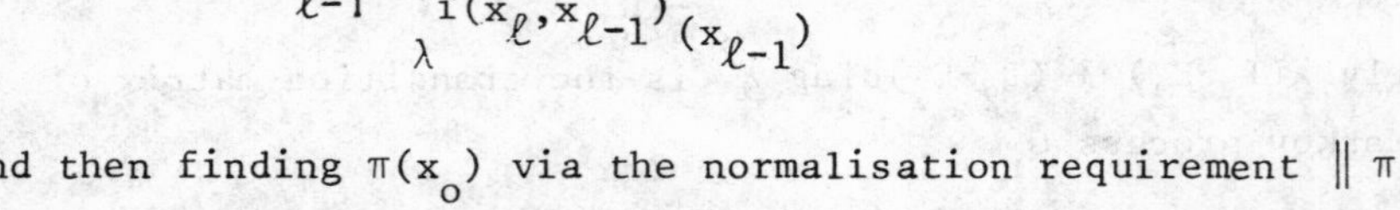

$$X_t = \begin{pmatrix}1\\0\\ \vdots\\0\end{pmatrix} \qquad X_t = \begin{pmatrix}0\\1\\0\\ \vdots\\0\end{pmatrix} \qquad X_t = \begin{pmatrix}0\\ \vdots\\0\\1\end{pmatrix}$$

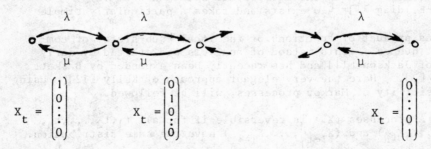

Fig 7

This in fact still holds for any queue with $\lambda(n)$ and $\mu(n)$

functions of the present queue length only. We can then rewrite (3) as :

$E_1 \operatorname{diag} \underline{\lambda}^1 . \pi = T_2 . \operatorname{diag} \underline{\lambda}^2 . \pi$ and equivalently (shifting indices)

$T_1 \operatorname{diag} \underline{\lambda}^1 . \pi = E_2 . \operatorname{diag} \lambda^2 . \pi$. The solution π is then easy to write :

$$\pi(n) = \frac{\prod_{j=0}^{n-1} \frac{\lambda(j)}{\mu(j+1)}}{\sum_{\ell=0}^{B} \prod_{j=0}^{\ell-1} \frac{\lambda(j)}{\mu(j+1)}} \tag{5}$$

where the empty product $\prod_{j=0}^{-1} (\ ) = 1$ by definition. If $B = \infty$, assume the denominator is finite. This guarantees equilibrium.

Since (Q_t) is reversible, (Q_{-t}) also represents a queue with an arrival occurring each time Q_t has a departure ($Q_{t+dt} - Q_t = -1 \Rightarrow Q_{-t} - Q_{-t-dt} = +1$) and vice versa. Assume now that the arrival process is Poisson, i.e. $\lambda(Q_{t-}) = \lambda$ ($B = \infty$), but allow a general $\mu(Q_{t-})$. Since (Q_t) and (Q_{-t}) have the same probabilistic description, the arrival process of Q_{-t}, and hence the departure process of Q_t, is a Poisson process with rate λ, assuming that the queue is in equilibrium. Moreover arrivals after time $-t$ of the reversed time system Q_{-t}, are independent of Q_{-t}. Hence departures of the original system before t, $\{D_s, s \leq t\}$, are independent of Q_t. The symmetric statement that future arrivals, $\{A_s, s \geq t\}$, are independent of Q_t is trivial in this case.

Consider now a series connection of queues as described above (with Poisson input to the first queue), starting in equilibrium. Then the input to the second queue is Poisson also, and hence its output is Poisson. Moreover Q_t^2 depends only on $\{D_s^1, s \leq t\}$ and the service times at the second queue, both of which are independent of Q_t^1. Hence Q_t^1 and Q_t^2 are independent in equilibrium :

$$\pi(Q_t^1 = n_1, Q_t^2 = n_2) = \pi(Q_t^1 = n_1).\pi(Q_t^2 = n_2) \tag{6}$$

Both marginal distributions are calculated for a queue with arrival rate λ(since $A_t^2 = D_t^1$ has rate λ) and departure rates $\mu^1(Q_t^1)$, μ^2 (Q_{t-}^2) respectively i.e.

$$\pi(Q_t^1 = n_1, Q_t^2 = n_2) = \frac{\prod_{j=1}^{n_1} \frac{\lambda}{\mu^1(j)}}{\sum_{n=o}^{\infty} \prod_{j=1}^{n} \frac{\lambda}{\mu^1(j)}} \cdot \frac{\prod_{j=1}^{n_2} \frac{\lambda}{\mu^2(j)}}{\sum_{n=0}^{\infty} \prod_{j=1}^{n} \frac{\lambda}{\mu^2(j)}} \qquad \ldots . \quad (7)$$

Finiteness of the infinite sums is a necessary and sufficient condition for existence of an equilibrium distribution. This is always assumed. This argument and the resulting product form equilibrium distribution is easily extended to several question in series.

Suppose now that at each queue customers are served in order of arrival, one at a time. The waiting time $W_t^{n,1} = T_n^D - T_n^A$ of a customer arriving at queue 1 at T_n^A and leaving for queue 2 at T_n^D, is equal to the waiting time of a customer in the reversed time system arriving at $-T_n^D$ and leaving at $-T_n^A$. This last quantity is clearly independent of arrivals to the reversed time system after $-T_n^D$. By reversibility the waiting time in the original system for customer n is independent of the departure prior to T_n^D. But the waiting time of customer n in the second queue can only be influenced by the behaviour of preceding customers and its own service time in the second queue. Hence $W_t^{n,1}$ and $W_t^{n,2}$ (and extensions to a series connection of queues) are independent under equilibrium assumptions:

$$\begin{aligned} P(W_t^{n,1} \leq x, W_t^{n,2} \leq y) &= P(W_t^{n,1} \leq x) \cdot P(W_t^{n,2} \leq y) \\ &= \left(1 - e^{-(\mu_1-\lambda)x}\right) \cdot \left(1 - e^{-(\mu_2-\lambda)y}\right) \end{aligned} \qquad \ldots . \quad (18)$$

(see Kleinrock [14,I, p.202])

The independence of queue lengths and waiting times extends to tree networks as in fig. 8, with all

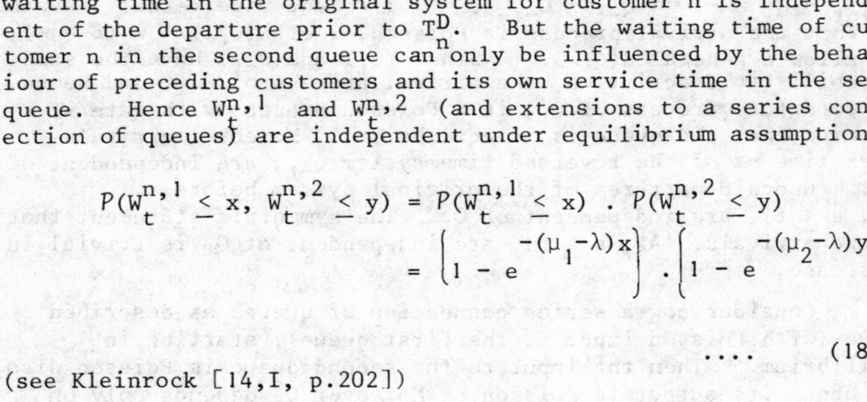

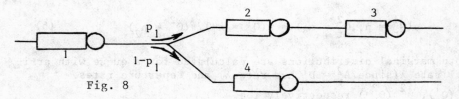

Fig. 8

queues M/M/1 and routing independent of previous behaviour of the queue. The argument for independence of queue lengths still holds if the outputs of queue 3 and 4 are combined to form the input for a 5th queue, or if the departure rates at each queue depend on the local queue length (e.g. multiple server). However one cannot prove independence of waiting times under these conditions because overtaking is then possible, i.e. the order in which customers are served is not the same in each queue.

If the service time is not exponentially distributed, the queue length Q_t is not Markovian. Nevertheless, by comparing directly (Q_t) and (Q_{-t}), it is possible to prove that (Q_t) is a reversible process for the following types of service behaviour (M/M/1 is called type 1):

type 2: M/G/1 queue with processor sharing service, i.e. if n customers are present they are all served at a rate $\frac{1}{n}$, with the total service requirements independent, identically distributed. This means that the remaining required service of each of the n customers decreases in $[t, t + dt]$ by $\frac{dt}{n}$.

type 3: M/G/∞ : since there are infinitely many servers, customers do not interact. This represents an independent, identically distributed delay for each customer.

type 4: M/G/1 with pre-emptive resume last come-first serve scheduling. This a is stack where the service starts working immediately on a newly arrived customer until completion or until the next arrival.

The time reversal argument is obvious for type 3 . For type 2 and 4 a proof, for slightly more general queues, can be found in Kelly [12, §3.3] .

Clearly, any type 1,2,3 or 4 queue has a Poisson process of rate λ as output, under stationarity assumptions. Hence series connections of such queues have, in equilibrium, independent queue lengths, that is (6) still holds but the marginal distributions may be different from those given in (7)

$$\text{type 3:} \quad \mathcal{P}(Q_t = n) = \frac{1}{n!} \left(\frac{\lambda}{\mu}\right)^n . e^{-\frac{\lambda}{\mu}} \tag{9}$$

It is important to note that while a single queue is reversible for each of the above types, the tandem connection of two queues is not reversible, as can easily be seen from the fact that (n_1 denoting queue length of first queue, n_2 of the second queue) $p((n_1, n_2) \to (n_1 - 1, n_2 + 1))$ is a non-zero transition probability, while $p((n_1 - 1, n_2 + 1) \to (n_1, n_2)) = 0$. This violates the necessary (and sufficient) condition for reversibility.

A series (or tree) connection of queues was assumed above because then the output of a queue cannot influence the future input. The absence of feedback hence led to the conclusion that the state Q_t of a queue is independent of its output $\{D_s, s \leq t\}$ up to t. This is no longer true in a general network which contains loops. Calculating equilibrium distributions for such queues with several classes and different types of queues becomes very complicated. For details and proofs the reader is referred to Kelly [12, §3] and Kleinrock [14,II,§4.12]. A brief outline follows.

Consider an arbitrary stationary Markov process (X_t) with countable state space and transition rates P(j,k) and equilibrium distribution $\pi(j)$. Then the transition rates of the reversed time process (X_{T-t}) are

$$P'(j,k) = \frac{\pi(k)P(k,j)}{\pi(j)}$$

and the reversed process has the same equilibrium distribution. In fact (Kelly [12, Thm. 1.13]) given (X_t) with transition rates P(j,k), then if the following equations have a solution for $(\pi(j))$ and P'(j,k) :

$$\sum_k P(j,k) = \sum_k P'(j,k) \tag{10}$$

$$\pi(j)P(j,k) = \pi(k)P'(k,j) \tag{11}$$

then $(\pi(j))$ is the equilibrium distribution of X(t) and of the reversed process (X_{T-t}) (which has P'(j,k) as transition rates). (10) and (11) is a larger set of equations than $P.\pi = 0$ (as in (2) with $\frac{d\pi_t}{dt} = 0$) but it has more structure and is easier to solve. The product form of the solution is obtained if in

$$\pi(j) = \frac{P'(k,j)}{P(j,k)} \cdot \pi(k)$$

the quotient has an easily defined structure, as for an M/M/1 queue.

This is true for a quasi-reversible queue, i.e. a queue with state (X_t), a stationary Markov process, such that X_t, for any t, is independent of future arrivals $(A_s, s \geq t)$ and past departures $(D_s, s \leq t)$. It is easy to see that arrival and departure rates are then constant, i.e. A_t and D_t are Poisson processes. Note that the state (X_t) may involve more than the queue length to make it Markovian. The assumption that the state is independent of the future input is very reasonable from a dynamical system point of view. The second condition is then imposed because the reversed time process (X_{-t}) should also be quasi-reversible. All the queueing models considered above are quasi-reversible.

Consider now a network of quasi-reversible queues. On leaving queue i, a customer goes to queue j with probability r(i,j), independent of his past behaviour. At each queue i there is an arrival process of rate λ_i, while r(i,N+1) denotes the probability that a customer completing service at queue i leaves the network altogether. The total arrival rate ν_i of customers at queue i, i = 1,...,N, is then found by solving the flow equations (" preservation of customers"):

$$\nu_i = \lambda_i + \sum_{j=1}^{N} \nu_j \, r(j,i) \qquad (12)$$

These equations have a unique solution because $\sum_{j=1}^{N} r(j,i) < 1$.

In an open network (all entering customers eventually leave) of quasi-reversible queues (in equilibrium) the states of the individual queues (and a fortiori the queue lengths) are independent, each behaving as if it had an independent Poisson process of rate ν_i as arrival process. This same joint distribution is also observed by each arriving customer (i.e. $\mathcal{P}(X_{T^A_{n-}})$). This distribution is given by

$$\mathcal{P}(Q^i_t = n_i, \; i = 1,\ldots,N) = \prod_{i=1}^{N} \mathcal{P}(Q^i_t = n_i) \qquad (13)$$

with $\mathcal{P}(Q^i_t = n_i)$ of the form (7) or (9) depending on the type of service discipline at queue i. Kelly [12, §3.2] proves these results by studying the transition probabilities of the reversed process, showing that the reversed network is also quasi-reversible and then solving equations (10) and (11). As a subsidiary result one finds that streams of customers leaving the network are Poisson and that the present state is independent of past departures.

If the individual queues are quasi-reversible but the network is closed (no customer ever leaves) or mixed (some customers remain in the network forever) the queue lengths cannot be independent. However the equilibrium distribution of queue lengths is still of the product form as in (13), except for a multiplicative factor which normalizes the distribution.

The above results also continue to hold if there are several classes of customers. For details, see Kelly [12], Kleinrock [14] or Baskett et al [2].

Remark: It is very important to observe that the ν_i arrival rates calculated from (12) depend on the arrival rates λ_i and routing probalilities $r(i,j)$ but not on the service rates or service discipline of each queue.

4. MARTINGALE ANALYSIS OF QUEUES

In this section some results of §3 will be rederived using the martingale model of §2 and filtering equations. The approach is as follows: let $\mathcal{D}_t = \sigma(D_s, s \leq t)$ represent the information available from observing the output (or several output streams) up to t; derive formulas for $\hat{X}_t = E(X_t|\mathcal{D}_t)$, the best estimate of the state X_t of the queue (or queueing networks); find conditions such that $\hat{X}_t = EX_t = \pi$, the equilibrium distribution; this implies that $\mathcal{D}_t$ and X_t are independent. This method was first used by Brémaud[7] and was also used by Walrand and Varaiya [18,19]. They derived a system of non-linear differential equations for $\hat{X}_t$. Here the linear Mortensen-Zakai equations for $d\widehat{L_t X_t}$ (as derived by Boel, Varaiya and Wong [3] and corrected by Brémaud [6]) will be used. These equations are similar to the Mortensen-Zakai equations discussed elsewhere in this volume, but with point processes of unknown rate (the signal) as observations.

Consider a single Markovian queue, arrival rate $\lambda(Q_{t-})$, departure rate $\mu(Q_{t-})$ modelled as a countable state Markov process X_t in §2. Then on the same measure space (Ω, F) we can define another probability measure P_o such that the rate of D_t becomes constant, equal to the average departure rate $\bar{\mu} = \sum_m \mu(n)\pi(n)$. In [3] it is shown that

$L_t = E_o\left(\frac{dP}{dP_o} \middle| F_t\right)$ is a well defined locally integrable (P_o, F_t)-martingale satisfying :

$$dL_t = L_{t-} \cdot \left(\frac{\underline{\lambda}^2 . X_{t-}}{\bar{\mu}} - 1\right)(dD_t - \bar{\mu} \cdot dt) \tag{14}$$

or

$$L_t = \prod_{T_n^D \leq t}\left(\frac{\underline{\lambda}^2 \cdot X_{T_n^D-}}{\bar{\mu}}\right) \cdot \exp\left(-\int_o^t (\underline{\lambda}^2 \cdot X_s - \bar{\mu}) \cdot ds\right) \quad \ldots . \tag{15}$$

From stochastic calculus rules one easily finds :

$$dL_tX_t = [(T_1-E_1)\text{diag }\underline{\lambda}^1 + (T_2-E_2)\text{diag }\underline{\lambda}^2]L_{t-}X_{t-}dt$$

$$+ (T_1-E_1)\text{diag }\underline{\lambda}^2.L_{t-}X_{t-}(dA_t - \underline{\lambda}^1X_{t-}dt)$$

$$+ [(T_2-E_2+I)\text{diag }\underline{\lambda}^2 - \overline{\mu}.I]\frac{1}{\overline{\mu}}.L_{t-}X_{t-}$$

$$.(dD_t-\overline{\mu}.dt)$$

Assume now that the departures are observed and find the optimal estimate $\hat{X}_t = E(X_t|\mathcal{D}_t)$ via Bayes' rule

$$\hat{X}_t = \frac{E_o(L_tX_t|\mathcal{D}_t)}{E_o(L_t|\mathcal{D}_t)} = \frac{E_o(L_tX_t|\mathcal{D}_t)}{\|E_o(L_tX_t|\mathcal{D}_t)\|_1} = \frac{\widehat{L_tX_t}}{\|\widehat{L_tX_t}\|_1}$$

Then using the following formulas of Brémaud [6] for point process observations, one obtains (using $E_1 = I$) :

$$d\widehat{L_tX_t} = [(T_1-I)\text{diag }\underline{\lambda}^1 + (T_2-E_2)\text{diag }\underline{\lambda}^2]\widehat{L_{t-}X_{t-}}.dt$$

$$+ [(T_2-E_2)\text{diag }\underline{\lambda}^2 + \text{diag }(\underline{\lambda}^2-\overline{\mu})] . \frac{1}{\overline{\mu}} . \widehat{L_{t-}X_{t-}}.(dD_t-\overline{\mu}.dt)$$

or

$$d\widehat{L_tX_t} = [(T_1-I)\text{diag }\underline{\lambda}^1 + \text{diag}(\underline{\lambda}^2-\overline{\mu})] . \widehat{L_{t-}X_t}.dt$$

$$+ [T_2.\text{diag }\frac{\underline{\lambda}^2}{\overline{\mu}} - I] . \widehat{L_{t-}X_{t-}} . dD_t \tag{18}$$

where $\lambda^2(0) = \mu(0) = 0$ has been used in $E_2.\text{diag }\underline{\lambda}^2 = I.\text{diag }\underline{\lambda}^2$. Equation (18) is an infinite system of linear differential equations in between departures, while at a jump

$$\widehat{L_{T_n^D}X_{T_n^D}} = (T_2.\text{diag }\frac{\underline{\lambda}^2}{\overline{\mu}}).\widehat{L_{T_{n-}^D}X_{T_{n-}^D}}$$

represents multiplication by an (infinite) matrix. Since $\widehat{L_tX_t}$ is well-defined as conditional expectation of an integrable random variable, it suffices to prove uniqueness of the solutions of (18) in $\ell_2 \supset \ell_1$. This can be done by verifying that the boundedness conditions of Arley and Borschsenius [1] are satisfied. Finally notice that all elements of $\widehat{L_t X_t}$ are positive, hence

$$\hat{L}_t = E_o(L_t|\mathcal{D}_t) = \| \widehat{L_t X_t} \|_1 = \underline{1}^T . \widehat{L_t X_t}$$

and the calculation of $\hat{X}_t$ involves linear operations only, except for one division($\underline{1}^T = (1,1,1,\ldots)$).

The above calculations can be repeated for the general queueing network modelled at the end of §2. Assume some transition counts, N_t^j, $j \in J \subset \mathbb{N}_+$, are being observed. Define the new probability measure $\mathcal{P}_o$ such that N_t^j has constant rate $\bar{\lambda}_j$, its long term average under $\mathcal{P}$, for $j \in J$, while other transition rates remain unchanged. Let L_t be the corresponding likelihood ratio; and let

$$\widehat{L_t X_t} = E_o(L_t X_t | N_s^j,\ j \in J,\ s \le t). \quad \text{Then}$$

$$d\widehat{L_t X_t} = \sum_j (T_j - E_j) \operatorname{diag} \underline{\lambda}^j . \widehat{L_t X_t}\, dt \tag{19}$$

$$+ \sum_{j \in J} (T_j \operatorname{diag} \underline{\lambda}^j - \operatorname{diag} \bar{\lambda}_j) \frac{\widehat{L_{t-} X_{t-}}}{\bar{\lambda}_j} . (dN_t^j - \bar{\lambda}_j\, dt)$$

Consider now a single stationary queue, i.e. let $\widehat{L_0 X_0} = \hat{X}_0 = \pi$ be the equilibrium distribution. Then one can show :

<u>Theorem</u>. For a single queue in equilibrium as described above, the following are equivalent

i) $\lambda(Q_{t-}) = \lambda, \quad \forall t$

ii) X_t and the observations $\mathcal{D}_t$ are independent

iii) $E(X_{T_n^D} | \mathcal{D}_{T_n^D}) = \pi, \quad \forall n$

iv) $E(X_t | \mathcal{D}_t) = \pi, \ \forall t$

v) the queue is quasi-reversible

If either of these conditions holds the output D_t is a Poisson process with rate λ.

Essentially one has to prove that $\hat{X}_t = \frac{\widehat{L_t X_t}}{\underline{1}^T . \widehat{L_t X_t}} = \pi$ is equivalent to the input process being Poisson. Premultiplying (18) by $\underline{1}^T$ it is clear that $d\underline{1}^T . \widehat{L_t X_t} = 0$ in between jumps while at a jump

$$\underline{1}^T . \widehat{L_{T_n^D} X_{T_n^D}} = \underline{1}^T . T_2 . \operatorname{diag} \frac{\underline{\lambda}}{\overline{\mu}} . \widehat{L_{T_{n-}^D} X_{T_{n-}^D}}$$

Then the unique solution of (18) on $[0, T_1^D]$ will be $\pi = \widehat{L_t X_t}$ if and only if (using $T_2 \operatorname{diag} \underline{\lambda}^2 . \pi = \operatorname{diag} \underline{\lambda}^1 . \pi$)

$$T_1 . \operatorname{diag} \underline{\lambda}^1 = \operatorname{diag} \overline{\mu}$$

or $\lambda(n) = \lambda = \overline{\mu}$ for all n. Then $\underline{1}^T . \widehat{L_{T_1^D} X_{T_1^D}} = 1$ remains unchanged. The argument can thus be extended to all intervals $[0, T_n^D]$: recursively.

Essentially the above argument shows that π is an eigenvalue of the linear operator on the right of (18), corresponding to the eigenvalue zero, if and only if the arrival process is Poisson.

The difficulty in extending these results to general networks, using (19), is in connecting the matrices T_i to the topology of the network. Intuitively one expects that if the components (function) of $\widehat{L_t X_t}$ corresponding to the state of a sub-queue are not influenced by artificially creating one or more of the observed transitions at a given time, then the state of the sub-queue should be independent of the observed transitions. The above independence can only hold if the customers making the observed transition never return to the sub-queue whose state is being estimated. Walrand and Varaiya [19] have made these heuristic arguments rigorous. Melamed [16] derived similar results via different methods. The main conclusions are that, if a network has type 1,2,3 or 4 servers as in §3, and external inputs are Poisson processes, then external outputs and more generally flows of customers on branches which are not part of a loop (no feedback) are Poisson processes with a rate given by the flow conservation equations (12). These flows observed up to time t, are independent of the state of that part of the network for which they can never return as an input.

Brémaud [8] has shown the opposite as well for the M/M/1 queue with feedback, fig 5. If A_t is a Poisson process then so is D_t, since it is the output of a queueing network satisfying the conditions of Walrand and Varaiya. However the flows E_t, F_t and $F_t + D_t = S_t$ cannot be Poisson for $0 < p < 1$ (and $p = 1$ requires the input rate $\lambda = 0$ for equilibrium). To prove this, consider the following model. Let B_n be a Bernouilli process, $\mathcal{P}(B_n=1) = p$, $\forall n$, with a value 1 indicating that the n-th departure from the queue is fed back. Otherwise the notation is as in §2, with $E_1 = I$, $\lambda(n) = \lambda$ and $\mu(n) = \mu$, $\underline{\lambda}^1 = (\lambda,\ldots\lambda)$, $\underline{\lambda}^2 = (0,\mu,\mu,\ldots)$. Then

$$dX_t = [\lambda(T_1-I) + (T_2-E_2)\text{diag}\,\underline{\lambda}^2.(1-B_{N_{t-}+1})] \,.\, X_{t-}.dt$$

$$+(T_1-I)X_{t-}(dA_t-\lambda.X_{t-}.dt)$$

$$+ (T_2-E_2)X_{t-}.(1-B_{N_{t-}+1}) \,.\, (dS_t-\underline{\lambda}^2.X_{t-}dt)$$

Notice that the last component can also be written as

$$(T_2-E_2)X_{t-}.(dD_t-(1-B_{N_{t-}+1}) \,.\, \underline{\lambda}^2.X_{t-}.dt)$$

The average rate of S_t is $\frac{\lambda}{1-p}$. Therefore define the liklihood ratio L_t^S:

$$dL_t^S = L_t^S(\frac{1-p}{\lambda} \,.\, \underline{\lambda}^2.X_{t-}-1)(dS_t - \frac{\lambda}{1-p}.dt)$$

which takes S_t into a Poisson process. Applying stochastic calculus rules gives

$$dL_t^S X_t = [\lambda(T_1-I) + (T_2-E_2)\text{diag}\,\underline{\lambda}^2.(1-B_{N_{t-}+1}) \, L_{t-}^S X_{t-}.dt$$

$$+(T_1-I)L_{t-}^S.X_{t-}.(dA_t-\lambda.X_{t-}.dt)$$

$$+[\frac{1-p}{\lambda}((1-B_{N_{t-}+1})(T_2-E_2)+I)\text{diag}\underline{\lambda}^2-I] \,.\, L_{t-}^S X_{t-}(dS_t - \frac{\lambda}{1-p}.dt)$$

and the best estimator of X_t, given observations $\{S_s, s \leq t\}$ is

$$\hat{X}_t = \frac{\widehat{L_t^S X_t}}{\| \widehat{L_t^S X_t} \|_1} \quad \text{with (noting that } E_o(B_{N_{t-}+1} \mid S_s, s \leq t) = p$$

and $(T_2 - E_2)\text{diag}\, \underline{\lambda}^2 = \mu(T_2 - E_2))$:

$$d\widehat{L_t^S X_t} = [\lambda(T_1 - I) + \mu(T_2 - E_2)(1-p)]\widehat{L_t^S X_t} \cdot dt$$
$$+ [\frac{(1-p)^2}{\lambda}\mu(T_2 - E_2) + \frac{1-p}{\lambda} \cdot \mu - 1] \cdot \widehat{L_{t-}^S X_{t-}}$$
$$\times (dS_t - \frac{\lambda}{1-p}\,dt)$$

The equilibrium distribution is

$$\pi(n) = \left(1 - \frac{\lambda}{(1-p)\mu}\right)\left(\frac{\lambda}{(1-p)\mu}\right)^n$$

and exists as long as $p < 1 - \frac{\lambda}{\mu}$. This follows from the results in §3. It is easily verified that $\widehat{L_o^S X_o} = \pi$ implies $\widehat{L_t^S X_t} \neq \pi$ for all $t > 0$ and in particular $\frac{\widehat{L_t^S X_t}(0)}{\widehat{L_t^S X_t}(1)} \neq \frac{\pi(0)}{\pi(1)} = \frac{(1-p)\mu}{\lambda}$,

unless $p = 0$. This insures that $\hat{X}_t \neq \pi$ and the rate of S_t, given observation $\{S_s, s \leq t\}$, is not constant, $\frac{\lambda}{1-p} \neq \underline{\lambda}^2 . \hat{X}_t$. Hence S_t is not a Poisson process. Similarly one can find likelihood ratios L_t^F and L_t^D to prove that F_t is not Poisson and that D_t is a Poisson process.

5. SEMI-MARKOV MODELS OF FINITE BUFFER QUEUES

The obvious disadvantage of the analysis in the preceding sections is the excessive size of the state space. One way out of this difficulty is to consider the network of queues only at times T_n when it enters certain important states such as buffer full, queue empty, completion of a non-exponential service times. The state space will then be much smaller, but the time between successive events $(T_n - T_{n-1})$ is no longer exponential. If one chooses the special states such that the "embedded process" X_{T_n} is Markovian, then one obtains a semi-Markov model. A complete description requires $P(X_{T_n} | X_{T_{n-1}})$ (usually easy to get) and $P(T_n - T_{n-1} \leq t | X_{T_n}, X_{T_{n-1}})$.

The martingale models will be useful in calculating this last distribution, as shown by the following example.

Consider an M/M/1/B queue and reduce the size of the state space by looking at the system only when the waiting room is full or empty ($Q_{T_n} = 0$ or B). We then want to calculate the joint distribution of the length of time between two visits to the boundary (0 or B) together with the number of customers served during this interval ($D_{T_n} - D_{T_{n-1}}$), conditioned on the initial state, say Q_o, and the final state Q_{T_1}. Equivalently we will derive the moment generating function. It is easy to see that on $[0,T_1]$:

$$dz_1^{A_t} z_2^{D_t} e^{-[(z_1-1)\lambda+(z_2-1)\mu]t} = z_1^{A_{t-}} \cdot z_2^{D_{t-}} \cdot e^{-[z_1-1)\lambda+(z_2-1)\mu]t}$$
$$\cdot [(z_1-1)(dA_t-\lambda dt)+(z_2-1)(dD_t-\mu.dt)]$$

is a martingale increment. Substituting $z_2 = \frac{z}{z_1}$, multiplying by $z_1^{Q_o}$ and substituting $z_1(s,z) = \frac{s+\lambda+\mu}{2\lambda} - \sqrt{\left(\frac{s+\lambda+\mu}{2\lambda}\right)^2 - \frac{z\mu}{\lambda}}$ one obtains the martingale

$$z_1(s,z)^{Q_t} z^{D_t} e^{-st} \quad \text{for } t \in [0,T_1].$$

It can be shown that T_1 is a regular stopping time. Hence by optional sampling

$$z_1(s,z)^{Q_o} = E[z_1(s,z)^{Q_{T_1}} . z^{D_{T_1}} e^{-st} | Q_o]$$
$$= z_1(s,z)^B . \frac{\left(\frac{\mu}{\lambda}\right)^{Q_o} - 1}{\left(\frac{\mu}{\lambda}\right)^B - 1} . E[z^{D_{T_1}} e^{-sT_1} | Q_o, Q_{T_1} = B]$$
$$+ \frac{\left(\frac{\mu}{\lambda}\right)^B - \left(\frac{\mu}{\lambda}\right)^{Q_o}}{\left(\frac{\mu}{\lambda}\right)^B - 1} . E[z^{D_{T_1}} e^{-sT_1} | Q_o, Q_{T_1} = 0]$$

A second equation in $E[z^{D_{T_1}} e^{-sT_1} | Q_o, Q_{T_1} = 0]$ and $E[z^{D_{T_1}} e^{-sT_1} | Q_o, Q_{T_1} = B]$ is obtained by a recursive (birth-and-death process) analysis of $E[z^{D_{T_1}} e^{-sT_1} | Q_o]$. This requires inverting a Jacobian Toeplitz matrix. Details will be given in a forth coming paper [5]. Related results have been derived by Kennedy [13].

6. DESIGN AND OPTIMAL CONTROL OF QUEUEING NETWORKS

The equilibrium distributions for quasi-reversible networks, obtained in §3, provide a reasonable model for optimal design. By Little's law average waiting times are proportional to average queue lengths, while the cost of a transmission link (server) can reasonably be modeled as an increasing function $f_i(\mu_i)$ of the service rate. Hence for a network as in fig.4 (with one class only considered here to simplify the notation) the following optimization problem will lead to an optimal choice of transmission capacities for given arrival rates λ_i and routing probabilities $r(i,j)$, assuming the total amount of resources available is F :

$$\min_{\sum_{i=1}^{N} f_i(\mu_i) \leq F} \sum_{i=1}^{N} E\, g_i(Q_t^i) \tag{20}$$

Because of the product form of the equilibrium solution and because the arrival rates ν_i at queue i (defined by (12)) are independent of the design parameters μ_i one obtains a separable nonlinear programming problem (assuming type 1, 2 or 4 servers)

$$\min_{\sum_{i=1}^{N} f_i(\mu_i) \leq F} \sum_{i=1}^{N} \sum_{k_i} g_i(k_i) \cdot \frac{\nu_i^{k_i}}{\mu_i^{k_i}} \left(1 - \frac{\nu_i}{\mu_i}\right)$$

This can be solved using Lagrange multipliers. If $f_i(\mu_i) = f_i \cdot \mu_i$ and $g_i(Q_t^i) = g_i \cdot Q_t^i$ then (Kelly [12, p.97])

$$\mu_i^{opt} = \nu_i + \frac{\sqrt{g_i f_i \nu_i}}{f_i} \cdot \frac{F-\Sigma f_j \nu_j}{\Sigma \sqrt{g_j f_j \nu_j}}$$

An important limitation of the design of networks as above, using equilibrium distributions of quasi-reversible networks, is that little can be said about the delay distributions. Moreover quasi-reversible queueing networks cannot deal with finite buffer queues. Therefore a further analysis (as attempted in §5 for example) is necessary to make a rational choice allocating available resources between increasing the service rate or increasing the size of the waiting room.

The model of a queueing network described in this paper can also be used for optimal control models, where the decisions are changes of the rates of the different types of transitions This may represent adjusting departure rates (change the number of servers) or arrival rates (e.g. by levying admission tolls); it may also represent changes in the priority of serving different types of customers or changes of the routing matrices. Let $\mathcal{Y}_t = \sigma\{N_s^i,\ i \in J \subset \mathcal{N}_+,\ s \leq t\} \subset \mathcal{F}_t$ represent the observations available for making a decision u_t at time t (i.e. u_t is a $\mathcal{Y}_t$ - adapted process). The values of this control law lie in a fixed set U and the transition rates $\lambda_t^i(X_{t-},u_{t-})$ can thus vary in a given range. The model of the system then becomes (cf.(1)):

$$dX_t = \sum_i (T_i - E_i).\text{diag}\,\underline{\lambda}^i(u_{t-}) \;.\; X_{t-}.dt + M_t$$

If a cost is associated to spending time in certain states (long queues, full buffer), to using certain control values (using many servers) and possibly also to changes in control value (e.g. a switchover cost associated to changing from one type of services to another) the following discounted cost should be minimized

$$\underset{u \text{ admissible}}{\text{Inf}} \; E_u \int_o^T e^{-\alpha t}\, C_t(X_t,u_t,u_{t-}).dt$$

The problem is now in the right form for applying dynamic programming. Let

$$V(t,u_{[0,t]},\mathcal{Y}_t) = \inf_{u_{[t,T)}} E_u\left(\int_t^T e^{-\alpha s} C_s(X_s,u_s,u_{s-})ds \,\middle|\, u_{[0,t]},\mathcal{Y}_t\right)$$

be the value function. Then

$$0 \leq \inf_u E_u \left(\frac{\partial V(t,u,\mathcal{Y}_t)}{\partial t} - \alpha \, V(t,u,\mathcal{Y}_t) + C_t(X_t,u,u_{t-}) \right. \qquad (22)$$

$$\left. + \sum_{i \in J} \lambda^i(u) \, \Delta_i \, V(t,u,\mathcal{Y}_t) \,\middle|\, \mathcal{Y}_t \right)$$

where $\Delta_i V$ represents the (discontinuous) change in V when a transition of type i is observed. A control law u is optimal if and only if equality holds in Bellman-Hamilton-Jacobi equation (22) [4].

Very few cases are known where (22) can be solved analytically. In an excellent survey paper Sobel [17] studies the problem for a single Markovian queue, with adjustable service rate $\mu(u) = u$, $\underline{\mu} \leq u \leq \bar{\mu}$, and cost

$$C_t(X_t,u_t,u_{t-}) = w(Q_t) + c(u_t)$$

where $\begin{cases} w \geq 0, & \text{convex, nondecreasing} \\ c \geq 0, & \text{nondecreasing} \end{cases}$

is assumed. Then if the queue length can be observed the control laws are of the form $u_t(Q_{t-})$. By reducing the problem to a discrete time Markov decision problem (splitting up the cost in $\sum \int_{T_n \wedge T}^{T_{n+1} \wedge T}$), and showing that present cost plus value function is monotone in u (by backwards induction) Sobel shows that there exists a deterministic function g_t such that

$$u_t = \begin{cases} = \bar{\mu} & \text{if } Q_{t-} > q_t \\ = \underline{\mu} & \text{if } Q_{t-} \leq q_t . \end{cases}$$

If $T = \infty$ then $q_t = q$.

In more complicated cases it will be necessary in order to solve an optimal control problem to assume an a priori structure for the control law, similar to the above bang-bang form. For example if M servers are available, with an operating cost and a starting cost for each server, one could imagine that a good control law will switch from m to m+1 servers active when $Q_t > \bar{q}_{m+1}$, and switch from m to m-1 servers as soon as $Q_t < \underline{q}_m$

($\underline{q}_m < \bar{q}_m$). One then has to determine the 2M numbers $\bar{q}_m$, $\underline{q}_m$. This could be done by analyzing, for arbitrary $\bar{q}_m$, $\underline{q}_m$, the semi-Markov process $(X_{T_n}, T_n - T_{n-1})$ where T_n is the time of the n-th switch in control value. Extensions of the methods of §5 would be useful for this purpose.

An important limitation to most optimal control results is the assumption of complete information, including the assumption that arrival rates are exactly known. If a queue is part of a large network, the arrival rate depends on the state of the rest of the network, which is usually not available to the controller. The arrival rate however determines the parameters of the control law, such as q_t above, and an estimate, obtained by filtering equations as in §4, will have to be used. Very little is known about properties of such adaptive controllers. The only reference dealing with unknown arrival rates is by Yadin and Zacks [20] who prove optimality for a control law with simple structure.

To find optimal control laws one could also attempt to solve (22) numerically. This is usually impossible because of the size of the state space. Again looking at the system only when it is in particular states (or grouping states) leads to semi-Markov models with much smaller state space. It may be the only numerically feasible way to obtain a good control law from (22).

REFERENCES

1. N. Arley and V. Borchsenius, On the theory of infinite systems of differential equations and their application to the theory of stochastic processes and the perturbation theory of quantum mechanics, Acta Mathematica, 76, (1945), p. 261-322
2. F. Baskett, M. Chandy, R. Muntz and J. Palacios, Open, closed and mixed networks of queues with different classes of customers, Journal of the A.C.M., 22 (1975), p. 248-260.
3. R. Boel, P. Varaiya and E. Wong, Martingales on jump processes; II : Applications, SIAM Journal on Controls, 13 (1975), p. 1022-1061.
4. R. Boel and P. Varaiya, Optimal control of jump processes, SIAM J. on Control and Optimization, 15 (1977), p. 92-119.
5. R. Boel, Martingale methods for the semi-Markov analysis of queues with blocking, in preparation.
6. P.Brémaud, La méthode des semi-martingales en filtrage quand l' observation est un processus ponctuel marqué Séminaire de Probabilités X, Lecture Notes in Mathematics, vol. 511, p. 1-18, Springer 1976.

7. P. Brémaud, On the output theorem of queueing theory, via filtering, J. Applied Probability, 15 (1978), p.387-405
8. P. Brémaud, Streams of a M/M/1 feedback queue in statistical equilibrium, z. Wahrscheinlichkeitstheorie u. verw. Geb., 45 (1978), p.21-33
9. P. Courtois, Decomposability - queueing and computer system applications, Academic Press, 1977.
10. W. Feller, An introduction to probability theory and its applications, volume 2, 2nd. ed., Wiley, 1971
11. J. Jackson, Networks of waiting lines, Operations Research, 5 (1957), p. 518-521.
12. F. Kelly, Reversibility and stochastic networks, Wiley, 1979
13. D. Kennedy, Some martingales related to cumulative sum tests and single-server queues, Stochastic Processes and their Applications, 4 (1976), p. 261-269
14. L. Kleinrock, Queueing systems, vol. I: Theory, 1975; vol. II: Computer applications, 1976, Wiley.
15. H. Kobayashi and A. Konheim, Queueing models for computer communications system analsyis, IEEE Trans. on Communications, COM - 25 (1977), p. 2129.
16. B. Melamed, Characterization of Poisson traffic streams in Jackson queueing networks, Advances in Applied Probability, 11 (1979), p. 422-439.
17. M. Sobel, Optimal operation of queues, in Mathematical Methods in Queueing Theory, A.B. Clarke, ed., Lecture Notes in Economics and Mathematical Systems, vol. 98, Springer, 1973.
18. J. Walrand and P. Varaiya, Interconnections of Markov chains and quasi-reversible queueing networks, to appear in : Stochastic Processes and their Applications, 10 (1980)
19. J. Walrand and P. Varaiya, Flows in queueing networks; a martingale approach; preprint, 1979.
20. M. Yadin and S. Zacks, Adaptation of the service capacity in a queueing system which subjected to a change in the arrival rate at unknown epoch, Technical Report no. 30, Dept. of Math. and Stat., Case Western Reserve University, 1977.

STATE SPACE MODELS FOR GAUSSIAN STOCHASTIC PROCESSES

Anders Lindquist and Giorgio Picci

University of Kentucky, Lexington, Kentucky
LADSEB-CNR, Padova, Italy

ABSTRACT: A comprehensive theory of stochastic realization for multivariate stationary Gaussian processes is presented. It is coordinate-free in nature, starting out with an abstract state space theory in Hilbert space, based on the concept of splitting subspace. These results are then carried over to the spectral domain and described in terms of Hardy functions. Each state space is uniquely characterized by its structural function, an inner function which contains all the systems theoretical characteristics of the corresponding realizations. Finally coordinates are introduced and concrete differential-equation-type representations are obtained. This paper is an abridged version of a forthcoming paper, which in turn summarizes and considerably extends results which have previously been presented in a series of preliminary conference papers.

1. INTRODUCTION

In recent years there has been a considerable interest in various versions of the so-called *stochastic realization problem* [1-31], which, loosely speaking, can be described as the problem of finding (a suitable class of) stochastic dynamical systems, called *realizations*, all having a given random process $\{y(t); t \in T\}$ as its output. (Here T is the index set, which usually is the real line R or the set Z of integers.) In the past it has often been

---

The first author was supported partially by the National Science Foundation under grant ECS-7903731 and partially by the Air Force Office of Scientific Research under grant AFOSR-78-3519.

M. Hazewinkel and J. C. Willems (eds.), Stochastic Systems: The Mathematics of Filtering and Identification and Applications, 169–204.

assumed that y is a stationary (or stationary increment) process with rational spectral density, thus insuring the existence of finite dimensional realizations.

The early papers on the subject [1-3] consider a deterministic version of the problem, the objective being to realize (in the deterministic sense [32,33]) the spectral factors of the given process which is defined up to second-order properties only. The probabilistic aspects of the stochastic realization problem were subsequently clarified in [8-10]. In all these papers the states of the realizations are represented in a fixed coordinate system to avoid trivial questions of uniqueness.

However, the most natural approach to the stochastic realization problem is coordinate free: Begin by constructing families $\{X_t;\ t \in T\}$ of *state spaces* which evolve in time in a Markovian manner. These state spaces should be as small as possible, but large enough to contain the essential information for determining the temporal evolution of the given process. Then, for each such family, concrete realizations can be obtained by introducing suitable bases in the state spaces. This line of study was initiated in [4-6], where a restricted version of the problem of this paper was studied, considering only state spaces contained in the closed span of the past (or, symmetrically, the future) of the given process. With such a state space approach we need not restrict the analysis to processes with rational spectral density, since the framework will also accommodate infinite dimensional state spaces.

During the last couple of years we have been developing a state space theory of stochastic realization which is now in a reasonably complete form. Part of our work has been reported in a series of preliminary conference papers [11-15]; a more complete account will appear in a forthcoming paper [16], which is now under preparation. Some results in the first phase of this work were obtained in cooperation with Ruckebusch [17], who parallelly developed his own geometric state space theory [20,21]. The present paper is an attempt to summarize the results presented in [16]. Due to page limitations, not all topics of [16] will be discussed. Also we have left out the proofs of the theorems, instead providing the reader with references for the proofs. For simplicity, only realizations of continuous-time stationary processes will be discussed, but it should be understood that our basic geometric theory holds also for stationary increment processes and discrete-time processes, and that the subsequent spectral theory can be appropriately modified to take care of these cases also.

2. PROBLEM FORMULATION

Let $\{y(t);\ t \in R\}$ be a real stationary m-dimensional Gaussian process which is purely nondeterministic, mean-square continuous and centered, and let H be the Gaussian space [34] generated by y, i.e. the linear span of the stochastic variables $\{y_k(t);\ t \in R, k = 1,2,\ldots,m\}$ closed in L_2 norm. The space H is a Hilbert space when endowed with the inner product $(\xi,\eta) = E\{\xi\eta\}$, where $E\{\cdot\}$ stands for mathematical expectation. For any two subspaces A and B of H (which are always taken to be closed), $A \vee B$ denotes the closed linear hull of A and B, E^A denotes the orthogonal projection on A, and $\bar{E}^A B$ signifies the closure of $E^A B$. Moreover, let $A^\perp$ denote the orthogonal complement of A in H, and let $A \ominus B$ be the orthogonal complement of B in A, implicitly implying that A is a subspace of B. Since y is a stationary process, there is strongly continuous group $\{U_t;\ t \in R\}$ of unitary operators $H \to H$ such that $y_k(t) = U_t y_k(0)$ for all t and $k = 1,2,\ldots,m$ [35]. In the sequel we shall also consider the two semigroups $\{U_t;\ t \geq 0\}$ and $\{\bar{U}_t;\ t \geq 0\}$ obtained by setting $U_t := U_t$ and $\bar{U}_t := U_{-t}$ for $t \geq 0$.

The first problem at hand is to determine families $\{X_t;\ t \in R\}$ of subspaces of H, such that

$$y_k(t) \in X_t \quad ; \quad k = 1,2,\ldots,m \tag{2.1}$$

for all t, which are *Markovian* in the sense that

$$E^{X_t^-}\lambda = E^{X_t}\lambda \qquad \text{for all } \lambda \in X_t^+ \tag{2.2}$$

where $X_t^- := \bigvee_{\tau \leq t} X_\tau$ and $X_t^+ := \bigvee_{\tau \geq t} X_\tau$, and which are *stationary*, i.e. satisfy the condition

$$X_t = U_t X_0 \qquad \text{for all } t \in R\ . \tag{2.3}$$

The subspaces $\{X_t\}$ will be called *state spaces*. This is a generalization of the following more concrete problem: Find a vector-valued stationary Markov process $\{x(t);\ t \in R\}$ and a matrix C so that $y(t) = Cx(t)$ for all $t \in R$. However, as we shall see in Section 9, (in a strict sense) the latter problem generally makes sense only if the spectral density of y is rational, in which case there are finite dimensional Markov processes x. To circumvent this difficulty we would have to consider *weak* Hilbert space-valued Markov processes. Problem formulation (2.1)-(2.3), on the other hand, is *coordinate free* and makes sense without further restrictions or modifications.

In view of condition (2.3), it is enough to require that (2.1) and (2.2) hold for one t, say t = 0; then they will automatically hold for all $t \in R$. For simplicity we shall drop the index and write X instead of X_0. Now define the *past space* H^- and the *future space* H^+ as the closed linear hulls in H of the stochastic variables $\{y_k(t);\ t \le 0,\ k = 1,2,\ldots,m\}$ and $\{y_k(t);\ t \ge 0,\ k = 1,2,\ldots,m\}$ respectively. Then, it follows from (2.1) that $H^- \vee X \subset X^-$ and that $H^+ \subset X^+$. Consequently applying the projection $E^{H^-\vee X}$ to (2.2) with t = 0 we obtain

$$E^{H^-\vee X}\lambda = E^X\lambda \quad \text{for all } \lambda \in H^+ . \tag{2.4a}$$

It is easy to show [15] that the symmetric condition

$$E^{H^+\vee X}\lambda = E^X\lambda \quad \text{for all } \lambda \in H^- \tag{2.4b}$$

is equivalent to (2.4a). A subspace satisfying one of conditions (2.4) is called a *splitting subspace*. Loosely speaking, such a subspace contains all information about the past needed in predicting the future, or, equivalently, all the information about the future required to estimate the past.

PROPOSITION 2.1. [14]. *Let* X *be a subspace of* H *and let* $S := H^- \vee X$ *and* $\bar{S} := H^+ \vee X$. *Then the family* $\{U_tX;\ t \in R\}$ *satisfies conditions (2.1) and (2.2) if and only if* X *is a splitting subspace such that*

$$\bar{U}_t S \subset S \quad \text{for all } t \ge 0 \tag{2.5a}$$

and

$$U_t\bar{S} \subset \bar{S} \quad \text{for all } t \ge 0 . \tag{2.5b}$$

In view of this proposition we shall say that a splitting subspace is *Markovian* if it satisfies conditions (2.5). Hence we can instead consider the problem of finding Markovian splitting subspaces. This problem formulation has the advantage of also covering situations which are not discussed in this paper, e.g. realization of processes with stationary increments and discrete-time processes, for which problem formulation (2.1)-(2.3) is too restrictive (since it does not allow for observation noise, for example). Hence the basic geometric theory developed below has a wider applicability, as we shall demonstrate in some subsequent papers.

Obviously the whole space H is a Markovian splitting subspace, and so are H^- and H^+, but they are too large for our purposes. Indeed, the whole idea is to obtain "data reduction." Therefore we shall be particularly interested in (Markovian) splitting subspaces X which are *minimal* in the sense that there is no proper subspace of X which is also a (Markovian) splitting subspace. The following proposition implies that a minimal Markovian splitting subspace is the same thing as a Markovian minimal splitting subspace.

PROPOSITION 2.2. [16]. *Any minimal Markovian splitting subspace is a minimal splitting subspace.*

When the Markovian splitting subspaces have been adquately characterized, there remains the problem of obtaining *concrete* dynamical representations of the given process y based on these state spaces, if possible of differential equation type. Note that we are only considering representations for which the state spaces are contained in the Hilbert space H generated by the given process (or its increments), i.e. *internal* realizations. Our theory could be modified to accommodate realizations containing exogeneous random elements (*external* realizations), but this is outside the scope of this paper, being a bit unnatural in the present setting.

3. THE GEOMETRY OF SPLITTING SUBSPACES

Our first problem will be to determine the set of all splitting subspaces, and, in particular, those which are minimal.

The *predictor space*

$$X_- := \bar{E}^{H^-} H^+ \tag{3.1}$$

is a splitting subspace. To see this, note that, for all $\lambda \in H^+$, $E^{H^-}\lambda \in X_-$; hence $E^{H^-}\lambda = E^{X_-}E^{H^-}\lambda = E^{X_-}\lambda$. Moreover, all splitting subspaces $X \subset H^-$ contain X_-. In fact, by (2.4a), $E^{H^-}H^+ = E^X H^+$ which is contained in X. Hence X_- is a minimal splitting subspace. By symmetry, we see that the *backward prediction space*

$$X_+ := \bar{E}^{H^+} H^- \tag{3.2}$$

is also a minimal splitting subspace. (The reader is urged to carefully distinguish between X_- and X_+ and X^- and X^+ defined in Section 2. The reason for the former notation will be clear from what follows.)

PROPOSITION 3.1. [14]. *The spaces* H, H^- *and* H^+ *have the orthogonal decompositions*

$$H = N^- \oplus H^{\square} \oplus N^+ , \tag{3.3a}$$

$$H^- = N^- \oplus X_- \quad \text{and} \quad H^+ = N^+ \oplus X_+ , \tag{3.3b}$$

where $H^{\square} := X_- \vee X_+$, $N^- := H^- \cap (H^+)^{\perp}$ *and* $N^+ := H^+ \cap (H^-)^{\perp}$.

The space $H^{\square}$ is called the *frame space* and N^- and N^+ are called the (past respectively the future) *junk spaces*. These notations are suggested by the following result.

PROPOSITION 3.2. [14]. *Let* X *be a minimal splitting subspace. Then*

$$H^- \cap H^+ \subset X \subset H^\square . \tag{3.4}$$

Hence the frame space $H^\square$ is the closed linear hull of all minimal splitting subspaces, and consequently it contains all "information" needed for state space construction. On the other hand, the junk spaces contain no useful information and could be discarded. It is not hard to see that the frame space is itself a (generally nonminimal) splitting subspace. The following proposition illustrates the importance of these concepts in filtering theory. It should be compared with the corresponding result in [20], which is weaker.

PROPOSITION 3.3. [16]. *Let* X *be a splitting subspace. Then*

$$E^{H^-} X = X_- \tag{3.5a}$$

if and only if $X \perp N^-$, *and symmetrically*

$$E^{H^+} X = X_+ \tag{3.5b}$$

if and only if $X \perp N^+$.

We shall say that the given process y is *noncyclic* if it has nontrivial junk spaces, i.e. $N^- \neq 0$ and $N^+ \neq 0$, and *strictly noncyclic* if N^- and N^+ are both full range. (A subspace $A \subset H$ is *full range* if the closed linear hull of $\{U_t A;\ t \in R\}$ in H is all of H.) In the scalar case (m = 1) strict noncyclicity is the same as noncyclicity. Clearly the problem of state space construction is not very interesting unless we have noncyclicity, since otherwise there will be no "data reduction," H^- and H^+ being minimal splitting subspaces.

In order to describe the set of splitting subspaces, we need to introduce the concept of *perpendicular intersection*. Two subspaces of H, A and B, are said to *intersect perpendicularly* if

$$\bar{E}^A B = A \cap B \tag{3.6a}$$

or equivalently [15]

$$\bar{E}^B A = A \cap B . \tag{3.6b}$$

If A and B together span all of H, we have the following characterization of perpendicular intersection.

PROPOSITION 3.4. [15]. *Let* A *and* B *be subspaces of* H *such that* $A \vee B = H$. *Then* A *and* B *intersect perpendicularly if and only if* $B^{\perp} \subset A$ *or, equivalently,* $A^{\perp} \subset B$.

Now, if H^- and H^+ intersect perpendicularly,† $H^{\square} = H^- \cap H^+$, and consequently there is a *unique* minimal splitting subspace, namely $H^- \cap H^+$ (Proposition 3.2). Hence the problem of finding the minimal splitting subspaces is trivial. In general, however, H^- and H^+ do not intersect perpendicularly, but, by appropriately extending H^- and H^+ so that the extended spaces intersect perpendicularly, we can still describe each splitting subspace as the intersection between two subspaces.

THEOREM 3.1. [15]. *The subspace* $X \subset H$ *is a splitting subspace if and only if*

$$X = S \cap \bar{S} \tag{3.7}$$

for some perpendicularly intersecting subspaces S *and* $\bar{S}$ *such that* $S \supset H^-$ *and* $\bar{S} \supset H^+$. *The correspondence* $X \leftrightarrow (S,\bar{S})$ *is one-one, the pair* $(S,\bar{S})$ *being uniquely determined by relations*

$$S = H^- \vee X \tag{3.8a}$$

and

$$\bar{S} = H^+ \vee X\ . \tag{3.8b}$$

COROLLARY 3.1. *A subspace* $X \subset H$ *is a splitting subspace if and only if there are subspaces* $S \supset H^-$ *and* $\bar{S} \supset H^+$ *such that one of the following four equivalent conditions hold*

$$X = \bar{E}^{S}\bar{S} \tag{3.9a}$$

$$X = \bar{E}^{\bar{S}}S \tag{3.9b}$$

$$X = S \ominus \bar{S}^{\perp} \tag{3.9c}$$

$$X = \bar{S} \ominus S^{\perp}\ . \tag{3.9d}$$

A subspace S such that $S \supset H^-$ ($S \supset H^+$) will be called an *augmented past (future) space*. Hence, each (minimal or nonminimal) splitting subspace is uniquely characterized by two perpendicularly intersecting subspaces S and $\bar{S}$, one being an augmented past space

---

†For a process y with rational spectral density Φ, H^- and H^+ intersect perpendicularly if and only if Φ has no zeros, i.e. y is a "purely autoregressive" process.

and one an augmented future space. We shall write $X \sim (S,\bar{S})$ to recall this correspondence. For example, $X_- \sim (S_-,\bar{S}_-)$ where $S_- = H^- \vee X_- = H^-$ and $\bar{S}_- = H^+ \vee X_- = (N^-)^\perp$ (Proposition 3.1). Likewise, $X_+ \sim (S_+,\bar{S}_+)$ where $S_+ = (N^+)^\perp$ and $\bar{S}_+ = H^+$.

To have X minimal we clearly need to make S and $\bar{S}$ as small as possible. Given an S, the smallest $\bar{S}$ which both contains H^+ and intersects S perpendicularly is

$$\bar{S} = H^+ \vee S^\perp \tag{3.10a}$$

(Proposition 3.4). Likewise, given $\bar{S}$,

$$S = H^- \vee \bar{S}^\perp \tag{3.10b}$$

is the smallest subspace containing H^- which intersects $\bar{S}$ perpendicularly. It is not hard to see that the two conditions not only characterize minimality but also the splitting property.

THEOREM 3.2. [15]. *Let* $S \supset H^-$ *and* $\bar{S} \supset H^+$ *be two subspaces, and set* $X = S \cap \bar{S}$. *Then* X *is a minimal splitting subspace if and only if both conditions (3.10) hold.*

In view of Proposition 3.2, (3.3a) and (3.8), it is clearly necessary that

$$S \subset (N^+)^\perp \tag{3.11a}$$

and that

$$\bar{S} \subset (N^-)^\perp \tag{3.11b}$$

in order that $X \sim (S,\bar{S})$ be minimal, but *not* sufficient; in fact, any subspace of $H^\square$ satisfies (3.11). However, in Theorem 3.2., conditions (3.10) can be replaced by (3.10a) + (3.11a) or by (3.10b) + (3.11b), as seen from the following pair of propositions.

PROPOSITION 3.5. [14]. *Let* $S \supset H^-$ *and* $\bar{S} \supset H^+$ *be two subspaces satisfying (3.10a). Then (3.10b) holds if and only if (3.11a) holds.*

PROPOSITION 3.6. [14]. *Let* $S \supset H^-$ *and* $\bar{S} \supset H^+$ *be two subspaces satisfying (3.10b). Then (3.10a) holds if and only if (3.11b) holds.*

It follows from Theorem 3.2 and Proposition 3.5 that X is a minimal splitting subspace if and only if $\bar{S} := H^+ \vee X$ is given by (3.10a) and $S := H^- \vee X$ satisfies

$$H^- \subset S \subset (N^+)^\perp , \tag{3.12}$$

in which case $X = \bar{E}^S H^+$, as can be seen from (3.9a) and (3.10a).

Consequently the minimal splitting subspaces are in one-one correspondence with subspaces S satisfying (3.12). For this reason, we shall call a subspace satisfying (3.12) a *minimal augmented past space*. The set of such subspaces form a complete lattice, where the partial ordering is induced by the $\subset$ operation. Consequently the set of minimal splitting subspaces also form a complete lattice, in which X_- is the minimum element and X_+ is the maximum. In fact, as we have seen above, $S_- = H^-$ and $S_+ = (N^+)^\perp$.

Symmetrically, there is a one-one correspondence between the minimal splitting subspaces and subspaces $\bar{S}$ such that

$$H^+ \subset \bar{S} \subset (N^-)^\perp \, . \tag{3.13}$$

We shall call such a subspace a *minimal augmented future space*. In terms of it, the minimal splitting subspace has the representation $X = \bar{E}^S H^-$.

4. OBSERVABILITY, CONSTRUCTIBILITY AND MINIMALITY

Relation (2.4a), defining a splitting subspace $X \sim (S,\bar{S})$, can be written

$$E^S|_{H^+} = E^S|_X \circ E^X|_{H^+} \, , \tag{4.1a}$$

where $|_A$ denotes restriction to the domain A. (Here the first operator on the right-hand side is merely an insertion map, insuring that the range spaces match.) Likewise, the alternative definition (2.4b) can be written

$$E^{\bar{S}}|_{H^-} = E^{\bar{S}}|_X \circ E^X|_{H^-} \, . \tag{4.1b}$$

Define $G^+ := E^S|_{H^+}$ and $G^- := E^{\bar{S}}|_{H^-}$. Then the splitting property (2.4) is equivalent to either of the two *Hankel operators* G^+ and G^- having a factorization through X described by the commutative diagrams

(4.2)

respectively, where $O := E^X|_{H^+}$, $C := E^X|_{H^-}$, $R = E^S|_X$ and $\bar{R} := E^{\bar{S}}|_X$. Such a factorization is said to be *canonical* [32] if the first factor (here R or $\bar{R}$) is *one-one* and the second factor (here O or C)

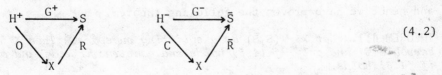

maps *onto* a dense subset of X; if the second factor maps onto X we say that the factorization is *exactly canonical*.

Since the insertion map R is trivially one-one, the first of diagrams (4.2) is canonical if and only if

$$\bar{E}^{X}H^{+} = X \, . \tag{4.3a}$$

A splitting subspace with this property is said to be *observable;* the mapping O is called the *observability operator*. Likewise, the second factorization (4.2) is canonical if and only if

$$\bar{E}^{X}H^{-} = X \, . \tag{4.3b}$$

If this condition holds, we say that X is *constructible;* we call C the *constructibility operator* [20]. If one of the factorizations is exactly canonical, the closure bar over the E in the correspondind relation (4.3) can be removed; then we say that X is *exactly observable* ($E^{X}H^{+} = X$) or *exactly constructible* ($E^{X}H^{-} = X$) respectively.

It follows from (3.9a) and the splitting property (2.4a) that a splitting subspace X is observable if and only if

$$\bar{E}^{S}\bar{S} = \bar{E}^{S}H^{+} \, . \tag{4.4}$$

But this condition is equivalent to

$$\bar{S} = H^{+} \vee S^{\perp} \, . \tag{4.5}$$

In fact, since $E^{S}\lambda = 0$ for $\lambda \in S^{\perp}$, it is easy to see that (4.5) implies (4.4). To see that (4.5) is a consequence of (4.4), first note that, since S and $\bar{S}$ intersect perpendicularly (Theorem 3.1), $\bar{S} \supset H^{+} \vee S^{\perp}$ (Proposition 3.4). But $Z := \bar{S} \ominus (H^{+} \vee S^{\perp}) \subset S \cap (H^{+})^{\perp}$, and therefore (4.4) cannot hold unless Z = 0. In the same way we see that a splitting subspace X is constructible if and only if

$$S = H^{-} \vee \bar{S}^{\perp} \, , \tag{4.6}$$

and hence we have proven the following theorem.

THEOREM 4.1. *Let* $X \sim (S,\bar{S})$ *be a splitting subspace. Then* X *is observable if and only if (4.5) holds and constructible if and only if (4.6) holds.*

We can now tie together the concepts of observability and constructibility with that of minimality, discussed in Section 3. It follows from Theorems 3.2 and 4.1 that a splitting subspace is minimal if and only if it is both observable and constructible. This point can also be illustrated in the following way: Apply the projector E^{H^-} to (2.4a) to obtain $E^{H^-}\lambda = E^{H^-}E^X\lambda$ for all $\lambda \in H^+$, i.e. the diagram

$$\begin{array}{ccc} H^+ & \xrightarrow{\;E^{H^-}|_{H^+}\;} & H^- \\ & {\scriptstyle O}\searrow \qquad \nearrow{\scriptstyle C^*} & \\ & X & \end{array} \qquad (4.7)$$

commutes if X is a splitting subspace, where $C^* = E^{H^-}|_X$ is the adjoint of the constructibility operator $C = E^X|_{H^-}$. It is not hard to see that C^* is one-one if and only if C maps onto a dense subset of X. (Cf. [36; p.89].) Consequently, in view of the equivalence between minimality and observability plus constructibility proven above, the factorization (4.7) is canonical if and only if X is minimal. Of course, in the same way, if X is a splitting subspace, the dual diagram

$$\begin{array}{ccc} H^- & \xrightarrow{\;E^{H^+}|_{H^-}\;} & H^+ \\ & {\scriptstyle C}\searrow \qquad \nearrow{\scriptstyle O^*} & \\ & X & \end{array} \qquad (4.8)$$

commutes, and it is canonical if and only if X is minimal.

Finally we summarize the connections between observability, constructibility and minimality provided by Theorems 3.2 and 4.1 and Propositions 3.5 and 3.6.

THEOREM 4.2. *Let* $X \sim (S,\bar{S})$ *be a splitting subspace. Then the following conditions are equivalent:*

(i) X *is minimal*

(ii) X *is observable and constructible*

(iii) X *is observable and* S *is minimal*

(iv) X *is constructible and* $\bar{S}$ *is minimal.*

5. MARKOVIAN SPLITTING SUBSPACES

Let $X \sim (S,\bar{S})$ be a splitting subspace. Then, by Proposition 2.1, X is *Markovian* if and only if the two conditions

$$\bar{U}_t S \subset S \quad \text{(left invariance)} \tag{5.1a}$$

and

$$U_t \bar{S} \subset \bar{S} \quad \text{(right invariance)} \tag{5.1b}$$

both hold. It is immediately seen that $X_- \sim (H^-,(N^-)^\perp)$ and $X_+ \sim ((N^+)^\perp,H^+)$ satisfy these conditions, and consequently X_- and X_+ are Markovian splitting subspaces. Hence all minimal Markovian splitting subspaces form a complete sublattice of the lattice defined in Section 3, with X_- being the minimum and X_+ the maximum element.

If (5.1a) holds, $\{\bar{U}_t|_S;\ t \ge 0\}$ is a strongly continuous semigroup on S, and the same holds true for the adjoints

$$U_t(S) := E^S U_t|_S \quad ; \quad t \ge 0 \tag{5.2a}$$

Similarly, if (5.1b) holds, $\{U_t|_{\bar{S}};\ t \ge 0\}$ and the adjoints

$$\bar{U}_t(\bar{S}) := E^{\bar{S}} \bar{U}_t|_{\bar{S}} \quad ; \quad t \ge 0 \tag{5.2b}$$

both form strongly continuous semigroups on $\bar{S}$. Operators of type (5.2) are called *compressions* of the shifts U_t and $\bar{U}_t$ respectively. Compressions with respect to subspaces other than S and $\bar{S}$ will be denoted analogously. It can be shown that a Markovian splitting subspace $X \sim (S,\bar{S})$ is invariant for $U_t(S)$ and $\bar{U}_t(\bar{S})$. More precisely we have:

PROPOSITION 5.1. [16]. *Let $X \sim (S,\bar{S})$ be a splitting subspace. Then the conditions (5.1a) and*

$$\bar{U}_t(\bar{S})X \subset X \tag{5.3a}$$

are equivalent. Similarly (5.1b) is equivalent to

$$U_t(S)X \subset X . \tag{5.3b}$$

Conditions (5.3) imply that for each $\xi \in X$, $U_t(S)\xi = U_t(X)\xi$ and $\bar{U}_t(\bar{S})\xi = \bar{U}_t(X)\xi$, where $U_t(X)$ and $\bar{U}_t(X)$ are defined as in (5.2).

The operators $U_t(X) : X \to X$ and $\bar{U}_t(X) : X \to X$ will play a very important role in what follows.

THEOREM 5.1. [16]. *The splitting subspace* X *is Markovian if and only if* $\{U_t(X);\ t \geq 0\}$ $[\{\bar{U}_t(X);\ t \geq 0\}]$ *is a strongly continuous semigroup.*

Hence, we shall call $\{U_t(X);\ t \geq 0\}$ the *forward Markov semigroup* and $\{\bar{U}_t(X);\ t \geq 0\}$ the *backward Markov semigroup* of the Markovian splitting subspace X. The following theorem describes how these shift operators intertwine the Hankel, observability and constructibility operators introduced in Section 4.

THEOREM 5.2. [16]. *Let* $X \sim (S,\bar{S})$ *be a Markovian splitting subspace. Then the following diagrams commute.*

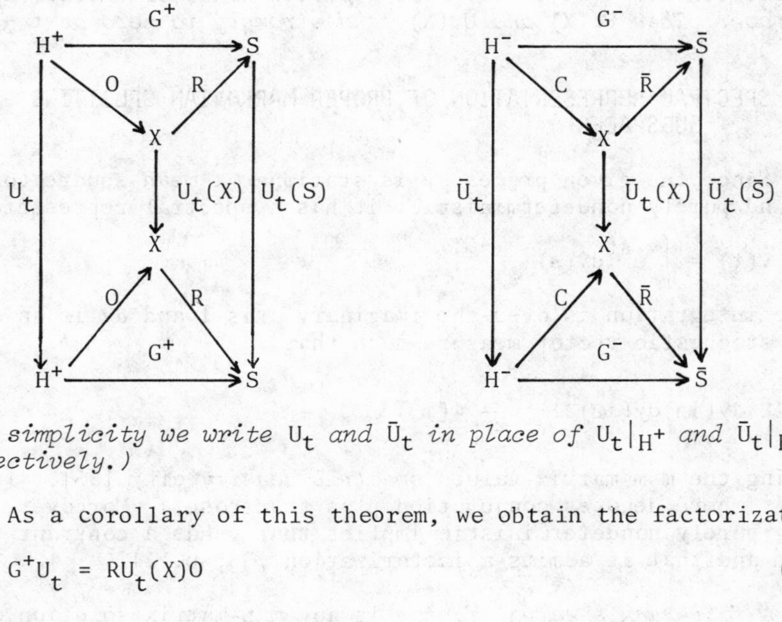

(For simplicity we write U_t *and* $\bar{U}_t$ *in place of* $U_t|_{H^+}$ *and* $\bar{U}_t|_{H^-}$ *respectively.)*

As a corollary of this theorem, we obtain the factorization

$$G^+U_t = RU_t(X)O \tag{5.4a}$$

and its backward counterpart

$$G^-\bar{U}_t = \bar{R}\bar{U}_t(X)C , \tag{5.4b}$$

which should be compared with the corresponding factorizations in deterministic realization theory [32]. Relations (5.4) will be used in Section 9.

A Markovian splitting subspace $X \sim (S,\bar{S})$ is said to be *proper* if both S and $\bar{S}$ are purely nondeterministic. [A subspace Z is *purely nondeterministic* if $\bigcap_{t\in R} U_t Z = 0$.] If X is proper, neither S nor $\bar{S}$ has a doubly invariant subspace, i.e. a subspace which satisfies both conditions (5.1), and, moreover, X is a proper subspace of both S and $\bar{S}$. In fact, if X = S (say), (3.7) implies that $S \subset \bar{S}$, and hence we must have $\bar{S} = H$, which contradicts the purely nondeterministic assumption. Consequently properness of X insures *effective data reduction.*

PROPOSITION 5.2. [14]. *Let y be strictly noncyclic. Then all splitting subspaces* $X \subset H^{\square}$ *(i.e., in particular, the minimal ones) are proper.*

Any proper splitting subspace is also purely nondeterministic [but the opposite is not true; $H^- \sim (H^-, H)$ could serve as a counterexample], and therefore the following result can be applied.

PROPOSITION 5.3. [16]. *Let X be a purely nondeterministic splitting subspace. Then* $U_t(X)$ *and* $\bar{U}_t(X)$ *tend strongly to zero as* $t \to \infty$.

6. SPECTRAL REPRESENTATION OF PROPER MARKOVIAN SPLITTING SUBSPACES

Since the given process y is stationary, mean-square continuous and purely nondeterministic, it has a spectral representation

$$y(t) = \int e^{st} d\hat{y}(s) \ , \tag{6.1}$$

where integration is over the imaginary axis I and $d\hat{y}$ is an orthogonal stochastic vector measure such that

$$E\{d\hat{y}(i\omega)d\hat{y}(i\omega)^*\} = \frac{1}{2\pi}\Phi(i\omega)d\omega \ , \tag{6.2}$$

Φ being the m×m matrix-valued *spectral density* of y [35]. (Asterisk (*) here denotes conjugation plus transpose.) Moreover, y being purely nondeterministic implies that Φ has a constant rank $p \le m$ and that it admits a factorization [35; p.114].

A *full-rank spectral factor* is any m×p-matrix solution of

$$W(s)W(-s) = \Phi(s) \tag{6.3}$$

such that rank W = p. To any such spectral factor we may associate a p-dimensional Wiener process on R

$$u(t) = \int \frac{e^{st} - 1}{s} d\hat{u}(s) \quad ; \quad d\hat{u} = W^{-L} d\hat{y} \ , \tag{6.4}$$

where $E\{d\hat{u}(i\omega)d\hat{u}(i\omega)^*\} = \frac{1}{2\pi} I d\omega$, and W^{-L} is a left inverse of W.

Despite the fact that, in general, W has more than one left inverse, it can be shown [16] that $d\hat{u}$, and hence u, is uniquely defined. Let $\mathcal{U}$ denote the class of all such Wiener processes, and let H(du), $H^-(du)$ and $H^+(du)$ be defined as the closed linear hulls in H of $\{u_k(t);\ t \in T,\ k = 1,2,\dots,p\}$, where T is R, $\{t \le 0\}$ and $\{t \ge 0\}$ respectively. As these notations suggest, we are merely interested in the increments of the processes $u \in \mathcal{U}$, the assumption $u(0) = 0$, contained in (6.4), being for convenience only.

It can be shown [14] that each $u \in \mathcal{U}$ spans all of H, i.e. H(du) = H. Consequently our basic Hilbert space H consists precisely of the random variables

$$\eta = \int_{-\infty}^{\infty} f(-t)\,du(t)\ , \tag{6.5a}$$

where f varies over the space $L_2^p(R)$ of p-dimensional row-vector functions square-integrable on the real line (with respect to the Lebesgue measure). This representation can be transformed to the spectral domain to read

$$\eta = \int \hat{f}(i\omega)\,d\hat{u}(i\omega)\ , \tag{6.5b}$$

where $\omega \mapsto \hat{f}(i\omega)$ is the Fourier-transform of f, defined in the L_2 sense. To conform with formulations prevalent in the systems sciences, formally we use the double-sided Laplace-transform, $\hat{f}$ belonging to the space $L_2^p(I, \frac{1}{2\pi}\,d\omega)$ of p-dimensional row-vector functions which are square-integrable on the imaginary axis I; we shall write $L_2^p(I)$ for short. Also, we shall adopt the convention of writing $\mathcal{F}f$ to denote $\hat{f}$. Then $\mathcal{F}$ is a unitary operator from $L_2^p(R)$ to $L_2^p(I)$.

For each $u \in \mathcal{U}$, (6.5b) defines an isomorphism between H and $L_2^p(I)$. Let $Q_u : H \to L_2^p(I)$ be the mapping

$$\eta \overset{Q_u}{\mapsto} \hat{f}\ . \tag{6.6}$$

Then it follows from the definition of stochastic integral that Q_u is a unitary operator. As is evident from (6.5a), $H^-(du)$ consists precisely of those η for which $f \in L_2^p(0,\infty)$, i.e. for which f vanishes on the negative real line. Likewise, $H^+(du)$ consists of those η for which $f \in L_2^p(-\infty,0)$. Then, defining the *Hardy spaces* $H_2^+ := \mathcal{F}L_2^p(0,\infty)$ and $H_2^- := \mathcal{F}L_2^p(-\infty,0)$, we clearly have $Q_uH^-(du) = H_2^+$ and $Q_uH^+(du) = H_2^-$.

The functions in H_2^+ can be extended to the right complex half-plane and can be seen to be analytic there [37-40]. Likewise, the

H_2^--functions, properly extended, are analytic in the open left half-plane. Therefore we shall say that a full-rank spectral factor is *stable* if all its rows belong to H_2^+ and *strictly unstable* if its rows belong to H_2^-. Let U^+ and U^- be the subclasses of U corresponding to stable and strictly unstable spectral factors respectively.

LEMMA 6.1. [14]. *There is a one-one correspondence between stable full-rank spectral factors* W *(determined modulo multiplication with a constant unitary matrix) and left invariant [i.e., satisfying (5.1a)] and purely nondeterministic subspaces* $S \supset H^-$. *The subspace* S *is related to* W *by*

$$S = H^-(du) \tag{6.7}$$

where $u \in U^+$ *is the Wiener process corresponding to* W.

LEMMA 6.2. [14]. *There is a one-one correspondence between strictly unstable full-rank spectral factors* $\overline{W}$ *and right invariant [i.e., satisfying (5.1b)] and purely nondeterministic subspaces* $\bar{S} \supset H^+$. *The subspace* $\bar{S}$ *is related to* $\overline{W}$ *by*

$$S = H^+(d\bar{u}) \tag{6.8}$$

where $\bar{u} \in U^-$ *is the Wiener process corresponding to* $\overline{W}$.

Since y is a purely nondeterministic process, H^- and H^+ are purely nondeterministic subspaces, and therefore the lemmas above apply. Hence there is $u_- \in U^+$ such that

$$H^-(du_-) = H^- . \tag{6.9}$$

This is the *innovation process* of y. Let W_- be the corresponding spectral factor. In view of (6.1) and the fact that $d\hat{y} = W_- d\hat{u}_-$, $Q_{u_-} ay(t) = e^{i\omega t} aW_-$ for all $t \in R$. But $Q_{u_-} H^- = H_2^+$, and therefore $\overline{sp}\{e^{i\omega t} aW_- \mid t \le 0, a \in R^m\} = H_2^+$, where $\overline{sp}\{\cdot\}$ denotes closed span in H. A function W with this property is called *outer* [37-40]; W_- is the unique outer spectral factor. Likewise, there is a $\bar{u}_+ \in U^-$ such that

$$H^+(d\bar{u}_+) = H^+ , \tag{6.10}$$

called the *backward innovation process* of y. The corresponding spectral factor $\overline{W}_+$ has the outer property with $t \le 0$ and H_2^+ exchanged for $t \ge 0$ and H_2^-. Such a function is called *conjugate outer*; $\overline{W}_+$ is the only spectral factor with this property.

By Lemmas 6.1 and 6.2 there is a one-one correspondence between proper Markovian splitting subspaces $X \sim (S,\bar{S})$ and pairs $(W,\overline{W})$ of full-rank spectral factors with W stable and $\overline{W}$ strictly unstable or, equivalently, pairs $(u,\bar{u})$ of Wiener processes with $u \in U^+$ and

$\bar{u} \in \mathcal{U}^-$. We shall call $W(\bar{W})$ the *forward (backward) spectral factor* of X and $u(\bar{u})$ the *forward (backward) generating process* of X. For each such pair $(W,\bar{W})$, we define a p×p matrix function

$$K = \bar{W}^{-L}W , \tag{6.11}$$

which we call the *structural function* of X. Although the left inverse $\bar{W}^{-L}$ is nonunique, it can be shown [16] that K is uniquely defined. In fact,

$$d\hat{\bar{u}} = Kd\hat{u} . \tag{6.12}$$

The structural function K will play a very important part in what follows. Due to certain similarities with the Lax-Phillips scattering operator [41], we shall alternatively call if the *scattering function*. It is not hard to see that $K(i\omega)$ is a unitary matrix for each $\omega \in R$. Next, we shall show that, in addition, K is bounded and analytic in the open right half-plane. A function with all these properties is called *inner* [37-40].

Now, in view of Theorem 3.1 and Lemmas 6.1 and 6.2, X is a proper Markovian splitting subspace if and only if

$$X = H^-(du) \cap H^+(d\bar{u}) \tag{6.13}$$

for some $u \in \mathcal{U}^+$ and $\bar{u} \in \mathcal{U}^-$ such that $H^-(du)$ and $H^+(d\bar{u})$ intersect perpendicularly. As pointed out above, the pair $(u,\bar{u})$ is unique, being the pair of generating processes of X.

LEMMA 6.3. [14]. *Let* $u \in \mathcal{U}^+$ *and* $\bar{u} \in \mathcal{U}^-$, *and let* W *and* $\bar{W}$ *be the corresponding spectral factors. Then* $S := H^-(du)$ *and* $\bar{S} := H^+(d\bar{u})$ *intersect perpendicularly if and only if* K, *defined by (6.11), is inner.*

The proof of this lemma, which can be found in [14], is based on the vector version of Beurling's Theorem [37-40].

By Corollary 3.1, (6.13) can be written $X = H^-(du) \ominus H^-(d\bar{u})$, the isomorphic image of which (under Q_u) is $Q_u X = \mathcal{H}_2^+ \ominus (\mathcal{H}_2^+ K)$. Consequently

$$X = \int (\mathcal{H}_2^+ K)^{\perp} d\hat{u} , \tag{6.14}$$

where $u \in \mathcal{U}^+$ is the Wiener process corresponding to W, and the superscript $\perp$ denotes orthogonal complement in $\mathcal{H}_2^+$. We collect these observations in the following theorem. Representation (6.14) should be compared with the deterministic solutions of [48,49].

THEOREM 6.1. [15,16]. *The subspace* X *is a proper Markovian splitting subspace if and only if (6.14) holds for some pair* $(W,\overline{W})$ *of full rank spectral factors such that* W *is stable,* $\overline{W}$ *is strictly unstable, and* $K := \overline{W}^{-L}W$ *is inner.*

In particular, if y is strictly noncyclic, all minimal Markovian splitting subspaces are given by Theorem 6.1 (Proposition 5.2). The minimum and maximum lattice elements X_- and X_+ correspond to the pairs $(W_-,\overline{W}_-)$ and $(W_+,\overline{W}_+)$, where W_- and $\overline{W}_+$ are the outer and conjugate outer spectral factors defined above. The corresponding pairs of generating processes are $(u_-,\bar{u}_-)$ and $(u_+,\bar{u}_+)$. The processes u_- and $\bar{u}_+$ are the forward and backward innovation processes respectively, and $\bar{u}_-$ and u_+ can be defined in terms of the junk spaces through the relations $H^-(d\bar{u}_-) = N^-$ and $H^+(du_+) = N^+$. The following result, which is a generalization to the vector case of a result found in [38], provides a test for noncyclicity in terms of the outer and conjugate outer spectral factor.

PROPOSITION 6.1. [14]. *The process* y *is strictly noncyclic if and only if there are inner functions* J_1, J_2, J_3 *and* J_4 *such that*

$$\overline{W}_+^{-L}W_- = J_1J_2^{-1} = J_3^{-1}J_4 . \tag{6.15}$$

In the scalar case $(m=1)$, the structural function K is invariant over the set of (proper) minimal Markovian splitting subspaces, but this is not so in the vector case. This point can be illustrated by a finite dimensional example: The Kronecker structure of a concrete differential-equation representation (of the type derived in Section 9) is uniquely determined by K, but this structure varies with different minimal X [16].

PROPOSITION 6.2. [14]. *Let* X *be a proper Markovian splitting subspace. Then* X *is finite dimensional if and only if its structural function* K *is rational.*

This proposition has two interesting corollaries.

COROLLARY 6.1. [15]. *Suppose that the spectral density* Φ *is rational. Then all splitting subspaces* X *contained in the frame space* $H^\square$ *are finite dimensional, and all minimal splitting subspaces have the same dimension.*

COROLLARY 6.2. [15]. *Suppose that* Φ *is rational. Then* y *is strictly noncyclic.*

For further discussion of the rational case we refer the reader to Section 7 in [15], where differential-equations representations are derived by factorization of the structural function K, using the ideas of [42].

7. SPECTRAL DOMAIN CRITERIA FOR OBSERVABILITY CONSTRUCTIBILITY AND MINIMALITY

Theorem 6.1 provides us with a procedure to find all proper Markovian splitting subspaces: All possible pairs $(W,\overline{W})$ of full-rank spectral factors with W stable, $\overline{W}$ strictly unstable, and $K = \overline{W}^{-L}W$ inner, inserted into (6.14), generate the whole family of such splitting subspaces. But how can we decide whether such a pair will provide an observable, or a constructible or a minimal splitting subspace? We need to translate the geometric criteria of Section 4 into spectral domain language.

To this end, first note that W is a stable full-rank spectral factor if and only if it can be written

$$W = W_-Q \tag{7.1}$$

where Q is an inner function and W_- is the unique outer spectral factor. Similarly, $\overline{W}$ is a strictly unstable full-rank spectral factor if and only if it has the representation

$$\overline{W} = \overline{W}_+\overline{Q}\ , \tag{7.2}$$

where $\overline{Q}$ is conjugate inner (i.e. $\overline{Q}^*$ is inner) and $\overline{W}_+$ is the unique conjugate outer spectral factor [37-40]. The observability condition (4.5) and the constructibility condition (4.6) can now be expressed in terms of the inner functions K, Q and $\overline{Q}^*$.

THEOREM 7.1. [15]. *Let* X *be a proper Markovian splitting subspace, let (7.1) and (7.2) be the corresponding spectral factors, and let* K *be the structural function (6.11). Then* X *is observable if and only if* K *and* $\overline{Q}^*$ *are left coprime and constructible if and only if* K *and* Q *are right coprime.*

Two inner functions are *left (right) coprime* if they have no common left (right) inner factor, except possibly for a constant unitary matrix. Hence, by Theorem 7.1, X is observable if and only if there is no nontrivial cancellation in the factorization $T = \overline{Q}K$. But according to [43,44] (also see [39]), this is the case if and only if

$$\text{cl}(\text{Im}H_T) = (H_2^+K)^\perp\ , \tag{7.3}$$

where $H_T : H_2^- \to H_2^+$ is the *Hankel operator* $H_Tf = P^{H_2^+}fT$ and $\text{Im}H_T$ denotes the range of H_T, cl the closure, and $P^{H_2^+}$ the orthogonal projection on H_2^+. In the same way, X is constructible if and only if there are no nontrivial cancellations in $\overline{T} = QK^*$, which statement is equivalent to

$$\mathrm{cl}(\mathrm{Im}H_{\bar{T}}) = (H_2^- K^*)^{\perp}, \tag{7.4}$$

where $H_{\bar{T}} : H_2^+ \to H_2^-$ is the Hankel operator $H_{\bar{T}} f = P^{H_2^-} f\bar{T}$ and $\perp$ now denotes orthogonal complement in H_2^-.

To clarify the nature of conditions (7.3) and (7.4), we shall take a closer look at the Hankel operators H_T and $H_{\bar{T}}$, along the lines of [14]. To this end, first note that H_T and $H_{\bar{T}}$ are related to the Hankel operators G^+ and G^- through the commutative diagrams

$$\begin{array}{ccc} H^+ & \xrightarrow{G^+} & S \\ \downarrow Q_{\bar{u}_+} & & \downarrow Q_u \\ H_2^- & \xrightarrow{H_T} & H_2^+ \end{array} \qquad \begin{array}{ccc} H^- & \xrightarrow{G^-} & \bar{S} \\ \downarrow Q_{u_-} & & \downarrow Q_{\bar{u}} \\ H_2^+ & \xrightarrow{H_{\bar{T}}} & H_2^- \end{array} \tag{7.5}$$

Then it follows from (4.2) that H_T and $H_{\bar{T}}$ factor according to the commutative diagrams

$$\begin{array}{ccccc} H_2^- & & \xrightarrow{H_T} & & H_2^+ \\ & \searrow^{\tilde{O}} & & \nearrow_{\tilde{R}} & \\ & & \mathcal{X} & & \end{array} \qquad \begin{array}{ccccc} H_2^+ & & \xrightarrow{H_{\bar{T}}} & & H_2^- \\ & \searrow^{\tilde{C}} & & \nearrow_{\tilde{\tilde{R}}} & \\ & & \bar{\mathcal{X}} & & \end{array} \tag{7.6}$$

where $\mathcal{X} := Q_u X$, $\bar{\mathcal{X}} := Q_{\bar{u}} X$, $\tilde{O}f = P^{\mathcal{X}} fT$, $\tilde{C}f = P^{\bar{\mathcal{X}}} f\bar{T}$, and $\tilde{R}$ and $\tilde{\tilde{R}}$ are the insertion maps $\tilde{R}f = f$ and $\tilde{\tilde{R}}f = f$.

Since $\tilde{R}$ is merely an insertion, $\mathrm{Im}H_T = \mathrm{Im}\tilde{O}$, and therefore (7.3) is precisely the observability condition (4.3a), transformed as in (7.5), for, by Theorem 6.1, $\mathcal{X} = (H_2^+ K)^{\perp}$. Likewise it is seen that (7.4) is the same as the constructibility condition (4.3b).

However, [43,44] also contain the stronger result that (7.3) holds with the closure operation removed if and only if K and $\bar{Q}^*$ are *strongly* left coprime, i.e. $\inf_{\mathrm{Re}(s)>0}\{|aK(s)| + |a\bar{Q}^*(s)|\} > 0$ for every $a \in R^p$. The analogous statement holds for (7.4) and K and Q. Hence we have the following strong version of Theorem 7.1.

THEOREM 7.2. *Let* X, K, Q *and* $\bar{Q}$ *be as in Theorem 7.1. Then* X *is exactly observable if and only if* K *and* $\bar{Q}^*$ *are strongly left coprime and exactly constructible if and only if* K *and* Q *are strongly right coprime.*

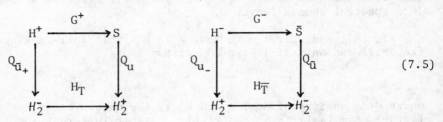

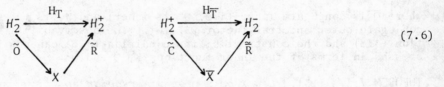

In order to apply conditions (iii) and (iv) of Theorem 4.2, we also need to characterize minimality of S and $\bar{S}$ in the spectral domain. We shall say that a stable (strictly unstable) full-rank spectral factor is *minimal* if it corresponds to a minimal augmented past (future) space via the correspondence of Lemma 6.1 (Lemma 6.2). Now, assume that y is strictly noncyclic. Then the spectral factors W_+ and $\overline{W}_-$ introduced in Section 6 are well-defined. Let Q_+ and $\overline{Q}_-$ be the inner factors in (7.1) and (7.2) respectively corresponding to these spectral factors. Moreover, let K_- and K_+ be the structural functions of X_- and X_+.

THEOREM 7.3. [14]. *Suppose* y *is strictly noncyclic. Let* W *be a stable full-rank spectral factor, and let* $J := W^{-L}W_+$. *Then,* J *is uniquely defined, and the following conditions are equivalent:*

(i) W *is minimal*

(ii) Q *is a left inner divisor of* Q_+, *i.e. there is an inner function* θ *such that* $Q\theta = Q_+$

(iii) J *is inner.*

If one of these conditions holds, J *and* K_+ *are right coprime.*

The fact that (i) and (ii) are equivalent was first proven in [19] for the scalar case. The dual version of Theorem 7.3 goes as follows.

THEOREM 7.4. [14]. *Suppose* y *is strictly noncyclic. Let* $\overline{W}$ *be a strictly unstable full-rank spectral factor, and let* $\overline{J} := \overline{W}^{-L}\overline{W}_-$. *Then,* $\overline{J}$ *is uniquely defined, and the following conditions are equivalent:*

(i) $\overline{W}$ *is minimal*

(ii) $\overline{Q}^*$ *is a right inner divisor of* $\overline{Q}_-^*$

(iii) $\overline{J}^*$ *is inner.*

If one of these conditions holds, $\overline{J}^*$ *and* K_- *are left coprime.*

Note that the coprimeness conditions of Theorems 7.3 and 7.4 are related to T and $\overline{T}$ as follows: $T = K_+J^*$ and $\overline{T} = K_-\overline{J}^*$. This has some significance for state space isomorphism [16].

8. ABSTRACT REALIZATION THEORY IN HARDY SPACE

We shall now translate the results of Section 5 to the spectral domain, thereby laying the ground work for concrete differential-equation type representations to be introduced in the next section.

In view of (6.1), any $\eta \in H$ can be written

$$\eta = \int f d\hat{y} \tag{8.1}$$

for some $f \in L_2^m(I,\Phi d\omega)$. Define the unitary operator Q_y : $H \to L_2^m(I,\Phi d\omega)$ by $Q_y\eta = \hat{f}$, and introduce the spaces $\mathcal{Y}^+ := Q_yH^+$ and $\mathcal{Y}^- := Q_yH^-$, which consist of all linear functionals of the future and past respectively of y. Then $\mathcal{Y}^+ = \overline{\text{sp}}\{ae^{i\omega t} \mid a \in R^m;\ t \geq 0\}$ and $\mathcal{Y}^- = \overline{\text{sp}}\{ae^{i\omega t} \mid a \in R^m;\ t \leq 0\}$. Clearly, for $t \geq 0$, $\mathcal{Y}^+(\mathcal{Y}^-)$ is invariant under multiplication by $e^{i\omega t}(e^{-i\omega t})$. For each $t \geq 0$, let $\Sigma_t : \mathcal{Y}^+ \to \mathcal{Y}^+$ and $\overline{\Sigma}_t : \mathcal{Y}^- \to \mathcal{Y}^-$ be the mappings $\Sigma_t f = e^{i\omega t}f$ and $\overline{\Sigma}_t f = e^{-i\omega t}f$ respectively. Now, for any proper Markovian splitting subspace $X \sim (S,\bar{S})$, let the operators $\hat{G}^+$ and $\hat{G}^-$ be defined in terms of the Hankel operators G^+ and G^- by the commutative diagrams

$$\begin{array}{ccc} H^+ & \xrightarrow{G^+} & S \\ {\scriptstyle Q_y}\downarrow & & \downarrow{\scriptstyle Q_u} \\ \mathcal{Y}^+ & \xrightarrow{\hat{G}^+} & H_2^+ \end{array} \qquad \begin{array}{ccc} H^- & \xrightarrow{G^-} & \bar{S} \\ {\scriptstyle Q_y}\downarrow & & \downarrow{\scriptstyle Q_{\bar{u}}} \\ \mathcal{Y}^- & \xrightarrow{\hat{G}^-} & H_2^- \end{array} \tag{8.2}$$

Then $\hat{G}^+f = P^{H_2^+}fW$ and $\hat{G}^-f = P^{H_2^-}f\overline{W}$, i.e. $\hat{G}^+$ and $\hat{G}^-$ are the Hankel operators corresponding to the input-output maps $f \mapsto fW$ and $f \mapsto f\overline{W}$ respectively.

For any $t \geq 0$ and subspace $Z \subset L_2^p(I)$, define the *compressions* $\Sigma_t(Z) := P^Z e^{i\omega t}|_Z$ and $\overline{\Sigma}_t(Z) := P^Z e^{-i\omega t}|_Z$, where we are slightly misusing notations by letting $e^{i\omega t}$ also denote multiplication by the function $\omega \mapsto e^{i\omega t}$. Then, setting $\mathcal{X} := Q_uX$ and $\overline{\mathcal{X}} := Q_{\bar{u}}X$, it is immediately clear that $Q_uU_t(X)Q_u^{-1} = \Sigma_t(\mathcal{X})$ and $Q_{\bar{u}}\overline{U}_t(X)Q_{\bar{u}}^{-1} = \overline{\Sigma}_t(\overline{\mathcal{X}})$ and that $\Sigma_t(H_2^+)$ and $\overline{\Sigma}_t(H_2^-)$ are analogously related to $U_t(S)$ and $\overline{U}_t(\bar{S})$ respectively. Note that $\{\Sigma_t(\mathcal{X});\ t \geq 0\}$ and $\{\overline{\Sigma}_t(\overline{\mathcal{X}});\ t \geq 0\}$ are strongly continuous semigroups (Theorem 5.1) which tend strongly to zero as $t \to \infty$ (Proposition 5.3). We shall call them the *forward* and the *backward spectral semigroups of* X; $\mathcal{X}$ and $\overline{\mathcal{X}}$ will be called the *forward* and *backward spectral images of* X respectively. Then, by isomorphism, the following proposition is a corollary of Theorem 5.2.

PROPOSITION 8.1. *Let* X *be a proper Markovian splitting subspace with generating processes* $(u,\bar{u})$ *and spectral images* $\mathcal{X} := Q_uX$ *and*

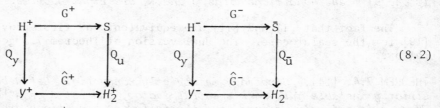

$\overline{X} := Q_{\bar{u}}X$. *Then the diagrams*

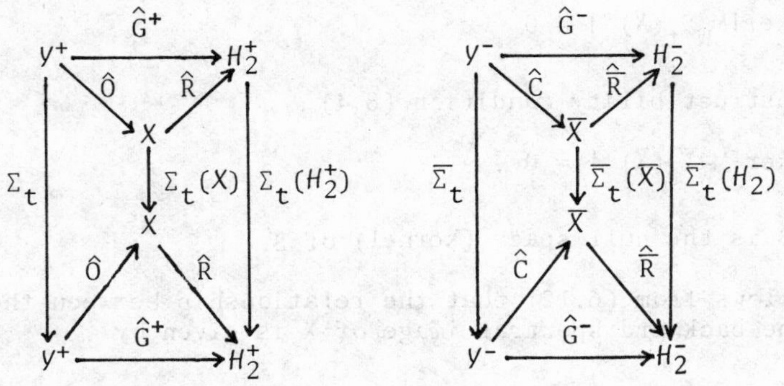

(8.2)

commute, where $\hat{O}f = P^X fW$, $\hat{C}f = P^{\overline{X}} f\overline{W}$ *and* $\hat{R}$ *and* $\hat{\hat{R}}$ *are the insertion maps* $\hat{R}f = f$ *and* $\hat{\hat{R}}f = f$.

By isomorphism, X is observable if and only $\text{cl}\{\text{Im}\hat{O}\} = X$ and constructible if and only if $\text{cl}\{\text{Im}\hat{C}\} = \overline{X}$. Clearly, since $aW \in X$ for all $a \in R^m$ (this is equivalent to $ay(0) \in X$, as can be seen by inserting $d\hat{y} = Wd\hat{u}$ into (6.1)), $\hat{O}e^{i\omega t}a = \Sigma_t(X)M_W a$ for $t \geq 0$, where $M_W : R^m \to X$ is the mapping $M_W a = aW$. Then, since $Y^+ = \overline{\text{sp}}\{ae^{i\omega t} \mid a \in R^m;\ t \geq 0\}$, $\text{cl}\{\text{Im}\hat{O}\} = \text{cl}\{\bigcup_{t\geq 0}\text{Im}[\Sigma_t(X)M_W]\}$. Likewise, noting that $a\overline{W} \in \overline{X}$ for all $a \in R^m$, $\text{cl}\{\text{Im}\hat{C}\} = \text{cl}\{\bigcup_{t\geq 0}\text{Im}[\overline{\Sigma}_t(\overline{X})M_{\overline{W}}]\}$, where $M_{\overline{W}} : R^m \to \overline{X}$ is defined as the mapping $M_{\overline{W}}a = a\overline{W}$. Hence we have proven

PROPOSITION 8.2. *Let* X *be a proper Markovian splitting subspace with spectral images* X *and* $\overline{X}$. *Then* X *is observable if and only if*

$$\text{cl}\{\bigcup_{t\geq 0}\text{Im}[\Sigma_t(X)M_W]\} = X \tag{8.3}$$

and constructible if and only if

$$\text{cl}\{\bigcup_{t\geq 0}\text{Im}[\overline{\Sigma}_t(\overline{X})M_{\overline{W}}]\} = \overline{X}\ . \tag{8.4}$$

Condition (8.3) is precisely the observability condition of a (deterministic) dynamical system with semigroup $\{\Sigma_t(X)^*;\ t \geq 0\}$ and read-out map (*observation operator*) $M_W^* : X \to R^m$ [45; p.211], and this is precisely the role these operators will play in the differential-equation type representations of Section 9. Likewise, (8.4) is the observability condition of a system with semigroup $\{\overline{\Sigma}_t(\overline{X});\ t \geq 0\}$ and observation operation $M_{\overline{W}}^* : \overline{X} \to R^m$. Since any linear operator A from one Hilbert space Y to another Z satisfies the condition

$$\text{cl}\{\text{Im}A\} \oplus \ker A^* = Z \tag{8.5}$$

we may alternatively write the observability condition (8.3) as

$$\bigcap_{t\geq 0} \ker[M_W^*\Sigma_t(X)^*] = 0 \tag{8.6}$$

and the constructibility condition (8.4)

$$\bigcap_{t\geq 0} \ker[M_{\overline{W}}^*\overline{\Sigma}_t(\overline{X})^*] = 0 \ , \tag{8.7}$$

where ker B is the null space (kernel) of B.

It follows from (6.12) that the relationship between the forward and the backward spectral image of X is given by

$$X = \overline{X}K \ , \tag{8.8}$$

where K is the structural function of X. Let $M_K : \overline{X} \to X$ be the mapping $f \mapsto fK$. Then M_K is a unitary operator. The following proposition described the interplay between forward and backward.

PROPOSITION 8.3. [16]. *Let X be a proper Markovian splitting subspace. Then the relations between its forward and backward spectral semigroups and its forward and backward observation operators is given by the commutative diagram*

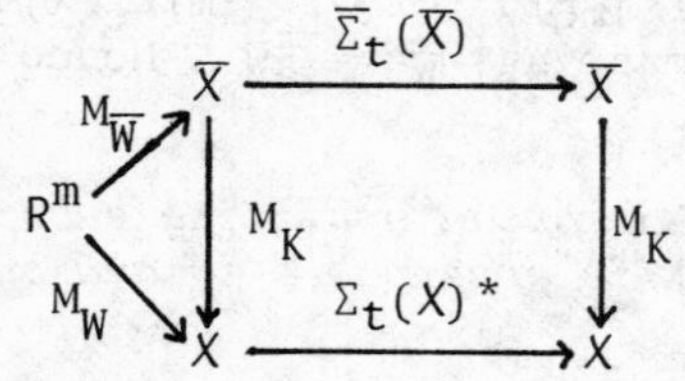

COROLLARY 8.1. *The constructibility condition (8.4) can be written*

$$\text{cl}\{\bigcup_{t\geq 0} \text{Im}[\Sigma_t(X)^* M_W]\} = X \tag{8.9}$$

or equivalently

$$\bigcap_{t\geq 0} \ker[M_W^*\Sigma_t(X)] = 0 \ . \tag{8.10}$$

In the sequel we shall express all relations only in its forward form, referring the reader to Proposition 8.3 for a recipe to obtain the backward counterpart.

This theory also provides us with a natural factorization of the autocorrelation function of y.

PROPOSITION 8.4. [16]. *The m×m matrix function* $\Lambda(t) := E\{y(t)y(0)'\}$ *has the factorizations*

$$\Lambda(t) = \begin{cases} M_W^*\Sigma_t(X)^*M_W & \text{for } t \geq 0 \\ M_W^*\Sigma_{-t}(X)M_W & \text{for } t \leq 0 \; . \end{cases} \tag{8.11}$$

Since M_W^* and $\Sigma_t(X)^*$ are bounded operators, they can be represented by matrices. Let $\{\hat{x}_1, \hat{x}_2, \ldots, \hat{x}_n\}$ be an arbitrary basis in X, where, in general, $n = \infty$. (Such a basis exists, since X is separable.) Then $\Sigma_t(X)\hat{x}_i = \sum_j \alpha_{ij}(t)\hat{x}_j$ for some numbers $\{\alpha_{ij}(t)\}$. Since $\Sigma_t(X)$ is a strongly continuous semigroup, it has a representation $\Sigma_t(X) = e^{At}$, where A is the infinitesimal generator [39,41,45]. Therefore we shall write e^{Ft} to denote the n×n-matrix $\{\alpha_{ij}(t)\}$; i.e.

$$\Sigma_t(X)\hat{x}_i = \sum_{j=1}^{n} (e^{Ft})_{ij}\hat{x}_j \; . \tag{8.12}$$

Note that, unless $n < \infty$, F should not be interpreted as a matrix, since, in general, A is not a bounded operator defined everywhere on X. Similarly, if $\{e_1, e_2, \ldots, e_m\}$ is the canonical (orthonormal) basis in R^m, we define the m×n-matrix H through the relation

$$M_W e_i = \sum_{j=1}^{n} H_{ij}\hat{x}_j \; . \tag{8.13}$$

Applying the operator Q_u^{-1} to this relation, we obtain

$$y_i(0) = \sum_{j=1}^{n} H_{ij}x_j \; , \tag{8.14}$$

where $\{x_1, x_2, \ldots, x_n\} := \{Q_u^{-1}\hat{x}_1, Q_u^{-1}\hat{x}_2, \ldots, Q_u^{-1}\hat{x}_n\}$ is a basis in X. Relation (8.14) illustrates the fact that H is a matrix of the observation operator. Define P to be the n×n covariance matrix with components

$$P_{ij} = E\{x_i x_j\} \; . \tag{8.15}$$

Then it is clear from (6.5b) that

$$P_{ij} = \langle \hat{x}_i, \hat{x}_j \rangle_X \; , \tag{8.16}$$

where $\langle \cdot, \cdot \rangle_X$ denotes inner product in the Hilbert space X.

We are now in a position to formulate the results of this section in matrix form. For $t \geq 0$, Proposition 8.4 yields

$$\Lambda_{ij}(t) = \langle e_i, M_W^* \Sigma_t(X)^* M_W e_j \rangle_{R^m}$$

$$= \langle \Sigma_t(X) M_W e_i, M_W e_j \rangle_X \tag{8.17}$$

But, combining (8.12) and (8.13), we have

$$\Sigma_t(X) M_W e_i = \sum_{j=1}^{n} (He^{Ft})_{ij} \hat{x}_j \ , \tag{8.18}$$

which inserted into (8.17) together with (8.13) yields

$$\Lambda(t) = He^{Ft}PH' \quad \text{for} \quad t \geq 0 \tag{8.19a}$$

after applying (8.16). In the same way, letting $e^{F't}$ denote the transpose of e^{Ft}, we have

$$\Lambda(t) = HPe^{-F't}H' \quad \text{for} \quad t \leq 0 \ . \tag{8.19b}$$

To obtain matrix representations for the backward operators, first note that $\{\hat{\bar{x}}_1, \hat{\bar{x}}_2, \ldots, \hat{\bar{x}}_n\}$, where $\hat{\bar{x}}_i = M_K^{-1}\hat{x}_i$ $(i = 1,2,\ldots,n)$, is a basis in $\bar{X}$. By applying M_K^{-1} to (8.13), we see that $M_{\bar{W}}$ has the matrix H with respect to this basis. However, for the semigroup the situation is a bit more complicated. For $t \leq 0$, define the $n \times n$-matrix $e^{\bar{F}t}$ by

$$\bar{\Sigma}_{-t}(\bar{X})\hat{\bar{x}}_i = \sum_{j=1}^{n} (e^{\bar{F}t})_{ij} \hat{\bar{x}}_j \ , \tag{8.20}$$

which, by Proposition 8.3, can be written

$$\Sigma_t(X)^* \hat{x}_i = \sum_{j=1}^{n} (e^{-\bar{F}t})_{ij} \hat{x}_j \tag{8.21}$$

for all $t \geq 0$. Then, taking (8.16) into account, inserting (8.12) and (8.21) in the defining relation

$$\langle \hat{x}_i, \Sigma_t(X)\hat{x}_j \rangle = \langle \Sigma_t(X)^* \hat{x}_i, \hat{x}_j \rangle$$

yields

$$e^{\bar{F}t} = Pe^{-Ft}P^{-1} \quad \text{for} \quad t \geq 0 \ , \tag{8.22}$$

which should be compared with the corresponding result in [8]. (Also see [46] for the original wide sense result.)

Finally, it is not hard to see that the observability criterion (8.6) has the matrix formulation

$$\bigcap_{t\geq 0} \ker(He^{Ft}) = 0 \tag{8.23}$$

and that the matrix version of the constructibility criterion (8.7) reads

$$\bigcap_{t\leq 0} \ker(He^{\bar{F}t}) = 0 \ . \tag{8.24}$$

9. DYNAMICAL REPRESENTATIONS OF MARKOVIAN SPLITTING SUBSPACES

At this point the abstract realization problem has been solved. All proper Markovian splitting subspaces have been determined, and the corresponding semigroups and observation operators have been characterized. The problem is now to what extent these abstract realizations can be represented by stochastic differential equations.

Let X be a proper Markovian splitting subspace with forward generating process u. We shall say that y *admits a regular (forward) realization with respect to* X if, for some Hilbert space $\mathcal{S}$, there are a strongly continuous semigroup $\{e^{At};\ t \geq 0\}$ on $\mathcal{S}$ and a bounded operator $B : R^p \to \mathcal{S}$ such that

$$X = \left\{ \sum_{k=1}^{p} \int_{-\infty}^{0} \langle c, e^{-A\sigma}Be_k\rangle_{\mathcal{S}} du_k(\sigma) \ \middle|\ c \in \mathcal{S} \right\}, \tag{9.1}$$

where $\langle\cdot,\cdot\rangle_{\mathcal{S}}$ denotes the inner product in $\mathcal{S}$ and $\{e_1,e_2,\ldots,e_p\}$ is the canonical basis in R^p. [Of course, in order that (9.1) be well-defined, the function $t \mapsto \langle c,e^{At}Be_k\rangle$ must belong to $L_2(0,\infty)$ for all $c \in \mathcal{S}$ and $k = 1,2,\ldots,m$, so this is implicitly assumed in the definition.] Then, since $y_i(0) \in X$ for each $i = 1,2,\ldots,m$, there are $c_1,c_2,\ldots,c_m \in \mathcal{S}$ such that

$$y_i(0) = \sum_{k=1}^{p} \int_{-\infty}^{0} \langle c_i, e^{-A\sigma}Be_k\rangle_{\mathcal{S}} du_k(\sigma) \ . \tag{9.2}$$

Define the operator $C : \mathcal{S} \to R^m$ in the following way (since we shall have no further use of the constructibility operator, we shall take the liberty to give a new meaning to C): Let Cf be the m-dimensional vector with components $\langle c_i,f\rangle_{\mathcal{S}}$, $i = 1,2,\ldots,m$. Then C is a bounded operator, and (9.2) can be written

$$y(0) = \int_{-\infty}^{0} Ce^{-A\sigma}Bdu(\sigma) \; . \tag{9.3}$$

We shall call such a representation a *(forward) regular realization of* y *with respect to* X, the word "regular" referring to the boundedness of the operators e^{At}, B and C and "with respect to X" referring to property (9.1). The Hilbert space S will be called the *range space* of the realization, X the *state space*. Applying the group $\{U_t;\ t \in R\}$ of shift operators to (9.3) we obtain

$$y(t) = \int_{-\infty}^{t} Ce^{A(t-\sigma)}Bdu(\sigma) \tag{9.4}$$

for each $t \in R$, as can be seen by a simple change of variables.

Formally we can write (9.4) as

$$\begin{cases} x(t) = \int_{-\infty}^{t} e^{A(t-\sigma)}Bdu(\sigma) \\ y(t) = Cx(t) \; , \end{cases} \tag{9.5}$$

where the *state* x(t) takes values in S. However, in doing so, we shall have to be careful. Unless $\dim S < \infty$, x(t) cannot in general be defined as a Hilbert space-valued random variable in the usual sense [45,47]. For this to be possible the covariance operator must be nuclear, and, as we shall see below, this is usually not the case. However, $\{x(t);\ t \in R\}$ can be interpreted as a *weak* Markov process, by using the theory for weak random variables developed by Balakrishnan [45]; for a discussion of this, see [16]. In any case, we may formally write the realization in the differential-equation form

$$\begin{cases} dx = Axdt + Bdu \\ y = Cx \; , \end{cases} \tag{9.6}$$

to be interpreted either by (9.4) or (9.5), the latter requiring the theory of [45].

From (6.4) and (6.5) we have

$$y(0) = \int_{-\infty}^{0} w(-\sigma)du(\sigma) \; , \tag{9.7}$$

where $w = F^{-1}W$ is the inverse Fourier transform of the spectral factor W. Comparing (9.3) and (9.7) it is seen that, in general, there is no regular realization with respect to an arbitrary X, since w must be continuous for this to be the case. Nevertheless, as we shall see below, the regular realizations are, in a sense, dense so that we can always find one that is an arbitrarily good approximation. Let us assume, for the moment, that y does admit a regular realization with respect to X. Then we may ask the question whether any deterministic realization $w(t) = Ce^{At}B$ which is regular (i.e. B and C are bounded [48,49]) would do the job. To answer this question, note that, for $t \geq 0$,

$$E^S y_i(t) = \sum_{k=1}^{n} \int_{-\infty}^{0} \langle e^{A^*t} c_i, e^{-A\sigma} Be_k \rangle_S \, du_k(\sigma) , \tag{9.8}$$

where e^{A^*t} denotes the adjoint of e^{At}, and that, by the splitting property (2.4), $E^S y_i(t) = E^X y_i(t)$. Therefore, forming the closed span of (9.8) over all $t \geq 0$ and $i = 1,2,\ldots,m$, it is seen that (9.1) is satisfied if $\overline{E^X H^+} = X$ and $\overline{\text{sp}}\{e^{A^*t}c_i;\ t \geq 0,\ i = 1,2,\ldots,m\} = S$. Hence, if X is observable *and* the pair (C,A) is observable in the deterministic sense [45], the answer to our question is "yes." For an arbitrary proper Markovian splitting subspace X, however, things are more complicated. This should be expected, since in general *two* spectral factors, namely W and $\overline{W}$, are needed to characterize X, whereas in the observable case $\overline{W}$ is uniquely determined from W via (4.5). Also it is reasonable to require that the range space is as small as possible. This is achieved by requiring that (A,B) is controllable [45]. Note, however, that the controllability condition has nothing to do with the splitting subspace X, but is desired merely in order to obtain a range space S "of the same size" as X.

In view of the results of Section 8, this suggests that we try to use the spectral image $\mathcal{X}$ of X as a range space. To this end, we first need to introduce the concept of *regular splitting subspace*. Let $\hat{C}$ be the class of all functions in $L_2^p(I)$ having a continuous inverse Fourier transform. (Note that $L_1(I) \cap L_2(I)$ is a proper subset of $\hat{C}$.) Then, in view of (6.5), $\mathcal{X} \subset \hat{C}$ if the regularity condition (9.1) holds, and therefore the operator $V_0 : \mathcal{X} \to R^p$ given by $V_0 f = (F^{-1}f)(0)$ is defined on all of $\mathcal{X}$. We shall say that X is a *regular splitting subspace* if V_0 is a bounded operator.

LEMMA 9.1. [16]. *Let X be a proper Markovian splitting subspace and let $\mathcal{X} := Q_u X$ be its spectral image. Then X is regular if and only if $\mathcal{X} \subset \hat{C}$.*

Clearly all finite dimensional X, if there are any, are regular. The following lemma is a corollary of Proposition 8.1.

LEMMA 9.2. [16]. *Let* $\hat{\xi} \in X$ *and let* $\check{\xi} := F^{-1}\hat{\xi}$. *Then, for each* $t \geq 0$,

$$\check{\xi}(t + \sigma) = (F^{-1}\Sigma_t(X)\hat{\xi})(\sigma) \tag{9.9}$$

for almost all σ *on* $[0,\infty)$.

Now, let X be a regular splitting subspace, and let $a \in R^m$. Then, since $aW \in X \subset \hat{C}$ (Lemma 9.1), $aw(t)$ is continuous and Lemma 9.2 yields

$$aw(t) = V_0\Sigma_t(X)M_W a \ . \tag{9.10}$$

[It can be seen [16] that (9.10) corresponds to the factorization (5.4a).] This leads to the factorization described by the commutative diagram

$$\begin{array}{ccc} R^m & \xrightarrow{\quad M_{w(t)} \quad} & R^p \\ {\scriptstyle M_W}\searrow & & \nearrow{\scriptstyle V_0} \\ X & \xrightarrow{\quad \Sigma_t(X) \quad} & X \end{array} \tag{9.11}$$

where $M_{w(t)}$ denotes multiplication from the right by the *matrix* $w(t)$. Since matrix multiplication from the left is the adjoint of multiplication by the same matrix from the right,

$$w(t)b = (V_0\Sigma_t(X)M_W)^*b \ . \tag{9.12}$$

Now, $\Sigma_t(X)$ and M_W are bounded operators, and regularity of X insures that V_0 is bounded also. Therefore (9.7) and (9.12) yield after applying the shift

$$y(t) = \int_{-\infty}^{t} M_W^*\Sigma_{t-\sigma}(X)^*V_0^*du(\sigma) \ . \tag{9.13}$$

This is a regular realization with respect to X, as is seen from the following theorem.

THEOREM 9.1. [16]. *The process* y *has a regular realization with respect to* X *if and only if* X *is a regular splitting subspace. In this case, the range space can be taken to be* $X := Q_u X$ *and* e^{At}, B *and* C *to be* $\Sigma_t(X)^*$, V_0^* *and* M_W^* *respectively.*

We shall call (9.13) the *standard (forward) realization* corresponding to X. We have already encountered the observation operator

M_W^* and the Markov semigroup $\Sigma_t(X)^*$ in Section 8, where conditions for observability and constructibility were given. It can be shown [16] that the standard realization is *spectrally minimal* [39], and that the pair $(\Sigma_t(X)^*, V_0^*)$ is exactly controllable. The covariance matrix of x(0) is the identity I, and therefore the corresponding representation (9.5) must be interpreted in the weak sense.

Obviously, if y admits a forward regular realization with respect to X, it also admits a backward one. In particular, the analysis above can be carried out in the backward setting also to yield the *standard backward realization*

$$y(t) = \int_t^\infty M_{\bar{W}}^* \bar{\Sigma}_{\sigma-t}(\bar{X})^* \bar{V}_0^* d\bar{u}(\sigma) \tag{9.14}$$

corresponding to X. The relationship between the operators in (9.13) and (9.14) is described by Proposition 8.3 and

$$V_0 = M_K \bar{V}_0 . \tag{9.15}$$

Also it is not hard to see that $\bar{x}(0) := M_K^{-1} x(0)$ is the state corresponding to the standard backward realization.

However, an arbitrary proper Markovian splitting subspace will in general not be regular, since $F^{-1}Q_u X$ may contain functions which are not continuous (Lemma 9.1), and in this case y will have no regular realization with respect to X (Theorem 9.1). (Note that this situation is unique to *continuous-time* processes; in the discrete-time setting all X are regular, since the evaluation operator V_0 is always bounded.) We shall show, however, that in each X there is a *dense* subset of random variables ξ such that each process $\xi(t) := U_t \xi$ admits a regular realization with respect to a subspace of X.

Our basic strategy will be to convolute each function in $F^{-1}Q_u X$ by a scalar L_2 function ϕ. The resulting functions will be continuous [50; p.398] and consequently the techniques described above can be applied. Hence define the subspace $X_\phi \subset H$ to be

$$X_\phi = Q_u^{-1} F[\phi * (F^{-1} Q_u X)] . \tag{9.16}$$

Also let $B_\phi : R^p \to \mathcal{X}$ be the bounded operator defined by $B_\phi b = P^{\mathcal{X}}(F\phi_- b)$, where $\mathcal{X} = Q_u X$ and $\phi_-(t) := \phi(-t)$.

PROPOSITION 9.1. [16]. *Let* $\phi \in L_2(0,\infty)$. *Then* $X_\phi \subset X$. *Moreover,*

$$X_\phi = \left\{ \sum_{k=1}^{p} \int_{-\infty}^{0} \langle \xi, \Sigma_{-\sigma}(X)^* B_\phi e_k \rangle_X du_k(\sigma) \;\middle|\; \xi \in X \right\} \tag{9.17}$$

where $\{e_1, e_2, \ldots, e_p\}$ *is the canonical basis in* R^p.

Here (9.17) should be compared with (9.1). Even if y does not admit a regular realization with respect to X, it can be approximated uniformly closely by another process which has a regular realization with respect to $X_\phi \subset X$.

PROPOSITION 9.2. [16]. *For any* $\varepsilon > 0$, *there is a* $\phi \in L_2(0,\infty)$ *such that, for all* $t \in R$ *and* $k = 1,2,\ldots,m$,

$$\| y_k(t) - y_k(t;\phi) \| < \varepsilon$$

where

$$y(t;\phi) = \int_{-\infty}^{t} M_W^* \Sigma_{t-\sigma}(X)^* B_\phi du(\sigma) . \tag{9.18}$$

Note that the only difference between (9.13) and (9.18) is the B-matrix. For the moment, let B denote either V_0^* or B_ϕ, and let $\{\tilde{e}_1, \tilde{e}_2, \ldots, \tilde{e}_p\}$ be the canonical basis in R^p, $\{e_1, e_2, \ldots, e_m\}$ being reserved for R^m as in Section 8. Let $h(t) := M_W^* \Sigma_t(X)^* B$. Then

$$\begin{aligned} h_{ij}(t) &= \langle e_i, M_W^* \Sigma_t(X)^* B \tilde{e}_j \rangle_{R^m} \\ &= \langle B^* \Sigma_t(X) M_W e_i, \tilde{e}_j \rangle_{R^p} . \end{aligned} \tag{9.19}$$

Now, let n be the (usually infinite) dimension of X, and let $\{\hat{x}_1, \hat{x}_2, \ldots, \hat{x}_n\}$ be the basis in X introduced in Section 8. Let e^{Ft} and H be the matrices of $\Sigma_t(X)^*$ and M_W^* respectively as defined by (8.12) and (8.13), and let G be the corresponding matrix of B, i.e.

$$B^* \hat{x}_i = \sum_{j=1}^{n} G_{ij} \tilde{e}_j . \tag{9.20}$$

Then it follows from (9.19) that $h(t) = He^{Ft}G$, i.e. (9.13) or (9.18) has the matrix representation

$$y(t) = \int_{-\infty}^{t} He^{F(t-\sigma)} B du(\sigma) , \tag{9.21}$$

which, taking due care to properly define the possibly infinite-dimensional state, can be written in differential-equation form.

REFERENCES

[1] B.D.O. Anderson, The inverse problem of stationary covariance generation, *J. Statistical Physics* 1(1969), pp. 133-147.

[2] P. Faurre, Realisations markoviennes de processus stationaires, Research Report No. 13, March 1973, IRIA (LABORIA), Le Chesnay, France.

[3] J.C. Willems, Dissipative dynamical systems, Part II: Linear systems with quadratic supply rates, *Archive for Rational Mechanics and Analysis* 45(1972), pp. 352-393.

[4] H. Akaike, Markovian representation of stochastic processes by canonical variables, *SIAM J. Control* 13(1975), pp. 162-173.

[5] H. Akaike, Stochastic theory of minimal realization, *IEEE Trans.* AC-19 (1974), pp. 667-674.

[6] G. Picci, Stochastic realization of Gaussian processes, *Proc. IEEE* 64(1976), pp. 112-122.

[7] Y.A. Rozanov, On two selected topics connected with stochastic systems theory, *Applied Mathematics & Optimization* 3(1976), pp. 73-80.

[8] A. Lindquist and G. Picci, On the stochastic realization problem, *SIAM J. Control and Optimization* 17(1979), pp. 365-389.

[9] M. Pavon, Stochastic realization and invariant directions of the matrix Riccati equation, *SIAM J. Control and Optimization* 18(1980), pp. 155-180.

[10] G. Ruckebusch, Representations Markoviennes de processus Gaussiens Stationaires, Ph.D. Thesis, Univ. of Paris VI, 1975.

[11] A. Lindquist and G. Picci, On the structure of minimal splitting subspaces in stochastic realization theory, *Proc. 1977 Conf. Decision and Control*, New Orleans, pp. 42-48.

[12] A. Lindquist and G. Picci, A state-space theory for stationary stochastic processes, *Proc. 21st Midwest Symposium on Circuits and Systems*, Ames, Iowa, August, 1978.

[13] A. Lindquist and G. Picci, A Hardy space approach to the stochastic realization problem, *Proc. 1978 Conf. Decision and Control*, San Diego, pp. 933-939.

[14] A. Lindquist and G. Picci, Realization theory for multivariate stationary gaussian processes I: State space construction, *Proc. 4th Intern. Symp. Math. Theory of Networks and Systems*, July 1979, Delft, Holland, pp. 140-148.

[15] A. Lindquist and G. Picci, Realization theory for multivariate gaussian processes II: State space theory revisited and dynamical representations of finite dimensional state spaces, *Proc. 2nd Intern. Conf. on Information Sciences and Systems*, Patras, Greece, July 1979, Reidel Publ. Co., pp. 108-129.

[16] A. Lindquist and G. Picci, Realization theory for multivariate gaussian processes (contains as a proper subset the material in [14,15]; under preparation).

[17] A. Lindquist, G. Picci, and G. Ruckebusch, On splitting subspaces and Markovian representations, *Math. Syst. Theory* 12(May 1979), pp. 271-279.

[18] G. Ruckebusch, Représentations markoviennes de processus gaussiens stationnaires, *C.R. Acad. Sc. Paris*, Series A, 282(1976), pp. 649-651.

[19] G. Ruckebusch, Factorisations minimales de densités spectrales et représentations markoviennes, *Proc. 1re Colloque AFCET-SMF*, Palaiseau, France, 1978.

[20] G. Ruckebusch, A state space approach to the stochastic realization problem, *Proc. 1978 Intern. Symp. Circuits and Systems*, New York.

[21] G. Ruckebusch, A geometric approach to the stochastic realization problem, Center Math. Appl. Ecole Polytechnique, Internal Report 41 (November 1978).

[22] G. Ruckebusch, Theorie geometrique de la representation markovienne, These de doctorat d'etat, Univ. Paris VI, 1980.

[23] J.H. van Schuppen, Stochastic filtering theory: A discussion of concepts, methods and results, Mathematisch Centrum, Amsterdam, Report BW 96/78, December 1978.

[24] C. van Putten and J.H. van Schuppen, On stochastic dynamical systems, Mathematisch Centrum, Amsterdam, Report BW 101/79, 1979.

[25] J.C. Willems and J.H. van Schuppen, Stochastic systems and the problem of state space realization, Mathematisch Centrum, Report BW 108/79, 1979.

[26] M. Gevers and W.R.E. Wouters, An innovations approach to the discrete-time stochastic realization problem, *Journal A* 19(1978), pp. 90-110.

[27] A.E. Frazho, On minimal splitting subspaces and stochastic realizations, (preprint).

[28] P. Faurre, M. Clerget and F. Germain, *Opérateurs Rationnels Positifs*, Dunod 1979.

[29] F. Badawi, A. Lindquist and M. Pavon, A stochastic realization approach to the smoothing problem, *IEEE Trans. Autom. Control* AC-24 (1979), pp. 878-888.

[30] F. Badawi, A. Lindquist and M. Pavon, On the Mayne-Fraser smoothing formula and stochastic realization theory for non-stationary linear stochastic systems, *Proc. 1979 Conf. on Decision and Control*, Fort Lauderdale.

[31] L. Finesso and G. Picci, On the structure of minimal spectral factors, *IEEE Trans. Automatic Control* (to be published).

[32] R.E. Kalman, P.L. Falb and M.A. Arbib, *Topics in Mathematical System Theory*, McGraw-Hill, 1969.

[33] R.W. Brockett, *Finite Dimensional Linear Systems*, Wiley 1970.

[34] J. Neveu, *Processus Aléatoires Gaussiens*, Presses L'Université de Montréal, 1968.

[35] Yu. A. Rozanov, *Stationary Random Processes*, Holden-Day, 1967.

[36] P.R. Halmos, *Finite-Dimensional Vector Spaces*, Springer-Verlag, 1974.

[37] K. Hoffman, *Banach Spaces of Analytic Functions*, Prentice-Hall, 1962.

[38] H. Dym and H.P. McKean, *Gaussian Processes, Function Theory, and the Inverse Spectral Problem*, Academic Press, 1976.

[39] P.A. Fuhrmann, *Linear Operators and Systems in Hilbert Space*, McGraw-Hill (to be published).

[40] H. Helson, *Lecture Notes on Invariant Subspaces*, Academic Press, 1964.

[41] P.D. Lax and R.S. Phillips, *Scattering Theory*, Academic Press, 1967.

[42] G.D. Forney, Jr., Minimal bases of rational vector spaces, with applications to multivariable linear systems, *SIAM J. Control*, 13(May 1975), pp. 453-520.

[43] R.G. Douglas, H.S. Shapiro and A.L. Shields, Cyclic vectors and invariant subspaces for the backward shift operator, *Ann. Inst. Fourier, Grenoble* 20, 1(1970), pp. 37-76.

[44] R.G. Douglas and J.W. Helton, Inner dilatations of analytic matrix functions and Darlington synthesis, *Acta Sci. Math.* 34(1973).

[45] A.V. Balakrishnan, *Applied Functional Analysis*, Springer-Verlag, 1976.

[46] L. Ljung and T. Kailath, Backwards markovian models for second-order stochastic processes, *IEEE Trans.* IT-22 (1976), pp. 488-491.

[47] I.A. Ibragimov and Y.A. Rozanov, *Gaussian Random Processes*, Springer-Verlag, 1978.

[48] J.S. Baras and R.W. Brockett, H^2 functions and infinite dimensional realization theory, *SIAM J. Control* 13(1975), pp. 221-241.

[49] J.S. Baras and P. Dewilde, Invariant subspace methods in linear multivariable-distributed systems and lumped-distributed network synthesis, *Proc. IEEE* 64(1976), pp. 160-178.

[50] E. Hewitt and K. Stromberg, *Real and Abstract Analysis*, Springer-Verlag, 1965.

ERGODIC THEORY OF LINEAR PARAMETER-EXCITED SYSTEMS

Volker Wihstutz

Forschungsschwerpunkt
Dynamische Systeme
Universität Bremen
Fachbereich Mathematik
Postfach 330 440
D-2800 Bremen 33
F R G

Abstract

Using Oseledec' multiplicative ergodic theorem we prove the existence of a fundamental system of solutions of $\dot{x} = A(t)x$, $A(\cdot)$ stationary, allowing a Floquet type decomposition into a stationary angular part and a growing radial one. For triangular matrices the decomposition is explicitely calculated as a functional of A.

1. THE PROBLEM

We consider the linear vector differential equation

(1) $\dot{x} = A(t)x$

where $A(\cdot)$ is "real noise", i.e. an n×n-matrix valued stochastically right continuous strict sense stationary (and ergodic) process with $E|A(o)|<\infty$. How does the solution $x_t = \Phi(t)x_0$, $\Phi(t)$ fundamental matrix, reflect the stationarity of the noise? In the deterministically degenerated case with constant $A(t,\omega) = A$ the fundamental matrix $\Phi(t) = \mathrm{Id}\exp(tA)$ is decomposed into a constant, thus stationary matrix Id and a part describing the growth; more generally, for deterministic periodic matrices $A(t) = A(t+T)$ by Floquet theory we have $\Phi(t) = P(t)\exp(tR)$ with differentiable periodic $P(t) = P(t+T)$ and constant R.

M. Hazewinkel and J. C. Willems (eds.), Stochastic Systems: The Mathematics of Filtering and Identification and Applications, 205–218.

Our stochastic problem is to find a fundamental system $\Phi(t)$ with an analogous decomposition such that

(2) $\Phi(t) = S(t)\exp(\Lambda t+o(t))$

where $S(\cdot)$ is a differentiable stationary matrix valued process and Λ is constant with a spectrum consisting of all possible orders of growth defined by the Lyapunov numbers

$$\lambda(x) = \overline{\lim_{t\to\infty}} \frac{1}{t} \log|\Phi(t)x|, \Phi(t) \text{ fundamental matrix with } \Phi(0) = Id$$

2. ORDER OF GROWTH

Let $(\Omega,\mathcal{F},P)$ be the canonical probability space of the stationary noise $A(\cdot)$, $(\Theta_t)_{t\in\mathbb{R}}$ the associated group of measure preserving shifts. Then the fundamental matrix $\Phi(t)$ is an A-cocycle, i.e. it is A-measurable and

(3) $\Phi(t+s,\omega) = \Phi(t,\Theta_s\omega)\Phi(s,\omega)$.

Moreover, since $A(t) \in L_1(\Omega,P)$, we have (see OSELEDEC [5])

(4) $\sup_{-1\leq s\leq 1} \log^+ ||\Phi(s)|| \in L_1(\Omega,P)$.

Theorem 1 (Multiplicative ergodic theorem of Oseledec) Under the above conditions we have P-almost surely:

(5) $\mathbb{R}^n = \bigoplus_{i=1}^{r(\omega)} E_i(\omega)$, $k_i(\omega) = \dim E_i(\omega)$,

with the properties:

(i) there are $r(\omega)$ Lyapunov numbers $\lambda_1(\omega)<\ldots \lambda_r(\omega)$ such that

$$\lambda_i(\omega) = \lambda_i(\omega,x) = \lim_{t\to\infty} \frac{1}{t} \log|\Phi(t,\omega)x|$$

uniformly for all $x \in E_i(\omega)$, $i = 1, \ldots, r(\omega)$;

(ii) the random variables r, k_i, λ_i are invariant with respect to (Θ_t) and $E_i(\Theta_t\omega) = \Phi(t)(\omega)E_i(\omega)$; thus if A is stationary and ergodic r, k_i and λ_i are constants.

For proofs see OSELEDEC [5], RAGHUNATHAN [6] and RUELLE [7].

In the following we always assume ergodicity of A . So if $x_0(\omega) \in E_i(\omega)$, by projection onto the unit sphere S^{n-1} we have $x_t = w_t|x_t| = w_t \exp(t\lambda_i+o(t))$, where

$w_t = x_t/|x_t|$ satisfies the angular differential equation associated with (1):

$$\dot{w}_t = (A(t) - q(A(t),w_t)\,Id)w_t \qquad (6)$$

$$\text{with } q(A(t),w_t) = w_t'\,\frac{1}{2}\,(A(t) + A'(t))w_t$$

where the prime denotes transposition.

3. FLOQUET TYPE DECOMPOSITION

In order to find the stationary part, we are looking for stationary angles $w_t \in E_i(\omega) \cap S^{n-1}$ using a generalization of KHASMINSKII's [4] criterion for stationarity.

Theorem 2

Let $\xi(\cdot)$ be a stochastically right continuous stationary process with state space $(\mathbb{R}^m, \mathcal{B}^m)$, and C a ξ-cocycle acting as measurable operator on $(\mathbb{R}^n, \mathcal{B}^n)$. Then, on a suitable probability space $(\tilde{\Omega}, \tilde{\mathcal{F}}, \tilde{P})$, there exists an orbit $x_t = C(t)x_0$ which is ξ-stationary (i.e.: $(x_t, \xi(t))$ is stationary) if and only if there exists an orbit $y_t = C(t)y_0$ which satisfies the tightness condition

$$\frac{1}{t}\int_0^t P(|y(s)|>r)\,ds \to 0 \quad \text{for} \quad r \to \infty$$

uniformly in t $(t \geq t_o$ or $t \leq -t_o)$.

Proof

The proof is somewhat shorter than that of KHASMINSKII [4], generalizing his arguments to non-differentiable operators. The necessity of the condition is obvious, take $y_t = x_t$.
Let y_t satisfy the tightness condition, let τ_n be a sequence of r.v.'s uniformly distributed on $[0,n]$ such that the τ_n's, y_0 and $\xi(0)$ are independent. Put

$$y_n(t,\omega) := y(t+\tau_n(\omega),\omega),\ \xi_n(t,\omega) := \xi(t+\tau_n(\omega),\omega)\ .$$

Then, by the assumptions on ξ and the tightness condition for y_t, the process $(y_n(o),\xi_n(t))$ is uniformly stochastically continuous and bounded (see KHASMINSKII [4] and ARNOLD/WIHSTUTZ [1]). Thus we can apply SKOROKHOD's convergence theorem [8], according to which it is possible to construct a probability space $(\tilde{\Omega}, \tilde{\mathcal{F}}, \tilde{P})$ and a subsequence $(\tilde{y}_{n_k}(o), \tilde{\xi}_{n_k}(t))$ with the same finite dimensional distributions as $(y_{n_k}(o), \xi_{n_k}(t))$, converging in probability to $(\tilde{x}_o, \tilde{\xi}(\cdot))$, where $\tilde{\xi}(\cdot)$ and $\xi(\cdot)$ define the same measure. To simplify notation we use only one subscript for the subsequence. By the ξ-cocycle property of C

$$C(s+t), \tilde{\xi}_k(\cdot))\tilde{y}_k(o) = C(s+t+\tilde{\tau}_k, \tilde{\xi}(\cdot))\tilde{y}(o) = \tilde{y}_k(s+t).$$

Now, as a consequence of the stochastic convergence of $(\tilde{y}_k(o), \tilde{\xi}_k(t))$ to $(\tilde{x}_o, \tilde{\xi}(t))$, the stationarity of $(\tilde{x}_t, \tilde{\xi}_t)$, $\tilde{x}_t := C(t, \tilde{\xi}(\cdot))\tilde{x}_o$ follows from

$$\begin{aligned}
& Ef(\tilde{x}(s_1+t), \ldots, \tilde{x}(s_r+t), \tilde{\xi}(s_1+t), \ldots, \tilde{\xi}(s_r+t)) \\
= \; & Ef(C(s_1+t, \tilde{\xi}(\cdot)\tilde{x}_o, \; . \; . \; . \qquad , \tilde{\xi}(s_r+t)) \\
= \; & \lim_{k\to\infty} Ef(C(s_1+t, \tilde{\xi}_k(\cdot))\tilde{y}_k(o), \; . \; . \; . \; , \tilde{\xi}_k(s_r+t)) \\
= \; & \lim_{k\to\infty} Ef(\tilde{y}_k(s_1+t) \qquad , \; . \; . \; . \; , \tilde{\xi}_k(s_r+t)) \\
= \; & \lim_{k\to\infty} \frac{1}{k}\int_o^k Ef(\tilde{y}(s_1+t+u) \qquad , \; . \; . \; . \; , \tilde{\xi}(s_r+t+u)) \, du \\
= \; & \lim_{k\to\infty} \frac{1}{k}\int_o^k Ef(\tilde{y}(s_1+v) \qquad , \; . \; . \; . \; , \tilde{\xi}(s_r+v)) \, dv \\
= \; & Ef(\tilde{x}(s_1), \ldots, \tilde{x}(s_r), \tilde{\xi}(s_1), \ldots, \tilde{\xi}(s_r)),
\end{aligned}$$

where f is bounded and continuous, $r \in \mathbb{N}$, $s_1 \ldots, s_r \in \mathbb{R}$.

Theorem 3 (Floquet type decomposition)

Let $A(t)$ be stochastically right continuous, stationary and ergodic with $E|A_o|<\infty$. Then the stochastic differential equation (1) has a fundamental system of solutions

(7) $\quad \Psi(t) = \mathbf{S}(t) \; \mathbf{exp} \; (t\Lambda + o(t))$

with constant diagonal matrix Λ, consisting of the Lyapunov numbers λ_i of (1) each repeated k_i times and a differentiable angular part $S(t) = (w_t^1, \ldots, w_t^n)$, $w_t^i = x_t^i/|x_t^i|$, such that to each Lyapunov number λ_i there is at least one solution w_t^i of (6) which is A-stationary.

Proof

1. The application of theorem 1 to the fundamental matrix $\Phi(\cdot)$ of (1) yields the state space decomposition (5).

By utilizing results of CASTAING/VALADIER [3] we can choose in each subspace $E_i(A(\cdot))$, $i=1,\dots,r$, an A-measurable selection of k_i orthonormal vectors; collect them as matrix $\Sigma_o(A(\cdot))$. Let $\Sigma_t(A(\cdot)) := \Sigma_o(\Theta_t A)$.

$$\hat{\Phi}(t,A(\cdot)) := \Sigma_t^{-1}(A(\cdot))\Phi(t,A(\cdot))\Sigma_o(A(\cdot)) = \begin{pmatrix} \boxed{\Phi_1(t,A(\cdot))} & & o \\ & \ddots & \\ o & & \boxed{\Phi_r(t,A(\cdot))} \end{pmatrix}$$

reduces the cocycle $\Phi(\cdot)$ to a "diagonal form" of $k_i \times k_i$-matrices Φ_i with invariant subspaces $\hat{E}_i$ which are fixed portions of $\mathbb{R}^n$, associated with the same Lyapunov number λ_i as $E_i(t,\omega)$ (Lyapunov homology, see OSELEDEC [5]). In general, $\hat{\Phi}(t)$ is not differentiable anymore.

2. The continuous operator $\hat{C}(t)(x) := \hat{\Phi}(t)x/||\hat{\Phi}(t)x||$ on $\mathbb{R}^n$ is an A-cocycle, since $\hat{C}(t,\Theta_s A(\cdot))C(s,A(\cdot))x =$

$$= \frac{\hat{\Phi}(t,\Theta_s A(\cdot))(\hat{\Phi}(s,A(\cdot))x/||\hat{\Phi}(s,A(\cdot))x||)}{||\hat{\Phi}(t,\Theta_s A(\cdot))(\hat{\Phi}(s,A(\cdot))x/||\hat{\Phi}(s,A(\cdot))x||)||}$$

$$= \frac{\hat{\Phi}(t,\Theta_s A(\cdot))\hat{\Phi}(s,A(\cdot))x}{||\Phi(t,\Theta_s A(\cdot))\hat{\Phi}(s,A(\cdot))x||}$$

$$= \hat{\Phi}(t+s,A(\cdot))x/||\hat{\Phi}(t+s,A(\cdot))x||$$

$$= \hat{C}(t+s,A(\cdot))(x)$$

and obviously $w_t = C(t)w_o$, $w_o \in S^{n-1}$, satisfies the tightness condition of theorem 2 . Thus, there is a probability space $(\tilde{\Omega},\tilde{\mathfrak{F}},\tilde{P})$, a $(\tilde{w}_k(o),\tilde{A}_k(\cdot))$, stochastically converging to $(\hat{w}_o,\tilde{A}(\cdot))$, such that the finite-dimensional distributions of these processes equal those of $(w_k(o),A_k(\cdot))$ and $\hat{w}_t(\tilde{\omega}) = \hat{C}(t,\tilde{A}(\cdot))\hat{w}_o(\omega)$ is $\tilde{A}$-stationary. If all components of $w_o(\omega)$ are non-zero, then by construction $\hat{w}_t(\omega)$ has non vanishing stationary components in all compact sets $\hat{E}_i \cap S^{n-1}$ $(i=1,\dots,r)$.

3. We remark that $\Phi(t) = \Sigma_t(\tilde{A}(\cdot))\hat{\Phi}(t, \tilde{A}(\cdot))\Sigma_0^{-1}(\tilde{A}(\cdot))$ is an $\tilde{A}$-cocycle with Oseledec subspaces $E_i(\tilde{A}(\cdot))$, dim E_i=I Lyapunov numbers λ_i and the stationary $\tilde{A}$-measurable selection $\Sigma_t(\tilde{A}(\cdot))$. The construction of a stationary angle $w_t(\tilde{\omega}) \in E_i(t, \tilde{A}(\cdot) \cap S^{n-1}$ by the following lemma completes the proof.

Lemma

Let $\hat{\Phi}(t) = \Sigma_t^{-1}\Phi(t)\Sigma_0$ be a transformation of $\Phi(t)$ with stationary A-measurable Σ_t. If $\hat{w}_t = \hat{\Phi}(t)\hat{w}_0/||\hat{\Phi}(t)\hat{w}_0||$ is A-stationary, then $w_t = \Phi(t)(\Sigma_0\hat{w}_0)/||\Phi(t)(\Sigma_0\hat{w}_0)||$ inherits this property.

Proof

If $\hat{w}_t$ is A-stationary, then so is $\Sigma_t(A(\cdot))\hat{w}_t = \Sigma_t\hat{\Phi}(t)\hat{w}_0/||\hat{\Phi}(t)\hat{w}_0||$. Thus the projection onto S^{n-1}

$$\frac{\Sigma_t\hat{\Phi}(t)\hat{w}_0/||\hat{\Phi}(t)\hat{w}_0||}{||(\Sigma_t\hat{\Phi}(t)\hat{w}_0/||\hat{\Phi}(t)\hat{w}_0||)||} = \frac{\Sigma_t\hat{\Phi}(t)\hat{w}_0}{||\Sigma_t\hat{\Phi}(t)\hat{w}_0||} = \frac{\Phi(t)(\Sigma_0\hat{w}_0)}{||\Phi(t)(\Sigma_0\hat{w}_0)||}$$

has this property, too.

Remarks 1. In general, the stationary angle $w_t(A(\cdot)) = C(t,A(\cdot))\Sigma_0(A(\cdot))\hat{w}_0(A(\cdot))$ is not A -measurable. (We drop the ~.)
2. If in the decomposition (7) Λ contains n different eigenvalues $\lambda_1, \ldots, \lambda_r$, $r = n$, the whole matrix $S(\cdot)$ is A-stationary. For $r < n$ several cases are possible: S may contain any number of stationary angles between r and n. Thus in general the stationarity properties of S cannot be improved.

Examples

a) For stationary, ergodic

$$A(t) = \begin{pmatrix} 0 & \beta(t) \\ 0 & 0 \end{pmatrix}, \quad b(t) = \int_0^t \beta(s)\, ds,$$

we have

$$\Phi_t = \begin{pmatrix} 1 & b(t) \\ 0 & 1 \end{pmatrix} = \begin{pmatrix} 1 & b(t)/\sqrt{1+b(t)^2} \\ 0 & 1/\sqrt{1+b(t)^2} \end{pmatrix} \exp\begin{pmatrix} 0 & 0 \\ 0 & \frac{1}{t}\log\sqrt{1+b(t)^2} \end{pmatrix} t\ .$$

There is one Lyapunov number equal to 0 if $b(\cdot)$ grows with order 0 . But we have two stationary solutions even of the original linear differential equation (1) if there is a random variable b_o such that $(b_o + b(t))$ is stationary, while there is only one stationary angle $w_t^1 \equiv \binom{1}{0}$ solving (6), any other converging to $\pm w_t^1$, if for instance $E\beta(t) \neq 0$.

b) For stationary, ergodic $A(t) = \begin{pmatrix} 0 & \alpha(t) \\ -\alpha(t) & 0 \end{pmatrix}$ we have

$$\Phi(t) = \begin{pmatrix} \cos a(t) & \sin a(t) \\ -\sin a(t) & \cos a(t) \end{pmatrix}, \quad a(t) := \int_0^t \alpha(s)\, ds ,$$

where again the Lyapunov number is 0 . Theorem 2 yields at least one stationary angle $w^1(t)$. But then $w^2(t) := Uw^1(t)$ where U is an arbitrary orthogonal (2×2) matrix $\neq \pm \mathrm{Id}$ is a second A-stationary solution of (6) which is linearly independent of $w^1(t)$.

4. SYSTEMS WITH TRIANGULAR MATRICES

Triangular matrices of the form

$$A(t) = \begin{pmatrix} a_{11}(t) & 0 & \cdots & 0 \\ a_{21}(t) & a_{22}(t) & 0 & \cdots \\ \vdots & \vdots & \ddots & 0 \\ a_{n1}(t) & a_{n2}(t) & \cdots & a_{nn}(t) \end{pmatrix} \text{ stationary, ergodic,}$$

allow the direct computation of the Lyapunov numbers and the stationary angles associated with each of them.

Put

$$R_k(t) := \frac{1}{t}\int_0^t a_{kk}(s)\, ds, \quad \lambda_k := E\, a_{kk}(0) = \lim_{t\to\infty} R_k(t), \quad 1 \le k \le n ,$$

using Birkhoff's ergodic theorem. Then

$$x_1(t) = e^{tR_1(t)}\, c_1,$$

$$x_j(t) = e^{tR_j(t)} \int_0^t e^{-sR_j(s)} (a_{j1}x_1(s) + \ldots + a_{j,j-1}x_{j-1}(s))\, ds + e^{tR_j(t)}\, c_j$$

with constants c_j, $j = 1,\ldots,n$. In order to compute the kth solution $x^{(k)}$ of the fundamental system $S(t)$ choose the constant vector $(0 , \ldots , 0 , 1 , c_{k+1,k} ,\ldots, c_{n,k})$. Then

$x_i^{(k)}(t) \equiv 0$ for $i < k$, $x_{k,k}(t) = 1\cdot\exp[tR_k(t)]$.
Now assume

$$x_k^{(k)}(t) = y_k^{(k)}(t)\exp[tR_k(t)]$$

$$\vdots$$

$$x_{j-1}^{(k)}(t) = y_{j-1}^{(k)}(t)\exp[tR_k(t)] \quad , \quad (j > k+1) ,$$

where $(y_k^{(k)}, \ldots, y_{j-1}^{(k)}, A_t)$ is A-measurable and stationary, thus $(u_{j-1,k}, \lambda_j, \lambda_k)$ is A-measurable and stationary with

$$u_{j-1,k}(t) := a_{jk}(t) + y_k^{(k)}(t)a_{j,k+1}(t) + \ldots + y_{j-1}^{(k)}(t)a_{j,j-1}(t).$$

Then we have

$$x_j^{(k)}(t) = e^{tR_j(t)} \int_0^t \exp((R_j(s)-R_k(s))(-s))u_{j-1,k}(s)\, ds + e^{tR_j(t)} c_{j,k}.$$

Now we make the choice of the constants $c_{j,k}$ depending on whether λ_j is greater or smaller than λ_k.
If $\lambda_j > \lambda_k$,

$$\text{(8a)} \quad c_{j,k} = -\int_0^\infty e^{[R_j(s)-R_k(s)](-s)} u_{j-1,k}(s)\, ds$$

exists and we get

$$\text{(8b)} \quad x_j^{(k)}(t) = \left(-\int_t^\infty \exp([R_j(t)-R_k(t)]t - [R_j(s)-R_k(s)]s) \times u_{j-1,k}(s)\right) ds\, e^{tR_k(t)}$$

$$=: \quad y_j^{(k)}(t)\exp[tR_k(t)] ,$$

where $y_j^{(k)}(t)$ is $(y_k^{(k)}, \ldots, y_{j-1}^{(k)}, A_t)$- stationary
The stationarity follows from the fact that $x_j^{(k)}(t)$ is a functional of the stationary process

$(u_{j-1,k}, a_{jj}, a_{kk})$ which is shift compatible, i.e.

$$x_j^{(k)}(t;u(\cdot+h), a_{jj}(\cdot+h), a_{kk}(\cdot+h)) = x_j^{(k)}(t+h; u(\cdot), a_{jj}(\cdot), a_{kk}(\cdot))$$

as one easily checks.

If $\lambda_j < \lambda_k$ we can define by similar arguments

$$\text{(9a)} \quad c_{j,k} := \int_{-\infty}^{o} e^{[R_j(s)-R_k(s)](-s)} u_{j-1,k}(s)\, ds$$

in order to get

$$\text{(9b)} \quad x_j^{(k)}(t) = \left(\int_{-\infty}^{t} \exp([R_j(t)-R_k(t)]t - [R_j(s)-R_k(s)]s \times u_{j-1,k}(s)\, ds \right) e^{tR_k(t)} =: y_j^{(k)}(t) \exp[tR_k(t)] .$$

If $\lambda_j = \lambda_k$ there is in general no component $x_j^{(k)}(t)$ growing with order λ_k and having a stationary part. For instance if $a_{jj}(t) = a_{kk}(t) = \text{const}$ the solution $x_j^{(k)}(t) = \exp[t\lambda_k]\cdot(c_{j,k} + \int_o^t u_{j-1}(s)\, ds)$ has a stationary angular part if and only if $u_{j-1,k}(t)$ is the derivative of a stationary process.

If all n Lyapunov numbers λ_i are different, by choice of the constants

$$\begin{pmatrix} 1 & 0 & 0 & \cdot & \cdot & \cdot & \cdot & 0 \\ c_{2,1} & \cdot & & & & & & \cdot \\ \vdots & & \cdot & & & & & \cdot \\ c_{k,1} & & & 1 & & & & \cdot \\ \vdots & & & c_{k+1,k} & \cdot & & & 0 \\ c_{n,1} & \cdot & \cdot & c_{n,k} & \cdot & \cdot & & 1 \end{pmatrix}$$

with $c_{j,k}$ $(j > k)$ defined in (8a) and (9a), resp., we obtain the desired Floquet type decomposition $\Phi_t = S(t) \exp(t\Lambda + o(t))$ with $\Lambda = \text{diag}(\lambda_1, \ldots, \lambda_n)$ and the A-measurable stationary matrix of angles

$$S(t) = \begin{pmatrix} 1 & 0 & \cdots & \cdots & 0 \\ y_2^{(1)} & \ddots & 0 & & \vdots \\ \vdots & & 1 & & \vdots \\ \vdots & & y_{k+1}^{(k)} & & 0 \\ y_n^{(1)} & \cdots & y_n^{(k)} & & 1 \end{pmatrix} \cdot \operatorname{diag}(\|y^{(j)}\|^{-1})_{1 \leq j \geq n}$$

where $$y^{(j)} := \begin{pmatrix} 0 \\ \vdots \\ 0 \\ 1 \\ y_{j+1}^{(j)} \\ \vdots \\ y_n^{(j)} \end{pmatrix}$$

satisfying (6). Here $y_j(t)$ $(j>k)$ is of form (8b) if $\lambda_j>\lambda_k$ or of form (9b) if $\lambda_j<\lambda_k$. In the first case the component $x_j^{(k)}(t)$ is anticipating and depends on the whole future of the process $A(\cdot)$, while in the latter case $x_j^{(k)}(t)$ is non-anticipating depending only on the past of $A(\cdot)$.

If some of the λ's are equal each equivalence class of indices, two indices i and j, $1 \leq i,j \leq n$ defined to be equivalent iff the corresponding λ_i and λ_j are equal, has a maximal element with index i_p. This index corresponds to a solution with growth λ_{i_p} and a stationary angle. In general this is the only one.

5. HIERARCHY OF SOLUTIONS. DOMAIN OF ATTRACTION

As usual we assume $\lambda_1 < \lambda_2 < \ldots < \lambda_r$, $r \leq n$,

and define a flag of linear subspaces by $V_i(\omega) := \bigoplus_{j=1}^{i} E_j($

$K_i := \sum_{j=1}^{i} \dim E_j(\omega) = \dim V_i(\omega)$, $i=1,2,\ldots,r$. Then we

have a hierarchy of solutions with respect to their domain of attraction or "the law of large numbers".

Theorem 4

Let $w_t^{(j)}$, $1 \le j \le n$ denote the angles of the Floquet type decomposition in Theorem 3 and let x_t be any solution of (1) with associated angle $w_t = x_t/||x_t||$.

Representing w_o by $w_o = \sum_{j=1}^{n} c_j w_o^{(j)}$ let

$H := \max \{j \mid c_j \neq 0\}$.

If $K_{i-1} < H \le K_i$, i.e. $w_o \in V_i \setminus V_{i-1}$ then

(i) $||x_t|| = \exp(\lambda_i t + o(t))$

(ii) w_t is attracted by the angle w_t^* of the solution x_t^* starting in

$$w_o^* = \sum_{j=K_{i-1}+1}^{H} c_j w_o^{(j)} \in E_i,$$

i.e. $\lim_{t\to\infty} (w_t - w_t^*) = 0.$

Proof

Let $\lambda(j)$ be the Lyapunov number of the solution $x_t^{(j)}$ starting in $w_o^{(j)}$, $1 \le j \le n$. (Caution: $\lambda(j) \neq \lambda_j$ in general!). We have $\lambda(H) = \lambda_i$.

Since $w_o = x_o/||x_o||$

and
$$\begin{aligned} w_t &= x_t/||x_t|| = \Phi(t)x_o/||\Phi(t)x_o|| \\ &= (\Phi(t)x_o/||x_o||) \quad ||\Phi(t)x_o/||x_o|||| \\ &= \Phi(t)w_o/||\Phi(t)w_o|| \end{aligned}$$

and $||\Phi(t)w_o^{(j)}|| = \exp(t\lambda(j)+o(t))$ we have

$$w_t = \sum_{j=1}^{H} c_j\Phi(t)w_o^{(j)} \Big/ ||\sum_{k=1}^{H} c_j\Phi(t)w_o^{(j)}||,$$

$$w_t^* = \sum_{j=K_{i-1}+1}^{H} c_j\Phi(t)w_o^{(j)} \Big/ ||\sum_{j=K_{i-1}+1}^{H} c_j\Phi(t)w_o^{(j)}||,$$

thus

$$(w_t - w_t^*) = \sum_{j=1}^{K_{i-1}} c_j\Phi(t)w_o^{(j)} \Big/ ||\sum_{j=1}^{H} c_j\Phi(t)w_o^{(j)}|| +$$

$$+ \sum_{j=K_{i-1}+1}^{H} c_j\Phi(t)w_o^{(j)} \Big/ \Big\| \sum_{j=1}^{H} c_j\Phi(t)w_o^{(j)} \Big\|$$

$$- \sum_{j=K_{i-1}+1}^{H} c_j\Phi(t)w_o^{(j)} \Big/ \Big\| \sum_{j=K_{i-1}+1}^{H} c_j\Phi(t)w_o^{(j)} \Big\|$$

$$= \sum_{j=1}^{K_{i-1}} c_j\Phi(t)w_o^{(j)} \Big/ \Big\| \sum_{j=1}^{H} c_j\Phi(t)w_o^{(j)} \Big\| +$$

$$+ \left(\sum_{j=K_{i-1}+1}^{H} c_j\Phi(t)w_o^{(j)} \right) \left(\Big\| \sum_{j=1}^{H} c_j\Phi(t)w_o^{(j)} \Big\|^{-1} - \Big\| \sum_{j=K_{i-1}+1}^{H} c_j\Phi(t)w_o^{(j)} \Big\|^{-1} \right)$$

Considering the norm of the first term we can estimate

$$\Big\| \sum_{j=1}^{K_{i-1}} c_j\Phi(t)w_o^{(j)} \Big\| \left(\Big\| \sum_{j=1}^{H} c_j\Phi(t)w_o^{(j)} \Big\| \right)^{-1}$$

$$\leqq \left(\sum_{j=1}^{K_{i-1}} \| c_j\Phi(t)w_o^{(j)} \| \right) \left(\Big\| \sum_{j=1}^{H} c_j\Phi(t)w_o^{(j)} \Big\| \right)^{-1}$$

$$= \left(\sum_{j=1}^{K_{i-1}} |c_j| \exp(t\cdot\lambda(j)+o(t)) \right) \left(\Big\| \sum_{j=1}^{H} c_j\Phi(t)w_o^{(j)} \Big\| \right)^{-1}$$

where the denominator contains an $\exp(t\lambda(H)+o(t))$ - term, thus the first term tends to zero for $t\to\infty$.

The remaining terms yield

$$\Big\| \sum_{j=K_{i-1}+1}^{H} c_j\Phi(t)w_o^{(j)} \Big\| \left(\Big\| \sum_{j=1}^{H} c_j\Phi(t)w_o^{(j)} \Big\|^{-1} - \Big\| \sum_{j=K_{i-1}+1}^{H} c_j\Phi(t)w_o^{(j)} \Big\|^{-1} \right)$$

$$= ||x_t^*|| \left(|| \sum_{j=K_{i-1}+1}^{H} c_j\Phi(t)w_o^{(j)} || - || \sum_{i=1}^{H} c_j\Phi(t)w_o^{(j)} || \right) \times$$

$$\times \left(|| \sum_{j=1}^{H} c_j\Phi(t)w_o^{(j)} || \; || \sum_{j=K_{i-1}+1}^{H} c_j\Phi(t)w_o^{(j)} || \right)^{-1}$$

$$\leq ||x_t^*|| \left(\frac{|| \sum_{j=1}^{K_{i-1}} c_j\Phi(t)w_o^{(j)} ||}{|| \sum_{j=1}^{H} c_j\Phi(t)w_o^{(j)} || \; || \sum_{j=K_{i-1}-1}^{H} c_j\Phi(t)w_o^{(j)} ||} \right)$$

which is asymptotically

$$\exp(\lambda_i t+o(t))\exp(\lambda_{i-1} t+o(t))(\exp(2\lambda_i t+o(t))^{-1} \xrightarrow[t\to\infty]{} 0.$$

Remark

If we know the set $V_i \setminus V_{i-1}$ we can calculate λ_i by

$$\lambda_i = \lim_{t\to\infty} \frac{1}{t} \int_0^t q(A_s, w_s)\, ds = \lim_{t\to\infty} \frac{1}{t} \int_0^t q(A_s, w_s^{(i)})\, ds ,$$

$$q(A,w) = w' \frac{A+A'}{2} w ,$$

w_s starting anywhere in $V_i \setminus V_{i-1}$ (see ARNOLD/WIHSTUTZ [1]) .

Example

The undamped linear oscillator $\ddot{y} + \alpha_t y = 0$ with

$A_t = \begin{pmatrix} 0 & 1 \\ -\alpha_t & 0 \end{pmatrix}$, $\alpha_t \leq a < 0$ stationary and ergodic,

has the two Lyapunov numbers $\lambda_1 = -\lambda < 0$ and $\lambda_2 = \lambda$ (see BENDERSKIJ/PASTUR [2]), thus linearly independent α-stationary angle processes $w^1(t)$ and $w^2(t)$. Since the domain of attraction of $w^2(t)$ contains the first quadrant (see ARNOLD/WIHSTUTZ [1]), one can compute λ by starting somewhere in this quadrant using the formula of the preceding remark.

REFERENCES

[1] Arnold, L. and Wihstutz, V.: On the stability and growth of real noise parameter-excited linear systems, in:
Kallianpur, G. and Kölzow, D.: Measure Theory, Application to Stochastic Analysis, Springer, Lecture Notes in Mathematics 695, 1978, 211-217

[2] Benderskii, M.M. and Pastur, L.A.: The asymptotic behavior of a second order differential equation with random coefficients, Teorija Funkcii Funkcional'nyi Analiz i ich Pričenija, Charkov, vol. 22 (1975), 3-14 (Russian)

[3] Castaing, Ch. and Valadier, M.: Convex Analysis and Measurable Multifunctions, Lecture Notes in Mathematics 580, Springer, Berlin 1977

[4] Khasminskii, R.S.: Stability of Systems of Differential Equations with Random Perturbation of their Parameters, Nauka, Moscow 1969 (Russian)

[5] Oseledec, V.I.: A Multiplicative Ergodic Theorem. Lyapunov Characteristic Numbers for Dynamical Systems, Trans. Moscow Math. Soc. 19 (1968), 197-231

[6] Raghunathan, M.S.: A Proof of Oseledec's Multiplicative Ergodic Theorem, Israel J. of Math. 32 (1979), 356-362

[7] Ruelle, D.: Ergodic Theory of Differentiable Dynamical Systems, Extrait des Publications Mathématiques no. 50 of IHES, Bures sur Yvette 91440 France, 1978

[8] Skorokhod, A.V.: Studies in the Theory of Random Processes, Addison-Wesley, Reading (Mass.), 1965

One service mathematics has rendered the human race; it has put common sense back where it belongs, on the topmost shelf next to the dusty cannister labelled "discarded nonsense".

E.T. Bell

Part 4

IDENTIFICATION

The last paper in this section uses ideas from nonlinear filtering theory to attack identification problems. It could have been placed in either section.

SYSTEM IDENTIFICATION

E.J. Hannan

The Australian National University

In the first section the introduction of parameters to describe multivariate ARMAX systems is described and some topological properties of the parameter spaces are considered that relate to parameter estimation. In particular the analytic manifold, $M(n)$, of all ARMAX structures of given order (McMillan degree) is considered. A very general discussion is given, in the second section, of the properties of maximum likelihood estimation of the point on $M(n)$, emphasising both minimal conditions on the inputs and the linear innovations and avoiding assumptions of a too restrictive nature concerning the true stochastic structure. In the third section the estimation of the order is discussed and a number of new results are presented. Other topics briefly considered are the construction of initial estimates for an iterative solution of the likelihood equations, time varying parameter models and non-linear models.

1. INTRODUCTION

Substantially linear time invariant systems, only, will be dealt with in this account. The theory for these has reached a certain stage of completeness but is elaborate and in the space available only an abbreviated account even of these can be given. The theory is asymptotic with the record length, T. Such theories can be deceptive and we shall try to indicate where deception can occur. References will be dealt with through bibliographic notes at the end of the account.

M. Hazewinkel and J. C. Willems (eds.), Stochastic Systems: The Mathematics of Filtering and Identification and Applications, 221–246.

2. PARAMETERS FOR LINEAR, TIME INVARIANT SYSTEMS

We may initially describe such systems in the ARMAX form. (ARMAX = autoregressive-moving average with exogenous components.)

$$\sum_{0}^{p} A(j)y(t-j) = \sum_{1}^{m} D(j)u(t-j)+\sum_{0}^{q} B(j)\epsilon(t-j), \quad A(0) = B(0) \qquad (2.1)$$

Here $y(t)$ is s-dimensional and $u(t)$ is r dimensional and both are observed for $t = 1,\ldots,T$. We at least require of the unobserved s-dimensional vector $\epsilon(t)$ that

$$E\{\epsilon(s)\epsilon(t)'\} = \delta_{st}\Sigma, \quad \Sigma > 0.$$

"True" values of parameter matrices such as Σ, $A(j)$ will be indicated by a zero subscript, Σ_0, $A_0(j)$. We introduce the z transforms, which are matrices of polynomials

$$\tilde{d}(z) = \Sigma D(j)z^j, \quad \tilde{g}(z) = \Sigma A(j)z^j, \quad \tilde{h}(z) = \Sigma B(j)z^j,$$

and put

$$k(z) = \tilde{g}(z^{-1})^{-1}\tilde{h}(z^{-1}), \quad \ell(z) = \tilde{g}(z^{-1})^{-1}\tilde{d}(z^{-1}).$$

These are matrices of rational functions. Then we formally have

$$y(t) = \ell(z)u(t) + k(z)\epsilon(t) \qquad (2.2)$$

where z^{-1} is interpreted as the backwards shift so that

$$k(z)\epsilon(t) = \sum_{0}^{\infty} K(j)\epsilon(t-j), \quad k(z) = \sum_{0}^{\infty} K(j)z^{-j}, \quad K(0) = I_s. \qquad (2.3)$$

If $\det(\tilde{g}) \neq 0$, $|z| \leq 1$ then certainly (2.3) converges almost surely (a.s.) and in mean square. Equally if, for example, $u(t)$ is generated by a stationary process with finite variance then, under this condition

$$\ell(z)u(t) = \sum_{1}^{\infty} L(j)u(t-j), \quad \ell(z) = \sum_{1}^{\infty} L(j)u(t-j) \qquad (2.4)$$

converges a.s. and in mean square. Of course we might also view (2.1) as having been initiated at $t = 0$ when no questions of convergence would arise in (2.3) and (2.4). It is also well known that, in the stationary case, $\epsilon(t)$ will be the linear innovation sequence for $y(t)$ if and only if $\det(\tilde{h}) \neq 0$, $z < 1$. Finally the system as prescribed will contain redundant elements unless $(\tilde{g},\tilde{h},\tilde{d})$ are left prime. By this is meant that if $\tilde{g} = v\tilde{g}_1$, $\tilde{h} = v\tilde{h}_1$, $\tilde{d} = v\tilde{d}_1$, where all components are matrices of polynomials, then $\det(v)$ is a non zero constant. It is thus natural to require that

$$\det(\tilde{g})\neq 0,\ |z|\leq 1;\ \det(\tilde{h})\neq 0,\ |z|<1;\ \tilde{g},\tilde{h},\tilde{d} \text{ are left prime.} \tag{2.5}$$

It is clear that any set for which the second part of (2.5) is as prescribed cannot be open. For this reason (2.5) is sometimes strengthened so that $\det(\tilde{h}) \neq 0$, $|z| \leq 1$. For brevity we shall call the strengthened version (2.5)'.

The set of all structures (2.1), subject to (2.5), for given m,p,q, is very complicated to parameterize. We may separate off a dense subset (see below) by further requiring that

$$[A(p)\vdots B(q)\vdots D(m)] \text{ is of rank } s,\quad A(0)=B(0)=I_s, \tag{2.6}$$

The set $M(p,q,m)$ of all structures satisfying (2.1), (2.5)', (2.6), for given Σ, may be mapped into an open set in Euclidean space of dimension $(p+q)s^2+mrs$ by using the elements of the $A(j)$, $B(j)$, $D(j)$, $j > 0$, as coordinates. Of course the set of all $\Sigma > 0$ is mapped by the on and above diagonal elements of the matrix into an open set in Euclidean space of dimension $s(s+1)/2$. It is convenient to distinguish the system parameters that coordinatise $M(p,q,m)$ from the variances and covariances in Σ.

It is an interesting and important fact that the set of all structures (1.1) can be parameterized with the introduction of only one integer parameter in place of the three in $M(p,q,m)$. (This is a somewhat misleading statement as we see below.) Of course integer parameters are troublesome as they have to be scanned in an optimisation routine. This parameterization may be introduced via the, alternative, state space form for (1.1) which we here present in prediction error form

$$\begin{aligned} x(t+1|t) &= Fx(t|t-1)+Gu(t)+K\epsilon(t) \\ y(t) &= Hx(t|t-1)+\epsilon(t). \end{aligned} \tag{2.7}$$

Here $x(t|t-1)$ is not observed. Under (2.5) F has eigenvalues of modulus less than unity. We have

$$k(z) = H(zI_n-F)^{-1}K+I_s,\quad \ell(z) = H(zI_n-F)^{-1}G$$

If (2.7) is made reachable and observable (see reference 23) then F has minimal dimension, which we henceforth call n, as above. This will be called the system order (or McMillan degree).

We may choose $x(t|t-1)$ canonically as follows. Put, in the stationary case,

$$y(t+j|t) = \sum_{j}^{\infty} K(a)\epsilon(t+j-a) + \sum_{j}^{\infty} L(a)u(t+j-a).$$

Then as we proceed from left to right along the row

$$y(t+1|t)', \; y(t+2|t)', \; \ldots \tag{2.8}$$

we shall meet a first linearly independent set of n elements

$$y_k(t+j|t), \; j=1,\ldots,n_k, \; \Sigma n_k = n. \quad \text{(The rank of (2.8) is } n\text{)}. \tag{2.9}$$

These may be chosen to constitute $x(t+1|t)$. These "dynamical indices" n_k also may be discovered from the Hankel matrix

$$\mathcal{H} = \begin{bmatrix} [K(1)\; L(1)] & [K(2)\; L(2)] & \cdots \\ [K(2)\; L(2)] & [K(3)\; L(3)] & \cdots \\ \vdots & \vdots & \end{bmatrix}$$

which also is of rank n and for which rows (j,k); $j=1,\ldots,n_k$; $k=1,\ldots,s$ constitute a (first) linearly independent set. Here row (j,k) is the kth row in the jth block of rows. The set of all structures (2.1), or equivalently (2.7), for these dynamical indices will be called $V(\alpha)$, where α symbolises a set $\{n_k, k=1,\ldots,s\}$. Using the canonical form for the minimal (2.7) induced by the canonical choice of $x(t)$ the freely varying elements in F,G,H,K (those not identically zero or unity) map (2.1), subject to (2.5)' into an open set in Euclidean space of dimension

$$n(r+s+1) + \sum_{j<i} \{\min(n_j,n_i) + \min(n_j,n_i+1)\}. \tag{2.10}$$

We may also consider all structures (2.1) for which (2.9) is a linearly independent subset of (2.8), whether or not they are the first such set met as you proceed from left to right. Call this set $U(\alpha)$. If $(k,\ell) \in U(\alpha)$ then we may always represent (k,ℓ) as $g^{-1}(z)(h(z),d(z))$ where g,h,d are left prime, $\det(g) \neq 0$, $|z| \geq 1$, $\det(h) \neq 0$, $|z| \geq 1$ (We assume (2.5)'), and

$$g_{ii} \text{ is a monic polynomial and degree } (g_{ij}) < \text{degree } (g_{jj}) = n_j, \quad j \neq i. \tag{2.11}$$

The $U(\alpha)$ overlap and indeed each $U(\alpha)$ is dense in their union, $M(n)$, in the topology we introduce below for that set. Of course $V(\alpha) \subset U(\alpha)$ but only for one case is $V(\alpha) = U(\alpha)$, namely that case where $n_1 = n_2 = \ldots = n_w = n_{w+1}+1 = \ldots = n_s+1$ so that $n = n_1 s+w$, $w < s$. Indeed this is the case where the first n linearly independent rows of $\mathcal{H}$ are the first n rows so that any point in $U(\alpha)$ is evidently also in $V(\alpha)$, for this α. We shall call this special $U(\alpha) = V(\alpha)$ by a special name, for ease of reference, which we choose to be $U_0(n)$. It is sometimes spoken of as the "generic" neighbourhood. All of the other $V(\alpha)$ are proper subsets of the corresponding $U(\alpha)$ and indeed are lower dimensional since, except for $U_0(n)$, (2.10) is less than $n(2s+r)$, which (see below) is the dimension of $M(n)$. A natural topology for $M(n)$, i.e. the set of all (k, ℓ) satisfying (2.5) for some $\tilde{g},\tilde{h},\tilde{d}$ (and also the last part of (2.1)), is obtained from the metric

$$||k,\ell|| = \sum_1^\infty ||[K(n),L(n)]||2^{-u}$$

where $||.||$, on the right is the Euclidean norm of the indicated matrix. Of course $M(n)$ is not complete in this topology. We may introduce coordinates in each $U(\alpha)$ which map $U(\alpha)$ in a one to one manner into Euclidean space of dimension $n(2s+r)$. One way to choose these coordinates is as follows. Let $(k,\ell) = g^{-1}(h,d)$ be the representation for a pair $(k,\ell) \in U(\alpha)$ described in and above (2.11). Let $\alpha_{ij}(u)$ be the coefficient of z^u in the (i,j)th element of g. Let $(\beta_{ij}(u),\delta_{ij}(u))$ similarly relate to h,d. Then we choose as coordinates

$$\alpha_{ij}(u),\ u=0,1,\ldots,n_j-1;\ i,j=1,\ldots,s;(\beta_{ij}(u),\delta_{ij}(u)),\ u=0,1,\ldots n_i-1;$$
$$i,j=1,\ldots,s.$$

Thus we have taken all $\alpha_{ij}(n)$ not prescribed to be 0 or 1 by (2.11) but we have taken only the first n_i of the $\beta_{ij}(n)$, $\delta_{ij}(n))$ for each j, any given i. (When $(k,\ell) \in V(\alpha)$ then we may, in addition to (2.11), require also degree $(g_{ij}) \leq$ degree (g_{ii}), $j \leq i$; degree $(g_{ij}) <$ degree (g_{ii}), $j > i$; degree $(h_{ij}) \leq$ degree (g_{ii}), degree $(d_{ij}) <$ degree (g_{ii}); $i,j = 1,\ldots,s$. Thus, remembering that $A(0) = B(0)$ so that $\alpha_{ij}(n_i) = \beta_{ij}(n_i)$, $j = 1,\ldots,s$, we see that no $\alpha_{ij}(n)$, $\beta_{ij}(n)$, $\delta_{ij}(n)$ is omitted from the coordinate vector unless it is identically 0 or 1.) The remainder are rational functions of the above chosen set of coordinates. There are evidently $n(2s+r)$ coordinates chosen above. Moreover these $\binom{n+s-1}{s-1}$ coordinate neighbourhoods cover $M(n)$ in such a way as to make it an analytic manifold. We collect these results together as a theorem.

Theorem 1. Let $M(n)$ be the set of all (k,ℓ) of order n satisfying (2.5)'. The $U(\alpha)$, $\Sigma n_j = n$, serve as coordinate neighbourhoods through which $M(n)$ may be topologized as an analytic manifold of dimension $n(2s+r)$. Each $U(\alpha)$ is dense in $M(n)$. The $V(\alpha)$ are disjoint and their union is $M(n)$. $V(\alpha) \subset U(\alpha)$ and the inclusion is proper unless $U(\alpha) = U_0(n)$.

Thus the problem of identifying a structure (2.1) from data has been reduced to the estimation of the appropriate n and then the choice of the appropriate point in $M(n)$, and the appropriate $\not{Z}$. We shall, in the next sections give a very nearly definitive account of certain "maximum likelihood" (M.L.) methods for this identification problem. However this account must be interpreted with care. In the first place there are many sets, α, of indices to be considered unless n, r are small, which is why the statement, that only one integer n will suffice, is somewhat spurious. We go ahead in the remainder of this section to discuss these problems. Part of the problem is related to the topological structure of $M(n)$ but not to the conditions (2.5). Hence it is convenient to consider the set, which we shall call $M_c(n)$, of all rational matrix pairs (k,ℓ), that are "proper", i.e. for which all elements have denominators no smaller in degree than their numerators. (From (2.3), (2,4) it is clear that all $(k,\ell) \in M(n)$ satisfy this condition). Precisely as before we may introduce dynamical indices via $\mathcal{H}$ and hence define neighbourhoods $U_c(\alpha)$. These define a neighbourhood system covering $M_c(n)$ with respect to which it is an analytic manifold. Of course $M(n)$ is an open submanifold of $M_c(n)$ and $U_c(\alpha)$ of $U(\alpha)$. Let ϕ_α map $U_c(\alpha)$ into $n(2s+r)$ dimensional space via the coordinates introduced above and let π_α be the inverse mapping. Thus $\pi_\alpha(g,h,d) = g^{-1}(h,d) = (k,\ell)$. We may also topologise $M_c(n)$ as a metric space so that convergence corresponds to the convergence element by element, of the $K(u)$, $L(u)$. Call this topology T_p. Finally we may topologise $U_c(\alpha)$ by means of the strongest topology for which π_α is continuous. This is called the quotient topology and we indicate it by T_Q. Thus now a set in $U_c(\alpha)$ is being called open if its pre-image under π_α is open. Of course $T_p \subset T_Q$. The reason why we consider T_Q is the following. As we shall see the M.L. estimators $(\hat{k},\hat{\ell})$ converge in T_p to k_0, ℓ_0, the true values, if $n = n_0$(= "true" value). For many purposes this is all that is needed, e.g. for prediction or filtering, which depend only on a knowledge of k_0, ℓ_0. However for other purposes (e.g. future modification of the system for control) the description in terms of g, h, d may be needed. Moreover the theorem about $\hat{k}$, $\hat{\ell}$ does not tell us how to compute these and for this purpose we clearly need coordinates. If we only know that $(\hat{k}, \hat{\ell})$ converge to (k_0, ℓ_0) in T_p we do not know about the behaviour of their coordinates. Of course, in any topology for which π_α is continuous, convergence of the coordinates implies convergence of $(\hat{k}, \hat{\ell})$ and thus it is relevant to consider the strongest topology for which π_α is continuous. If T_Q is strictly stronger than T_p then convergence of $(\hat{k}, \hat{\ell})$

in T_p does not imply convergence in T_Q and if this does not happen certainly the coordinates of $(\hat{k},\hat{\ell})$ cannot converge. By $\overline{M_c(n)}$ we mean the closure of $M_c(n)$ in the T_p topology. We say $\beta \leq \alpha$ if $\beta = (n_1',\ldots,n_s')$ and $n_j' \leq n_j$, $= 1,\ldots,s$.

Theorem 2. Let $\alpha = (n_j,\ldots,n_s)$, $\Sigma n_i = n$. Then

(i) $\overline{U_c(\alpha)} = \overline{M_c(n)} = \bigcup_{i\leq n} M_c(i)$

(ii) $\overline{\phi_\alpha U_c(\alpha)} = \mathbb{R}^{n(2s+r)}$

(iii) $\pi_\alpha \overline{\phi_\alpha U_c(\alpha)} = \bigcup_{\beta\leq\alpha} U_c(\beta)$

(iv) $U_c(\alpha)$ is the largest subset of $\rho_\alpha \overline{\phi_\alpha U_c(\alpha)}$ on which $T_p = T_Q$.

(iv) shows that $(\hat{k}, \hat{\ell})$ may not converge in T_Q. Even if $n_0 = n$ and $k_0,\ell_0 \in U(\beta)$ and we choose α so that $\beta \not\leq \alpha$ then (iii) shows that $\phi_\alpha(k_0,\ell_0)$ lies at infinity and any program using these coordinates must overflow or give a spurious estimate. The last part of Theorem 1 makes this last situation extraordinarily unlikely but clearly even if (k_0,ℓ_0) lies in $U(\alpha)$ it may be near the edge of the coordinate neighbourhood we are using.

Another difficulty with $M(n)$ comes from the use of canonical g,h,d. These correspond to canonical $\tilde{g},\tilde{h},\tilde{d}$ where the jth row of the latter is the jth row of $(g(z^{-1}), h(z^{-1}), d(z^{-1}))$ multiplied by z raised to a power that is the greatest found in the jth row of (g,h,d). There may have been constraints on some initial $\tilde{g},\tilde{h},\tilde{d}$ but it will be extraordinarily difficult to translate these into constraints on the canonical forms. One may therefore prefer to remain with the initial prescription and trust that (2.6) holds. The nature of this choice can be seen from the following. We may consider the set of all $\tilde{g},\tilde{h},\tilde{d}$ satisfying (2.5)' and (2.11) plus degree $(g_{ij}) \leq n_i$, $j \leq i$; $< n_i$; $j > i$, degree $(h_{ij}) < n_i$. In order that the corresponding $(k,\ell) \in V(\alpha)$ it is necessary and sufficient that the matrix, $J(\alpha)$, whose jth row is composed of the coefficients of the highest power of z that occurs in the jth row of $[\tilde{g},\tilde{h},\tilde{d}]$, should be of rank s. Thus $M(p,p,p) = U_0(ps)$ and hence is dense in $M(ps)$, which from Theorem 1(i) is dense in the union of all $M(i)$, $i \leq sp$. Thus restricting oneself to $M(p,q,m)$ cannot be costly. Of course a linear system may originally be prescribed in state space form (anterior to (2.7))

$$x(t+1) = Fx(t)+Gu(t)+\xi(t+1), \quad \mathcal{E}\{\xi(s)\xi(t)'\} = \delta_{st}Q$$
$$y(t) = Hx(t)+\eta(t), \quad \mathcal{E}\{\eta(s)\eta(t)'\} = \delta_{st}R. \tag{2.12}$$
$$\mathcal{E}\{\xi(s)\eta(t)\} = \delta_{st}S.$$

In this case constraints will be on initial F,G,H, and the covariance matrices and none of the manifolds introduced in this section will be of much use.

Finally it must be remembered that the data is most unlikely to be precisely generated by any system (2.1).

3. IDENTIFICATION WHEN THE ORDER IS PRESCRIBED

As previously said estimation will be by maximum likelihood as if the $\epsilon(t)$ were Gaussian. However this is done only to construct the likelihood and the properties of the M.L. estimates will be shown to hold more generally. For that matter it will not be assumed that the true order is as prescribed or even that the data is generated by a system (2.1). We shall explicitly discuss only $M(n)$ for brevity but all results extend to $M(p,q,m)$, for example.

The $u(t)$ will be treated as part of an infinite sequence of sure vectors satisfying

$$\lim_{T\to\infty} \frac{1}{T}\sum_{1}^{T-j} u(t)u(t+j)' = \Gamma_u(j), \quad \frac{1}{T}\sum_{1}^{T} u(t) = 0 \tag{3.1}$$

Thus mean corrections have been made. It is an immediate consequence of (3.1) that

$$\Gamma_u(j) = \int_{-\pi}^{\pi} e^{ij\omega} dF_u(\omega),$$

where $F_u(\omega) = \Gamma_u(0)-F_u(-\omega)$ at all points of continuity and $F_u(\omega)$ is Hermitian non-negative and increasing with ω. It is assumed that $u(t)$ is "persistently exciting" in the sense that for any matrix of polynomials $P(z)$,

$$\mathrm{tr}\{\int_{-\pi}^{\pi} P(e^{i\omega})dF_u(\omega)P^*(e^{i\omega})\} \geq c\ \mathrm{tr}\{\int_{-\pi}^{\pi} P(e^{i\omega})P^*(e^{i\omega})d\omega\}, \quad c > 0.$$

Such conditions as this are needed for an

asymptotic treatment. Of course a signal could fail to be persistently exciting and still give information about $\ell(z)$, for example if $u(t)$ was transient.

The likelihood is constructed as if the $\epsilon(t)$ were Gaussian with the covariance properties stated below (2.1) and as if (2.1) held for $t \geq 1$ with $y(0),\ldots,y(-p+1)$, $u(0),\ldots,u(-m+1)$ unknown constants that will have to be estimated. The estimates of these initial values will not be consistent and they will be omitted from the discussion below. Put $\theta = (\sigma^2,\mu)$ where σ^2 varies over the manifold of all $\Sigma > 0$ and μ varies over $M(n)$. Of course the maximum may be in $\overline{M(n)}$ and not $M(n)$ but σ^2 will always be an interior point of its manifold, at least for T large enough. We shall not write down the likelihood here since proofs are not being given. To compute the estimate an iteration will be needed and for a practicable procedure an initial estimate of μ will be needed. We shall later indicate how this might be got.

Let us call $\tilde{M}(n)$ the "manifold with edge points" obtained from $M(n)$ by allowing $\det\{h(z)\}$ to be zero for some z with $|z| = 1$. We do not assume $\mu_0 \varepsilon \tilde{M}(n)$ but <u>rather</u> $\mu_0 \varepsilon \tilde{M}(i)$, $i \leq n$. Of course $\tilde{M}(i)$, $i \leq n$ is contained in $\overline{M(n)}$ because of Theorem 2. We assume that $\epsilon(t)$ is generated by a stationary ergodic stochastic process with $E\{\epsilon(s)\epsilon(t)'\} = \delta_{st}\Sigma_0$, $\Sigma_0 > 0$.

Theorem 3. <u>If the conditions just stated hold and $\hat{k}, \hat{\ell}, \hat{\Sigma}$ are the M.L. estimates then these converge a.s. to k_0, ℓ_0, Σ_0.</u>

<u>Remarks</u>. (i) We could prescribe that θ_0 lies in some subset of the parameter space defined by a fixed, known, set of continuous constraining functions, $\psi(\theta) = 0$. The theorem would still hold.

(ii) The theorem will hold in an extended sense even if no model (2.1) holds. Thus if, for example, let $u(t)$ be generated by a stationary process, independent of $\epsilon(t)$, with spectral function $F_u(\omega)$, and let

$$y(t) = \sum_0^\infty K_0(j)\epsilon(t-j) + \sum_1^\infty L_0(j)u(t-j), \quad K_0(0) = I_s$$

with

$$\sum_0^\infty \|K_0(j)\| < \infty, \quad \sum_1^\infty \|L_0(j)\| < \infty.$$

If $\mathcal{L}_T(\theta)$ is the likelihood and $L_T(\theta) = -2T^{-1}\log \mathcal{L}_T(\theta)$ then $L_T(\theta)$ will converge, a.s., to

$$\log\det \mathcal{Z}_\theta + \frac{1}{2\pi}\int_{-\pi}^{\pi} \mathrm{tr}[(k_\theta \mathcal{Z}_\theta k_\theta^*)^{-1} k_0 \mathcal{Z}_0 k_0^* \, d\omega +$$

$$\{k_\theta^{-1}(\ell_0-\ell_\theta)dF_u(\omega)(\ell_0-\ell_\theta)^* k_0^{*-1}\}] \tag{3.2}$$

Assume that this is minimised over $\overline{M(n)}$, and with respect to $\mathcal{Z}_\theta$, at $\mathcal{Z}_1, k_1, \ell_1$. Then $\hat{\mathcal{Z}}, \hat{k}, \hat{\ell}$ will converge a.s. to $\mathcal{Z}_1, k_1, \ell_1$. Of course under the conditions of the theorem $\mathcal{Z}_1 = \Sigma_0$, $k_1 = k_0$, $\ell_1 = \ell_0$.

(iii) The conditions on $u(t)$ are rather restrictive since $u(t)$ could not contain a trend, for example. Relaxations of these conditions are possible and we give some references in the Bibliographic notes.

We indicate briefly how the theorem may be proved. It may be shown that $L_T(\theta)$ may be replaced by

$$\log\det \mathcal{Z}_\theta + \frac{1}{2\pi}\int \mathrm{tr}\{(h_\theta \mathcal{Z}_\theta h_\theta^*)^{-1} w(e^{i\omega}) w(e^{i\omega})^*\} d\omega \tag{3.3}$$

$$w(e^{i\omega}) = \frac{1}{\sqrt{T}} \sum_1^T \{g_\theta y(t) - d_\theta u(t)\} e^{it\omega} \tag{3.4}$$

where $g_\theta y(t) = \Sigma A_\theta(j) y(t-j)$, for example, and $L_T(\theta)$ is being (approximately) evaluated in a particular coordinate system. The expression (3.3) is plausible since $\epsilon_\theta(t) = h_\theta^{-1}\{g_\theta y(t) - d_\theta u(t)\}$ is the estimate of $\epsilon(t)$ based on θ and hence the second term in (3.3) is approximately $\mathrm{tr}(\Sigma_\theta^{-1}\tilde{\Sigma})$ where

$$\tilde{\Sigma} = \frac{1}{T}\sum_1^T \epsilon(t)\epsilon(t)'. \tag{3.5}$$

That (3.3) converges to (3.2) may be proved by approximating matrices such as $(k_\theta \mathcal{Z}_\theta^{-1} k_\theta^*)^{-1}$ by the Cesaro sum of their Fourier series and thus reducing the proof to that for the convergence of

$$\frac{1}{T}\sum_1^T \epsilon(t)\epsilon(t+j)', \quad \frac{1}{T}\sum_1^T u(t)u(t+j)', \quad \frac{1}{T}\sum_1^T \epsilon(t)u(t+j)'. \tag{3.6}$$

Ergodicity and (3.1) establish the result for the first two and a simple argument (using the method of subsequences) establishes the result for the third. The technical details are quite difficult because $\det(k_\theta)$ may be zero, the coordinate neighbourhoods map into unbounded regions in Euclidean space, there are many of them and while $\mathcal{Z}_0 > 0$ we have not bounded it away from zero.

For a CLT we need to assume that $\mu_0 \varepsilon \bar{M}(n)$. (More elaborate theorems would be possible for $\mu_0 \varepsilon \bar{M}(n)/M(n)$ but no published results seem to be available.) Again this CLT may be reduced to that for (centred, scaled forms of) the quantities (3.6). Let F_t be the σ-algebra of events determined by $\epsilon(s)$, $s \leq t$. If

$$E\{\epsilon(t)|F_{t-1}\} = 0 \tag{3.7}$$

then the best linear predictor is the best predictor and conversely. Thus without (3.7) the formulation of the system as in (2.1) does not seem to have much meaning. A little more is needed. If

$$E\{\epsilon(t)\epsilon(t)'|F_{t-1}\} = \Sigma_0 \tag{3.8}$$

then the CLT holds. This condition is not so plausible. Of course it holds if the $\epsilon(t)$ are independent, for example Gaussian. A much weaker set of conditions, implied by (3.8), are

$$\lim_{k\to\infty} E\{\epsilon(t)\epsilon(t)'|F_{t-k}\} = \Sigma_0 \quad \text{a.s.}$$
$$\lim_{\ell\to\infty} E\{\epsilon_i(t)\epsilon_j(t)\epsilon_k(t-a)|F_{t-\ell}\} = E\{\epsilon_i(t)\epsilon_j(t)\epsilon_k(t-a)\}, \quad a > 0, \text{ a.s.} \tag{3.9}$$

The a.s. limits in (3.9) exist in any case (since, for example, $b_k = E\{\epsilon(t)\epsilon(t)'|F_{t-k}\}$ is a reverse martingale). Asserting that the limits are the constants shown is a weak form of non determinacy condition. We shall call the right side, of the second part of (3.9), $\sigma_{ijk}(a)$. If to (3.9) is adjoined

$$\sigma_{ijk\ell}(a,b) = E\{\epsilon_i(t)\epsilon_j(t)\epsilon_k(t-a)\epsilon_\ell(t-b)\} \text{ exists}, \quad a, b > 0, \text{ all } i,j,k,\ell, \tag{3.10}$$

then the CLT holds. Of course (3.10) is also implied by (3.8).

Theorem 4. _If $\mu_0 \varepsilon M(n)$ and (3.7), (3.8) hold or (3.7), (3.9), (3.10) hold then the ML estimates of μ_0 obey the central limit theorem._

Remarks. (i) If (3.8) holds the asymptotic distribution is precisely the same as that which would obtain were the $\epsilon(t)$ Gaussian.

(ii) If constraints are applied and the constraining $\psi(\theta)$ do not mix system parameters with covariances and are twice continuously differentiable on $M(n)$ then the CLT continues to hold for $\hat{\mu}$ under the conditions of the theorem

(iii) If (3.9), (3.10) hold but not (3.8) the covariance matrix of $\hat{\mu}$ in the limiting distribution is more complicated. As the $\sigma_{ijk}(a)$ approach zero and the $\sigma_{ijk\ell}(a,b)$ approach $\delta_{ab}\sigma_{ij}\sigma_{k\ell}$ (the values, in each case, when (3.8) holds) then this covariance matrix approaches that for the Gaussian case.

(iv) If $E\{\epsilon_j(t)^4\} < \infty$, $j = 1,\ldots,s$, the CLT also holds for the covariance estimators.

A key part in the proof of this theorem is played by the elimination of the "squared epsilons" i.e. terms of the form $\epsilon_i(t)\epsilon_j(t)$. In the scalar case, when $m = 0$ (i.e. $u(t)$ does not occur) the reason for this can be seen by considering the gradient of the second term in (3.3) evaluated at μ_0, namely (in this case)

$$\partial_\mu \int_{-\pi}^{\pi} f_\theta(\omega)^{-1} I(\omega)d\omega|, f_\theta(\omega)=\frac{\sigma^2}{2\pi}|k_\mu(e^{i\omega})|^2, I(\omega)=\frac{1}{T}|\sum_1^T y(t)e^{it\omega}|^2. \tag{3.11}$$

Since the first term is, evaluated at μ_0,

$$-\partial_\mu^2 \int f_\theta(\omega)^{-1} I(\omega)|_{\bar{\mu}}(\hat{\mu}-\mu_0),$$

where $\bar{\mu}$ lies between $\hat{\mu}$ and μ_0, and hence converges to μ_0 by Theorem 3, we need to find the distribution of

$$T^{\frac{1}{2}}\partial_\mu \int f_\theta(\omega)^{-1} I(\omega)d\omega|_0 \tag{3.12}$$

However

$$\partial_\mu \int f_\theta(\omega)^{-1}|k_0|^2 d\omega|_0 = \sigma^{-2}\partial_\mu \int \log f_\theta(\omega)d\omega|_0 = 0$$

since $(2\pi)^{-1}\int \log\{2\pi f_\theta(\omega)\}d\omega = \log \sigma^2$ and is independent of μ. Thus $T^{\frac{1}{2}}\partial_\mu \int f_\theta(\omega)^{-1}|k_0|^2 \tilde{Z} d\omega$ may be subtracted from (3.12) and this eliminates most of the squared terms. (See (3.5) for $\tilde{Z}$.) This argument also shows why constraints that mix system parameters with covariances affect the theorem.

Finally we consider, briefly, in this section how to obtain an initial estimate of μ. If that can be done by means of a procedure that both gives a consistent estimator and is one that can always be completed (as compared to an iterative or search procedure that might never converge) then, in theory at least, the inferential procedure is rather complete. We say in theory because, while the method we present is consistent and even gives estimates with errors that are $O(T^{-\frac{1}{2}})$ yet it gives inefficient initial estimates and some experience suggest that with T not large the initial estimators may not be satisfactory. This is mainly because they commence from an estimator of $\tilde{g}(z)$ that need not satisfy (2.5), though the estimator will satisfy (2.5) if T is large enough, when the model is valid. Here we restrict ourselves to the ARMA case for brevity. The methods we deal with are described for $M(p,q)$ (i.e. what was earlier called $M(p,q,0)$) in reference (12) and we shall not repeat them here but shall deal with $M(n)$. (In reference (12) it was asserted that the method is always consistent but that is not true as can be seen from the paper by B. Hanzon delivered at this conference. However, a priori, the method is almost sure to be consistent in the sense that if an element (of $M(p,q)$ is chosen according to any continuous probability distribution on that manifold then with probability unity the consistency will hold.) It will be rather difficult to construct initial estimates when $k \varepsilon U(\alpha)$ unless α corresponds to $U_0(n)$. This is partly because only then is $V(\alpha) = U(\alpha)$ and partly because only then are all of the elements of the coefficient matrices of $\tilde{g}$, $\tilde{h}$, that are not identically zero or unity, freely varying. (See above Theorem 1 in section 2.) However if $M(n)$ is to be the basis for estimation then it is almost certain that it will be commenced on the assumption that $k \varepsilon U_0(n)$ which assumption again must be virtually costless since $U_0(n)$ is dense in $M(n)$. Only if iteration becomes troublesome (because the coordinate neighbourhood, $U_0(n)$, is unsuitable since k_0 is near the "edge") will a new system of coordinates be sought and then further iteration will presumably commence from the last point reached in $U_0(n)$. These arguments to some extent justify the consideration only of $U_0(n)$. Put $n = sn + w$, $w < s$. For $U_0(n)$ we have $A(0) = B(0)$ and

$$A(0) = \begin{bmatrix} I & 0 \\ A_{21}(0) & I \end{bmatrix}, \quad A(1) = \begin{bmatrix} A_{11}(1) & 0 \\ A_{21}(1) & A_{22}(1) \end{bmatrix}$$

where the partition is after the row and column numbered w. The first w row of $\tilde{g}(z)$, $\tilde{h}(z)$ are of degree n_s+1 and the last $(s-w)$ rows are of degree n_s. Put

$$A(j) = \begin{bmatrix} A_1(j) \\ A_2(j) \end{bmatrix}$$

with the partition again after row w. Then consider the equations of estimation

$$\begin{bmatrix}\hat{A}_{11}(1)\\ \hat{A}_{21}(0)\end{bmatrix}\tilde{C}(-k) + \sum_{1}^{n_s-1}\begin{bmatrix}\hat{A}_1(j+1)\\ \hat{A}_2(j)\end{bmatrix}C(j-k) + \begin{bmatrix}\hat{A}_1(n_s+1)\\ \hat{A}_2(n_s)\end{bmatrix}\tilde{C}(n_s-k) \tag{3.13}$$

$$= -\begin{bmatrix}[I_w\,\vdots\,0] & C(-k+1)\\ [0\,\vdots\,I_{s-w}] & C(-k)\end{bmatrix}, \quad k = n_s+1,\ldots,2n_s+1.$$

Here

$$C(n) = \frac{1}{T}\sum_{1}^{T-n} y(t)y(t+n)'$$

and $\tilde{C}(-k)$ is $C(-k)(= C(k)')$ with the last $(s-w)$ rows removed and $\tilde{C}(n_s-k)$ is $C(n_s-k)$ with the last $(s-w)$ columns removed. These estimates will converge a.s. to the true values if $\mu_0 \in U_0(n)$ and the conditions of Theorem 1 hold, provided the matrix Γ_n^n which we now describe, is non singular. This matrix is made up of the blocks

$$\Gamma(t) = E\{y(s)y(s+t)'\} = \int_{-\pi}^{\pi} e^{it\omega} f(\omega)d\omega, \quad f(\omega) = \frac{1}{2\pi}\, k\Sigma k^*.$$

Let Γ_∞^∞ be the infinite Hankel matrix

$$\Gamma_\infty^\infty = [\Gamma(1-j-k)]_{j,k=0,1} \tag{3.14}$$

Then Γ_n^n is obtained by choosing rows (and columns) as follows. We index these as (j,u) (and (k,v)) where j, u is the uth row in the jth block, $j = 0,1,\ldots;$ $u = 1,\ldots,s$. $(k = 0,1,\ldots;$ $v = 1,\ldots,s)$. Then we choose rows (j,u), $j = 0,1,\ldots,n_u-1;$ $u = 1,\ldots,s$ $((k,v), k = 0,1,\ldots,n_v-1;$ $v = 1,\ldots,s)$. (If $C(1-j-k)$ replaces $\Gamma(1-j-k)$ in (3.14) then the matrix on the left side of the system (3.13) results). Γ_n^∞ (i.e. having all the columns of (3.14) but only the rows chosen as above) is of rank n and if we institute any prior continuous probability distribution on $U_0(n)$ then Γ_n^n will have that rank with probability one.

Forming

$$\hat{v}(t) = \sum_{0}^{n_s+1}\hat{A}(j)y(t-j), \quad v(t) = \sum_{0}^{n_s+1} B_0(j)\epsilon(t-j)$$

we may use $\hat{v}(t)$ in place of $v(t)$ to estimate the $B_0(j)$ by a standard consistent procedure. We refer to the notes for further details. These procedures yield strongly consistent estimates. Analogous procedures are available for any of the spaces $V(\alpha)$.

4. ESTIMATION OF ORDER

The previous section indicates that once the system order is known the asymptotic inferential theory is rather complete. In this section we deal with the estimation of the system order. Our procedures, and indeed all procedures so far as we know, depend on the examination of $\det \hat{\Sigma}(n)$ where $\hat{\Sigma}(n)$ is the ML estimator over $\overline{M(n)}$. The criteria considered are of the form

$$\log \det \hat{\Sigma}(n) + n\, C_T, \quad C_T > 0, \quad n \leq N \tag{4.1}$$

where the C_T are a prescribed sequence of constants. The estimator, $\hat{n}$, is chosen so as to minimise (4.1). Some bound, N, on n is assumed imposed, a priori. The second term in (4.1) is needed since the first is a monotone decreasing function of n. The most commonly used rule for C_T is $2(2s+r)/T$, i.e. so that nC_T is $2T^{-1}$ by the number of system parameters. We discuss various choices below. We confine our comments, for brevity, to $M(n)$ though, for example, $M(p,q,m)$ could be treated in the same way, for example using

$$\log \det \hat{\Sigma}(p,q,m) + 2\{(p+q)s^2+sr\}T^{-1}, \quad p,q,m \leq N.$$

Other criteria have been suggested. Some of these are essentially variants of (4.1). However one important idea is to make C_T depend upon the estimation procedure. Thus if we put

$$I_\mu(\theta) = -\mathcal{E}\{\partial^2 \log \mathcal{L}_T(\mu,\sigma^2)/\partial\mu\partial\mu'\}$$

then this is the part of the information matrix relevant to $\hat{\mu}$. Then $I_\mu(\hat{\theta})^{-1}$ estimates the covariance matrix of $\hat{\mu}$. It has been suggested that the normalised trace of $I_\mu(\hat{\theta})^{-1}$ be considered i.e.

$$\text{sp}\,\{I_\mu(\hat{\theta})^{-1}\} = \{n(2s+r)\}^{-1}\,\text{tr}\{I_\mu(\hat{\theta})^{-1}\}. \tag{4.2}$$

For $n = n_0$ (or $n < n_0$) where n_0 is the true order this is $O(T^{-1})$ and $\text{sp}\{I_\mu(\hat{\theta})^{-1}\}$, or some multiple of this, might not unreasonably be used as C_T in (4.1). The reasoning is that when $n > n_0$ then $T\,\text{sp}\{I_\mu(\hat{\theta})^{-1}\}$ will become infinite and force the choice $\hat{n} = n_0$. Consider, for example, the very simplest case where $n = 1 = s$, $m = 0$. Then

$$T^{-1}I_\mu(\hat\theta) \to \sigma^2 \begin{bmatrix} \dfrac{1}{1-\alpha_0^2} & \dfrac{-1}{1-\alpha_0\beta_0} \\ \dfrac{-1}{1-\alpha_0\beta_0} & \dfrac{1}{1-\beta_0^2} \end{bmatrix}$$

so that

$$T\,\mathrm{sp}\{I_\mu(\hat\theta)^{-1}\} \to \frac{1}{2\sigma^2}\{(2-\alpha_0^2-\beta_0^2)(1-\alpha_0\beta_0)^2\}/(\alpha_0-\beta_0)^2. \quad (4.3)$$

When $n = 0$ then $\alpha_0 = \beta_0$ and evidently the right side of (4.3) tends to infinity as $(\alpha_0-\beta_0) \to 0$. Of course to investigate the asymptotic behaviour of (4.2) when $n = 0$ is a more delicate matter. It has also been suggested that $\det\{I_\mu(\hat\theta)^{-1}\}$ be considered but the use of this to replace C_T gives a term that is $O(T^{-2})$ when $n = n_0$ but, in the case considered in (4.3), is still only $O((\alpha_0-\beta_0)^{-2})$ as a function of $(\alpha_0-\beta_0)$. There is no asymptotic theory available for the use of statistics like (4.2) but there is some evidence from experience that they work well. Their use has not necessarily been formally through (4.1) but rather through an examination both of statistics such as (4.1) as well as (4.2).

We begin with a theorem relating to the parameterization of $M(n)$ when $n_0 < n$. We recall the definition $\beta \le \alpha$ above Theorem 2. We shall write $|\alpha|$ for Σn_i when the n_i are the components of α.

Theorem 5. Let $(k_0,\ell_0)\varepsilon U_c(\beta)$, $\beta \le \alpha$. Then the (k_0,ℓ_0) equivalence class in $\phi_\alpha U_c(\alpha)$ is an affine subspace of dimension $(n-n_0)s$, where $n_0 = |\beta|$ and $n = |\alpha|$.

This theorem relates to the current problem for the following reason. We know from Theorem 2(iii) that in examining $M(n)$ only neighbourhoods $U(\alpha)$ for $\beta \le \alpha$ need to be considered, at least if we a priori bound the length of the parameter vector in that coordinate system. (We will also need to bound the zeros of $\det(h)$ away from the unit circle because we are not concerned with $M_c(n)$ but with $M(n)$.). Then we may choose coordinates in $U(\alpha)$ so that $(n-n_0)s$ of these coordinatise the complementary space. Put $\phi_\alpha\mu = (\psi_\alpha,\chi_\alpha)$ where ψ_α coordinatises the affine space and χ_α its complement. The coordinates making up ψ_α may be chosen by considering $h_\alpha-g_\alpha k_0$ where h_α, g_α are the canonical factors described in (2.11), for $U(\alpha)$. Then

$$h_\alpha-g_\alpha k_0 = \sum_{-\infty}^{m} \Psi_\alpha(j)z^j, \quad m \le \max(n_j)-1$$

where we choose the elements of the coefficient matrices as co-ordinates, taking a maximum linearly independent set of these. Now $(\hat{k},\hat{\ell}) \to (k_0,\ell_0)$ so that $\hat{\chi}_\alpha \to 0$ a.s., for $\alpha \geq \beta$. However $\hat{\psi}_\alpha$ does not converge, apparently, to any limit. For the theorem we quote below we restrict ourselves to the case $m = 0$, though there is no doubt it may be extended to cover the case where $u(t)$ is present. We illustrate the result first by taking $n_0 = 0$, $n = 1$, $s = 1$. In this case all vectors in ϕU are bounded but we bound β, the parameter specifying h, by $|\beta| \leq 1-\delta$, $\delta > 0$. (Of course there is now only one coordinate neighbourhood.) Then $\chi = \beta-\alpha$ and $\psi = \frac{1}{2}(\beta+\alpha)$, may be chosen as the coordinates. It is sufficient to bound only β, away from $|\beta| = 1$, since $\hat{\chi} \to 0$ so that eventually $\hat{\alpha}$ will also be so bounded. Consider, in relation to (4.1), $-T\log\{\hat{\sigma}^2(1)/\hat{\sigma}^2(0)\}$. This has to be compared to TC_T. If it is greater than TC_T then $n = 1$ is preferred and if it is less then $n = 0$ is preferred. Because $\hat{\chi} \to 0$ then we see that $-T\log\{\hat{\sigma}^2(1)/\hat{\sigma}^2(0)\}$ eventually becomes a function only of $\hat{\psi}$. It can be shown that eventually it is of the form

$$\sup_{\psi}[T^{-\frac{1}{2}}\frac{(1-\psi^2)^{\frac{1}{2}}}{\sigma^2}\sum_1^T\{\epsilon(t)\sum_0^\infty \psi^k\epsilon(t-k-1)\}]^2 + o(1), \quad \text{a.s.} \tag{4.4}$$

where we mean by this that the $o(1)$ term converges a.s. to zero. The sup is over $|\psi| \leq 1-\delta$ and $\hat{\psi}$ is, essentially, the optimising value. Now under (3.7), (3.8) and for $E\{\epsilon(t)^4\} < \infty$, it may be shown that the expression in square brackets in (4.4) converges weakly (narrowly) to a Gaussian random process on $[-1+\delta, 1-\delta]$. If the change of variable $\tau = \log\{(1+\psi)/(1-\psi)\}$ is made then this process becomes Gaussian and stationary with absolutely continuous spectrum and density $(\cosh\omega\pi)^{-1}$. Call this process $w(\tau)$. If TC_T increases sufficiently fast it is apparent, then, that $n = 0$ will be, correctly, preferred. The dividing line is given by the law of the iterated logarithm, which however needs special proof in this case since the sum involved, namely $(1-\psi^2)^{\frac{1}{2}}\Sigma\{\epsilon(t)\Sigma\psi^k\epsilon(t-k-1)\}$, depends on a parameter and its square is being maximised with respect to that. Nevertheless the law holds so that

$$\limsup_{T\to\infty}\sup_{\psi}\ [\frac{1}{(2T\ell n\ell nT)^{\frac{1}{2}}}\frac{(1-\psi^2)^{\frac{1}{2}}}{\sigma^2}\sum_1^T \epsilon(t)\{\sum_0^\infty\psi^k\epsilon(t-k-1)]^2 = 1 \quad \text{a.s.}$$

Thus if

$$\text{lim inferior } TC_T/\{2\ell n\ell nT\} > 1 \tag{4.5}$$

then, eventually and correctly, $n = 0$ will be preferred whereas if the right side of (4.5) is <1 a.s. convergence cannot take place. We can also evaluate $P\{\hat{n}=1\}$ as $T \to \infty$ because it is just the probability that $\sup w(\tau)$ exceeds TC_T, the sup being over $|\tau| \leq \ln\{(2-\delta)/\delta\}$. If

$$\limsup TC_T < \infty \tag{4.6}$$

then

$$\lim_{\delta\to 0} \lim_{T\to\infty} P\{\hat{n}>n_0\} = 1, \tag{4.7}$$

so that the wrong decision is almost certainly made. It is also apparent that as $\delta \to 0$, $\hat{\psi}$ will tend to ± 1, since the maximising value $\hat{\tau}$ will get larger and larger.

These partial results hold in general as the following theorem shows.

Theorem 6. <u>Let $\hat{n}$ be obtained by minimising (4.1) for $N \geq n_0$. Assume (3.7), (3.8) and</u>

$$\det(k_0)\neq 0,\ |z|>1-\delta, \delta>0,\ \|\phi_\alpha \mu_0\| \leq \rho<\infty,\ \ \mathcal{E}\{|\epsilon_j(t)|^\gamma\}<\infty,\ j=1,\ldots,s. \tag{4.8}$$

<u>Then $\hat{n}$ cannot be consistent, in general, unless $C_T \to 0$. If $\gamma=4$ and the $\epsilon(t)$ are independent then $\hat{n} \to n_0$, a.s. if (4.5) holds but a.s. convergence does not hold if the inequality sign in (4.5) is reversed. If $\gamma>4$ and $TC_T\uparrow\infty$ then $\hat{n}$ converges in probability to n_0. If (4.6) holds then (4.7) holds, provided $N > n_0$.</u>

<u>Remarks</u>. (i) The fact that, when $n = 1$, $n_0 = 0$, $\psi \to \pm 1$ as $\delta \to 0$ shows how necessary it is to bound the zeros of $\det(k)$ a priori, i.e. of det h, for it shows that otherwise the optimum will be found near $\alpha = \beta = 1$ or $\alpha = \beta = -1$. Moreover the likelihood is very complicated here as is shown by the need to transform by $\tau = \log\{(1+\psi)/(1-\psi)\}$ for this stretches out the line $\alpha = \beta$ by extending the parts near ± 1 greatly, so that the correlation between neighbouring values of the process being transformed into $w(\tau)$ was very low near these points. Without such bounding, if $n > n_0$, the computer will almost certainly overflow. Of course there may be no true n_0 but situations near to that dealt with in the theorem may arise.

(ii) The last result needs to be read with caution for two reasons. In the first place it is probable that T needs to be very large before it is relevant (and δ very small). The mass in the distribution of $\sup w(\tau)$ will be concentrated near $2\ln\{2\ln(2-\delta)/\delta\}$. Even for $\delta = 0.01$ this is only about 4.7. Thus δ needs to be very small, and T then needs to be very large, for the result to be relevant. Moreover the use of

$TC_T = 4s$ is, as earlier said, a commonly used procedure. In fact there will be no true n_0 and we are trying to find an $\hat{n}$ that will give a rational transfer function model that is suitable, say for prediction. Nevertheless this result in the theorem must be of some relevance.

(iii) The occurrence of $\ell n \ell n T$ as TC_T in providing a dividing line between cases where a.s. convergence does and does not hold should not lead to the use of, say, $TC_T = 2.1 \ell n \ell n T$. Indeed $\ell n \ell n T$ changes too slowly with T for this to be very meaningful. Of course it does show that $TC_T = C \ell n T$ will give a.s. convergence and this rule has been recommended and used.

(iv) There is no doubt that the first part of the theorem will hold when the $\epsilon(t)$ are not independent but merely obey (3.7), (3.8) and (possibly) $\gamma > 4$ in (4.8). However this has not been proved. However if $TC_T \sim \ell n T$ (see (iii) above) then it has been shown that $\hat{n}$ does converge a.s. to n_0 under these conditions.

There are three further sets of observations relating to the determination of the order, n. The first of these relates to methods introduced by Akaike, again for the ARMA case. He was concerned to estimate the dynamical indices, $\alpha = (n_1, n_2, \ldots, n_s)$ and to obtain a first estimate of the parameters specifying the canonical form (sometimes called the echelon form) for elements of $V(\alpha)$. As described in (2.9) the dynamical indices define a canonical set of predictions of the $y(t)$. Let $y_{(\alpha)}(t)$ be a vector comprised of the $y_k(t+j)'$, $j = 1, \ldots, n_k$; $k = 1, \ldots, r$. (See 2.9). The $y_k(t+j|t)^k$ are (in this ARMA case) obtained by projecting the components of this vector on the space spanned by the $y(s)$, $s \geq t$. Of course with a finite amount of data this cannot be done but one might argue that if the zeros of $\det(k)$ are well away from the unit circle then only $y(s)$ for $t-\nu+1 \leq s \leq t$, for some suitable ν, would matter in this projection. Thus consider the vector $y_{(\nu)}(t)$ comprised of the $y_j(s)$, $t-\nu+1 \leq s \leq t$. Of course ν will be larger than $n = \Sigma n_j$. The technique of canonical correlation is concerned with the rank of the matrix, B let us say, of coefficients in the regression of $y_{(\alpha)}(t)$ on $y_{(\nu)}(t)$. Thus $B = \mathcal{E}\{y_{(\alpha)}(t) y_{(\nu)}(t)'\} [\mathcal{E}\{y_{(\nu)}(t) y_{(\nu)}(t)'\}]^{-1}$. In practice B would be estimated as $\hat{B}$ from $y_{(\alpha)}(t)$, $y_{(\nu)}$, $t = \nu, \nu+1, \ldots, T-\max(n_k)$. If some n_k is too large, i.e. the true indices are n_{0j} with $n_{0j} = n_j$, $j \neq k$, $n_{0k} = n_k - 1$, then there will be a vector α so that $\alpha' B = 0$. This vector will indeed describe the linear dependence in the rows of $\mathcal{H}$ (see below 2.9) that makes row (k, n_k+1) dependent on earlier rows. By examining the estimated canonical correlations (which estimate the eigenvalues of B, in the metrics induced by $\mathcal{E}\{y_{(\nu)}(t) y_{(\nu)}(t)'\}$ and $\mathcal{E}\{y_{(\alpha)}(t) y_{(\alpha)}(t)'\}$) the dynamical indices, n_{0k}, can also be estimated. How well these procedures will work in practice remains to be seen. They do not generalise readily to the ARMAX case and seem to demand T very large.

A further procedure will also be described for the scalar ARMA case, though this has been generalised to the vector ARMAX case. The idea here is to construct a prior estimate of the prediction variance, σ^2, as a benchmark against which to compare the $\hat{\sigma}^2(n)$, or $\hat{\sigma}^2(p,q)$, obtained in the process of ARMA fitting. (We use $\hat{\sigma}^2(n)$, $\hat{\sigma}^2(p,q)$ in preference to $\hat{\Sigma}(n)$, $\hat{\Sigma}(p,q)$ in the scalar case. See 4.1 and below that equation for the latter quantities.) Recall the definition of $I(\omega)$ in (3.11). We use these for $\omega_j = 2\pi j/T$, omitting $j = 0$ and $j = T/2$ if T is even. (The former amounts to making the mean correction and the latter is omitted for simplicity here; it can be included in a modified definition.) Thus there are $s = [\frac{1}{2}(T-1)]$, j values $j = 1,\ldots,s$. Form

$$\hat{\sigma}^2(1) = \exp[\frac{1}{s}\sum_1^s\{\log I(\omega_j) + \gamma\}], \quad \gamma = -\int_0^\infty e^{-x}\log x\, dx.$$

Then under reasonable conditions $\hat{\sigma}^2(1) \to \sigma^2$, a.s. Moreover $\hat{\sigma}^2(1)$ is based on minimal smoothness assumptions for $f(\omega)$ (see 3.11) so that, if $\hat{\sigma}^2(n)$, or $\hat{\sigma}^2(p,q)$, is near to $\hat{\sigma}^2(1)$, no great improvement in prediction variance can be expected from raising n, or p,q. We refer the reader to the references for further details. This brings up a further point which needs emphasis. As said before, the prediction variance is not the only criterion of excellence for an estimator that needs to be considered. It may be that what is most relevant is some minor feature of the spectrum, $f(\omega)$, near a critical frequency. If, for example, $f(\omega)$ is the spectrum of measurements of the diameter of a wave guide then most power will be at low frequencies since the diameter will be near to constant but small variations in diameter at high frequencies may be important. These will hardly affect σ^2, however, so that ARMA models may entirely fail to describe this feature and, for that matter, $\hat{\sigma}^2(1)$ will also only very weakly reflect it.

Finally we emphasise again that in these discussions we have been concerned with the situation where there is a true n_0. In practice this will not be so but the truth may be near to such a situation and it is in this light that the results must be interpreted. However there is some considerable merit in a discussion where no finite true n_0 is assumed. Consider the scalar ARMA case, for example, and the fitting of an autoregression (i.e. $q = m = 0$ in $M(p,q,m)$ and $s = 1$). Assume that

$$\sum_0^\infty \alpha(j)y(t-j) = \epsilon(t), \quad \sum_0^\infty|\alpha(j)| < \infty, \quad \alpha(0) = 1, \qquad (4.8)$$

with infinitely many $\alpha(j)$ non null, where the $\xi(t)$ are Gaussian and independent with $E(\epsilon(t)) = 0$ $E(\epsilon(t)^2) = \sigma^2$. Say we use (4.1) with $nC_T = 2p/T$ (see below 4.1). Then it can be shown that the resulting $\hat{p}_T$ minimising (4.1) is optimal in the following sense. Let N depend on T so that $N/T \to 0$ as $T \to \infty$. Let p_T be such as to minimise the error of prediction of a series $\tilde{y}(t)$, totally independent of $y(t)$ but with the same stochastic structure, when prediction is made on the basis of an estimated autoregression of order p using $y(1),\dots,y(T)$. Then $\hat{p}_T$ asymptotically gives such an error of prediction whose ratio, to that minimising error, converges to unity. Thus in this context, with (4.1) chosen in this way (due to Akaike), an optimal estimate of order is obtained.

5. OTHER CONSIDERATIONS

We shall briefly here indicate some further developments.

The first of these relates to time varying parameter models. Some special models have been successfully investigated theoretically and via simulations. The furthest advanced of these seems to be the autoregressive model of the type

$$\sum_{0}^{p} \{A(j) + B_t(j)\}y(t-j) = \epsilon(t), \qquad (5.1)$$

wherein the $y(t)$ and $\epsilon(t)$ are vectors of s components and the $B_t(j)$ are matrices of random variables totally independent of the $\epsilon(t)$ sequence and independent for different values of t. It is assumed that the $\epsilon(t)$ are a sequence of independent random vectors. Conditions for stability, in the sense of convergence of the means and covariances of the $y(t)$ when (5.1) is initiated at $t = 0$ with arbitrary initial values, have been obtained and a fairly complete inferential theory derived. Such models could be useful though random variation of the coefficients from time to time is not fully realistic.

The other type of situation that has been considered, almost entirely for $s = 1$, is that where in (2.12) H is dependent on t but is directly observed. Thus

$$y(t) = h(t)'x(t)+\eta(t), \quad x(t+1) = Fx(t)+\xi(t) \qquad (5.2)$$

where $\xi(t)$, $\eta(t)$ have covariance matrices as prescribed in (2.12). Often F would be taken to be I_n and $S = 0$. The model is useful since $h(t)$ has the interpretation of a set of observed variables in a regression relation and $x(t)$ is a time varying set of regression coefficients. The model is clearly a useful one. There is no definitive discussion, so far as I know, of the problem of determining just what <u>can</u> be known about Q and R (or F for that matter).

Non-linear models have also received attention, again mainly for $s = 1$ and $m = 0$. The most complete theoretical treatment available is probably that for a model of the form

$$y(t) = f(y(t-1),\ldots,y(t-p)) + \epsilon(t) \tag{5.3}$$

wherein $\epsilon(t)$ is Gaussian and serially independent with variance σ^2. The stability of such systems have been discussed. (Essentially f must be bounded by a linear function). One way that people have tried to handle models such as (5.3) is via approximating f by a piecewise linear function. Thus as values of the $y(t-j-\tau)$, $j = 1,\ldots,p$ at some prior time, τ, pass through a suitably defined "threshold" the structure changes to a new linear (autoregressive) structure. With $p = 1$ and one threshold, algorithms have been written but in the general case the procedure seems too unwieldy, with too much scanning, for a practicable algorithm. A preferable approach seems to be a suitably economical parametric family of functions on the right of (5.3).

A third type of model is that of the kind (again for $s = 1$ and $m = 0$)

$$\sum_0^p \alpha_j y(t-j) = \sum_{k=0}^{Q} \sum_{\ell=1}^{P} \gamma_{k,\ell}\,\epsilon(t-k)y(t-\ell) + \sum_0^q \beta_j \epsilon(t-j) \tag{5.4}$$

where $\epsilon(t)$ is as below (5.3) and

$$\alpha(z) = \Sigma\alpha_j z^j, \quad \beta(z) = \Sigma\beta_j z^j, \quad \gamma(z_1,z_2) = \Sigma\Sigma\gamma_{k,\ell} z_1^k z_2^\ell.$$

We assume $\alpha(z) \neq 0$, $|z| \leq 1$. A solution is sought in the form

$$y(t) = \sum_{r=1}^{\infty} \Sigma_r d_r(j_1,\ldots,j_r)\epsilon(t-j_1)\ldots\epsilon(t-j_r). \tag{5.5}$$

It is easy to show that

$$\delta_r(z_1,\ldots,z_r) = \Sigma_r d(j_1,\ldots,j_r) z_1^{j_1}\ldots z_r^{j_r}$$

satisfies

$$\delta_r(z_1,\dots,z_r) = (-)^r \frac{\gamma(z_1,z_2z_3\dots z_r)\gamma(z_2,z_3\dots,z_r)\dots\gamma(z_{r-1},z_r)\beta(z_r)}{\alpha(z_1\ z_2\dots z_r)\alpha(z_2\ z_3\dots,z_r)\dots\alpha(z_{r-1}\ z_r)\alpha(z_r)}$$

and hence, when the first term on the right side of (5.4) requires $k < \ell$ then a necessary and sufficient condition for the mean square convergence of (5.5) is that

$$\sum_{r=1}^{\infty} \left(\frac{\sigma^2}{2\pi}\right)^r \int_{-\pi}^{\pi}\dots\int |\delta_r(e^{i\omega_1},e^{i\omega_2},\dots,e^{i\omega_r})|^2 \prod_1^r d\omega_j < \infty$$

so that a sufficient condition is

$$\frac{\sigma^2}{2\pi}\int_{-\pi}^{\pi}\left|\frac{\gamma(e^{i\omega},e^{i\phi})}{\alpha(e^{i(\omega+\phi)})}\right|^2 d\omega \leq c < 1, \quad \phi\varepsilon[-\pi,\pi].$$

Individual cases may be more easily treated. For example if

$$y(t) = \gamma\epsilon(t-k)y(t-\ell)+\epsilon(t), \quad k,\ell > 0 \tag{5.6}$$

then (5.5) will converge a.s. if and only if

$$\ell n|\gamma| + E\{\ell n|\epsilon(t)|\} < 0 \tag{5.7}$$

assuming only that $|E\{\ell n|\epsilon(t)|\}| < \infty$ and the $\epsilon(t)$ are stationary and ergodic with $P\{\epsilon(t)=0\} = 0$.

A more difficult question is that as to whether $\epsilon(t)$ in (5.4), or (5.6) in particular, is measurable with respect to the σ-algebra of events determined by $y(s)$, $s \leq t$, for unless this is true $\epsilon(t)$ cannot be the sequence of (non-linear) innovations and also unless this is true the likelihood for these models will be difficult to construct and the models will have little meaning. If (5.7) holds as well as the conditions below that and $\ell n|\gamma| + E\{\ell n|y(t)|\} \leq 0$ then the $\epsilon(t)$ are the linear innovations but there seems to be no treatment of the general case. In any case this condition is rather unsatisfactory since it is expressed in terms of $y(t)$ not the intrinsic parameters, γ, σ^2.

6. BIBLIOGRAPHIC NOTES

Theorem 1 first came to my notice through (5). The structure of $M(n)$ discussed below that theorem is widely known and understood. Some references are (6), (23). $M(p,q,m)$, essentially, was introduced in (10). Theorem 2 is proved in (6). For the basic spectral theory used see, for example, (11). Theorems 3 and 4 are proved in (16). Some discussion of the relaxation of the conditions on $u(t)$ is given in (6). See also (11). In these discussions trends are accounted for by a preliminary trend removal by regression. A more general treatment for a scalar autoregressive model is discussed in (8). The asymptotic distribution theory for the central limit theorem is discussed in (13). The results relating to (3.13) are of the same kind as those derived in (12) but a proof for the case given in (3.13) has not yet been published. Criteria of the form of (4.1) appear first to have been suggested in (1). For $C_T = 2(2s+r)/T$ Akaike appears to have had in mind that $\mu_0 \notin M(n)$ for any n but the structure was well approximated by some $\mu \varepsilon M(n)$ for some n. In case $\mu_0 \varepsilon M(n)$ for some n he suggested $C_T = (2s+r)\log T$ and the same suggestion was made also in (22). The canonical correlation technique for the dynamical indices is due to Akaike (2). The statistic $\hat{\sigma}^2(1)$ is discussed in (4) and references therein. The discussion based on (4.8) is due to Shibata (24). The use of $\mathrm{sp}\{I_\mu(\hat{\theta})^{-1}\}$ was suggested in (26) and of $\det\{I_\mu(\hat{\theta})^{-1}\}$ in (21). Theorem 5 is proved in (6) and Theorem 6 in (14), (15). A discussion of (5.1) for the scalar case is given in (3) and for the vector case in (20). Models of the form of (5.2) have been widely considered. See (18) for some discussion and references. Systems of the form of (5.3) are considered in (17). A discussion of their linearised approximations together with references is given in (25). Models of the form of (5.4) are discussed in (9) but the analysis of their stability properties is due to the author and B.G. Quinn and has not yet been published. See also (19).

REFERENCES

(1) Akaike, H.: 1969, "Fitting autoregressive models for prediction", Ann. Inst. Statist. Math. 21, pp. 243-247.

(2) Akaike, H.: 1976, "Canonical correlation analysis of time series and the use of an information criterion". In *System Identification: Advances and Case Studies* (R.K. Mehra and D.G. Lainiotis, eds) pp. 27-96. Academic Press, New York.

(3) Andel, J.: 1976, "Autoregressive series with random parameters", Math. Operationsforsch. u. Statist. 7, pp. 735-741.

(4) Cameron, M.A.: 1978 "The prediction variance and related statistics for stationary time series", Biometrika 65, pp. 283-296.

(5) Clark, J.M.C.: 1976, "The consistent selection of parameterizations in system identification". Paper presented at JACC, Purdue University.

(6) Deistler, M., and Hannan, E.J.: 1979, "Some properties of the parameterization of ARMA systems with unknown order". To be published.

(7) Forney, D.G.: 1975, "Minimal bases of rational vector spaces with applications to multivariable linear systems". SIAM J. Control 13, pp. 493-520.

(8) Fuller, W.A., Hasza, D.P., and Goebel, J.J.: 1979, "Estimation of the parameters of stochastic difference equations", Res. Report. Dept of Stats. Iowa State Univ.

(9) Granger, C.W.J., and Anderson, A.D.: 1978, *An Introduction to Bilinear Time Series Models*. Vandenhoeck and Ruprecht. Göttingen.

(10) Hannan, E.J.: 1969, "The identification of vector mixed autoregressive-moving average systems", Biometrika 57, pp. 223-225.

(11) Hannan, E.J.: 1970, *Multiple Time Series*. Wiley, New York.

(12) Hannan, E.J.: 1975, "The estimation of ARMA models", Ann. Statist. 3, pp. 975-981.

(13) Hannan, E.J.: 1979, "The central limit theorem for time series regression", Stoch. Proc. and their Appn. 9, pp. 281-289.

(14) Hannan, E.J.: 1980, "The estimation of the order of an ARMA process", Ann. Statist. 8,

(15) Hannan, E.J.: 1980, "Estimating the dimension of a linear system". To be published.

(16) Hannan, E.J., Dunsmuir, W.T., and Deistler, M.: 1980, "Estimation of vector ARMAX models", J. Multivariate Anal. To appear.

(17) Jones, D.A.: 1978, "Non-linear autoregressive processes", Proc. Roy. Soc. London A, 360, pp. 71-95.

(18) Pagan, A.R.: 1980, "Some identification and estimation results for regression models with stochastically varying coefficients", J. Econometrics 13,

(19) Pham, Tuan D. and Tran, Lanh T.: 1980, "Quelques résultats sur les modèles bilinéaires de séries chronologiques", C.R. Acad. Sc. Paris, 290, Série A, pp. 330-338.

(20) Quinn, B.G. and Nicholls, D.F.: 1981, "The stability of autoregressive models with random coefficients", J. Multivariate Anal. 11,

(21) Rissanen, J.: 1976, "Minimax entropy estimation of models for vector processes". In *Systems Identification: Advances and Case Studies*, Eds. R.K. Mehra and D.G. Lainiotis. Academic Press, New York, pp. 97-120.

(22) Rissanen, J.: 1978, "Modeling by shortest data description", Automatica 14, pp. 468-471.

(23) Rosenbrock, H.H.: 1970, *State Space and Multivariable Theory*. Nelson, London.

(24) Shibata, R.: 1980, "Asymptotically efficient selection of the order of the model for estimating parameters of a linear process. Ann. Statist. 8, pp. 147-164.

(25) Tong, H., and Lim, K.S.: 1980, "Threshold autoregression, limit cycles and cyclical data. J. Roy. Statist. Soc. B, 42, pp.

(26) Young, P., Jakeman, A., and McMurtrie, R.: 1980, "An instrumental variable method for model order identification", Automatica 16,

RECURSIVE IDENTIFICATION

Lennart Ljung

Dept of Electrical Engineering
Linköping University, S-581 83 Linköping, Sweden

Notes compiled by

G. Bengtsson, S. Bittanti, G. Picci and A. C. M. van Swieten

Abstract

These lectures give a survey of recursive identification. The major approaches taken are briefly reviewed. A common framework, "Recursive prediction error methods" is chosen, within which many methods appear as special cases or as certain approximative variants.

Special attention is paid to asymptotic convergence properties. The family of recursive prediction error methods is shown to converge to a local minimum of a certain criterion function. If a stochastic Gauss-Newton search direction is used in the algorithm, it produces estimates with the same asymptotic distribution as an off-line version. Thus, asymptotically, no loss of information is encountered, despite the recursiveness of the method.

Some user aspects and choices are also briefly discussed, and implications for adaptive control are commented upon.

M. Hazewinkel and J. C. Willems (eds.), Stochastic Systems: The Mathematics of Filtering and Identification and Applications, 247–281.

1. INTRODUCTION

1.1. What is identification?

Identification can be defined in the following way

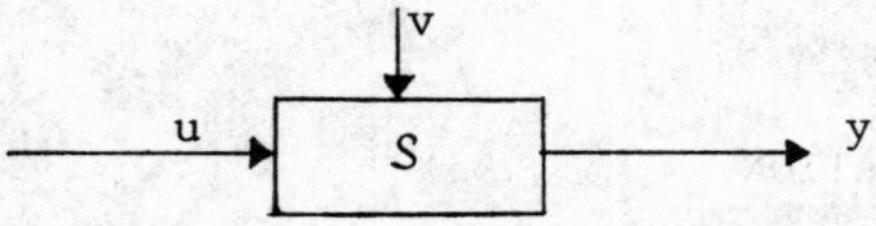

We have a dynamical system with input u and output y and disturbed by a stochastic disturbance v. We observe the sequences of inputs, u^N, and outputs, y^N, and we would like to infer a model for the system S based on the observations. So, identification is a mapping from input-output sequences to a model of the system,

$$u^N,\ y^N \to M(\hat{\Theta}_N)$$

where $\{M(\Theta)\}$ is a set of models parametrized by Θ.

1.2. Recursive identification

By recursive identification we mean that we would like to have the estimate of the parameters Θ, <u>on-line</u> i.e. recursively in time, as the process develops, and produces its sequences of inputs and outputs. More precisely, we say that <u>recursive identification</u> is a mapping of the following sort

$$\left.\begin{array}{l}\hat{\Theta}(t-1)\\ S(t-1)\\ y(t)\\ u(t)\end{array}\right\} \rightarrow \begin{array}{l}\hat{\Theta}(t)\\ S(t)\end{array}$$

At time t-1 we have available the previous model $\hat{\Theta}(t-1)$, and also a "memory vector", $S(t-1)$. At time t we receive the new measurements of input, u(t), and output, y(t), and from these quantities we infer a new model $\hat{\Theta}(t)$. We also need to update the memory vector in some way. For this algorithm to be recursive we would like the memory vector to be of finite and non increasing dimension so that we can condense the information as we get more and more data. We would like the calculations corresponding to this mapping to be finite and non increasing in time, in order to be usable on line.

1.3. Why is recursive identification interesting?

Essentially there are two situations in which recursive identification is of interest.

- One must somehow form on-line decisions related to the process. A typical situation for a control system is to design the next input, a decision that has to be taken in real time. (That is called adaptive control, if the decision is to be based on the current information of the system dynamics). Other situations could be monitoring the system to find out if some failure occurs. It could also be that one wants to identify the system only up to some degree of accuracy and stop when this accuracy is achieved.

- Another reason for recursive identification is that it is a convenient way of data reduction and summorises into the current model all past data.

1.4. Do there exist recursive identification methods?

The answer is yes, this has been the main topic of research for 15-20 years and as a result lots of methods have been produced.

It is indeed not very easy to get a good overview of the various approaches and methods which have been proposed. So if there is a problem associated with recursive identification we would list,

1 - that there are too many methods around, and it is difficult to get an overview over what possibilities one has,

2 - that many of these methods have been devised on an "ad hoc" basis and the convergence properties have been fairly unclear for a long time.

3 - We could say that the practical efficiency of the algorithms is an unsolved problem now. It might be that even if an algorithm has good asymptotic properties we might have to wait an exceedingly long time before a reasonable estimate of the parameters is obtained. So, to have a good practical choice we have to make sure that an algorithm has a good convergence rate.

4 - We are still lacking practical experience to really decide which algorithms are the best ones.

2. APPROACHES TO RECURSIVE IDENTIFICATION

Many of the different approaches that have been taken to recursive identification can be put under one of the following four headings

A - Modification of "off-line" methods
B - Nonlinear filtering
C - Stochastic approximation
D - Model reference/Adaptive observers (Pseudo-linear regression)

Before starting this dicussion we will just give some details about two input-output models that will be used repeatedly as examples in the following.

The first one is the well known and very simple difference equation model,

$$y(t)+a_1y(t-1)+...+a_ny(t-n)=b_1u(t-1)+...+b_nu(t-n)+v(t) \qquad (2.1)$$

Here u(t) and y(t) are the input and output of the system at time t and v(t) is a disturbance term taking account of the fact that there is no exact linear relationship between them.

We will sometimes use the following notation (q^{-1} is the delay operator, $q^{-1}y(t)=y(t-1)$)

$$A(q^{-1}):=1+a_1q^{-1}+...+a_nq^{-n},$$

$$B(q^{-1})=b_1q^{-1}+b_2q^{-2}+...+b_nq^{-n},$$

so that the model (2.1) can be also rewritten in polynomial notation as

$$A(q^{-1})y(t) = B(q^{-1})u(t) + v(t). \qquad (2.2)$$

We shall also rewrite the model in the following way. Collect all the parameters of the difference equation $\{a_i, b_i\}$, into a single parameter vector Θ

$$\Theta^T = (a_1, a_2, ..., a_n, b_1, ..., b_n) \qquad (2.3)$$

and collect the corresponding data into another vector $\varphi(t)$,

$$\varphi^T(t)=\left(-y(t-1),...,-y(t-n),u(t-1),...,u(t-n)\right) \qquad (2.4)$$

With the notation the difference equation (2.1) can be rewritten as

$$y(t) = \Theta^T\varphi(t) + v(t). \qquad (2.5)$$

This representation of the difference equation is convenient because it coincides with the well known linear regression model in statistics. Sometimes we will refer to this model as the "Least Squares" model (L.S.), for that reason. In most cases in real applications it is sufficient to deal with models of this type. The only drawback of the L.S. model is that we are not capable of modelling the character of the noise term. If we like to go further and get a more complicated model it would be natural to give some structure to the noise term v(t) and model it for example as a moving average

$$v(t) := e(t) + c_1 e(t-1) + \dots + c_n e(t-1) \tag{2.6}$$

where $\{e(t)\}$ is white noise.

With the above polynomial notation we shall write now the difference equation as

$$A(q^{-1})y(t) = B(q^{-1})U(t) + C(q^{-1})e(t) \tag{2.7}$$

which will be called an ARMAX model.

2.1. Modification of off line methods

If we take the L.S. model, a typical criterion to use with it would be to minimize the discrepancies between the actual outputs and the one produced by the model

$$\min_{\Theta} \sum_{k=1}^{t} [y(k) - \Theta^T \varphi(k)]^2$$

which is the usual least squares criterion for the linear regression model (2.5), the only complication being here that the vector $\varphi(t)$ is data dependent (i.e. it depends on the previous inputs and outputs up to time t). Since the criterion is quadratic in Θ it is easy to write down what is the minimizing Θ,

$$\hat{\Theta}(t) = \left[\frac{1}{t} \sum_{1}^{t} \varphi(k)\varphi^T(k) \cdot \right]^{-1} \cdot \frac{1}{t} \sum_{1}^{t} \varphi(k)y(k). \tag{2.8}$$

We form the matrix

$$R(t) := \frac{1}{t} \sum_{1}^{t} \varphi(k)\varphi^T(k) \tag{2.9}$$

which consists of correlations of inputs and outputs and write the least squares estimate as

$$\hat{\Theta}(t) = R^{-1}(t) \cdot \frac{1}{t} \sum_{1}^{t} \varphi(k)y(k).$$

Now with a very simple manipulation we can convert this estimate to an exact recursive form

$$\hat{\Theta}(t)=\hat{\Theta}(t-1) + \frac{1}{t} R^{-1}(t)\varphi(t)[y(t)-\hat{\Theta}^T(t-1)\varphi(t)] \tag{2.10a}$$

$$R(t)=R(t-1)+ \frac{1}{t}[\varphi(t)\varphi^T(t)-R(t-1)] \tag{2.10b}$$

The second equation updates the "memory matrix" R(t-1) and we just use our data at time t to update the estimate $\hat{\Theta}(t-1)$ in (2.10a). This is the recursive least squares algorithm (R.L.S.).

Normally we do not like to compute the inverse of R(t) at each step and rather update the inverse

$$P(t): = \frac{1}{t} R^{-1}(t) \tag{2.11}$$

by using the matrix inversion lemma.

This is the simplest example of a modification of an off-line method to obtain a recursive identification method. The modification was purely algebraic and we did not need to make any approximation. Normally given any off-line criterion, to transform it into a recursive method we would have to introduce approximations, but we will not discuss this point further here.

2.2. Recursive identification as non linear filtering

The second approach to recursive identification is to pose it as a non linear filtering problem. In this case we take a Bayesian viewpoint of the statistical problems and we assume that the "true value" Θ of the system parameters is a random variable (not dependent on time) and we seek the posterior density of that random variable, $p(\Theta;y^t,u^t)$ after having observed y^t and u^t, or the conditional expectation.

$$\hat{\Theta}(t) = E[\Theta|y^t u^t].$$

Let us take a simple example. Suppose we have the linear model (2.5) where $\varphi(t)$ is a function of y^{t-1}, u^{t-1} (it could be a more general non linear function of the previous data). If we suppose that the error term v(t) is white Gaussian noise and the prior density of Θ is also Gaussian, then the posterior density can be computed exactly via Bayes formula (here we assume the input deterministic i.e. independent of Θ). The result of this calculation is

$$p(\Theta;y^t u^t) \sim N\left(\hat{\Theta}(t),P(t)\right) \tag{2.12}$$

where the conditional mean $\hat{\Theta}(t)$ and variance P(t) are evaluated in the following way

$$\hat{\Theta}(t) = \hat{\Theta}(t-1) + K(t)[y(t)-\hat{\Theta}^T(t-1)\varphi(t)] \qquad (2.13a)$$

$$K(t) = P(t)\varphi(t) \qquad (2.13b)$$

$$P(t) = P(t-1) - \frac{P(t-1)\varphi(t)\varphi^T(t)P(t-1)}{1+\varphi^T(t)P(t-1)\varphi(t)} \qquad (2.13c)$$

This algorithm is identical to the R.L.S. algorithm (2.10) which was derived by straightforward algebraic manipulations. Of course we should use the matrix inversion lemma for P(t) defined in (2.11)).

Notice that for the R.L.S. scheme we need some starting point, i.e. initial conditions $\hat{\Theta}(0)$ and P(0), which can now be interpreted as the a priori knowledge of the parameter Θ. The algorithm (2.13) is identical to the Kalman filter if we make a ficticious state model

$$\begin{cases} \Theta(t+1) = \Theta(t) \\ y(t) = \varphi^T(t)\Theta(t)+e(t) \end{cases} \qquad (2.14)$$

This corresponds in standard notations, to F being the identity matrix $H=\varphi^T(t)$ and no process noise. So the conditional expectation is still given by the Kalman filter. There is a complication here, since the observation matrix is not only time varying but acutally data dependent, but this does not affect the derivation of the Kalman fitler formulas. For this simple model the Bayesion approach actually leadsus to linear filtering.

If we go to more general examples it becomes more difficult. Suppose we are faced with the standard state space model

$$\begin{aligned} x(t+1) &= F(\Theta)x(t) + G(\Theta)u(t) + w(t) \\ y(t) &= H(\Theta)x(t) + e(t) \end{aligned} \qquad (2.15)$$

where we are to estimate the parameters for the coefficients of the matrices F, G and H as well as the noise covariance matrices. One way to approach this problem is to create an extended state

$$\xi(t) := \begin{pmatrix} x(t) \\ \Theta(t) \end{pmatrix} \qquad (2.16)$$

and rewrite eq. (2.15) in terms of $\xi(t)$ instead. We would get in this way a non linear state equation of the form

$$\begin{cases} \xi(t+1) = \bar{F}(\xi(t),u(t)) + \bar{w}(t) \\ y(t) = \bar{H}(\xi(t)) + e(t) \end{cases} \tag{2.17}$$

Now if we want to find the conditional expectation of $\Theta(t)$, starting from the model (2.15), we end exactly with the non linear filtering problem for the model (2.17). By choosing any of the (approximate) non linear filtering algorithms we are able to generate a big family of recursive identification methods. Especially the Extended Kalman Filter (EKF) has been used as a parameter estimator in this context.

2.3. Stochastic approximation

Stochastic approximation has an important role in connection with recursive identification. Especially at the end of the 60's there have been several papers on recursive identification based on stochastic approximation.

To explain what stochastic approximation is, let us again consider the simple model (2.5).

We could say that we would like to choose Θ in the model (2.5) by minimizing the expected variance $V(\Theta)$ of the "unexplained" part of the observations (i.e. the error)

$$V(\Theta): = E[y(t) - \Theta^T\varphi(t)]^2, \tag{2.18}$$

and again, since this criterion is quadratic in Θ we can obtain the minimizing Θ by solving the equation

$$0 = V'(\Theta) = E\{\varphi(t)[y(t)-\Theta^T\varphi(t)]\} \tag{2.19}$$

where $V'(\Theta)$ is the gradient of $V(\Theta)$. So we are up to solve an equation like (2.19). Now, since we cannot observe expectations, somehow we have to deal with it by approximations. Here is where the classical literature on stochastic approximation can be used, because the Robbins-Monro shceme addresses exactly the problem of solving an equation where the right hand side is an expectation,

$$0 = f(\Theta) = E\, Q(\Theta,v). \tag{2.20}$$

What we have here is a function Q of a parameter Θ and a random variable v and the rule again is as follows. For any $\bar{\Theta}$ we choose, "nature", chooses a $\bar{v}$ and we can only observe the sample values $Q(\bar{\Theta},\bar{v})$. To solve (2.20) we update Θ in the following way

$$\hat{\Theta}(t) = \hat{\Theta}(t-1) + \gamma(t)Q(\hat{\Theta}(t-1),v(t)) \tag{2.21}$$

where the "correction term" $\gamma(t)Q\{\hat{\Theta}(t-1),v(t)\}$ is based on the past observation of Q at the current estimate $\hat{\Theta}(t-1)$, and $\gamma(t)$ is chosen as a sequence of positive scalar starting to zero (because we have to pay less and less attention to the last observations, which are random — the noise could destroy the estimate $\hat{\Theta}(t-1)$ constructed in the previous steps). The time index in v(t) is here simply a "numbering" of the realizations of the random variable v.

If we apply the Robbins Monro scheme to eq. (2.19) we obtain

$$\hat{\Theta}(t) = \hat{\Theta}(t-1) + \gamma(t)\varphi(t)[y(t)-\hat{\Theta}^T(t-1)\varphi(t)] \tag{2.22}$$

The term $\varphi(t)[y(t)-\hat{\Theta}^T(t-1)\varphi(t)]$ is the observation of the negative gradient of $[y(t)-\Theta^T\varphi(t)]^2$ at the current estimate $\hat{\Theta}(t-1)$. We will call this algorithm a <u>stochastic gradient</u> algorithm because the update is obtained by using a "stochastic" observation of the gradient of $V(\Theta)$. This is a kind of random search scheme, the search for the minimum being in the direction of the negative gradient of the criterion function (or what we <u>believe</u> to be the gradient). The gradient search is not very good even for deterministic problems and we would think we would get a more efficient algorithm if we change the gradient direction to the <u>Newton direction</u>, the last being determined by the inverse of the Hessian matrix

$$V''(\Theta) = E\ \varphi(t)\varphi(t)^T \tag{2.23}$$

Since the criterion was quadratic, the Hessian is acutally independent of Θ and we have a very natural approximation of the Hessian, based on observations, namely the sample variance

$$R(t) := \frac{1}{t}\ \Sigma\ \varphi(k)\varphi^T(k). \tag{2.24}$$

So the <u>stochastic Newton direction</u> would be

$$R^{-1}(t)\varphi(t)[y(t) - \Theta^T\varphi(t)] \tag{2.25}$$

and if we take this direction in place of the gradient direction in the stochastic approximation algorithm (2.22) we get

$$\hat{\Theta}(t) = \hat{\Theta}(t-1) + \gamma(t)R^{-1}(t)\varphi(t)\varepsilon(t) \tag{2.26a}$$

$$\varepsilon(t) = y(t) - \hat{\Theta}^T(t-1)\varphi(t) \tag{2.26b}$$

$$R(t) = R(t-1) + 1/t\{\varphi(t)\varphi^T(t)-R(t-1)\}.$$

We will call this algorithm a <u>stochastic Newton algorithm</u>. Now if we choose $\gamma(t)=1/t$ we get a third interpretation of the R.L.S. algorithm as a stochastic Newton algorithm.

Normally there is a vast difference in convergence rates between the gradient and Newton algorithms, in general by a factor of ten. It actually depends on what the covariance of $\varphi(t)$ looks like. If R(t) is closed to the identity there is no big differance between the two algorithms but if (as it normally is in identiciation) R(t) is far from the identity, then there is a vast difference that can be noticed in practice.

Notice that in general the Hessian matrix will be dependent on Θ and then more severe approximations than the sample covariance matrix would have to be made for the Hessian.

2.4. Adaptive observers

Suppose we have the following general state space model

$$\begin{aligned} x(t+1) &= F(\Theta)x(t) + G(\Theta)u(t) + w(t) \\ y(t) &= H(\Theta)x(t) + e(t) \end{aligned} \tag{2.27}$$

A lot of effort has been spent to find representations of this model in which the observation equation between the observed output and some kind of "state" vector $\tilde{x}(t)$ looks <u>linear</u> i.e.

$$\begin{aligned} \tilde{x}(t+1) &= \tilde{F}(\Theta)\tilde{x}(t) + \tilde{G}(\Theta)u(t) + w(t) \\ y(t) &= \Theta^T\tilde{x}(t) + e(t) \end{aligned} \tag{2.28}$$

The <u>adaptive observer</u> approach consists in treating the last equation as a <u>linear regression</u> equation, and apply R.L.S. techniques. But this equation is <u>not</u> a linear regression because $\tilde{x}(t)$ relates in a non linear way to the parameter Θ. If we neglect this dependence on Θ in $\tilde{x}(t)$ we get to <u>pseudo lienar regression</u> (P.L.R.) methods. (Solo (1978)).

To give an example of PLR let us consider the ARMAX model (2.7) and write it in the form

$$y(t) = \Theta^T\varphi(t) + e(t) + c_1 e(t-1)+\ldots+c_n e(t-n) \tag{2.29}$$

Now we could still try to treat this as a linear regression by including in the parameter vector the unknown constants $c_1,\ldots,c_n$,

$$\bar{\Theta}^T: = (\Theta^T,\ c_2,\ldots,c_n) \tag{2.30}$$

and including in a new φ vector the term e(t-i) i=1,...,n,

$$\bar{\varphi}_0(t): = [\varphi(t),e(t-1),\ldots,e(t-n)], \tag{2.31}$$

thereby rewriting the model (2.29) as

$$y(t) = \bar{\Theta}^T \bar{\varphi}_0(t) + e(t). \tag{2.32}$$

Now we could in principle use the RLS scheme for this model although there is one problem since $\bar{\varphi}_0(t)$ contains the unmeasurable noise terms $\{e(t-i)\}$. It is very natural to replace the terms $e(t-1),\ldots,e(t-n)$, by some estimates, and there are very natural estimates available, namely the current residuals

$$\hat{e}(t) := y(t) - \hat{\Theta}^T(t)\bar{\varphi}(t), \tag{2.33}$$

corresponding to some estimate of $\bar{\Theta}$, $\hat{\Theta}(t)$ available at time t. If we now replace all the e:s in $\bar{\varphi}_0(t)$ by these estimates, we can define a new vector $\bar{\varphi}(t)$

$$\bar{\varphi}(t) := [\varphi(t);\ \hat{e}(t-1),\ldots,\hat{e}(t-n)] \tag{2.34}$$

which can be used instead of $\bar{\varphi}_0$ in the R.L.S. algorithm. This algorithm is known as the Extended Least Squares (E.L.S.) algorithm and is very much used.

3. A GENERAL FRAMEWORK

One reason for the fact that there are so many different methods for recursive identification is that there are several different models that could be used and somehow it is felt that there are some direct ties between models and methods. A quick list of different models that have been used is reported below together with the algorithms that have been most commonly associated with them.

Table 3.1. Some common models in recursive identification

$A(q^{-1})y(t)=B(q^{-1})u(t)+e(t)$	[Least squares, Instrumental variable]
$A(q^{-1})y(t)=B(q^{-1})u(t)+C(q^{-1})e(t)$	[Extended least squares, Recursive maximum likelihood]
$A(q^{-1})y(t)=B(q^{-1})u(t) + \frac{1}{D(q^{-1})} e(t)$	[Generalized least squares]

Table 3.1 continued

$$A(q^{-1})y(t)=B(q^{-1})u(t) + \frac{C(q^{-1})}{D(q^{-1})} e(t)$$ [Extended Matrix method]

$$y(t) = \frac{B(q^{-1})}{F(q^{-1})} u(t) + e(t)$$ [Model reference, output error]

$$y(t) = \frac{B(q^{-1})}{F(q^{-1})} u(t) + \frac{C(q^{-1})}{D(q^{-1})} e(t)$$ [Refined instrumental variable method]

$$x(t+1)=F(\Theta)x(t)+G(\Theta)u(t)+v(t)$$
$$y(t)=H(\Theta)x(t)+e(t)$$ [Extended Kalman Filter]

From this list it is apparent that if we want to have a hope fora general framework we have to ask the underlying ideas to be independent of model structure and this implies that we must first find and discuss a fairly general model structure that would include the previous models.

3.1. Prediction error identification methods

To start this discussion we will have to say a word about the so called prediction-error (P.E.) identication method because from this method it will be possible to develop a general framework for prediction error recursive identification methods. We will show how this approach relates to or contains most of the methods that we have been discussing so far. Let us recall what the P.E. identification method is in the off-line case.

Suppose we have collected input-output data up to time t

$$y^t: = \{y(0),\ldots,y(t)\}$$

$$u^t: = \{u(0),\ldots,u(t)\},$$

we now need a model, or a model set to relate these data. But what is a model? We claim that a model is something that, given the observations up to time t, enables us to make some kind of prediction of what is to come at the next time instant. This general philosophical view agrees with the user models of dynamical systems (difference equations or state space models) commonly considered, since we can always look at these models as mathematical objects which can always be transformed into predictors in a unique way. In this general framework we will just define a model as a (deterministic) mapping which transforms the input-output data into the output space. The model will in general depend (on time and) on a finite dimensional parameter Θ. We shall use the notation

$$\hat{y}_{M}(t|\Theta) = g_{M}(\Theta;t,y^{t-1},u^{t-1}) \tag{3.1}$$

where the function $g_M(\Theta;t,\cdot,\cdot)$ is a function in the model class M of predictors and $\hat{y}_M(t|\Theta)$ is the predicted value of the output at time t. The hat has no "stochastic" significance, it is just a notation. Besides the model and the data we have a criterion to select that member of the model set M that best fits to data. In particular, based on the model and the observed data, we can form the prediction error

$$\varepsilon(t,\Theta): = y(t) - \hat{y}_{M}(t|\Theta) \tag{3.2}$$

associated with the prediction $\hat{y}_M(t|\Theta)$ which can be calculated once we have observed the actual output y(t). It is reasonable to form the criterion on some scalar measure of fit, e.g. by taking some scalar function of ε like

$$V_N(\Theta): = \frac{1}{N}\sum_{t=1}^{N} \varepsilon^2(t,\Theta)$$

or more generally as

$$V_N(\Theta): = \frac{1}{N}\sum_{1}^{N} \ell\{t,\Theta,\varepsilon(t,\Theta)\},$$

where ℓ is an arbitrary scalar function, and minimizing $V_N(\Theta)$ w.r. to Θ in order to obtain the estimate. This is off-line P.E. identification based on N data. This general form of the criterion contains several well known methods. For example if we choose the scalar criterion function ℓ as minus the log likelihood,

$$\ell(t,\Theta,\varepsilon): = -\log f(t,\Theta,\varepsilon),$$

where f is the probability density function of the innovations, we obtain the maximum likelihood estimate of Θ. If the model is linear in Θ and the criterion is quadratic we are back to L.S. methods. If the model does not explicitly depend on y we have the so called output-error methods, where we compute the prediction error based only on the inputs. If we choose ℓ such that

$$\ell(t,\Theta,\varepsilon)/\varepsilon^2 \to 0 \quad \text{as } |\varepsilon| \to \infty$$

we have the so called "robust" methods (i.e. robust w.r. to large errors).

The asymptotic properties that have been proven for P.E. methods are of the following type. Under fairly general conditions it can be proven that the estimate $\hat{\Theta}(t)$, that minimizes the P.E. criterion converges w.p.1 to a value Θ^*

$$\lim_{t\to\infty} \hat{\Theta}(t) \to \Theta^* \qquad \text{w.p.1} \tag{3.6}$$

where Θ^* minimizes the expected value of the criterion, $E\,\ell(t,\Theta,\varepsilon)$ (i.e. the model $M(\Theta^*)$ is the "best" approximation to the system, this even if there is a "true" system corresponding to a higher dimensional parametrization). This is the best we can hope for. Also the way in which the estimate $\Theta(t)$ approaches the limiting value Θ^* is asymptotically normal i.e.

$$\sqrt{t}\left(\hat{\Theta}(t)-\Theta^*\right) \in AsN(0,P) \tag{3.7}$$

and we can also write explicit expressions for the asymptotic covariance matrix P.

This is for off-line P.E. methods. We will see that for recursive P.E. we can obtain asymptotic results that are not much worse than the off-line.

3.2. Recursive prediction error methods

In the general approach to recursive identification we also need a general model description, which describes the abstract underlying ideas independent of the model structure. A model is a predictor, which somehow computes from current input and output data the future value of the output. We restrict the class of models to the linear finite dimensional ones.

$$\begin{aligned} \varphi(t+1,\Theta) &= F(\Theta)\varphi(t,\Theta)+G(\Theta)\begin{pmatrix} y(t) \\ u(t) \end{pmatrix} \\ \hat{y}(t|\Theta) &= H(\Theta)\varphi(t,\Theta) \end{aligned} \tag{3.8}$$

The restriction to finite-dimensional models is quite natural because otherwise we do not have time for the operation of the recursive algorithm.

The restriction to linear models is not substantial although it makes the analysis easier. The system itself may be not linear. However the models are in practice always linear. All the models discussed above are linear. Let

$$\frac{d}{d\Theta}\,\hat{y}^T(t|\Theta) = \psi(t,\Theta) \tag{3.9}$$

$\psi(t,\Theta)$ equals the negative gradient of the prediction error

$$\psi(t,\Theta) = -\frac{d}{d\Theta}\varepsilon(t,\Theta)$$

This is an $n_\Theta \times n_y$ matrix. It is now possible to introduce the following state space description, which can be obtained by differentiating φ and combination with (3.8)

$$\xi(t+1,\Theta) = A(\Theta)\xi(t,\Theta) + B(\Theta)\begin{pmatrix} y(t) \\ u(t) \end{pmatrix} \tag{3.11}$$

$$\begin{bmatrix} \hat{y}(t|\Theta) \\ \mathrm{col}\,\psi(t|\Theta) \end{bmatrix} = C(\Theta)\xi(t,\Theta)$$

This description shows how the output y(t) and the input u(t) produce the predicted output $\hat{y}(t)$ and the gradient of the prediction error. This will normally be quite high dimensional. This description is only used for the conceptual organising of the ideas.

Note. If we differentiate the model of (3.8) with respect to Θ, we retain the eigenvalues. The matrix A contains zero eigenvalues, but it must have the nonzero eigenvalues of F. As long as we have a stable predictor i.e. $|\lambda_{i,F}|<1$, $i=(1,...)$ the model (3.11) will be stable.

The basic idea of identification methods is to minimize the prediction error $\varepsilon(t,\Theta)=y(t)-\hat{y}(t|\Theta)$

$$\mathrm{Min}\ \frac{1}{2}E\{\varepsilon^2(t,\Theta)\} = V(\Theta) \tag{3.12}$$

We like to find that value of Θ which minimizes the sum of ε^2. It is possible to use more general criterion functions, but this does not affect the theory. The solution of this minimization problem can be obtained from the following equation

$$0 = \frac{\partial}{\partial\Theta}V(\Theta) = \psi(t,\Theta)\varepsilon(t,\Theta) \tag{3.13}$$

In order to obtain the value of Θ which is minimizing we apply the Robbins-Monroe scheme

$$\tilde{\Theta}(t) = \tilde{\Theta}(t-1) + \gamma(t)R^{-1}\psi\{t,\tilde{\Theta}(t-1)\}\varepsilon\{t,\tilde{\Theta}(t-1)\} \tag{3.14}$$

Unfortunately, this algorithm is not recursive, although it looks like it. In order to calculate $\hat{\Theta}_{t-1}$ we need all the previous data points. We have to plug in the latest estimates in the model set and start iterating the model from zero through all the data. We therefore have to replace the $\psi\{t,\hat{\Theta}(t-1)\}$ and $\varepsilon\{t,\hat{\Theta}(t-1)\}$ by their recursively computed estimates. We replace by another model where everywhere $\tilde{\Theta}_{t-1}$ is replaced by the estimate $\hat{\Theta}_{t-1}$ based on the current data. The algorithm has the following form

$$\hat{\Theta}(t) = \hat{\Theta}(t-1) + \gamma(t)R^{-1}(t)\psi(t)\varepsilon(t)$$

$$\varepsilon(t) = y(t) - \hat{y}(t)$$

$$\xi(t+1) = A\{\hat{\Theta}(t)\}\xi(t) + B\{\hat{\Theta}(t)\}\begin{pmatrix} y(t) \\ u(t) \end{pmatrix} \quad (3.15)$$

$$\begin{pmatrix} \hat{y}(t) \\ \psi(t) \end{pmatrix} = C\{\hat{\Theta}(t)\}\xi(t)$$

This equation contains a quantity R^{-1}. R^{-1} modifies the directions of the gradient in order to get better convergence rate. If we choose R to be the identity then we have the stochastic gradient algorithm. We could also choose R as an approximation of the second derivative of the criterion function, the Hessian. With the criterion $E\{\varepsilon^2\}$ a suitable such choice would be

$$R(t) \sim \frac{1}{t} \sum_{k=1}^{t} \psi(k)\psi^T(k)$$

which can be written recursively as

$$R(t) = R(t-1) + \gamma(t)\{\psi(t)\psi^T(t) - R(t-1)\}.$$

We will call this choice of R-matrix a <u>stochastic Gauss-Newton algorithm</u>.

This stochastic approximation approach leads to the same algorithm as the non linear filtering approach and the off-line algorithm. Several classes of algorithms can be brought in this form: which we call <u>recursive prediction error methods</u>.

Examples:

1) RML

$$A(q^{-1})y(t) = B(q^{-1})u(t) + C(q^{-1})e(t) \quad (3.16)$$

(ARMAX) (Åström, Söderström (1973))

$$A(q^{-1})y(t) = B(q^{-1})u(t) + \frac{1}{D(q^{-1})}e(t) \quad (3.17)$$

Recursive generalized least squares RGLS: R is blockdiagonal. Hasting-James and Sage (1969))

Maximum likelihood,ML: Gertler-Banyasz (1974).

For the Box-Jenkins structure Young (1979) has obtained the instrumental variable technique.

2)

$$y(t) = \frac{B(q^{-1})}{F(q^{-1})} u(t) + \frac{C(q^{-1})}{D(q^{-1})} e(t) \tag{3.18}$$

Refined IV: R is blockdiagonal.

Modified Extended Kalman filter (Ljung (1979a)).

3)

$$x(t+1) = A(\Theta)x(t) + B(\Theta)u(t) + k(\Theta)\varepsilon(t)$$

$$y(t) = C(\Theta)x(t) + \varepsilon \tag{3.19}$$

If we apply an Extended Kalman Filter to a state space model we get almost a recursive prediction error method. There is however one term missing, namely the coupling term between the Kalman gain and the parameters. That missing term spoils the convergence property of the Kalman fitler. With that coupling term included a prediction error method is obtained. This term is missing in the standard Extended Kalman filter because we do not model the system in innovations form. The dependence of the Kalman gain on Θ is then not explicit. Thus, the family of methods corresponding to the recursive prediction error algorithms is quite rich.

3.3. Pseudo linear regression methods

This class of algorithms starts from the adaptive observer point of view. We start with a linear model

$$\varphi(t+1,\Theta) = F(\Theta)\varphi(t,\Theta) + G(\Theta) \begin{bmatrix} y(t) \\ u(t) \end{bmatrix} \tag{3.20}$$

$$\hat{y}(t|\Theta) = \Theta^T\varphi(t,\Theta)$$

The idea of adaptive observers was to choose some representation, which allows a linear relationship between the parameter Θ and the data vector φ. All usual input/output models can be represented in this way with certain choices of parameters. The idea behind the pseudo linear regression is to treat this algorithm as a linear regression forgetting about the implicit dependence of φ on Θ. The "pseudo linear regression model" looks like

$$y(t) = \Theta^T\varphi(t,\Theta) + e(t) \tag{3.21}$$

This is treated in least squares sense. Replace φ_t by a approximation which is estimated recursively. It is easy to see how this method relates to the recursive prediction error. Recursive prediction error methods differ only in that they use ψ as the gradient in ε instead of φ. So the pseudo linear regression can be seen as a approximation of the recursive prediction error algorithm in that the non-linear dependence on Θ in $\varphi(t,\Theta)$ is neglected when forming the gradient. It is an obvious approximation. I may or may not be a successful one. The algorithm has the form

$$\hat{\Theta}(t) = \hat{\Theta}(t-1) + \gamma(t)R^{-1}(t)\varphi(t)\varepsilon(t)$$

$$\varepsilon(t) = y(t) - \hat{\Theta}^T(t-1)\varphi(t) \tag{3.22}$$

$$\varphi(t+1) = F\left(\hat{\Theta}(t)\right)\varphi(t) + G\left(\hat{\Theta}(t)\right)\begin{bmatrix} y(t) \\ u(t) \end{bmatrix}$$

Examples of the pseudo linear regression methods are

1) Extended least squares (ELS) (Young (1968), Panuska (1968)).

$$A(q^{-1})y(t) = B(q^{-1})u(t) + C(q^{-1})e(t) \tag{3.23}$$

2) Extended Matrix Method. (Talman and van den Boom (1973))

$$A(q^{-1})y(t) = B(q^{-1})u(t) + \frac{C(q^{-1})}{D(q^{-1})}e(t) \tag{3.24}$$

3) Landau's Modified reference method, Landau (1976).

$$y(t) = \frac{B(q^{-1})}{F(q^{-1})}u(t) + e(t). \tag{3.25}$$

The conclusion of Section 3 is that many or perhaps even "most" recursive identification models can be seen as the result of applying the recursive prediction error method or an approximation of it to a particular model set. In that sense we have constructed a general framework which tells us how we should understand the differences between the algorithms. With this overview we may hope to get some guidance how to choose a particular algorithm.

4. ANALYSIS OF RECURSIVE IDENTIFICATION METHODS

This part is organized as follows. In Section 4.1, we discuss what kind of problems are associated with convergence analysis. Then (Section 4.2) the different approaches to convergence analysis suggested in the literature are briefly discussed. The particular approach we will deal with, i.e. the ODE approach, is presented in Section 4.3. The convergence properties of recursive prediction error algorithms and of pseudo-linear regression recursive algorithms are then dealt with in Section 4.4 and Section 4.5 respectively. Finally, the asymptotic distribution of the recursive prediction error algorithms is discussed in Section 4.6.

4.1. Problems associated with convergence analysis

In a typical recursive algorithm, the new estimate is related to the current one by the error of the current model as follows

$$\hat{\Theta}(t) = \hat{\Theta}(t-1) + K(t)\varepsilon(t).$$

The error $\varepsilon(t)$ is computed by some time-varying filter fed by the observations. Thus, $\varepsilon(t)$ will depend upon all the previous estimates of the parameter. Consequently, the algorithm, which is recursive from the user's point of view, is not recursive from the analyst's point of view.

4.2. Approaches to the convergence analysis

The following approaches to convergence analysis have been taken. The first one consists in associating a deterministic differential equation with the recursive stochastic algorithm. The convergence properties of the algorithm are then studied in terms of the stability properties of the associated equation. This approach will be the subject of the subsequent sections. It has been suggested and worked out by Ljung (1977). Many results have been rederived by Kushner (1978) on the basis of more advanced probabilistic arguments. A second approach stands on martingale theory and it has been treated by Moore and Ledwich (1979), Solo (1978), Goodwin, Ramadge and Caines (1980). Finally, the convergence analysis has also been carried on by studying the explicit expression of the parameter estimate (Hannan 1978).

4.3. The ODE approach to convergence analysis

By means of heuristic arguments, we will now show how an ordinary differential equation (ODE) can be associated with a recursive identification algorithm. Let the algorithm be defined by the following equations

$$\hat{\Theta}(t) = \hat{\Theta}(t-1) + \gamma(t)\psi(t)\varepsilon(t) \tag{4.1}$$

$$\psi(t+1) = A\{\hat{\Theta}(t)\}\psi(t) + B\{\hat{\Theta}(t)\}\begin{pmatrix} y(t) \\ u(t) \end{pmatrix} \tag{4.2}$$

$$\varepsilon(t) = C\{\hat{\Theta}(t)\}\psi(t) \tag{4.3}$$

where $\gamma(t)$ tends to zero as t tends to infinity. Since $\gamma(t)$ is small for large t, the term $\gamma(t)\psi(t)\varepsilon(t)$ in (4.1) is small, and becomes smaller and smaller as t increases. Consequently, $\hat{\Theta}(t)$ will change slowly. If the filter (4.2), (4.3) is stable, the influence of old parameter estimates on the current $\psi(t)$ and $\varepsilon(t)$ is less relevant. Thus, we can approximatively replace $\hat{\Theta}(t)$ in (4.2), (4.3) by some nominal constant value Θ^o. This leads to the following equations

$$\bar{\psi}(t+1),\Theta^o) = A(\Theta^o)\bar{\psi}(t,\Theta^o) + B(\Theta^o)\begin{pmatrix} y(t) \\ u(t) \end{pmatrix} \tag{4.4}$$

$$\bar{\varepsilon}(t,\Theta^o) = C(\Theta^o)\bar{\psi}(t,\Theta^o) \tag{4.5}$$

The solutions of (4.2), (4.3) will be asymptotically close to the solutions of (4.4), (4.5)

$$\psi(t) \approx \bar{\psi}(t,\Theta^o)$$

$$\varepsilon(t) \approx \bar{\varepsilon}(t,\Theta^o)$$

Therefore, the updating equation will approximatively coincide with

$$\hat{\Theta}(t) = \hat{\Theta}(t-1) + \gamma(t)\bar{\psi}(t,\Theta^o)\bar{\varepsilon}(t,\Theta^o). \tag{4.6}$$

If we assume that the system (4.4), (4.5) is stable, its solutions are stationary. Then, in view of the law of large members, $\bar{\psi}(t,\Theta^o)\bar{\varepsilon}(t,\Theta^o)$ will approach

$$f(\Theta^o) \triangleq E[\bar{\psi}(t,\Theta^o)\bar{\varepsilon}(t,\Theta^o)].$$

Up to some random error, eq. (4.6) becomes

$$\hat{\Theta}(t) = \hat{\Theta}(t-1) + \gamma(t)f(\Theta^o).$$

Therefore, it seems plausible that $\hat{\Theta}(t)$ should asympotically follow the trajectories of the following ODE

$$\dot{\Theta} = f(\Theta).$$

Following these intuitive arguments, the following result can be derived under fairly general conditions.

Theorem 4.1.

Let

$$D_s = \{\Theta | A(\Theta) \text{ is stable}\}.$$

- If $\hat{\Theta}(t) \in D_s$ infinitely often w.p.1 and the ODE is globally asymptotically stable with a stationary point Θ^*, then

  $\hat{\Theta}(t) \to \Theta^*$ w.p.1.

- If $\hat{\Theta}(t) \to \Theta^*$ with positive probability, then Θ^* is a stable stationary point of the ODE.

Remarks.

4.1 Condition $\hat{\Theta}(t) \in D_s$ is not restrictive in identification. Indeed, this condition is connected with the predictor stability, which is to be met in practice to avoid the explosion of the identification algorithm. The condition can be satisfied by some suitable parameter projection into the stability region.

4.2 The second part of the above theorem may be used to prove lack of convergence. In fact, if a stationary point is not stable, then the algorithm will converge to it with zero probability.

4.4. Convergence analysis of recursive prediction error algorithms

Consider a Newton-type algorithm

$$\hat{\Theta}(t) = \hat{\Theta}(t-1) + \gamma(t)R(t)^{-1}\psi(t)\varepsilon(t).$$

Denoting by $\varepsilon(t,\Theta)$ the prediction error of the model whose parametrization is Θ, we know that $\varepsilon(t)$ is an approximation of $\varepsilon(t,\Theta)$ and $\psi(t)$ is an approximation of the gradient $\psi(t,\Theta)$ of $\varepsilon(t,\Theta)$.

The associated ODE has the form

$$\dot{\Theta} = R^{-1}f(\Theta) \tag{4.6}$$

where

$$f(\Theta) = E[\psi(t,\Theta)\varepsilon(t,\Theta)]. \tag{4.7}$$

Letting

$$V(\Theta) = E\left[\frac{1}{2}\,\varepsilon(t,\Theta)^2\right], \tag{4.8}$$

we have

$$f(\Theta) = -E\left[\left(\frac{d}{d\Theta}\,\varepsilon(t,\Theta)\right)\varepsilon(t,\Theta)\right]^T = -\left(\frac{d}{d\Theta}\,V(\Theta)\right)^T.$$

Consequently, $V(\Theta)$ can be used as a Lyapunov function for the ODE. Therefore $\hat{\Theta}(t)$ will converge to a local minimum of $V(\Theta)$ w.p.1 as t tends to infinity.

Since (4.8) is the off-line criterion, one could not hope for a better result. Also, notice that it has not been assumed that the true system belongs to the model set. Thus, we have a strong robustness property, which ensures convergence to the best approximating model in the given family. If the true system belongs to the model set, then the true parametrization will result in a global minimum point of $V(\Theta)$.

4.5. Convergence analysis of pseudo-linear regressions

In this section, we consider pseudo-linear regressions

$$\hat{y}(t|\Theta) = \Theta^T\varphi(t,\Theta).$$

Then,

$$\varepsilon(t,\Theta) = y(t) - \Theta^T\varphi(t,\Theta)$$

is the prediction error. Now, for several particular model sets, one can show that

$$\varepsilon(t,\Theta) = \tilde{\varphi}(t,\Theta)^T(\Theta-\Theta_0) + e(t);$$

here Θ_0 denotes the true parametrization, $e(t)=\varepsilon(t,\Theta_0)$ is the true system residual and $\tilde{\varphi}(t,\Theta)$ is obtained by filtering $\varphi(t,\Theta)$ through some filter which is associated with the true system

$$\tilde{\varphi}(t,\Theta) = H_0(q^{-1})\varphi(t,\Theta) \tag{4.9}$$

In view of (4.7), by letting

$$\tilde{G}(\Theta) = E[\varphi(t,\Theta)\varphi^T(t,\Theta)],$$

the ODE associated with a Newton-type algorithm is given by (4.6) with

$$f(\Theta) = \tilde{G}(\Theta)(\Theta_0-\Theta).$$

The stability properties of the ODE are then tied to the positive definiteness of $\tilde{G}(\Theta)$. By making reference to $(\Theta-\Theta_0)^T R(\Theta-\Theta_0)$ as Lyapunov function, it can be shown that Θ_0 is globally stable if $\tilde{G}(\Theta)-1/2$ is positive definite. Should the algorithm be a stochastic gradient algorithm, then the same conclusion would hold under the positive definiteness of $\tilde{G}(\Theta)$. From (4.9) it follows that $\tilde{G}(\Theta)-1/2[\tilde{G}(\Theta)]$ is positive definite if $H_0(q^{-1})-1/2[H_0(q^{-1})]$ is positive real. In view of the first part of Th. 4.1. This is a sufficient condition for the asymptotic convergence of pseudo-linear regression algorithms. The condition is not necessary though.

The second part of Th. 4.1 can be applied to derive lack of convergence conditions. Any convergence point must be a stable stationary point of the ODE. Thus, if the system obtained by linearizing the ODE around a true parameter Θ_0 is not stable, then Θ_0 cannot be a convergence point of the algorithm. In some cases, it is possible to explicitly compute the eigenvalues of this linearized system. Then, explicit conditions of lack of convergence can be derived.

Example.
Consider the Extended Least Square Algorithm applied to an ARMA-process

$$A(q^{-1})y(t) = C(q^{-1})e(t).$$

It can be shown (Holst, 1979) that the eigenvalues of the linearized ODE are given by -1 with miltiplicity equal to the degree of polynomial $C(q^{-1})$ and by $-1/C(\alpha_k)$, where α_k is any pole of the system.

Consequently, if $C(q^{-1})$ is not positive real and, for some α_k, $C(\alpha_k)<0$, then the algorithm will not converge to the true parametrization. As there is a unique stationary point of the ODE associated with an ARMA-process, the algorithm will not converge to any parameter value.

□

4.6. Asymptotic distribution of prediction-error recursive algorithms

The classical asymptotic distribution results for off-line methods are based on the following argument. As the estimate $\hat{\Theta}_N$ minimizes the estimation criterion $V_N(\Theta)$, then

$$V'_N(\Theta_N) = 0$$

(prime denotes differentiation w.r. to Θ). Assuming that Θ_0 is a true parametrization, from mean-value theorem, it follows that

$$V'_N(\Theta_0) + V''_N(\xi_N)(\Theta_N-\Theta_0) = 0 \tag{4.9}$$

where ξ_N is a suitable point in the parameter space and double-prime denotes double differentation w.r. to Θ. If the Hessian matrix is invertible, then

$$(\hat{\Theta}_N-\Theta_0) = [V''(\xi_N)]^{-1}V'_N(\Theta_0).$$

From this expression, by some central limit theorem, the asymptotic distribution results are derived. We turn now to the on-line case to show that the same argument applies to prediction error recursive algorithms. Precisely, consider a Gauss-Newton type algorithm

$$\hat{\Theta}(t) = \hat{\Theta}(t-1) + \bar{R}(t)^{-1}\psi(t)\varepsilon(t).$$

From this expression, defining

$$\tilde{\Theta}(t) = \hat{\Theta}(t) - \Theta_0,$$

we have

$$\bar{R}(t)\tilde{\Theta}(t) = \bar{R}(t)\tilde{\Theta}(t-1) + \psi(t)\varepsilon(t).$$

Since

$$\bar{R}(t) = \bar{R}(t-1) + \psi(t)\psi(t)^T"$$

one can also write

$$\bar{R}(t)\tilde{\Theta}(t)=\bar{R}(t-1)\tilde{\Theta}(t-1)+\psi(t)\left(\psi(t)^T\tilde{\Theta}(t-1)+\varepsilon(t)\right).$$

Solving this equation w.r. to $\bar{R}\tilde{\Theta}$ and adding and subtracting the true system residual e(t), one finally obtains

$$\bar{R}(t)\tilde{\Theta}(t) = \bar{R}(0)\tilde{\Theta}(0) + \sum_1^t {}_k\psi(k)e(k) +$$

$$+ \sum_1^t {}_k\psi(k)\left(\psi(k)^T\tilde{\Theta}(k-1) + \varepsilon(k)-e(k)\right). \tag{4.10}$$

Notice that

$$\varepsilon(k)-e(k)=\varepsilon(k,\Theta)-\varepsilon(k,\Theta_0) \approx \left(\frac{d\varepsilon}{d\Theta}\right)^T (\Theta-\Theta_0) \approx$$

$$-\psi(k)^T\tilde{\Theta}(k-1).$$

Therefore the last term appearing in (4.10) is asymptotically close to zero. Since $\psi(k)$ is close to the opposite of the gradient of ε at Θ_0, (4.10) will asymptotically coincide with (4.9). This means that the stochastic Gauss-Newton prediction error recursive algorithm gives use to the same asymptotic distribution as off-line prediction error algorithms. Such a conclusion is somewhat surprising. See Solo (1978) and Ljung (1980) for further details. However, we remark htat these results hold asymptotically. It might happen that for recursive algorithms to approach the asymptotic behaviour a longer transient is necessary than for off-line methods.

Conclusions.

We can summarize part 4 as follows

- Recursive prediction error methods converge with probability one to a local minimum of the expected value of the chosen criterion. So, they have the same convergence properties as the off-line prediction error methods.

- The asymptotic distribution of the recursive prediction error methods of the stochastic Gauss-Newton type coincide with the asymptotic distribution of the off-line methods.

- The convergence of pseudo-linear regression algorithms depend on the positive realness of certain transfer functions associated with the true system. Hence, in general, the convergence of the algorithm cannot be a priori guaranteed.

5. USER CHOICES

The user of an identification method has to make a choice of which algorithm he will use. This choice includes the following features

1. The model set: $A(\cdot)$, $B(\cdot)$ and $C(\cdot)$
2. "experimental condition": $u(\cdot)$
3. Criterion: $\ell(\cdot)$
4. Gain sequence: $\gamma(\cdot)$
5. Search direction: $R(t)$
6. Gradient approximation: $\psi(t) \to \varphi(t)$
7. Initial condition: $\xi(0)$, $\hat{\Theta}(0)$
8. Implementation aspects: "fast algorithms".

5.1. The model set

In order to obtain a good fit of the data a large dimension of the parameter vector is desirable. However, one has to pay a statistical penalty for large dimensions of the parameter vector. It causes less accuracy of the estimates. The accuracy has to be interpreted as the expected variance of the prediction error. One can prove that the accuracy is of the order $\dim \Theta/N$, where N is the number of data points. What we want is a flexible model set, that has the relevant information with a small number of parameters.

Other aspects that are relevant for the choice of the model set are the existence of local minima and the model order. The strategy to tackle the local minima is: first apply the robust least squares method, since this method guarantees the global minimum. If you are not satisfied you can apply one of the other algorithms. We will not discuss the second point, model order, here.

5.2. "Experimental condition" $u(\cdot)$

5.3. The criterion $\ell(\cdot)$

The asymptotic variance matrix P of the prediction error depends on the criterion $\ell(\cdot)$. It is easy to show that P achieves its minimum for

$$\ell(\varepsilon) = -\log f(\cdot)$$

where $f(\cdot)$ is the probability density function of the innovations.

This is the Cramér-Rao lower bound. Independently of the method we use, this is asymptotically the lowest bound that we can reach. Under various assumptions on the distributions of the innovations, certain choices of the criterion function can be made. For instance, a quadratic criterion function corresponds to gaussian distributed innovations.

The choice of the criterion is also important for the robustness of the method.

5.4. The gain sequence $\gamma(\cdot)$

The influence of the gain sequence on the convergence rate may be illustrated by the following situation. A second order system is used. The signal to noise ratio was 1. The model was ARMAX with b parameter. Fig. 5.1 shows the estimated value of parameter $\hat{c}_1$ as a function of the number of iterations for two different gain sequences.

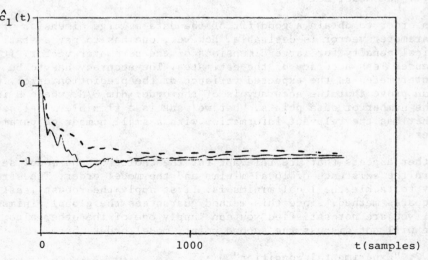

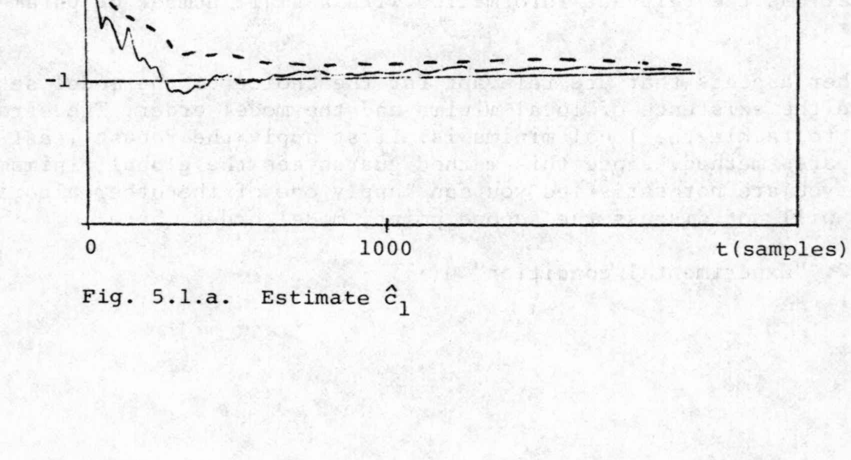

Fig. 5.1.a. Estimate $\hat{c}_1$

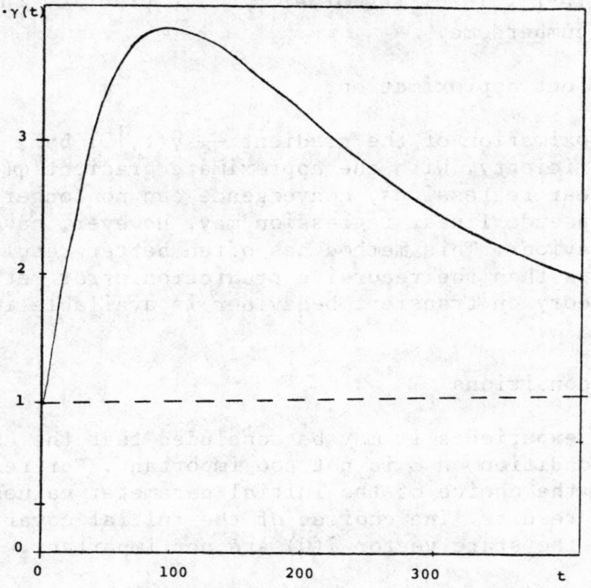

Fig. 5.1.b Gain sequence γ(t)

The change in $\mu(t)$ has a dramatic effect on the convergence rate. The reason for this appears to be the discounting of older measurements. For the least squares algorithm this can be shown as follows. The criterion function which is minimized is

$$\sum_{k=1}^{t} \beta(t,k)\,|y(k)-\Theta^T\varphi(k)|^2$$

where

$$\beta(t,k) = \frac{\gamma(t-1)}{\gamma(t)} \prod_{i=k}^{i=t} \big(1-\gamma(i)\big)$$

For $\gamma(t)$ as given, $\beta(t,k)$ will typically be increasing in k.

Asymptotically t $\gamma(t)$ should converge to 1, since for that value the optimal variance of the estimates is obtained.

5.5. The search direction R(t)

The Gauss-Newton algorithm gives strictly the best asymptotic estimates in terms of their covariance matrix. It usually gives an order of magnitude faster convergence rate than a straightforwardly applied gradient method. The number of operations for updating R^{-1} directly is of the order n_Θ^2. In practice this could sometimes be cumbersome.

5.6. The gradient approximation

Only the approximation of the gradient $\frac{d}{d\Theta}\hat{y}(t,|\Theta)$ by ψ gives asymptotic efficienty. With the approximate gradient $\varphi(t)$ used in pseudo-linear regressions, convergence can no longer be guaranteed. The pseudo linear regression may, however, have a good transient behaviour. This method has often better results within 200 à 300 steps than the recursive prediction error method. However no theory on transient behaviour is available in the literature.

5.7. Initial conditions

From personal experience it may be concluded that the choice of the initial condition on Θ is not too important. For recursive least squares the choice of the initial parameter values does not influence the results. The choices of the initial covariance matrix P(0) and the state vector $\xi(0)$ are not important.

5.8. Algorithmic aspects

Normally in the recursion formula $P(t) \triangleq \gamma(t) R^{-1}(t)$ is updated directly (n_Θ^2 operations) or with use of a square root algorithm P(t)1/2, since this algorithm has better numerical properties. We can update in a faster way by defining

$$K(t) \triangleq P(t)\psi(t) = \left[\sum_{k=1}^{k=t} \psi(k)\psi^T(k)\right]^{-1} \psi(t)$$

because we do not need P(t). This update costs only n_Θ operations. For fast algorithms see Ljung et al (1978).

6. APPLICATION OF RECURSIVE IDENTIFICATION TO ADAPTIVE CONTROL

The application of recursive identification to adaptive control will now be discussed. In connection with this concept the model reference adaptive control scheme is considered. Finally the convergence of adaptive control schemes will be analysed.

6.1. Adaptive control based on recursive identification

The problem in adaptive control is that in order to get a good regulator you have to know the system parameters. If the parameters are unknown, they should be estimated from the current data vector.

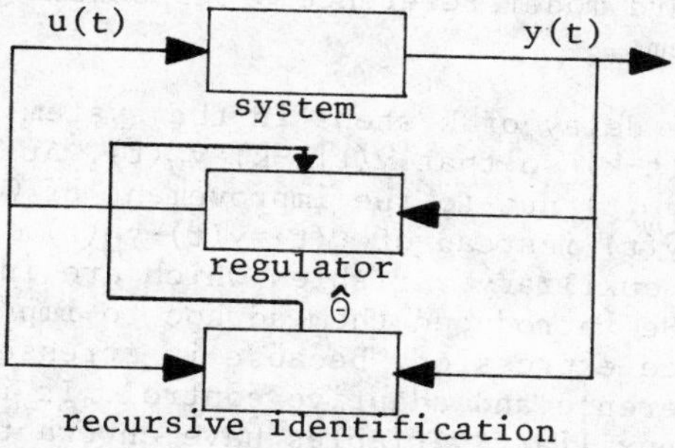

Fig. 6.1 Adaptive control scheme based on recursive identification

In Fig. 6.1 the adaptive control is visualized. The control follows from the following equations

$$\hat{\Theta}(t) = \hat{\Theta}(t-1) + \gamma(t)R^{-1}(t)\psi(t)\varepsilon(t)$$

$$\varepsilon(t) = y(t) - \hat{y}(t)$$

$$u(t) = h\left[\hat{\Theta}(t),\varphi(t)\right].$$

6.2. The model reference methods

In model references methods the control objective is to follow a reference output $y_M(t)$. The control u(t) has to be chosen in such a way that the actual output y(t) equals $y_M(t)$. The solution to this problem can be formulated as follows: choose u(t-1) in such a way that the predicted output y(t) equals the reference output $y_M(t)$. The prediction error in the parameter updating algorithm is then

$$\varepsilon(t) = y(t) - y_M(t)$$

If we redraw the scheme of fig. 6.1 for the model reference method the following scheme results.

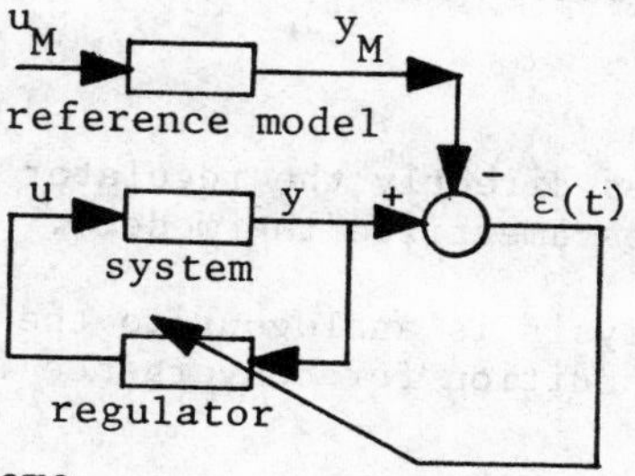

Fig. 6.2.
Model reference adaptive control scheme.

This is the standard model reference block scheme for the adaptive control system.

If there is a time delay of k steps in the system the solution changes: Choose u(t-k) so that $\hat{y}(t|t-k)=y_M(t)$. At time t-1 $\hat{y}(t)$ will differ from $y_M(t)$ due to the improvement of $\hat{\Theta}$. It is better to use $\varepsilon(t)=y(t)-\hat{y}(t)$ instead of $\varepsilon(t)=y(t)-y_M(t)$. This is, however, exactly the auxiliary variables which are introduced by Monopoli (1974). He introduced them ad hoc to improve the convergence. It is a nice expression, because it stresses the relation between model reference and adaptive control. It is interesting to know that the auxiliary variables have an exact interpretation in the sense of prediction error. This interpretation is due to Egardt (1979).

We illustrate the method with an ARMAX model.

$$Ay = Bu + Ce$$

The desired transfer function is

$$y = \frac{B}{\alpha} y_{ref} + \varepsilon$$

Use a regulator

$$u = \frac{T(q^{-1})}{R(q^{-1})} y_{ref}(t) - \frac{S(q^{-1})}{R(q^{-1})} y(t)$$

where R, S and T are determined from

$$T = \hat{C}$$

$$\alpha\hat{C} = \hat{A}R + \hat{B}S$$

The $\hat{A}$, $\hat{B}$ and $\hat{C}$ are estimated with the recursive identification method. Since the A, B and C are uniquely related with R, S and T, we drop the model and update directly R, S and T.

The control u(t) is determined such that

$$\hat{\Theta}(t)\varphi(t) = r(t+k)$$

This scheme estimates directly the regulator parameter. (We have essentially just reparametrized the model).

The convergence analysis is analogous to the extended least squares case. The condition for convergence is the positive realness of $\frac{1}{c} - 1/2$.

6.3. The convergence analysis of adaptive control methods

In the adaptive control scheme the estimated parameter vector $\hat{\Theta}$ affects ψ and ε in two ways: first as before via the identification scheme and secondly via the control law $u(t)=h(\hat{\Theta}(t),\psi(t))$. The control u depends on $\hat{\Theta}$ and hence the output y.

In a similar way as with the convergence analysis for the recursive identification method we define

$$f(\Theta) = E\{\psi(t,\Theta)\varepsilon(t,\Theta)\}$$

taking into account the dependencies on Θ in ψ and ε. Consequently, $\psi(t,\Theta)$ and $\varepsilon(t,\Theta)$ are the data and the residuals that would be obtained with a constant feedback Θ and with constant "predictor filters" Θ. The convergence of the adaptive control sheeme follows from the stability of the ordinary differential equation.

$$\dot{\Theta} = R^{-1}f(\Theta)$$

however the stability region of this differential equation is not known.

Example 6.1.
We estimate the parameters A, B and C in the model

$$A(q^{-1})y(t) = B(q^{-1})u(t) + C(q^{-1})e(t)$$

using extended least squares.

If the current estimates are used in any adaptive control law, the convergence analysis given previously holds without changes.

Hence, if $\frac{1}{C}$ -1/2 is positive real and the boundedness condition holds, then0

$$\hat{\Theta}(t) \rightarrow \{\Theta|\varepsilon(t,\Theta) - e(t)\} \text{ w.p.1.}$$

regardless of the control.

Unfortunately we cannot extend this result to the recursive prediction error method (RPEM) because of the existence of local minima. We used the criterion function itself as a Lyapunov function. This is no longer true in the adaptive control situation since ε depends in two ways of Θ. ψ does not account for the feedback part. Therefore

$$\frac{d}{d\Theta}\varepsilon(t,\Theta) \neq \psi(t,\Theta).$$

We conclude with another example

Example 6.2 Suppose the true system is given by

$$A_0(q^{-1})y(t)=q^{-k}B_0(q^{-1})u(t)+C_0(q^{-1})e(t)$$

We use the zero for indicating the true system. There is a delay of k steps. The control objective is to follow the reference signal r(t) as closely as possible. If A, B, and C were known, we would proceed as follows. First solve the equation

$$C = AF + q^{-k}S \quad \text{for F and S}$$

and use the control law

$$u(t) = \frac{Cr-Sy}{BF.}$$

This gives a closed loop system output

$$y(t) = r(t) + F(q^{-1})e(t)$$

The last part is a k-1-step moving average process. One possibility of adaptive control would be to do what we have done in Exampel 6.1. Use extended least squares, apply the corresponding adaptive control law. If the positive realness condition holds, we get convergence.

An alternative approach is to use a k-step ahead predictor.

$$y(t+k) = \hat{\Theta}^T\varphi(t)$$

where $\varphi(t)$ is some linear combination of past input and output data and future references values

$$\varphi^T(t)=[y(t),\ldots,y(t-n),u(t),\ldots,u(t-p),\ r(t+k-1),\ldots,r(t+k-n)]$$

In a similar way as before we obtain the adaptive control algorithm

$$\hat{\Theta}(t+k) = \hat{\Theta}(t+k-1) + \frac{1}{t}R^{-1}(t)\varphi(t)\varepsilon(t+k)$$

$$\varepsilon(t+k) = y(t+k) - \hat{\Theta}^T(t+k-1)\varphi(t)$$

$u(t)$ determined from $\hat{\Theta}^T(t-k)\varphi(t)=r(t)$.

An analysis of this scheme shows that it converges if $1/C_0(q^{-1})-$ $-1/2$ is positive real, provided the boundedness condition holds. □

7. CONCLUSIONS

The conclusions of presented material can be summarized as

- it is possible to give a general approach to the problem of recursive identification. ("Recursive prediction error methods")
- with a proper approximation, the recursive identification methods have asymptotically the same properties as the off-line methods
- more work has to be done on the transient behaviour of the estimates in order to improve the convergence rate
- recursive identification is the key to adaptive control.

REFERENCES

Åström, K. J. and P. Eykhoff (1971). System Identification - a survey. Automatica 7, pp 123-162.

Blum, H. (1954. Multidimensional stochastic approximation methods. Ann. Math. Stat. 25, pp 737-744.

Egardt, B. (1979). Unification of some condinuous-time adaptive control schemes. IEEE Trans AC-24, pp 558-592.

Gertler, J. and Cs. Bânyâsz (1974). A recursive (on-line) maximum likeoihood identification method. IEEE Trans AC-19, pp 816-820.

Goodwin, G. C., P. J. Ramadge and P. E. Caines (1980). Discrete time stochastic adaptive control. SIAM J on Control and Optimization, To appear.

Hannan, E. J. (1976). The convergence of some recursions. Ann. Statistics, Vol 4, pp 1258-1270.

Hannan, E. J. (1978). Recursive estimation based on ARMA models. To be published.

Hastings-James, R. and M. W. Sage (1969). Recursive generalized least squares procedure for on-line identification of process parameters. IEE Proc. 116, pp 2057-2062.

Holst, J. (1980). Local convergence of some recursive stochastic algorithms. Proc. 5th IFAC Symposium on Identification and System Parameter Estimation, Darmstadt, pp 1139-1146.

Isermann, R. U., W. Bamberger Bauer, P. Kenpo and H. Sieberg (1974). Comparison of six on-line identification and parameter estimation methods. Automatica 10, pp 81-103.

Kushner, H. J. and D. S. Clark (1978). Stochastic Approximation Methods for Constrained and Unconstrained Systems. Springer Verlag.

Landau, I. D. (1976). Unbiased recursive identification using model reference adaptive techniques. IEEE Trans. AC-21, pp 194-202.

Ljung, L. (1976). Consistency of the least squares identification method. IEEE Trans. Vol AC-21, No 5, pp 779-781.

Ljung, L. (1977a). Analysis of recursive stochastic algorithms. IEEE Trans AC-22, pp 551-575.

Ljung, L. (1977b). On positive real functions and the convergence of some recursive schemes. IEEE Trans AC-22, pp 539-551.

Ljung, L. (1978a). Convergence analysis of parametric identification methods. IEEE Trans. AC-23, pp 770-783.

Ljung, L. (1978b). On recursive prediction error identification algorithms. Report LiTH-ISY-I-0226, Dept of Electrical Engineering, Linköping University, Sweden. To appear in Automatica January 1981.

Ljung, L. (1978c). Convergence of an adaptive filter algorithm. Int. J. Control, Vol 27, No 5, pp 673-693.

Ljung, L., M. Morf and D. Falconer (1978d). Fast calculation of gain matrices for recursive estimation schemes. Int J Control Vol 27, No 1, pp 1-19.

Ljung, L. (1979). The extended Kalman filter as a parameter estimator for linear systems. IEEE Trans. AC-24, pp 36-50.

Ljung, L. (1980). Asymptotic gain and search direction for recursive identification algorithms. IEEE Conference on Decision and Control, Albuqurque.

Monopoli, R. V. (1974). Model reference adaptive control with an augmented error signal. IEEE Trans. AC-19, pp 474-484.

Moore, J. B. and H. Weiss (1979). Recursive prediction error methods for adaptive estimation. IEEE Trans. SMC-9, pp 197-204.

Moore, J. B. and G. Ledwich (1979). Multivariable adaptive parameter and state estimators with convergence analysis. J. Australian Math. Soc.

Panuska, V. (1968). A stochastic approximation method for identification of linear systems using adaptive filtering. Proc. JACC.

Panuska, V. (1969). An adaptive recursive least squares identification algorithm. Proc IEEE Symp on Adaptive Processes, Decision and Control.

Saridis, G. N., Z. J. Nikolic and K. S. Fu. Stochastic approximation algorithms for system identification, estimation and decomposition of mixtures. IEEE Trans. SSC-5, pp 8-15.

Saridis, G. N. (1974). Comparison of six on-line identification algorithms. Automatica 10, pp 69-79.

Solo, V. (1968). Time series recursions and stochastic approximation. Ph.D.-thesis Australian National University, Canberra.

Söderström, T. (1973). An on-line algorithm for approximate maximum likelihood identification of linear dynamic systems. Report 7308, Dept of Automatic Control, Lund Inst of Tech, Lund, Sweden.

Söderström, T., L. Ljung and I. Gustavsson (1978). A theoretical analysis of recursive identification methods. Automatica, 14, pp 231-244.
Talmon, J. L. and A. J. W. van den Boom (1973). On the estimation of transfer function parameters of process and noise dynamics using a single stage estimator. Proc. 3rd IFAC Symposium on Identification and System Parameter Estimation, The Hague/Delft.
Tsypkin, Ya. Z. (1971). Adaptation and Learning in Automatic Systems. Academic Press.
Young, P. C. (1968). The use of linear regression and related procedure for the identification of dynamic processes. Proc. 7th IEEE Symposium on Adaptive Processes, UCLA.
Young, P. C. and A. Jakeman (1979). Refined instrumental variable methods of recursive time series analysis. Int J Control, Vol 24, pp 1-30.

THE PROPERTIES OF THE PARAMETERIZATION OF ARMAX SYSTEMS IN STRUCTURAL SPECIFICATION AND THEIR RELEVANCE FOR ESTIMATION

M. Deistler

Technical University of Vienna

Topological and geometrical properties of the parameterization of ARMAX systems and their consequences for estimation are considered. Thereby we deal with the case when a-priori information about the parameters is available (structural specification).

1. Introduction

In dealing with inference in ARMAX systems, in addition to the purely algebraic problem of identifiability, some topological and geometrical properties of the parameterization turn out to be important.

In many applications there is a-priori information concerning the ARMAX parameters available and the ARMAX parameters have direct "physical" interpretation. In this case we use the (econometric) term "structurally" specified systems to distinguish from the case where there are no such restrictions and where we have "full" equivalence classes.

Here we only deal with systems in structural specification. Thereby genuine differences and difficulties occur. Analogous results for the nonstructural case have been derived in Deistler and Hannan (1979). Partly similar problems have been investigated by Kalman (1974), Clark (1976), Hazewinkel (1977) and Deistler, Dunsmuir and Hannan (1978).

Consider the ARMAX system

$$(1)\quad \sum_{i=0}^{p} A(i)y(t-i) = \sum_{i=1}^{r} D(i)x(t-i) + \sum_{i=0}^{q} B(i)\varepsilon(t-i)$$

M. Hazewinkel and J. C. Willems (eds.), Stochastic Systems: The Mathematics of Filtering and Identification and Applications, 283–289.

where $A(i)$, $B(i) \in \mathbb{R}^{s \times s}$, $D(i) \in \mathbb{R}^{s \times m}$ are parametermatrices and where

$$g(z) := \sum_{i=o}^{p} A(i)z^i; \; h(z) := \sum_{i=o}^{q} B(i)z^i; \; d(z) := \sum_{i=1}^{r} D(i)z^i$$

The assumption $\det g(z) \not\equiv o$ is part of the definition of an ARMAX system. $x(t)$ and $y(t)$ are observed inputs and outputs respectively and $x(t)$ is assumed to be nonstochastic, for simplicity. $(\varepsilon(t))$ is unobserved white noise:

$$E\varepsilon(t) = o; \quad E\varepsilon(s)\varepsilon'(t) = \delta_{st}.\Omega$$

Furthermore let

$$k(z) = \sum_{s=o}^{\infty} K(i)z^i : = g^{-1}(z).h(z); \; l(z) = \sum_{s=o}^{\infty} L(i)z^i : = g^{-1}(z).d(z)$$

denote the errors- and input transfer-functions respectively.

In the case of structurally specified ARMAX systems we usually have an a-priori prescription of the maximal lags with can occur for every variable, i.e. we have an a-priori prescription $\alpha = (p_1, \ldots, p_s, q_1, \ldots, q_s, r_1, \ldots, r_m)$ of the maximal degrees of the columns $g_1, \ldots, g_s$, $h_1, \ldots, h_s$, $d_1, \ldots, d_r$ of (g,h,d). Furthermore, in the case of structural specification we have additional restrictions on the parameters. As this second point is not so important for our results, for the sake of simplicity we will assume throughout

$$A(o) = B(o) = I$$

and that the parameters which are not explicitely restricted are "free" parameters. We also assume that $k(z)$ and $l(z)$ are uniquely determined from $(x(t))$ and $(y(t))$. For $l(z)$ this is implied by a suitable persistent exciting condition on $(x(t))$. $k(z)$ is uniquely defined, e.g. by the following assumptions:

$$\Omega \text{ is nonsingular}; \det g(z) \neq o \quad \forall |z| \leq 1; \; \det h(z) \neq 0 \quad \forall |z| < 1$$

We will not impose the last two assumptions explicitely here; their consequences for the results obtained are easily seen and these assumptions could also be replaced by more general ones.

A polynomial matrix (g,h,d) such that $(k,l) = g^{-1}(h,d)$ is called a <u>(left)matrix fraction</u> description (MFD) of the transfer-functions (k,l). The set of all MFD's corresponding to the transfer-functions (k,l) is called <u>(k,l)-equivalence class</u>. These equivalence classes, also called classes of <u>observational equivalence</u>, correspond to the equivalence kernel of the mapping π defined by $\pi(g,h,d) = g^{-1}.(h,d)$. A class of MFD's (or of ARMAX systems) is called <u>identifiable</u> if

π restricted to this class is injective, i.e. if any two observationally equivalent MFD's are identical.

2. Some Properties of the Parameterization

In most cases we assume that the MFD's are (relatively) left prime (i.e. they have no nonunimodular common left divisor). If (g,h,d) is left prime then a non necessarely left prime MFD $(\bar{g},\bar{h},\bar{d})$ is observationally equivalent to (g,h,d)if and only if there exists a polynomial matrix u such that

(2) $(\bar{g},\bar{h},\bar{d}) = u.(g,h,d)$

If $(\bar{g},\bar{h},\bar{d})$ is left prime too, then u must be unimodular (Hannan 1971).

Let α be the a-priori specified column degrees, let $a_i(j)$, $b_i(j)$ and $d_i(j)$ denote the coefficients of z^j in g_i, h_i and d_i respectively and finally let

$$C(-i): = (a_1(p_1-i),\ldots, a_s(p_s-i), b_1(q_1-i),\ldots, b_s(p_s-i), d_1(r_1-i),\ldots, d_m(r_m-i))$$

where $a_i(j) = b_i(j) = o,\ \ j < o;\ d_i(j) = o;\ \ j \leq o$

If we assume, that (g,h,d) are left prime and that the corresponding "column-end-matrix" C(o) has full rank s then, as easily can be seen, the matrix u in (2) must be constant. Therefore, the set of all left prime MFD's with prescribed column degrees α where C(o) has rank s (and where A(o) = I) is identifiable (Hannan 1971).

For prescribed α, every (g,h,d) may be characterized by a vector $\tau \in \mathbb{R}^n$, where $\tau: = \text{vec}\ (a_1(1),\ldots,a_1(p_1),\ldots, a_s(1),\ldots, a_s(p_s), b_1(1),\ldots, b_s(q_s), d_1(1),\ldots, d_m(r_m))$ ("vec" is understood rowwise and gives a row vector) and where $n: = s.(p_1+\ldots+p_s+q_1+\ldots+q_s+r_1+ +\ldots+r_m)$. We will identify (g,h,d) and τ. For given α let Θ_α be the set of all $\tau \in \mathbb{R}^n$ which are left prime and where the column-end-matrix has rank s. Thus Θ_α is the parameter-space under our identifiability assumptions. By $U_\alpha: = \pi(\Theta_\alpha)$ we denote the corresponding set of transfer-functions. Due to identifiability, the restriction of π to Θ_α , $\pi: \Theta_\alpha \to U_\alpha$ is bijective; $\psi: = \pi^{-1}: U_\alpha \to \Theta_\alpha$ is called the parameterization of U_α.

Of course, the natural topology for Θ_α is the relative Euclidian topology. From the point of view of (maximum-likelihood-) estimation, the natural topology for sets of transfer-functions (k,l) corresponds to the relative topology in the product space $(\mathbb{R}^{s\times(s+m)})^{\mathbb{Z}_+}$ for the power series coefficients $(K(i), L(i))_{i\in\mathbb{Z}_+}$. We call this topology the pointwise topology T_{pt}.

The next theorem clearifies the structure of the parameter space. The proof is given in Deistler (1980)

<u>Theorem 1:</u>

(i) Θ_α is open and dense in $\mathbb{R}^n$

(ii) For every $(k,l) \in \pi(\mathbb{R}^n - \Theta_\alpha)$, the corresponding (k,l)-equivalence class $\pi^{-1}(k,l) \subset \mathbb{R}^n - \Theta_\alpha$ is an affine subspace of $\mathbb{R}^n$ with dimension

$$n_{(k,l)} := s.(l\text{-rank}\begin{pmatrix} C(o), & o, & \dots\dots\dots, & o \\ C(-1), & C(o), & o, \dots\dots, & o \\ \dots & \dots & \dots & \dots \\ C(-l+1), & \dots\dots & \dots\dots, & C(o) \end{pmatrix}) > o$$

where $l := s.p$ and the $C(i)$ correspond to an arbitrary left prime $\tau \in \pi^{-1}(k,l)$.

Let the $(s.p + s.q + m.r) \times s(q + s.p)$ matrix $\tilde{K}$ be defined by

$$\tilde{K} := \begin{array}{l} \text{p blocks} \\ \\ \text{q blocks} \\ \\ \text{r blocks} \end{array} \left(\begin{array}{c|c} \begin{matrix} -I & -K(1), & \dots\dots\dots \\ 0 & -I & -K(1) \dots\dots \\ & \ddots & \ddots \\ 0 \,.\, .\, 0 & -I & -K(1) \dots \end{matrix} & \begin{matrix} -K(q) \dots\dots\dots -K(q+s.p-1) \\ -K(q-1) \dots\dots -K(q+s.p-1) \\ \dots\dots\dots\dots\dots\dots \\ -K(q-p) \dots\dots -K((s-1)p+q) \end{matrix} \\ \hline \begin{matrix} I, 0, \dots\dots\dots\dots, 0 \\ \ddots \\ 0 \dots\dots\dots\dots, 0, I \end{matrix} & 0 \\ \hline 0 & 0 \end{array} \right)$$

q blocks

and let the (s.p+s.q+m.r) $\times$ r(s.p+r) matrix $\tilde{L}$ be defined by

$$
\tilde{L} := \begin{array}{l} \\ p \text{ blocks} \\ \\ q \text{ blocks} \\ \\ r \text{ blocks} \\ \\ \end{array}
\left(\begin{array}{c|c}
0,-L(1)\ldots -L(p) & \ldots\ldots\ldots\ldots -L(r+s.p-1) \\
\vdots & \vdots \\
0\ldots\ldots 0,-L(1) & \ldots\ldots\ldots\ldots -L((s-1)p+r) \\
\hline
0 & 0 \\
\hline
I,0,\ldots\ldots\ldots,0 & \\
\ldots\ldots\ldots\ldots & 0 \\
\underbrace{0,\ldots\ldots\ldots,0,I}_{r \text{ blocks}} &
\end{array}\right)
$$

Let C_α denote the matrix consisting of the following rows of $(\tilde{K},\tilde{L})$: $1,1+s,..1+(p_1-1).s,2,2+s,...2+(p_2-1).s,...,s.2s,...p_s.s$, $s.p+1$, $s.p+1+s,...s.p+1+(q_1-1)s,..,s.p+q_s.s,s(p+q)+1,s(p+q)+1+r,...s(p+q)+1+(r_1-1).m,..,s(p+q)+m,...s(p+q)+r_m.m$.

Then the equation

(3) $(h,d) = g(k,1)$

may be written as (Deistler 1980):

(4) $c = \tau.(I_s \otimes C_\alpha)$

where

$$c := \text{vec}(K(1),K(2),\ldots,K(sp+q),L(1),L(2),\ldots,L(sp+r))$$

and where

$$I_s \otimes C_\alpha := \begin{pmatrix} C_\alpha,0\ldots 0 \\ 0, C_\alpha\cdots 0 \\ \hline 0,\ldots,0,C_\alpha \end{pmatrix}$$

The likelihood function only depends on the transfer-functions (and on Ω and the initial values) but not explicitely on the structural parameters τ. It is shown in Hannan, Dunsmuir and Deistler (1980) that if α has been specified and if the true (k,1) are in $\bar{U}_\alpha$ (where $\bar{U}_\alpha$ denotes the T_{pt}-closure of U_α in the set of all transfer-functions) then (under certain additional assumptions) their maximum likelihood estimators converge almost surely in T_{pt} to (k,l). Thereby the maximization is performed over $\bar{U}_\alpha$; i.e. we cannot exclude the possibility that estimates (k_t, l_t) are in $\bar{U}_\alpha - U_\alpha$ (even if the true transfer-functions are in U_α). This is the reason why the boundary of U_α has to be investigated. For the proof of the two theorems following, we again refer to Deistler (1980).

Theorem 2:

(i) $(k,l) \in U_\alpha$ if and only if $I_s \otimes C_\alpha$ has rank n and c is in the linear span of the rows of $I_s \otimes C_\alpha$.

(ii) $(k,l) \in \pi(\mathbb{R}^n - \Theta_\alpha)$ if and only if $I_s \otimes C_\alpha$ has rank less than n and c is in the linear span of the rows of $I_s \otimes C_\alpha$.

(iii) If $(k,l) \in \bar{U} - \pi(\bar{\Theta}_\alpha)$ then c is not in the linear span of the rows of $I_s \otimes C_\alpha$ and $I_s \otimes C_\alpha$ has rank less than n.

Let $\beta = (\bar{p}_1, \ldots, \bar{r}_m)$ be prescription of column degrees; we write $\beta \leq \alpha = (p_1, \ldots, r_m)$ if $\bar{p}_1 \leq p_1; \ldots, \bar{r}_m \leq r_m$

Theorem 3:

(i) $\psi_\alpha : U_\alpha \to \Theta_\alpha$ is a $(T_{pt}-)$homeomorphism

(ii) U_α is open in $\bar{U}_\alpha$

(iii) $\pi(\bar{\Theta}_\alpha) = \bigcup_{\beta<\alpha} U_\beta$

(iv) $\pi(\bar{\Theta}_\alpha) \subseteq \bar{U}_\alpha$ and equality holds for s=1.

3. Implications for Estimation

We now shortly discuss the implications of the preceeding results for estimation and specification. For more details see Deistler (1980). Let (k_t, l_t) be an estimate of $(k,l) \in \bar{U}_\alpha$ converging to (k,l). Then three different cases can be distinguished:

(a) First consider the case $(k,l) \in U_\alpha$. By theorem 3 (i) and (ii) (k_t, l_t) will be in U_α from a certain t_0 onwards and $\psi(k_t, l_t) \to \psi(k,l)$ $(t \geq t_0)$. Thus in this case the maximum likelihood estimators are (strongly) consistent for the true parameters $\tau = \psi(k,l)$ and we will say that α has been correctly specified. Note that for every rational transfer-function (k,l) there exists an α such that $(k,l) \in U_\alpha$ (Deistler, Dunsmuir and Hannan (1978)). The correct specification of α can be checked from the estimates of C_α (see theorem 2).

(b) Let $(k,l) \in \pi(\bar{\Theta}_\alpha) - U_\alpha$. To exclude the possibility $(k_t, l_t) \in \bar{U}_\alpha - \pi(\bar{\Theta}_\alpha)$

(see Deistler, Dunsmuir and Hannan (1978)) we have to introduce suitable prior bounds on the structural parameters in this case. Then the parameter-estimates will converge to the equivalence class $\pi^{-1}(k,l)$ (but not necessarely to a fixed point within this class). As indicated by theorem 3 (iii), there is a correct specification $\beta < \alpha$, i.e. a $\beta < \alpha$ such that $(k,l) \in U_\beta$. β can be found using theorem 2 (ii).

(c) The worst case is when $(k,l) \in \bar{U}_\alpha - \pi(\bar{\Theta}_\alpha)$. From theorem 2 (iii) we see that even if $(k_t,l_t) \in U_\alpha$ $\forall t$, $(k_t,l_t) \rightarrow (k,l)$ implies $\| \psi(k_t,l_t) \| \rightarrow \infty$, i.e. the computer will overflow. By an investigation of the linear dependence relations of the rows of the matrix $(\tilde{K},\tilde{L})$ a $\beta \neq \alpha$ can be found such that $(k,l) \in U_\beta$.

References

(1) Clark, J.M.C.:"The Consistent Selections of Parameterizations in Systems Identification." Paper presented at JACC (1976).

(2) Deistler, M.:"The Properties of the Parameterization of ARMAX Systems and Their Relevance for Structural Estimation and Dynamic Specification." Paper presented at the 4th World Congress of the Econometric Society, Aix-en-Provence 1980.

(3) Deistler, M., W. Dunsmuir and E.J. Hannan:"Vector Linear Time Series Models: Corrections and Extensions." Adv.Appl.Prob. 10, (1978), 360-372.

(4) Deistler, M. and E.J. Hannan:"Some Properties of the Parameterization of ARMA Systems with Unknown Order." Mimeo (1979).

(5) Hannan, E.J.:"The Identification Problem for Multiple Equation Systems with Moving Average Errors." Econometrica 39 (1971), 751-767.

(6) Hannan, E.J., W. Dunsmuir and M. Deistler:"Estimation of Vector ARMAX Models." To appear in J. Multivariate Analysis (1980).

(7) Hazewinkel, M.:"Moduli and Canonical Forms for Linear Dynamical Systems II: The Topological Case." Math. Systems Theory 10, (1977), 363-385.

(8) Kalman, R.E.:"Algebraic Geometric Description of the Class of Linear Systems of Constant Dimension." 8th Ann. Princeton Conf. on Information Sciences and Systems, Princeton, N.J. 1974.

TESTS OF ADEQUACY FOR ARMA MODELS AND TESTS OF SEPARATED HYPOTHESES

Guégan Dominique

C.S.P. Université Paris-Nord

The purpose of the NATO-ASI "Stochastic systems" is to study the important advances in the last years in stochastic system theory, particularly identification, recursive linear and non linear filtering and adaptive control. When we have identified a model we need to test its adequacy : it is this subject we want to discuss here. In a first part we are going to give some developments on the main tests used in the Arma case to verify the model. In a second part we present an asymptotic test constructed in the case of separated families, we then give some thoughts on the use of this test.

I. Tests of adequacy for models

Let us consider, for example an Arma (p,q) process X_t , $t \varepsilon Z$ following the equation $\Phi(B)X_t = \Theta(B)\varepsilon_t$ where ε_t is a gaussian white noise, B is the linear operator defined in the L^2 space generated by the family (ε), Φ and Θ the polynomials

$$\Phi(B) = 1 - \phi_1 B - \dots - \phi_p B^p \ , \quad \Theta(B) = 1 - \theta_1 B - \dots - \theta_q B^q$$

whose roots are supposed of modulus greater than 1. We suppose p and q known and from an observation $X^N = (X_1,\dots,X_N)$ we deduce estimates $\hat{\phi}_i$, $\hat{\theta}_i$, $\hat{\sigma}^2_\varepsilon$ of the other parameters. As we are interested in the tests of hypotheses on the coefficients of the model, a null hypothesis which is particularly important is the independence of the random terms. There exists a non parametric approach without assuming the gaussian hypothesis. For this point of view, we can find a review in [14].
In the parametric case, we can consider different approaches. Suppose we have identified two models from the observed data and we want to decide the best one, a usual way is to study the cor-

M. Hazewinkel and J. C. Willems (eds.), Stochastic Systems: The Mathematics of Filtering and Identification and Applications, 291–297.

relation of the estimated residuals of the models. First we study the variance of these coefficients of correlations , cf [5]. In effect, we need their variance to be small ; we know that it corresponds to the log-likelihood [1], and it appears in the calculus of the errors of the estimated values of the parameters. Generally this first analysis is not sufficient and we can mistake if we use it alone, cf [3], so we need global tests.
For that, consider the following statistic :

$Q(\hat{r}) = N \sum_{k=1,m} \hat{r}_k^2$, with $\hat{r}_k = \Sigma\, \hat{\varepsilon}_t \hat{\varepsilon}_{t-k} / \Sigma\, \hat{\varepsilon}_t^2$, m represents the number of considered correlations, (for example $m = 20$ for $N = 200$). Box and Pierce in [10] show that the asymptotic distribution of $\hat{r}_k$ is gaussian, and Box and Jenkins in [9] show that, in the case of an Arma (p,q), $Q(\hat{r}) \sim \chi^2_{m-p-q}$ for $N >> m+p+q$. Then we consider the following test : for a given level a_N , we define a reject region $\{Q(\hat{r})>C_N\}$ where C_N is such that $P(\chi^2_{m-p-q} > C_N) = a_N$. It is the Portmanteau's test. In fact this test is not a very good one : it turns out that using this test nearly always we accept the model. Its global formulation doesn't take into account the repartition of the correlations. For example:
- If all correlations are near zero, except only one which is big enough, we can obtain a value for the χ^2 which permits acceptance of the model, even if looking at the correlogram it is obvious we must reject it. - On other side, if all the correlations are very small but nonzero , and if the value of the statistic $Q(\hat{r})$ is too big when we compare it to the theoretical value of the χ^2 , we must reject the model, when a direct observation shows that we may accept it. In the object to take better into account the variance, Ljung and Box in [19] propose the use of another statistic $\tilde{Q}(\hat{r}) = N(N-2) \sum_{k=1,N} (N-k)^{-1} \hat{r}_k^2$ and they show that $\tilde{Q}(\hat{r}) \sim \chi^2_{E\tilde{Q}(\hat{r})}$ with $E\tilde{Q}(\hat{r}) = mN(N+2)^{-1}(1-(m+1)(2N)^{-1})$.

They show that numerically this approximation is better than the previous one. So in the Portmanteau's test, it seems natural to replace the statistic $Q(\hat{r})$ by $\tilde{Q}(\hat{r})$.
Two other tests are mainly used too, the Durbin Watson's test and the test of periodogram. The first one is based on the following statistic, cf [13] and [14],

$d = \sum_{i=1,N-1} (\hat{\varepsilon}_{i+1} - \hat{\varepsilon}_i)^2 / \sum_{i=1,N} \hat{\varepsilon}_i^2$ computed from a model of regression $X_t = \beta_1 U_1 + \ldots + \beta_k U_k + \varepsilon_t$, (or $X = U\beta + \varepsilon$ in the matricial form), where the U_i's are independent fixed variables.

So this test is constructed when the matrix U is fixed, but U is random as soon as we work in the Arma case and this produces errors at using this test. The other one, cf [6] and [12], is based on the cumulated periodogram estimated by

$$s_j = \sum_{r=1,j} p_r / \sum_{r=1,m} p_r \quad \text{with} \quad p_r = 2N^{-1} \left| \sum_{t=1,N} \hat{\varepsilon}_t \, e^{2i\pi rt/N} \right|^2$$

By this test we define an upper and a lower bound and we observe the graph $j \longrightarrow s_j$. If this graph stays between these two bounds, we accept the model, but as soon as the graph crosses one of these bounds, we reject it. Before to stop with this review we just want to note the existence of some tests which permit to decide the order of a model, for example Anderson's test [4], the corner's test [7], Hannan's test and Quenouille's test [17] and [20]. Nevertheless this rapid review permits to see that in the identification of a model, actually the ways to test its adequacy are not really satisfactory.

II. Tests of separated hypothesis

Let us consider a sup / sup test from an asymptotical point of view, generalising the results of A.M. Walker [21]. For the stationary process X_t, $t \in Z$, we want to test the composite hypothesis Θ_0 against Θ_1. For $\alpha \in \Theta_0$, we denote the spectral density by $f(\lambda,\alpha)$, and for $\beta \in \Theta_1$, by $g(\lambda,\beta)$. To these densities corresponds a family of distributions P_0 under Θ_0 and P_1 under Θ_1, for an appropriate space of sampling $\mathfrak{X}$. We suppose that the space of parameters is compact and the hypotheses are asymptotically separated in the Kullback sense, i.e.

$J(\Theta_0,\Theta_1) = \inf_{\alpha\varepsilon\Theta_0,\beta\varepsilon\Theta_1} K(\alpha,\beta) > 0$, where $K(\alpha,\beta)$ is the Kullback's information whose value in the present case is given by :

$$K(\alpha,\beta) = (2\pi)^{-1} \left[\int \log f(\lambda,\alpha) g^{-1}(\lambda,\beta) d\lambda + 1 - \int f(\lambda,\alpha) g^{-1}(\lambda,\beta) d\lambda \right]$$

We denote by $L_N(X,\alpha)$ the log likelihood of the process under Θ_0 and $L_N^*(X,\beta)$ under Θ_1. Under the hypothesis $(A_1$-$A_6)$ (given below), we use the following approximation for the likelihood (cf [15]), under Θ_0 :

$$L_N(X,\alpha) = N(2\pi)^{-1} \left[\int \log f(\lambda,\alpha) d\lambda + \int I_n(\lambda) f^{-1}(\lambda,\alpha) d\lambda \right]$$

$I_N(\lambda)$ is the periodogram of the process. We define in the same way the log-likelihood under Θ_1, $L_N^*(X,\beta)$. If $\hat{\alpha}_N$ and $\hat{\beta}_N$ are the estimates of the maximum likelihood, we define $\beta_\alpha \in \Theta_1$, as the nearest point from Θ_0, by the relation :

$$\int f(\lambda,\alpha)(g'_\beta \cdot g^{-1})(\lambda,\beta_\alpha) d\lambda = 0$$

(the derivative is taken w. r. t. β).

The hypotheses used are : (all the derivatives are taken w.r.t.λ).
A_1 - $f(\lambda,\alpha)$, $f'(\lambda,\alpha)$, $f''(\lambda,\alpha)$ belong to $L^2(d\lambda)$,
A_2 - the function $f(\lambda,\bullet)$ is cont. derivable at the second order,
A_3 - the function $g(\lambda,\beta)f^{-1}(\lambda,\alpha)$ is integrable.
We denote $b(\lambda,\alpha) = f^{-1}(\lambda,\alpha)$. Its first, second and third derivatives w.r.t.λ exist,
A_4 - $b'(\lambda,\alpha)$ is absolutely continuous in (λ,α),
A_5 - $b''(\lambda,\alpha)$ and $b'''(\lambda,\alpha)$ are continuous in (λ,α),
A_6 - m_1 and m_2 are constants such that: $\forall\alpha\ \forall\lambda\quad 0 < m_1 < f(\lambda,\alpha)$ and $|f''_\alpha(\lambda,\alpha)| < m_2$.
A_7 - let us fix two neighbourhoods of α and β_α such that $P_0(\hat{\alpha}_N \notin V_\alpha) = O(1)$ and $P_1(\hat{\beta}_N \notin V_\beta) = O(1)$; then the functions $f'f^{-1}(\lambda,\alpha')$, $f'(\lambda,\alpha')(g'g^2)^{-1}(\lambda,\beta')$, $f''(\lambda,\alpha')h(\lambda)$ where $h(\lambda) = f^{-1}(\lambda,\alpha) - g^{-1}(\lambda,\beta)$ are majorated by functions belonging to $L^2(d\lambda)$, for all $\alpha' \in V_\alpha$ and $\beta' \in V_\beta$.
We have then the following results : cf [15].

Theorem 1
Under the hypothesis A_1-A_2 $\hat{\alpha}_N \longrightarrow \alpha$ P_0 - a.s.
Under the hypothesis A_1-A_4 $\hat{\beta}_N \longrightarrow \beta_\alpha$ P_0 - a.s.
Under the hypothesis A_1-A_5 $(\hat{\alpha}_N,\hat{\beta}_N)$ is asymptotically normally distributed, under Θ_0 and Θ_1.

So now we want to construct a test on the following statistic :

$$Z_N = \sup_\alpha L_N(X,\alpha)/\sup_\beta L^*_N(X,\beta) = L_N(X,\hat{\alpha}_N)/L^*_N(X,\hat{\beta}_N)$$

K being finite, the hypotheses Θ_0 and Θ_1 are not separated at finite distance. But we know that under Θ_0, Z_N tends a.s. to a constant A/B with $A-B = K(\alpha,\beta_\alpha) > 0$, so that Θ_0 and Θ_1 are asymptotically separated. In other words, if we consider the statistic $\hat{L} = L_N(X,\hat{\alpha}_N) - L^*_N(X,\hat{\beta}_N)$, under Θ_0, $\hat{L}$ has the same behaviour as $NK(\alpha,\beta_\alpha)$ and tends a.s. to $+\infty$, symmetrically under Θ_1, $\hat{L}$ tends a.s. to $-\infty$. The hypotheses Θ_0 and Θ_1 are asymptotically separated, and the test based on this statistic is coherent. But the statistic $\hat{L}$ is not free under Θ_0 and we can't construct a reject region independent of α. In the object to have an asymptotically free distribution let us modify the statistic Z_N. We propose to norm it and to center it : let us study the statistic T_N

$$T_N = (\hat{L} - E_{\hat{\alpha}_N}\hat{L})(\mathrm{var}_{\hat{\alpha}_N}\hat{L})^{-1/2}$$

where $E_{\hat{\alpha}_N}$ and $\mathrm{var}_{\hat{\alpha}_N}$ represent the expectation and the variance taken under $P_{\hat{\alpha}_N}$. We can show the following result :

Theorem 2

Under the hypotheses A_1-A_7, the asymptotic distribution of T_N under Θ_0 is a centered gaussian law $N(0,\Gamma(\alpha))$ where

$$\Gamma(\alpha) = (2\pi)^{-1} K'_\alpha(\alpha,\beta_\alpha) I^{-1}(\alpha) V^{-1}(\alpha) \quad \text{with}$$

$$I(\alpha) = \int (f'^2 f^{-2})(\lambda,\alpha)\,d\lambda$$

$$K'_\alpha(\alpha,\beta_\alpha) = \int f'(\lambda,\alpha) g^{-1}(\lambda,\beta)\,d\lambda \quad \text{and}$$

$$V(\alpha) = 2\pi\left[1-2\int f(\lambda,\alpha) g^{-1}(\lambda,\beta_\alpha)\,d\lambda + \int f^2(\lambda,\alpha) g^{-2}(\lambda,\beta_\alpha)\,d\lambda\right]$$

For details see [15] and [16]. We can remark that we obtain for T_N a distribution which is not free, it depends on α. To avoid that, we are going to estimate $\Gamma(\alpha)$ by $\Gamma(\hat{\alpha}_N)$. In $\Gamma(\alpha)$ there are only constants which depend of the information obtained by simple passage to the limit under the integral sign. For example :

$$I(\hat{\alpha}_N) = \int (f'^2 f^{-2})(\lambda,\hat{\alpha}_N)\,d\lambda = \int (f'^2 f^{-2})(\lambda,\alpha)\,d\lambda + (\hat{\alpha}_N-\alpha)\,\ell_N(\alpha)$$

where $\ell_N(\alpha)$ is a bounded function, $(\hat{\alpha}_N-\alpha)$ is an $o(N^{-1/2})$, so the error made by this kind of approximation is of order $N^{-1/2}$, and we have $\mathrm{Var}\, T_N = \Gamma(\alpha) + O(N^{-1})$.

Theorem 3

Under the hypotheses A_1-A_7,

$$T_N \times \Gamma(\hat{\alpha}_N)^{-1} \quad \text{tends to a reduced centered gaussian law.}$$

In the same way, we can compute the distribution of the statistic T_N under Θ_1, and we can show (cf [15], [16]) the following result :

Theorem 4

Under the hypotheses A_1-A_7, the distribution of T_N, under Θ_1, tends to $+\infty$.

So the last two theorems permit to construct a consistant test based on the statistic T_N appropriately normed and centered, whose distribution in law is free with a reject region independent of α whose form is $\{T_N > C_N\}$ corresponding at a level a_N defined by $P(\Phi > C_N) = a_N$. Then the power of the test is near $\phi[C_N - N\Gamma(\hat{\alpha}_N) S^{-1}(\beta)]$ where ϕ represents the distribution of the law $N(0,1)$ and where :

$$S(\beta) = 2\pi\left[\int \frac{g^2(\lambda,\beta)}{f^2(\lambda,\alpha_\beta)}\,d\lambda + 1 - \int \frac{g(\lambda,\beta)}{f(\lambda,\alpha_\beta)}\,d\lambda\right].$$

α_β is defined as the same way as β_α.

III. Remarks and conclusions

The present test is a test of separated hypothesis. We can test for example an autoregressive model of order 1, $X_t-\alpha X_{t-1} = \varepsilon_t$ where $\alpha > \alpha_0 \neq 0$ against a moving average model of order 1, $X_t = \varepsilon_t + \beta\varepsilon_{t-1}$ with $\beta > \beta_0 \neq 0$. But if the two families have a common point, particularly if the true value of the parameter is near this point, then we are not in the case of separated hypothesis, but in the case of contiguous hypothesis. An article which seems decisive in this domain is the article of K.O. Dzaparidze [11]. He studies the distribution of the ratio of likelihood under Θ_0 and Θ_1 under the hypothesis of Le Cam [18]. In fact he tests under Θ_0, $f(\lambda,\alpha,o)$ against $f(\lambda,\alpha,\beta)$ under Θ_1 with $\beta = O(N^{-1/2})$ so $\beta = 0$ against $\beta \neq 0$. In this case, K.O. Dzaparidze shows that the asymptotic distribution of

$$- 2 \log L_N(\lambda,\hat{\alpha},O)L_N^{-1}(\lambda,\alpha, \frac{\beta}{\sqrt{N}}) \quad \text{is a} \quad \chi'^2$$

This point of view permits to consider the case of embedded hypotheses. To be complete on this problem, we have to note the point of view of Akaike [2] completed by Bouaziz [8] who study the asymptotic distribution of the likelihood ratio for estimates they suppose asymptotically gaussian and convergent to a same value. We can represent this by the figure below :

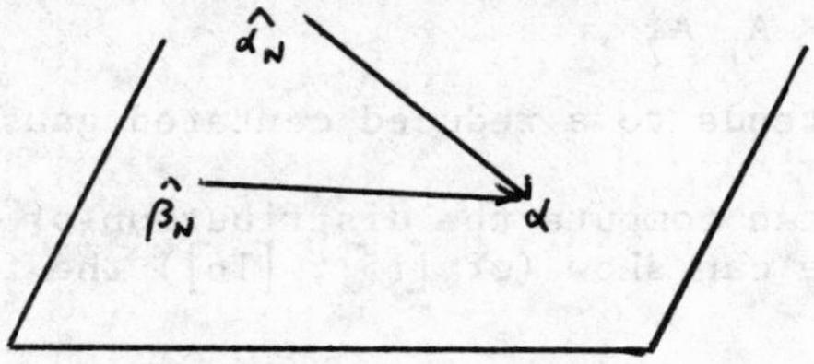

M. Bouaziz shows then, if p is the dimension of the parameters's space Θ_0 and r the dimension of the parameters's space Θ_1, the law of the likelihood's radio is then a $\chi^2(p-r)$. At last we can note that if we now consider an Armax (= Auto Regressive Moving Average with Exogenous Variables) model,

$$\sum_{j=0,p} \alpha_j X_{t-j} = \sum_{j=1,\infty} \gamma_j U_{t-j} + \sum_{j=0,\infty} \beta_j \varepsilon_{t-j}$$

as those used by Ljung, the test presented above is always consistent.

Bibliography

[1] H. Akaike, On the likelihood of a time series model. Inst. of Stat. Conference on time series analysis. Cambridge University, 1978.

[2] H. Akaike, Statistical predictor identification. Ann. Inst. Stat. Math. 22, p.203-217, 1970.

[3] O.D. Anderson, Time series analysis and forecast. The Box-Jenkins approach. Butterwortts, 1976.

[4] T.W. Anderson, The statistical analysis of time series. John Wiley, 1971.

[5] M.S. Bartlett, On the theoretical specification and sampling properties of autocorrelated time series. J.R.S.S. Serie B.8, p. 27-41, 1946.

[6] M.S. Bartlett, Problème de l'analyse spectrale des séries temporelles stationnaires. Publi. Stat. Université Paris 3, p. 119-134, 1957.

[7] J.M. Beguin, C. Courieroux, A. Montfort, Identification of a mixed autogressive-moving average process : the corner method. Doc de travail 7902, INSEE, 1979.

[8] M. Bouaziz, Identification de l'ordre de dépendance dans les séries temporelles. Thèse de 3ème cycle, Université Paris-Sud Orsay, 2450, 1978.

[9] G.E.P. Box, G.M. Jenkins, Time series analysis. Forecasting and control, Holden day, 1976.

[10] G.E.P. Box, D.A. Pierce, Distribution of residual autocorrelations in ARIMA time series models, J.A.S.A. , Vol. 65, p. 332, 1970.

[11] K.O. Dzaparidze, Tests of composite hypotheses for random variables and stochastic processes. Theory Prob. Its Appl. 22, 2, p. 104-120, 1977.

[12] J. Durbin, Tests of serial independance based on the cumulated periodogram. Bull. Inst. Stat. 42, 1947.

[13] J. Durbin, G.S. Watson, Testing serial correlation in least square regression II. Biometrika, 38, p. 159-178, 1951.

[14] D. Guegan, Tests d'adéquation pour des processus stationnaires, 10, Pré-publications Math. Université Paris 13, 1979.

[15] D. Guegan, Loi limite de la statistique de tests de vraisemblance, Thèse 3ème cycle, Université Paris 13, 2201, 1979.

[16] D. Guegan, Tests d'hypothèses séparées pour des processus stochastiques. C.R.A.S. 289, serie A, p. 241, 1979.

[17] E.J. Hannan, Time series analysis, London, 1960.

[18] L. Le Cam, Théorie asymptotique de la décision statistique. Montréal University, Press, 1969.

[19] L.A. Ljung, G.E.P. Box. Biometrika. 64, n°3, p. 297, 1978.

[20] M.H. Quenouille, A large sample test for the goodness of fit of autoregressive schemes J.R.S.S. 110, Serie A, p. 123-129, 1947.

[21] A.M. Walker. Biometrika. 54, 1 and 2, p. 39-68, 1967.

SOME NONLINEAR FILTERING PROBLEMS ARISING IN RECURSIVE IDENTIFICATION

P.S. Krishnaprasad* and Steven I. Marcus+

*University of Maryland, College Park, Maryland 20742
+University of Texas, Austin, Texas 78712

Abstract: In this paper, we outline an intrinsic formulation of the identification problem of linear system theory. The nonlinear filtering problems which appear in this way essentially fall into four distinct classes, distinguished by their estimation algebra. In principle, it is possible to explicitly solve the identification problem in the 'hyperbolic cases' using classical methods from the theory of partial differential equations. This is illustrated by an example which indicates the required sufficient statistics for solving the identification problem.

1. INTRODUCTION

Consider the linear stochastic differential system,

$$dx_t = A(p)x_t dt + b_1 u_t dt + b_2 dw_t$$
$$dy_t = \langle q, x_t \rangle\, dt + dv_t \tag{1.1}$$

where u_t denotes a known input function, $\{w_t\}$ and $\{v_t\}$ are independent Brownian motion processes, $\{x_t\}$ and $\{y_t\}$ are respectively the state and measured output processes. For reasons of identifiability we let,

$$A(p) = \begin{bmatrix} 0 & 1 & & & \\ & & 1 & & \\ & & & \cdot & \\ & & & & \cdot \quad 1 \\ -p_1 & , -p_2, & \ldots & & -p_n \end{bmatrix}$$

M. Hazewinkel and J. C. Willems (eds.), Stochastic Systems: The Mathematics of Filtering and Identification and Applications, 299–304.

the rational canonical form associated with $p=(p_1,\ldots,p_n)'\in\mathbb{R}^n$ and we let, $q=(q_1,\ldots,q_n)'\in\mathbb{R}^n$. The vectors b_1 and b_2 are known and fixed. When p and q are known, the state-estimation problem for (1.1) has the well-known solution -- the Kalman-Bucy filter. By the identification problem we shall mean the problem of jointly estimating the state and the parameters -- in other words it is the nonlinear filtering problem for the extended system with state $z_t=(x_t,p_t,q_t)$ defined by,

$$\begin{aligned} dx_t &= A(p_t)x_t\,dt + b_1u_t dt + b_2 dw_t \\ dp_t &= 0 \\ dq_t &= 0 \ , \end{aligned} \tag{1.3}$$

$$dy_t = \langle q_t, x_t\rangle dt + dv_t . \tag{1.4}$$

More precisely, in solving the identification problem one seeks a solution to the Kushner-Stratonovitch eqn. [1] satisfied by the conditional density p(t,z) given the observations y_s, $0\le s\le t$. Although this point of view goes back to Kushner [2], progress along these lines has been impeded by the nonlinearity of the Kushner equation.

More recently, it has been recognized [3,4,5] that an understanding of the evolution equation satisfied by the so-called unnormalized conditional density, $\psi(t,z)$ is essential for further progress in nonlinear filtering theory. Knowing $\psi(t,z)$, the conditional density is determined by the normalization

$$p(t,x) = \psi(t,z)/\int\psi(t,z)\,dz. \tag{1.5}$$

In the general situation when $\{z_t\}$ is a diffusion process with observation (semimartingale) $\{y_t\}$ of the form

$$dy_t = h(z_t)\,dt + dv_t \tag{1.6}$$

it is known that, $\psi(t,z)$ satisfies a linear stochastic partial differential equation (Mortenson-Duncan-Zakai equation) of the Ito type,

$$d\psi(t,z) = \mathcal{L}\psi(t,z)\,dt + h(z)\,\psi(t,z)dy_t \tag{1.7}$$

where $\mathcal{L}$ is the Kolmogorov forward operator associated with the diffusion $\{z_t\}$. (See the expository paper by Davis and Marcus in these proceedings). Now, as regards questions related to the complexity of a nonlinear filtering problem geometric ideas play a crucial role and one looks at the Stratonovitch version of (1.7) written formally as,

$$\partial\psi/\partial t(t,z)=[\mathcal{L}-\tfrac{1}{2}h^2(z)]\psi(t,z)+h(z)dy/dt. \tag{1.8}$$

In particular, the operator Lie algebra $\tilde{G}$ generated by $\mathcal{L}-\frac{1}{2}h^2$ and h, known as the estimation algebra [3], has been emphasized by Brockett, Mitter, Ocone and others as an object of central interest. In what follows we use the estimation algebra to classify identification problems and investigate special cases.

2. ESTIMATION ALGEBRAS:

The fact that the estimation algebra is invariant under change of coordinates [6], makes it useful as a classifying tool. In particular the choice of canonical forms is not crucial. Essentially there are four cases:

Case (1) $b_2=0$; $b_1=0$. Then $\mathcal{L}=-\langle Ax,\partial/\partial x\rangle-tr(A)$. Define $A_0=\mathcal{L}-\frac{1}{2}\langle q,x\rangle^2$; $A_1=\langle q,x\rangle$. Then the estimation algebra $\tilde{G}=\{A_0,A_1\}_{L.A.}$ is defined by the structure equations, $[A_0,A_j]=A_{j+1}$; $[A_j,A_k]=0$ $k,j\geq 1$ where, $A_j=(-1)^{j-1}\langle q,A^{j-1}x\rangle$ $j=1,2\ldots$. We have a sequence of abelian ideals $\tilde{G}_n=\text{span}\{A_j;j=n,n+1,..\}n\geq 1$, with finite codimension (a feature of potential value in connection with approximation problems).

Case (2) $b_1=e_n$ say; $b_2=0$. The presence of deterministic inputs does not alter the structure of the estimation algebra. Note $\mathcal{L}=-\langle Ax,\partial/\partial x\rangle-tr(A)-u_t\langle b_1,\partial/\partial x\rangle$. Define $A_0=\mathcal{L}-\frac{1}{2}\langle q,x\rangle^2$ and $A_j=(-1)^{j-1}\langle q,A^{j-1}x\rangle+(-1)^{j-1}\langle q,A^{j-2}b_1\rangle u_t$, $j\geq 2$. Then once again $[A_0,A_j]=A_{j+1}$; $[A_j,A_k]=0$; for $j,k\geq 1$.

Case (3) $b_1=e_n$; $b_2=0$ and the parameters p are known. $A_0,A_1,\ldots$ etc are as in the previous case. But by the Cayley-Hamilton theorem $A^n(p)=-\sum_{i=1}^{n} p_i A^{i-1}(p)$. Hence the operators A_k for $k>n+1$ are linearly dependent on the operators A_j for $j\leq N+1$. Then the estimation algebra $G=\{A_0,A_1\}_{L.A.}$ is finite dimensional. In fact using tensor products it can be shown that the underlying filtering problem is linear.

Case (4) $b_2\neq 0$. The presence of driving noise drastically alters the structure of the Lie algebra. The general situation is not unlike the example below:

$$dx_t = \alpha x_t dt + dw_t$$
$$d\alpha_t = 0$$
$$dy_t = x_t dt + dv_t$$

$$\mathcal{L}=-\alpha x\,\partial/\partial x-\alpha+\tfrac{1}{2}\partial^2/\partial x^2;\ A_0=\mathcal{L}-\tfrac{1}{2}x^2;\ A_1=x.$$

Define $A_{2n+1} = (\alpha^2+1)^n x$; $A_{2n+2} = (\alpha^2+1)^n(\partial/\partial x-\alpha x)$ with $n=0,1,2,\ldots$. Also let $B_k = -(\alpha^2+1)^k$, $k=0,1,2,\ldots$. Then the structure equations are,

$$[A_0,A_{2n+1}]=A_{2n+2};\quad [A_0,A_{2n+2}]=A_{2n+3}\quad [A_{2n+1},A_{2m+1}]=0;$$
$$[A_{2n+2},A_{2m+2}]=0;\ [A_{2n+1},A_{2m+2}]=B_{m+n}.\ [B_j,A_k]=0;\ [B_j,B_k]=0.$$

It is possible to write down filtrations of $\tilde{G}$ by sequences of ideals as before. In fact in each case above the algebra $\tilde{G}$ is a profinite dimensional filtered Lie algebra (see Hazewinkel-Marcus [10]). All the known nonlinear filtering problems that admit finite dimensionally computable statistics have Lie algebras of this type.

It is however important to note that the identification problem in our formulation is tractable in the cases (1), (2), (3) above where there is no driving noise. In these cases, the Stratonovitch form of the evolution equation for the unnormalized conditional density is given by,

$$\partial\psi/\partial t = (A_0 + A_1\dot{y})\psi \tag{2.1}$$

where, $A_0 = -\langle Ax, \partial/\partial x\rangle - tr(A) - u_t\langle b_1, \partial/\partial x\rangle - \frac{1}{2}\langle q,x\rangle^2$, and $A_1 = \langle q,x\rangle$. In principle, equation (2.1) can be solved by the method of characteristics. See below.

3. AN EXAMPLE

Consider the special case of (1.3)-(1.4) given by

$$\begin{aligned} dx_t &= -\alpha x_t dt + u_t dt \\ dy_t &= x_t dt + dv_t \end{aligned} \tag{3.1}$$

Then (2.1) reduces to,

$$\partial\psi/\partial t = (\alpha x - u_t)\partial\psi/\partial x + (\alpha - x^2/2 + x\dot{y})\psi. \tag{3.2}$$

Let the initial condition be given by $\psi(t,x,\alpha)|_{t=0} = \psi_0(x,\alpha)$. In the 4-dimensional (t,x,α,z) space, we want to pass an integral hyper-surface S: $z=\psi(t,x,\alpha)$ of the equation (3.2) through the 2-dimensional manifold Γ (Cauchy data) given parametrically by $x=s_1$, $\alpha=s_2$, $t=0$, $z=\psi_0(s_1,s_2)$. The characteristics passing through points (s_1,s_2) in Γ sweep out S and are given by the system of (characteristic) differential equations:

$$\begin{aligned} dx/d\tau &= -(\alpha x - u_t) \\ d\alpha/d\tau &= 0 \\ dt/d\tau &= 1 \\ dz/d\tau &= (\alpha - x^2/2 + x\dot{y}_t)z \end{aligned} \tag{3.3}$$

Solving (3.3), we obtain a parametric representation of the characteristic curves; $\alpha=s_2$; $t=\tau$; $x=X(s_1,s_2,\tau)= e^{-s_2\tau}s_1 + \int_0^\tau e^{-s_2(\bar{\tau}-\sigma)}u_\sigma d\sigma$ and finally,

$$z=\psi_0(s_1,s_2)\cdot\exp(s_2\tau)\cdot\exp\left(\int_0^\tau X(s_1,s_2,\sigma)\dot{y}_\sigma d\sigma - \tfrac{1}{2}\int_0^\tau X^2(s_1,s_2,\sigma)d\sigma\right). \tag{3.4}$$

Equation (3.4) for z is nothing but a parametric representation of the solution ψ we are seeking. It is easy to see that given x,t, and α the parameters s_1, s_2 and τ can be eliminated and

$$X(s_1,s_2,\sigma) = e^{\alpha(t-\sigma)}x - \int_\sigma^t e^{-\alpha(\sigma-\theta)} u_\theta \cdot d_\theta . \quad (3.5)$$

Substitution in (3.4) gives the explicit representation of $\psi(t,x,\alpha)$ for given input and output functions. The last exponential factor in (3.4) corresponds to the well-known likelihood ratio formula ([7], [8]).

4. SUFFICIENT STATISTICS

In Eqn. (3.4), only the term $\int_0^t X(s_1,s_2,\sigma)\dot{y}_\sigma d_\sigma$ inside the exponential depends on measured outputs and explicitly,

$$\int_0^t X(s_1,s_2,\sigma)\dot{y}_\sigma d_\sigma = x\sum_{k=0}^{\infty} \alpha^k \int_0^t \frac{(t-\sigma)^k}{k!} dy_\sigma - \sum_{k=0}^{\infty} (-\alpha)^k \int_0^t \gamma_k(t,\sigma) dy_\sigma \quad (4.1)$$

where $\gamma_k(t,\sigma) = \int_\sigma^t (\sigma-\theta)^k/k! u_\theta d_\theta$. The two sequences

(a) $\beta_k(t) = \int_0^t (t-\sigma)^k/k!\ dy_\sigma \quad k = 0,1,2 ..$

and

(b) $\omega_k(t) = \int_0^t \gamma_k(t,\sigma) dy_\sigma \quad k = 0,1,2, ...$

may be viewed as sufficient statistics for the problem. Each β_k can be computed as the output of a finite dimensional system driven by y_t. The same holds true for the ω_k's. We must mention that the statistics ω_k are similar to the sufficient statistics determining the likelihood ratio given by Giorgio Picci [9].

ACKNOWLEDGEMENT

This work was supported in part by Air Force Office of Scientific Research grant AFOSR 79-0025 and by the Department of Energy Contract DEAC01-79-ET-29363.

REFERENCES

[1] H.J. Kushner, "Dynamical Equations for Optimal nonlinear filtering", J. Differential Equations, Vol. 3, 1967, pp. 179-190.

[2] H.J. Kushner, "On the differential equations satisfied by conditional probability densities of Markov processes, with applications," RIAS Tech. Report No. 64-6, March 1964.

[3] R.W. Brockett and J.M.C. Clark, "The geometry of the conditional density equation," Proc. of the Oxford Conference

on Stochastic Systems, Oxford, England, 1978.
[4] S.K. Mitter, "On the analogy between mathematical problems of nonlinear filtering and quantum physics," Richerche di Automatica, 1980.
[5] E. Pardoux, "Stochastic partial differential equations and filtering of diffusion processes," Stochastics, Vol. 3, 1979, pp. 127-167.
[6] R.W. Brockett, "Classification and equivalence in estimation theory," Proc. IEEE Conference on Decision and Control, Dec. 1979, Ft. Lauderdale.
[7] T.T. Kadota and L.A. Shepp, "Conditions for absolute continuity between a certain pair of probability measures," Z. Wahrscheinlichkeitstheorie und Verw. Gebiete, 16, (1970) pp. 250-260.
[8] T. Kailath and M. Zakai, "Absolute Continuity and Radon-Nikodym derivatives for certain measures relative to the Wiener measure", Ann. of Math. Statistics, 42 (1971), pp. 130-140.
[9] G. Picci, "Some connections between the theory of sufficient statistics and the identifiability problem," SIAM J. Appl. Math., Vol. 33, No. 3, 1977, pp. 383-398.
[10] H. Hazewinkel and S.I. Marcus, "On the relationship between Lie algebras and nonlinear estimation," Proc. of the IEEE Conference on Decision and Control, Dec. 1980, Albuquerque.

philtre: a love - potion or
love - charm

Oxford English Dictionary

Part 5

LINEAR FILTERING

SOME TOPICS IN LINEAR ESTIMATION†

Thomas Kailath
Department of Electrical Engineering
Stanford University

INTRODUCTION

It must be with some trepidation that one ventures to speak about the problems of linear estimation to an audience already well familiar with the overwhelmingly more difficult nonlinear filtering problem. However, perhaps to compensate for this spectacle, the organizers have given me the opportunity to speak first, with considerable latitude in the choice of my topics.

For such an audience, there will be no need to present a tutorial on linear filtering, especially of the Kalman-Bucy type. I chose, therefore, to focus on some aspects generally less familiar to those with a 'modern' control theory background, i.e., largely a state-space background. In particular, we shall begin with a discussion of integral equations and of the important Wiener-Hopf technique. We shall specialize this to stationary processes over infinite intervals, and then describe some alternative, often computationally better, solution methods of Ambartzumian-Chandrasekhar and Krein-Levinson for finite-interval problems. For nonstationary processes, we start first with state-space models and build up to a brief description of the scattering theory framework for linear estimation. This will then lead us to nonstationary versions of the Ambartzumian-

† This work was supported in part by the Air Force Office of Scientific Research, Air Force Systems Command, under Contract AF49-620-79-C-0058; by the U.S. Army Command Office under Contract DAAG29-79-C-0215; and by the National Science Foundation, under Grant ENG78-10003.

M. Hazewinkel and J. C. Willems (eds.), Stochastic Systems: The Mathematics of Filtering and Identification and Applications, 307–350.

Chandrasekhar and Krein-Levinson equations, which we shall only consider very briefly, providing detailed references for further reading. We shall conclude with a remark on a possible implication for nonlinear filtering.

Our presentation is confined to continuous-time processes; a recent survey of the discrete-time case can be found in Kailath (1980).

In writing this chapter, it was a great help to have a carefully prepared preliminary reduction of the actual lectures, contributed by B. Hanzon, B. Ursin and D. Ocone. It is a pleasure also to thank M. Hazewinkel for these and several other organizational touches that made for an outstanding symposium.

TABLE OF CONTENTS

1. The Integral Equations of Smoothing and Filtering

Our estimation problems will be discussed in the context of the following model for the observed random process $\{y_t\}$:

$$y_t = z_t + v_t \quad , 0 \le t \le T \tag{1}$$

where z_t and v_t are R^p-valued stochastic variables with mean zero, and such that †

$$E\, v_t v_s' := I_p \delta(t-s) \ ,$$

where δ is the Dirac-delta distribution, and

$$E(z_t z_s' + z_t v_s' + v_t z_s') := K(t,s) \ ,$$

a continuous function on $[0,T]\times[0,T]$. This model describes a situation in which a signal, z_t, can only be observed with additive white noise v_t. We note that $K(t,s)$ does not have to be a covariance function; it is necessary only that

$$R(t,s) := E\, y_t y_s' = I_p\, \delta(t-s) + K(t,s)$$

be a covariance function.

It will be useful to consider two special cases:

(i) $v \perp z_s$ for all s,t with $0 \le s \le T$, $0 \le t \le T$. In this case $K(t,s) = E\, z_t z_s'$ is a covariance function

(ii) $v_t \perp z_s$ for all $t > s$. This possibility allows *causal dependence* of z on y (feedback!).

1a. The Smoothing Problem: Fredholm Equations

The *smoothing problem* is as follows. Given the observations $\{y_s : 0 \le s \le T\}$, find

† No special notation will be used to indicate matrix on vector quantities. Primes will denote transposes and E denotes expectation.

$$\hat{z}_{t|T} = \int_0^T H(t,s) y_s \, ds \tag{2a}$$

such that $E(z_t - \hat{z}_{t|T})'(z_t - \hat{z}_{t|T})$ is minimum, where the minimization is taken over all matrix-valued functions $H(t,s)$ of s, with t fixed, in the Hilbert space $L^2[0,T]$. It is well known that the following holds.

Theorem: A necessary and sufficient condition for the solution of this smoothing problem is

$$E(z_t - \hat{z}_{t|T})y_s' = 0 \text{ for } all \;\; s \,\varepsilon[0,T] \; . \tag{2b}$$

Put differently: every component of $z_t - \hat{z}_{t|T}$ must be orthogonal to every component y_s, for all $s\,\varepsilon[0,T]$, where the orthogonality is induced by the inner product $(a,b):=E\,ab$, a and b scalar stochastic variables.

Suppose now further that the special case (i) holds, namely that z_t is orthogonal to v_s for all s,t with $0 \le s \le T$, $0 \le t \le T$. Then the condition (2b) leads to the equation

$$H(t,s) + \int_0^T h(t,\tau)K(\tau,s)\, d\tau = K(t,\tau),\; 0 \le s,t \le T \; . \tag{3}$$

This is a Fredholm equation of the second kind (see, e.g., Courant and Hilbert, Vol. I, Ch. III) and the solution *H(t,s)* is called the ***Fredholm resolvent*** of *K(t,s)* on $[0,T] \times [0,T]$. Introduce the following operator notation:

$$HK \textit{ is defined as } (HK)(t,s) = \int_0^T H(t,\tau)K(\tau,s)\, d\tau \; ,$$

and I is the identity operator with kernel $I(t,s) = I_p \, \delta(t-s)$. In this notation the equation (3) can be written as

$$H + HK = K \tag{4a}$$

or in the equivalent forms

$$(I - H)(I + K) = I = (I + K)(I - H) . \tag{4b}$$

Clearly $I\text{-}H$ is the inverse, in the sense of the "operator multiplication" that we have just defined, of I+K . Note that in this case of complete orthogonality of z_t and v_s, the smoothing filter is precisely the resolvent of K.

How can the resolvent be computed? One answer is provided by the so-called Mercer expansion of $K(t,s)$ (see, e.g., Riesz and Nagy, p. 245; we use an extension to the vector case):

$$K(t,s) = \sum_{i=1}^{\infty} \lambda_i \varphi_i(t) \varphi_i'(s) \tag{5a}$$

where the φ_i are vector-valued orthonormal eigenfunctions of the operator K with eigenvalue φ_i:

$$\int_0^T K(t,\tau)\varphi_i(\tau)\, d\tau = \lambda_i \varphi_i(t) , \quad i = 1,2,\ldots \quad ,0 \le t \le T \tag{5b}$$

Then it can be seen easily that the Fredholm resolvent of K can be written as

$$H(t,s) = \sum_1^{\infty} (\lambda_i / 1 + \lambda_i) \varphi_i(t) \varphi_i'(s) \tag{6}$$

1b. The Filtering Problem: Wiener-Hopf Equations

In the special case that $T=t$, the smoothing problem becomes what is known as a *filtering problem*. We shall assume further that we are in one of the special cases (i) and (ii), viz., that v_t is either orthogonal to z_s for all s,t or just for all $s<t$. The filtered estimate can be written

$$\hat{z}_{t|t} = \int_0^t h(t,s) y_s\, ds \tag{7}$$

where $h(t,s)$ satisfies

$$h(t,s) + \int_0^t h(t,\tau) K(\tau,s)\, d\tau = K(t,s) , \quad 0 \le s \le t \le T , \tag{8}$$

Note that for each fixed t, we have a smoothing problem. The point is now that we have a collection of Fredholm integral equations, one for each value of t, and unlike as in (3) is more than just an indexing parameter in the family of equations. The filtering integral equation is said to be of "Wiener-Hopf type" rather than of "Fredholm type" and the solution can not be as simply expressed in terms of Mercer expansions as in the smoothing problem.

In one sense then, smoothing appears to be "easier" than filtering, a statement counter to the intuition current in the Kalman-Bucy state-space theory (see, e.g., Meditch (1969) and also the discussion following (15b) below). However, the following facts give some justification to this claim:

1. In the Wiener theory of estimation of stationary processes over infinite (smoothing) or semi-infinite (filtering) intervals, the smoothing solution is readily determined by Fourier transformation, while the filtering solution requires the more difficult Wiener-Hopf technique (further elaborated below).
2. In estimation given a fixed time-interval, smoothing can be implemented with *time invariant* filters (convolutions or fast Fourier transforms, see Lévy, Kailath, Ljung and Morf (1979)), while this will never be true for filtering.

1c. The Generalized Wiener-Hopf Technique

Wiener-Hopf equations first appeared in astrophysics and radiative transfer theory around 1900. In 1931, Wiener and Hopf invented an ingenious method for solving the equation, and it has since borne their name. Their so-called Wiener-Hopf technique plays a central role in linear filtering theory, and we will present it briefly here, both as a framework for later discussion and as a service to our state-space friends who might conceivably have never seen it! The technique was originally developed for difference (or convolution) kernels R; here we describe a generalized form (for arbitrary kernels) that captures the main idea.

To focus on the main idea, the treatment will leave aside technical issues (hypotheses on kernel functions, specification of function spaces, etc.) that are needed to build a rigorous theory (see, for example, Devinatz and Shinbrot (1967) and Gohberg and Feldman (1974)). To describe the technique, we first develop an operator notation for (8) that expresses the constraint $s \le t$. Thus if L is an integral operator associated with the kernel $L(t,s)$, define

$$\{L\}_+ f(t) := \int \{L(t,s)\}_+ f(s)\, ds$$

with

$$\{L(t,s)\}_+ := \begin{cases} L(t,s) & \text{if } s \le t \\ 0 & \text{if } s > t \end{cases}$$

Accordingly, define $\{I\}_+ := I$. $\{L\}_+$ is called the *causal part* of L. With this notation the Wiener-Hopf equation (8) becomes

$$\{hR\}_+ = \{K\}_+ \,, \tag{9}$$

As only the values of $h(t,s)$ for $s \le t$ play a role in this problem, $h(t,s)$ can be (and will be) taken equal to its causal part: $h = \{h\}_+$. We assume that $R = I + K$, $R = R'$, R is positive definite as a kernel, and K does not contain I term (alternatively: $I_p\, \delta(t-s)$ does not appear in $K(t,s)$).

The key idea of the method of Wiener and Hopf (1931) is to assume that R can be suitably factored. In our case, as

$$R = R^{1/2} R^{*/2} \,,$$

where $R^{1/2}$ is a causal and causally invertible operator, that is, $R^{1/2} = \{R^{1/2}\}_+$ and $R^{-1/2} := [R^{1/2}]^{-1}$ exists and satisfies $R^{-1/2} = \{R^{1/2}\}_+$. Here $R^{*/2}$ denotes the adjoint of $R^{1/2}$, that is $R^{*/2}(t,s) = R^{1/2}(s,t)'$. Such an $R^{1/2}$ is called the *canonical factor* of R, and when it exists it will be unique as a consequence of the causal and causally invertible requirement. [Observe that, despite the notation, $R^{1/2}$ is not the traditional operator-theoretic (symmetric) square root of a positive operator.]

Now make the simple but crucial observation that if h solves (9), there must exist some function g such that

$$\{g\}_+ = 0 \text{ and } hR = K + g \ . \tag{10}$$

Here g is *strictly* anti-causal, i.e., it does not have any I component. Multiplying (10) on the right by $R^{-*/2}$, we have

$$hR^{1/2} = KR^{-*/2} + gR^{-*/2} \ . \tag{11}$$

Now apply $\{\ \}_+$ to both sides of (11). Since h is causal and $R^{1/2}$ is causal and since the composition of two causal operators is again causal, $\{hR^{1/2}\}_+ = hR^{1/2}$. Likewise the composition of a strictly anti-causal operator (g) with an anti-causal operator $(R^{-*/2})$ is strictly anti-causal; $(R^{-*/2} = (R^{-1/2})^*$ is anti-causal since $R^{-1/2}$ is causal). Thus $\{gR^{-*/2}\}_+ = 0$. The end result is $hR^{1/2} = \{KR^{-*/2}\}_+$, or

$$h = \{KR^{-*/2}\}_+ R^{-1/2} \ . \tag{12}$$

This is the solution of the Wiener-Hopf equation.

However, we have not really so far used the assumption that R has the special form $R = I + K$. For this case further important results are available from canonical factorization. First observe that, since $K = R - I$,

$$\begin{aligned} h &= \{(R - I)R^{-*/2}\}_+ R^{-1/2} \\ &= \{R^{1/2} - R^{-*/2}\}_+ R^{-1/2} \\ &= I - \{R^{-*/2}\}_+ R^{-1/2} \ . \end{aligned} \tag{13}$$

However, when $R=I+K$, $R^{1/2}$ must have the form $R^{1/2}=I+h$, where h is some strictly causal operator, and so also $R^{-1/2}$ must have the form $R^{-1/2}=I+l$, where l is strictly causal. Therefore, $\{R^{-*/2}\}_+=\{I+l^*\}_+=I$, since l^* is strictly anti-causal. Hence (13) reduces to

$$h = I - R^{-1/2} \tag{14}$$

This striking formula has several interesting implications.

First, and most important, it shows that for this problem canonical factorization and filtering are *equivalent* problems; h is *immediately* determined if $R^{-1/2}$ is known and vice versa.

Secondly, if $R^{-1/2}$ is applied as a filter to y, we have

$$R^{-1/2}y = (I-h)y = y - \hat{z}\,.$$

Since $y=z+v$, $\hat{z}$ is the estimate of y_t given $\{y_s,\, s<t\}$, so that it is reasonable to expect that $\{y_t-\hat{y}_t\}$, the *new information* or *innovation* process, is a white noise process, consistent with the calculation

$$<R^{-1/2}y, R^{-1/2}y> = R^{-1/2}<y,y>R^{-*/2} = R^{-1/2}RR^{-*/2} = I\,.$$

This result can be rigorously established under quite general conditions (see Kailath (1968), Kailath (1971), Meyer (1972)).

1d. A Resolvent Identity relating Smoothing and Filtering

Recall the Fredholm resolvent of K, defined by $I-H=(I+K)^{-1}=R^{-1}$. By virtue of (14), we can write

$$I - H = R^{-*/2}R^{-1/2} = (I-h^*)(I-h) \tag{15a}$$

which immediately yields the nice formula

$$H = h^* + h - h^*h\,. \tag{15b}$$

This is actually an old identity, known by the early 1950's when it was discovered, in a differential version, independently by Siegert, Krein, Bellman and others (see references in Kailath (1974)).

Now when the signal and noise are completely uncorrelated, we saw earlier (cf. (4b)) that the Fredholm resolvent H is just the smoothing filter; the identity (15) then shows that the causal filter h determines the smoothing filter. This seems to contradict the remarks we made in Section 2b about the relative difficulties of smoothing and filtering as they appeared from thinking of the Wiener-Hopf equation as an infinite family of Fredholm equations. In the approach via canonical factorization, it would appear that filtering comes first, and then smoothing. In Sec. 3d, which describes a scattering theory approach, this sequence will again be reversed.

We shall illustrate these different relationships between filtering and smoothing by considering several specific examples in the next two sections.

2. Some Examples - Stationary Processes

2a. Scalar Stationary Proceses Over Infinite Intervals

These problems were studied by Wiener, Kolmogorov and Krein. For a concise exposition of Wiener's work, see a paper by N. Levinson (1947), reprinted as Appendix C of Wiener (1949). The papers of Kolmogorov and Krein are reprinted in Kailath (1977).

We suppose y_t, z_t, v_t to be scalar and stationary; then $R(t,s)=R(t-s)$. We assume the existence of

$$S_y(\omega) := F\{R(t)\}(\omega) = \int_{-\infty}^{\infty} R(t)e^{-j\omega t}\, dt ,$$

the Fourier transform of $R(t)$, where j is the imaginary unit. $S_y(\omega)$ is nonnegative (for real ω) and is known as the power spectral density of the process y. We assume further that

$$\int_{-\infty}^{\infty} \frac{|\ln S_y(\omega)|}{1+\omega^2} d\omega < \infty . \tag{16}$$

If this is not the case, then Kolmogorov and Wiener showed that y_t can be predicted perfectly from its own past (see, e.g., Doob (1953), Ch. 12).

Under this assumption the canonical factorization of $R(t)$ over $(-\infty,\infty)$ corresponds to a factorization of $S_y(\omega)$ as

$$S_y(\omega) = S_y^+(\omega) S_y^-(\omega) , \tag{17}$$

where $S_y^+(\sigma+j\omega)$ is analytic and bounded in the right half plane $\sigma>0$, $S_y^-(\omega)$ is analytic in the left half plane and

$$S_y^-(\sigma + j\omega) = \overline{S_y^+(\sigma - j\omega)} . \tag{17a}$$

It can be shown that (see, e.g., Solodovnikov (1960))

$$S_y^+(\omega) = \sqrt{S_y(\omega)}\, e^{j\vartheta(\omega)} \, , \tag{18}$$

where $\vartheta(\omega)$ is the Hilbert transform of $\ln\sqrt{S_y(\omega)}$.

In the case that $S_y(\omega+i\sigma)$ is rational, $S_y^+(\omega+i\sigma)$ can be found as follows:

$$S_y^+(\omega + j\sigma) = \text{constant x Monic polynomial of left half plane zeros}) \times \text{(Monic polynomial of all left half plane poles)}^{-1}. \tag{19}$$

This follows immediately from the fact that $S_y^+(\sigma+j\omega)$ and $1/S_y^+(\sigma+j\omega)$ must be analytic in the upper half plane $\sigma>0$, and from (17a). Note that (17a) implies $S_y^-(\omega)=\overline{S_y^+(\omega)}=S_y^+(-\omega)$, because $R(t)$ is real. Therefore,

Now the canonical causal factor (see the text between (9) and (10)) can be found as

$$R^{1/2} = F^{-1}\{S_y^+(\omega)\} \, , \tag{20}$$

the inverse Fourier transform of $S_y^+(\omega)$. The optimal filter (see (14)) is then equal to

$$h = F^{-1}\left\{1 - \frac{1}{S_y^+(\omega)}\right\} \tag{21}$$

How About Smoothing?

Consider the Fourier transform of the smoothing filter H (see (15)):

$$F\{H\} = F\{h + h^* - h^*h\} = \tag{22}$$

$$= \left[1 - \frac{1}{S_y^+(\omega)}\right] + \left[1 - \frac{1}{S_y^+(-\omega)}\right] - \left[1 - \frac{1}{S_y^-(\omega)}\right]\left[1 - \frac{1}{S_y^+(\omega)}\right] =$$

$$= 1 - \frac{1}{S_y^+(\omega)S_y^+(-\omega)} = 1 - \frac{1}{S_y(\omega)}$$

This is a well known formula, easy to derive directly from the equality $H=I-R^{-1}$. Note that smoothing does not require special factorization, so that the smoothing solution is easier to find from the given data than the filtering solution.

Important Remark: When $S_y(\omega)$ is rational, $S_y^+(\omega)$ is rational and so is $F\{h\} = 1 - \frac{1}{S_y^+(\omega)}$. This can be readily implemented in (a variety of) state-space forms, so that z can be "recursively computed", as noted by Whittle ((1963), p. 35) and others, independently of the direct state-space formulation of Kalman and Bucy (1961).

2b. Finite Intervals - The Ambartzumian-Chandrasekhar Equations

We shall next talk about the more difficult case of filtering stationary, scalar processes defined on finite intervals. This may be considered the first natural extension of Wiener's work and it began to be studied in the engineering literature around 1950, shortly after Wiener's work became public (see, e.g., Zadeh and Ragazzini (1950)). The finite interval case presented a new challenge because spectral factorization, in its traditional sense, does not work, and thus researchers tried various other methods to find the solution (see, e.g., Solodovnikov (1962)).

However, the astrophysicists V. A. Ambartzumian (USSR) and S. Chandrasekhar (USA) had already studied such problems in the mid-1940's, independently of engineers, and had demonstrated that the Wiener-Hopf equation could be replaced by an equivalent Riccati equation. Their results greatly simplified the numerical computation of solutions, and since computation in those days meant calculation by hand, they were considered to be a great success. The Ambartzumian-Chandrasekhar theory (see Chandrasekhar (1950)) assumes that the kernel of the Wiener-Hopf equation has of the form

$$K(t-s) = \int_0^1 e^{-\alpha|t-s|} w(\alpha)\, d\alpha \ , \tag{23}$$

in which $w(\alpha)$ is some known function. This form arose from the physical situation they considered, in which light incident at an angle α is propagating through a medium. If we assume that the light is incident at a finite number of values α, so that $w(\alpha)$ is a sum of δ-functions, the process y wll have a rational spectral density

$$S_y(\omega) = 1 + \sum_{i=1}^{n} \frac{2w_i \alpha_i}{\alpha_i^2 + \omega^2} , \tag{24}$$

The first result of the Ambartzumian-Chandrasekhar theory is that to solve the Wiener-Hopf equation it suffices to find the solution $Q(t,\alpha,\beta)$ to a Riccati-type partial differential equation

$$\frac{\partial}{\partial t} Q(t,\alpha,\beta) = P + k\, Q(t,\alpha,\beta) + \int_0^1 Q(t,\alpha,\beta')w(\beta')\,d\beta' \tag{25}$$

$$+ \int_0^1 Q(t,\alpha',\beta)w(\alpha')\,d\alpha' + \int_0^1\int_0^1 Q(t,\alpha,\beta')Q(t,\alpha',\beta)w(\alpha')w(\beta')\,d\alpha' d\beta' .$$

Q can be computed by discretization of this equation to obtain a finite dimensional system of ordinary Ricatti differential equations. This has great computational advantages, though since it may be required to compute $Q(t,\alpha,\beta)$ for t, α and β ranging over a large set of values, this is still burdensome.

However, using certain physical *invariance* arguments, Ambartzumian (1943) was able to show that Q could actually be computed in terms of two functions $X(t,\gamma)$ and $Y(t,\gamma)$, of *two* rather than *three* variables. Then Chandrasekhar (1947) derived a pair of differential equations for X and Y, considerably simpler than the original Riccati equation (25):

$$\frac{\partial X(t,\gamma)}{\partial t} = -\,Y(t,\gamma) \int_0^1 Y(t,\beta)w(\beta)\,d\beta \tag{26a}$$

$$\frac{\partial Y(t,\gamma)}{\partial t} = -\,Y(t,\gamma) - X(t,\gamma) \int_0^1 Y(t,\beta)w(\beta)\,d\beta \tag{26b}$$

with

$$X(0,\gamma) = Y(0,\gamma) = 1 , \quad 0 \le \gamma \le 1 .$$

Astrophysicists were quick to recognize the value of the recursive solution of the equations (26) (see, e.g., Sobolev (1965), p. 79). The ideas of Ambartzumian and Chandrasekhar were brought to the attention of applied mathematicians by the extensive work of Bellman, Kalaba and their colleagues on what they called *invariant imbedding* (see, e.g., Bellman and Wing (1975)). The equations (26) were first introduced into the estimation literature by Casti, Kalaba and Murthy (1972). Their extension to nonstationary processes was made by Kailath (1973).

2c. Sobolev's Identity and The Krein-Levinson Equations

Fundamental work on the finite interval, stationary case did not end with Ambartzumian and Chandrasekhar. In particular Sobolev (1965) went on to address the problem of arbitrary $K(t-s)$, for which a representation such as (23) is not given, and he succeeded in developing a much more direct approach. His idea was to exploit the Toeplitz structure of $K(t,s)=K(t-s)$ more deeply than in the previous theory. By this approach, he established a very powerful constraint on the Fredholm resolvent (smoothing kernel) $H(t,s;T)$ of a Toeplitz kernel:

If $A(T;t)$ is defined via the equation

$$A(T;t) + \int_0^T A(T;u)K(u-t)\,du = K(T-t)\,, \quad 0 \le t \le T \tag{27}$$

and $B(T;t)$ via

$$B(T;t) + \int_0^T B(T;u)K(u-t)\,du = K(-t)\,, \quad 0 \le t \le T \tag{28}$$

then Sobolev showed that

$$\left[\frac{\partial}{\partial t} + \frac{\partial}{\partial s}\right] H(t,s;T) = A'(T;t)A(T;s) - B'(T;t)B(T;s) \tag{29}$$

with

$$A(T;t) = H(T,t;T) = H'(t,T;T) \tag{30a}$$

and

$$B(T;t) = H(0,t;T) = H'(t,0;T) . \tag{30b}$$

(Note that these equations are written for the general case of matrices K, H, A and B.) Sobolev's identity shows that the resolvent $H(s,t;T)$ is determined for $(t,s) \in [0,T]\times[0,T]$ by its values on the boundaries of $[0,T]\times[0,T]$, i.e., by two functions of one variable.

Sobolev's identity is even more striking in its integrated form, which, when translated into operator form, is

$$I - H = (I - a^*)(I - a) - b^*b \tag{31}$$

where a and b are not only causal, but also Toeplitz. This means, for example, that the filter determind by $\mathbf{a}$,

$$(ay)_t = \int_0^t a(t-s)y_s \, ds$$

is *time-invariant*. Equation (31) is a useful modification of the formula (15a): $(I-H)=(I-h^*)(I-h)$, because it expresses H only in terms of time invariant (causal and anticausal) operators, whereas h, even for Toeplitz K, is not generally time invariant. Since time invariant filters are much easier to implement than time-variant ones, it is reasonable to use a instead of h, whenever h appears. Of course, an error is then incurred, but the remarkable implication of (31) is that if a replaces h in (15a), the correction can again be made with a time invariant filter b. Explicit formulas for a and b are as follows:

$$a(t) = A(T;t) \tag{32a}$$

$$b(t) = B(T;T-t) . \tag{32b}$$

For more on the applications of this identity to smoothing and other problems, see Lévy et al. (1979) and Kailath et al. (1978).

Sobolev's identity also has important implications for fast computation of the resolvent kernel $H(t,s;T)$ because we need only develop an efficient, recursive method for updating the boundary values $A(T;t)$ and $B(T;t)$ of $H(t,s;T)$. In fact, Krein (1955) had obtained such equations, which we shall present here in the special case of scalar processes:

$$\left[\frac{\partial}{\partial t}+\frac{\partial}{\partial s}\right]A(T;s) = -A(T;T-s)A(T;0) . \tag{36}$$

To see what this means, consider the naive discretization, $T=N\cdot\Delta$ and

$$A(T+\Delta;i\Delta+\Delta) = A(T;i\Delta) - A(T;T-i\Delta)A(T;0)\Delta \tag{36}$$

which propagates as illustrated in the figure

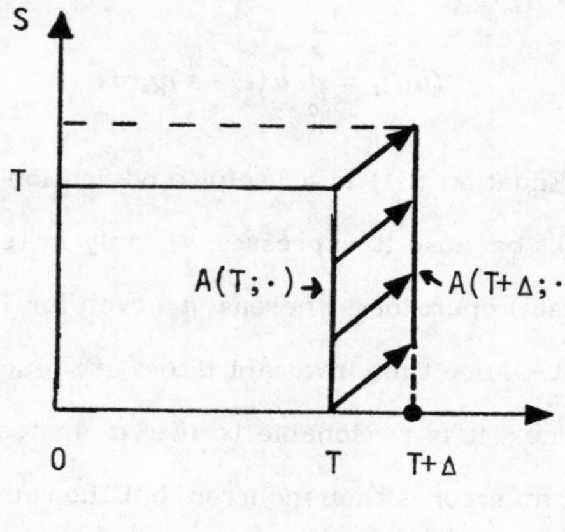

The one point not picked up by this scheme is $A(T+\Delta;0)$ and so it is computed by using the integral equation (27):

$$\begin{aligned} A(T+\Delta;0) &= \int_0^{T+\Delta} A(T+\Delta;u)K(u)\,du + K(T+\Delta) \\ &\approx \sum_{i=1}^{T+\Delta} A(T+\Delta;i\Delta)K(i\Delta)\Delta + K(T+\Delta) . \end{aligned} \tag{37}$$

How fast is this algorithm? If $T=N\cdot\Delta$, it takes $N+1$ multiplications to go from T to $T+\Delta$. Therefore, to generate the boundary out to T, takes $1+2+\ldots+N=N(N+1)/2=0(N^2)$ multiplications. Without the Toeplitz structure, we would need $O(N^3)$ operations to compute $H(t,s;T)$.

The above method of solving (36) via discretization leads to recursions very similar to those introduced for the prediction of discrete-time processes by Levinson (1947) and since then widely studied in the signal processing literature. We therefore call (36) a *Krein-Levinson* equation. Kailath, Ljung and Morf (1978) have extended these techniques to nonstationary processes.

We should also mention that when a representation of $K(t-s)$ in the exponential form (23) is available, the Krein-Levinson equations can in fact be reduced to the Ambartzumian-Chandrasekhar equations (cf. Kailath, Ljung, Morf (1976)).

3. Some Examples – Nonstationary Processes

3a. State Space Process Models - Kalman-Bucy Filters

The most common approach to nonstationary process estimation is via state-space models. It is assumed that the signal process z_t can be described as

$$z_t = H_t x_t \tag{38a}$$

$$\dot{x}_t = F_t x_t + G_t u_t \ , \quad t \geq t_0 \ , \tag{38b}$$

where x_t is an $n \times 1$ "state" vector, z_t is a $p \times 1$ vector and u_t is an $m \times 1$ "white noise" vector. The observed process is

$$y_t = z_t + v_t \tag{38c}$$

where is the observation noise v_t such that†

$$E\left\{\begin{bmatrix} u_t \\ v_t \end{bmatrix} [u_s' v_s']\right\} = \begin{bmatrix} I & C_t \\ C_t' & I \end{bmatrix} \delta(t-s) \ . \tag{38d}$$

The initial state x_{t_0} is assumed to be random, with

$$E\, x_{t_0} = 0 \ , \ E\, x_{t_0} x_{t_0}' = \Pi_0 \ , \tag{38e}$$

and furthermore, it is also assumed that

$$E\, u_t x_{t_0}' = 0 \ ; \ E\, v_t x_{t_0}' = 0 \ \text{for } all \ \ t \geq t_0 \ . \tag{38f}$$

Finally, the matrices F_t, G_t, H_t, Π_0 are all assumed to be known. The above assumptions ensure that

$$E\, v_t z_s' = 0 \ \text{if} \ s \leq t \ .$$

They also ensure that x_t is a Markov process, so that the signal z_t is modeled

---

† In the literature it is common to assume $E\, u_t u_s' = Q_t \delta(t-s)$ and $E\, v_t v_s' = R_t \delta(t-s)$, $R_t > 0$, but without loss of generality we can make the convenient normalizations of (38d).

as a so-called *projection* of the Markov state process x_t. Such descriptions had been widely used by physicists (see, e.g., the papers in Wax (1954)) and mathematicians (see, e.g., Doob (1944), (1948)) for stationary processes with rational power spectral densities. The extension to models with time-variant coefficients, as in (38) above, is at least in retrospect fairly natural and it began to be made in estimation theory in the late fifties by Laning and Battin (1958, p. 304), Stratonovich (1959), (1960), and most notably by Kalman (1960) and Kalman and Bucy (1961).

The celebrated Kalman-Bucy filter for the model (38) is

$$\hat{z}_t = H_t \hat{x}_t, \quad t \geq t_0 \tag{39a}$$

$$\dot{\hat{x}} = F_t \hat{x}_t + K_t \nu_t \tag{39b}$$

$$\hat{x}_{t_0} = 0 \ . \tag{39c}$$

Here ν_t is the "innovation process",

$$\nu_t = y_t - H_t \hat{z}_t , \tag{39d}$$

which is known to be white with

$$E\, \nu_t \nu_s' = I\, \delta(t-s) \tag{39e}$$

The $n \times p$ matrix K_t, which we shall call the "Kalman gain", can be computed as

$$K_t = P_t H_t' + G_t C_t \tag{40a}$$

where the $n \times n$ matrix P_t is the covariance matrix of the errors,

$$P_t = E\, \tilde{x}_t \tilde{x}_t' , \ \tilde{x}_t := x_t - \hat{x}_t , \tag{40b}$$

Kalman and Bucy showed that P_t could be recursively determined as the unique solution of the nonlinear Riccati-type differential equation

$$\dot{P}_t = F_t P_t + P_t F_t' + G_t G_t' - K_t K_t' , \ P_{t_0} = \Pi_0 . \tag{40c}$$

This is all very well known to estimation theorists by now. It is perhaps not so widely known that the recognition of the importance of state-space models and Markov processes in signal estimation problems is due independently to Stratonovich, who actually studied the nonlinear filtering problem, and, using "Gaussian approximation" methods, derived what was later called the "extended Kalman filter". For the linear case, Stratonovich gave an exact solution which is exactly the Kalman-Bucy filter. However, no stability analysis was given and no intensive study of the Riccati equation was made; these were the vital contributions of Kalman and Bucy.

It is often said that the reason for the wide applicability of the Kalman-Bucy filter (in particular, to general time-variant models) is that giving a state-space model avoids the difficult problem of "spectral factorization". This is totally wrong! Canonical spectral, or in the nonstationary case "covariance", factorization, results in a model that is not only causal but also causally invertible. This is clearly not the case for the assumed model (38): knowledge of $y=\{y_s,\ t_0\le s\le t\}$ does *not* allow us to reconstruct u, v and x_{t_0}. Therefore to find the filter, canonical factorization still has to be done in one way or another.

In fact, as we noted in Section 2c (cf. (14)), knowledge of the canonical spectral factor should immediately determine the filter *and vice versa*. Here we have the filter and could ask how to obtain the canonical factor. The answer is simple: just rewrite (39) in the form

$$\dot{\hat{x}}_t = F_t\hat{x}_t + K_t\nu_t \tag{41}$$

$$y_t = H_t\hat{x}_t + \nu_t\ ,\quad \hat{x}_{t_0} = 0\ ,$$

We see that (41) describes a causal and causally invertible relation between a white noise input ν and the desired process y. We call it the *innovations representation* (IR) of y. Thus the above equations determine the true canonical factor. But to find this factor, we have to do quite a bit of work, viz., solve the Riccati equation. In other words, assuming a state-space model does not in any sense allow us to avoid determining the canonical filter (unless, of course, we start with such a model).

A Remark on Derivations

By now numerous proofs of the Kalman-Bucy equations are available. In the present context, it may be interesting to note that if

$$\hat{x}_t = \int_{t_0}^{t} h_x(t,s) y_s \, ds$$

then $h_x(t,s)$ obeys the Wiener-Hopf equation

$$h_{xy}(t,s) + \int_0^t h_{xy}(t,\tau) K(\tau,s) \, d\tau = K(t,s), \quad t_0 \leq s < t$$

where $K(t,s)$ can be readily computed from the given model (38)--in fact, see (44) below. Some calculation will then show that

$$\frac{\partial}{\partial t} h_{xy}(t,s) = [F(t) - h_{xy}(t,t)] h_{xy}(t,s), \quad s < t$$

so that

$$\frac{d}{dt}\hat{x}_t = \int_0^t [F(t) - h_{xy}(t,t)H(t)] h_{xy}(t,s) y_s \, ds + h_{xy}(t,t) y_{ty}$$
$$= F(t)\hat{x}_t + h_{xy}(t,t)[y_t - H(t)\hat{x}_t],$$

which will be the Kalman-Bucy equation (39a) when we show that $h_{xy}(t,t) = K_t$ as given by (40). We shall omit these calculations.

3b. State Space Covariance Models - Recursive Wiener Filters

While many different models $\{F_t, G_t, H_t, Q_t\, G_t, \Pi_0\}$ can yield the same covariance function

$$R(t,s) = I \cdot \delta(t-s) + K(t,s),$$

the impulse response of the canonical factor is known to be uniquely determined by $R(t,s)$. Therefore one would expect that it is possible to compute the Kalman gain K_t (which should not be confused with the term $K(t,s)$ in the covariance funtion) *directly* from the parameters of the covariance function $R(t,s)$. This can in fact be done (Kailath and Geesey, 1971), as can be seen by examining the special form of the covariance function associated with a state-space model of the form (38).

Let $\Phi(t,s)$ be the state transition matrix, which is the unique solution of the linear differential equation

$$\frac{d\Phi(t,s)}{dt} = F_t \Phi(t,s)\ ;\ \Phi(s;s) = I$$

Let Π_t be the $n \times n$ covariance matrix of x_t,

$$\Pi_t = E\, x_t x_t' \ . \tag{42}$$

It is easy to check that Π_t is the solution to the Lyapunov-type equation

$$\dot{\Pi}_t = F_t \Pi_t + \Pi_t F_t' + G_t G_t' \ , \tag{43}$$

with given initial value Π_0. By direct calculation we have

$$R(t,s) = \begin{cases} I \cdot \delta(t-s) + H_t \Phi(t,s) N_s & \text{if } t \geq s \\ I \cdot \delta(t-s) + N_t' \Phi'(s,t) H_s' & \text{if } t \leq s \end{cases} \tag{44}$$

where

$$N_t = \Pi_t H_t' + G_t C_t \ . \tag{45}$$

We will now assume that (only) H_t, F_t and N_t are known. Define

$$\Sigma_t := E\, \hat{x}_t \hat{x}_t' \ . \tag{46}$$

If we recall that $P_t := E\, \tilde{x}_t \tilde{x}_t'$, where $\tilde{x}_t = x_t - \hat{x}_t$, then the orthogonality of $\hat{x}_t$ and $\tilde{x}_t$ immediately yields the identity

$$\Pi_t = P_t + \Sigma_t \tag{47}$$

Therefore we can now rewrite the Kalman gain (cf. (39f)) as

$$K_t := P_t H_t' + G_t C_t = \Pi_t H_t' + G_t C_t - \Sigma_t H_t = N_t - \Sigma_t H_t' \tag{48}$$

Moreover, note that

$$\dot{\Sigma}_t = \dot{\Pi}_t - \dot{P}_t = F_t \Sigma_t + \Sigma_t F_t' + K_t K_t' \; ; \; \Sigma_{t_0} = 0 \; . \tag{49}$$

The equations, (48) and (49), determine K_t completely in terms of the parameters $\{H_t, F_t, N_t\}$ of the covariance function $R(t,s)$.

We note that we have effectively also obtained a recursive form of the solution to the Wiener-Hopf integral equation (8) for estimating the signal z_t from y_t, provided the covariance function $R(t,s)$ is given in the (factored) form (44). This provides a nice answer to the efforts of several investigators in the mid-1950's, attacking Wiener-Hopf equations with nonstationary kernels (see, e.g., Dolph and Woodbury (1952), Shinbrot (1956), Zadeh and Miller (1956), Laning and Battin (1958)). We may thus call our solution (39), (48)-(49) a ***recursive Wiener filter*** as compared to the Kalman-Bucy filter (39)-(40).

The close connection we noted in Sec. 1c between the solution of the filtering problem and of the canonical factorization problem means that the above results, especially (41) and (48)-(49), also give us an expression of the canonical factor (innovations representation) directly in terms of the parameters of the covariance function. The notation $\hat{x}$ for the state of the model (40) is, of course, just a reflection of the original noncanonical model (38) that we started with; x has no particular significance if we are only given a process y with covariance $R=I+K$, and therefore it is preferable to rewrite (40) in a different notation. Following Faurre (1969), we shall write the canonical innovations model as

$$\dot{x}_* = F_t x_* + K_* \nu_* \; , \; x_{*0} = 0 \tag{50a}$$
$$y_t = H_t x_* + \nu_* \tag{50b}$$

where

$$K_{*} = N_t - \Sigma_{*} H_t' \tag{50c}$$

and Σ_* obeys the Riccati equation

$$\dot{\Sigma}_{*} = F_t \Sigma_{*} + \Sigma_{*} F_t' + K_{*} K_{*}' \ , \ \Sigma_{*0} = 0 \ . \tag{50d}$$

The value of our particular derivation of the IR is that it shows that of all (causal) models of a given covariance triple $\{F_t, H_t, N_t\}$, this one has the "smallest" state variance matrix, because its state, x_{*}, is the projection (on the space spanned by $\{y_s, s<t\}$) of the state, x_t, of any other model.

There has been some interest in studying the set of all causal models associated with a given covariance triple; in particular, one might ask by analogy with the above for a *maximum-variance* causal model. It turns out that this model can be defined via a certain *smoothing* problem, rather than a *filtering* problem as for the minimum-variance model. We refer to Kailath and Ljung (1981) for a discussion of this result, and its implications for the so-called *stochastic realization problem* of studying all causal models of a given process. There have been a number of recent papers on this, see especially, the book of Clerget, Faurre, Germain (1978), the thesis of Ruckebusch (1979), and Lindquist and Picci (1981), which contains references to several of their earlier works.

3c. Orthogonal Decomposition of the Space of Random Variables

Here we shall continue instead with another aspect of the interplay between smoothing and filtering theory, as brought out by a recent result of Weinert and Desai (1980), which we can formulate as follows.

Given the state-space model (38), it is easy to see that we cannot recover the "input" random variables $\{u_t, v_t, x_{t_0}\}$ just from knowledge of the "output" random variables $\{y_t\}$, unless we have additional information. The nice observation of Weinert and Desai is that this additional information can be provided by a model 'adjoint' to (38) in a certain sense. In particular, define $\{\eta_t, \Theta\}$ by the equations

$$\dot{\chi}_t = -F_t'\chi_t - H_t'v_t \ , \ \chi_{t_f} = 0 \tag{51a}$$

$$\eta_t = -G_t'\chi_t + v_t \tag{51b}$$

and

$$\Theta = -\Pi_0\chi_{t_0} + x_{t_0} \tag{51c}$$

Then we can check by direct calculation that

$$E\,\Theta y_t' = 0 \ , \ E\,\chi_t y_t' = 0 \tag{52}$$

and that the sets

$$\{u_t, v_t, x_{t_0}\} \text{ and } \{y_t, \eta_t, \Theta\} \tag{53}$$

can each be recovered from the other by causal linear operations. In fact, this latter equivalence can be seen by first combining the equations (38) and (51) into the set

$$\begin{bmatrix}\dot{x}_t\\ \dot{\chi}_t\end{bmatrix} = \begin{bmatrix}F_t & 0\\ 0 & -F_t'\end{bmatrix}\begin{bmatrix}x_t\\ \chi_t\end{bmatrix} + \begin{bmatrix}G_t & 0\\ 0 & -H_t'\end{bmatrix}\begin{bmatrix}u_t\\ v_t\end{bmatrix} \tag{54a}$$

$$\begin{bmatrix}\eta_t\\ y_t\end{bmatrix} = \begin{bmatrix}F_t & -G_t'\\ 0 & 0\end{bmatrix}\begin{bmatrix}x_t\\ \chi_t\end{bmatrix} + \begin{bmatrix}u_t\\ v_t\end{bmatrix} \tag{54b}$$

Then, by elimination, we can write

$$\begin{bmatrix}\dot{x}_t\\ \dot{\chi}_t\end{bmatrix} = \begin{bmatrix}F_t & -G_t G_t'\\ -H_t'H_t & -F_t'\end{bmatrix}\begin{bmatrix}x_t\\ \chi_t\end{bmatrix} + \begin{bmatrix}G_t\eta_t\\ -H_t'y_t\end{bmatrix} \tag{55a}$$

with the "two-point" boundary value conditions

$$\chi_{t_f} = 0 \ , \ x_{t_0} - \Pi_0\chi_{t_0} = \Theta \tag{55b}$$

Given $\{\Theta, y_t, \eta_t\}$, we can solve this linear two-point boundary value problem for $\{x_t, \chi_t\}$ and then calculate $\{u_t, v_t, x(0)\}$ from (51b,c).

We can summarize the above discussion by saying that the result of Weinert and Desai shows that the linear space of random variables spanned by $\{u_t, v_t, x_0\}$ can, at each t, be *orthogonally decomposed* into the space spanned by $\{y_t\}$ and $\{\eta_t, \Theta\}$.

[The following remarks might help explain the origin of the above equations. The point is that given $\{y_t,\ t_0 \le t \le t_f\}$, we cannot reconstruct $\{u_t, v_t, x_0\}$ but at best only their smoothed estimates $\{\hat{u}_t, \hat{v}_t, \hat{x}_0\}$. For a full reconstruction we must augment $\{y_t\}$ by the errors $\{\tilde{u}_t, \tilde{v}_t, \tilde{x}_0\}$. Now some simple calculation, which is facilitated and in fact illuminated by the use of operator notation, will show that the $\{\eta_t, \Theta\}$ as defined above span the same space as the $\{\tilde{u}_t, \tilde{v}_t, \tilde{x}_0\}$. Hence ...]

Of the several interesting implications of the above results, we shall here focus on just one.

3d. The Hamiltonian Equations and a Scattering Framework for Estimation Theory

Consider the projection of the random variables in the equations (55) onto the space spanned by $\{y_t,\ t_0 \le t \le t_f\}$.

Now, if we define

$$\lambda_{t|t_f} = \hat{\chi}_{t|t_f} \tag{56}$$

and note that the orthogonality properties (52) imply that $\hat{\eta}_{t|t_f} \equiv 0$ and $\hat{\Theta}_{|t_f} \equiv 0$, the equations (55) reduce to

$$\begin{bmatrix} \dot{\hat{x}}_{t|t_f} \\ -\dot{\lambda}_{t|t_f} \end{bmatrix} = \begin{bmatrix} F_t & -G_t G_t' \\ H_t' H_t & F_t' \end{bmatrix} + \begin{bmatrix} 0 \\ H_t' \end{bmatrix} y_t \tag{57a}$$

with boundary conditions

$$\lambda_{t_f|t_f} = 0\,,\ \hat{x}_{t_0|t_f} = \Pi_0 \lambda_{t_0|t_f}\ . \tag{57b}$$

These are the so-called *Hamiltonian equations* for linear estimation. They were first obtained by Bryson and Frazier (1963) in exploring a calculus of variations approach (maximizing a 'likelihood' function) to least-squares estimation. Verghese, Friedlander and Kailath (1980) used them to develop a so-called scattering theory approach to linear estimation. To introduce this, we shall for convenience change notation in (57):

$$t_0 \to \tau, \ t \to s, \ t_f \to t \ .$$

Then using Euler discretization, e.g.,

$$\dot{\lambda}_{s|t} = [\lambda_{s+\Delta|t} - \lambda_{s|t}]/\Delta + o(\Delta)$$

and the approximation

$$\int_s^{s+\Delta} y_\sigma \, d\sigma = y_s \Delta + o(\Delta) \, ,$$

we can obtain the following discrete approximation to (57):

$$\begin{bmatrix} \hat{x}_{s+\Delta|t} \\ \lambda_{s|t} \end{bmatrix} = \begin{bmatrix} I+F_s\Delta & G_s G_s'\Delta \\ -H_s'H_s\Delta & I+F_s'\Delta \end{bmatrix} \begin{bmatrix} \hat{x}_{s|t} \\ \lambda_{s+\Delta|t} \end{bmatrix} + \begin{bmatrix} 0 \\ H_s'y_s\Delta \end{bmatrix} \tag{58}$$

where we have, and shall in the future, consistently omit all $o(\Delta)$ terms. Note the arguments of the $\lambda_{\cdot|t}$, which are reversed from those of $\hat{x}_{\cdot|t}$. Because of this, we can depict (58) graphically as in Fig. 1, which suggests that we can regard $\hat{x}(\cdot\,|t)$ as a ***forward*** wave and $\lambda(\cdot\,|t)$ as a ***backward*** wave travelling through an incremental section at s of some *scattering* medium specified by the quantities

$I + F_s\Delta =$ the incremental *forward transmission coefficient*

$I + F_s'\Delta =$ the incremental *backward transmission coefficient*

$-H_s'H_s\Delta =$ the incremental *left reflection coefficient*

$G_s G_s'\Delta =$ the incremental *right reflection coefficient* and

$H_s' y_s \Delta =$ the incremental *internal backward source excitation.*

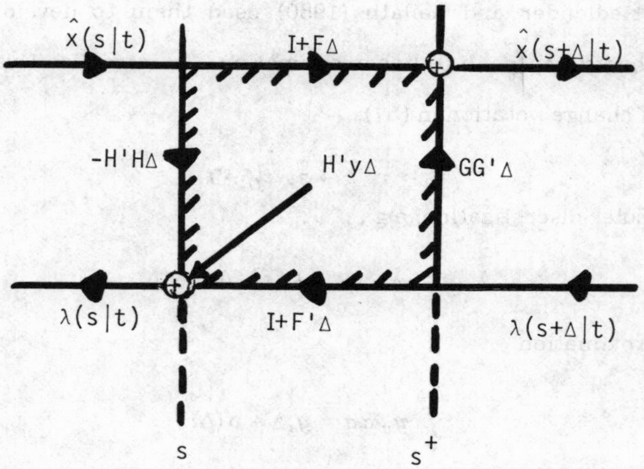

Figure 1. An incremental scattering layer corresponding to Eq. (58).

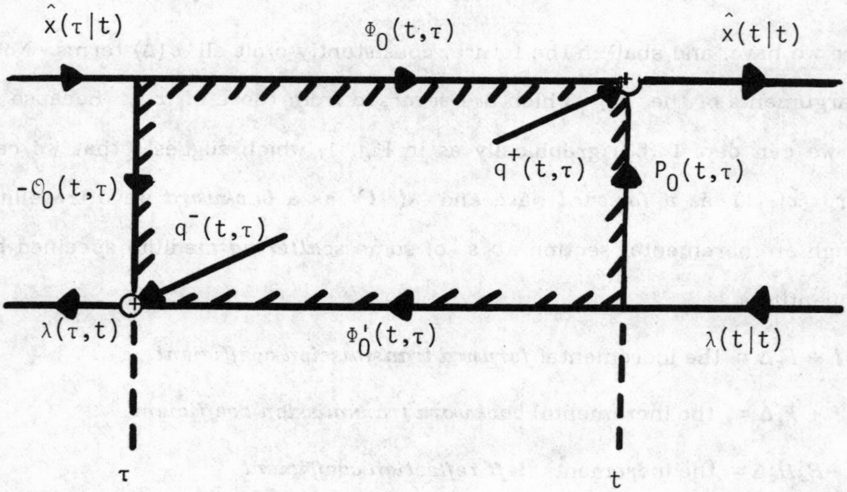

Figure 2. A macroscopic scattering section

We can put together such incremental sections to get a macroscopic section of the scattering medium from say $s=\tau$ to $s=t$. This is shown in Fig. 2, where, for reasons that will be clear very soon, we have denoted

$\Phi_0(t,\tau)$ = the *forward* transmission operator

$\Phi_0'(t,\tau)$ = the *backward* transmission operator

$P_0(t,\tau)$ = the *right* reflection operator

$\mathbf{O}_0(t,\tau)$ = the *left* reflection operator

$q_0^+(t,\tau)$ = the *forward* internal source (i.e., $y(.)$) contribution

$q_0^-(t,\tau)$ = the *backward* internal source contribution

The reasons for this notation can be seen by considering the effect of adding an incremental section from t to $t+\Delta$, as shown in Fig. 3. By tracing paths through the figure, we can write

$$\begin{aligned}\Phi_0(t+\Delta,\tau) &= (I+F\Delta)\Phi_0(t,\tau) - (I+F\Delta)P_0H'H\Delta\Phi_0(t,\tau)\\ &\quad + (I+F\Delta)P_0H'HP_0H'H\Delta^2\Phi_0(t,\tau) - \cdots\\ &= (I+F\Delta)(I-P_0H'H\Delta + o(\Delta)\Phi_0(t,\tau)\end{aligned} \tag{59}$$

where $o(\Delta)$ denotes terms that go to zero faster than Δ as $\Delta\to 0$. Then we see that

$$\lim \frac{\Phi_0(t+\Delta,t) - \Phi_0(t,\tau)}{\Delta} = (F - P_0H'H)\Phi_0(t,\tau) \tag{60}$$

which identifies $\Phi_0(t,\tau)$ as the state-transition matrix of $F-P_0H'H$.

Similarly, we can see by tracing paths through Figure 3 that

$$P_0(t+\Delta,\tau) = GG'\Delta + (I+F\Delta)(P_0 - P_0H'HP_0\Delta + o(\Delta))(I+F'\Delta) \tag{61}$$

so that

$$\lim \frac{P_0(t+\Delta,\tau) - P_0(t,\tau)}{\Delta} = GG' + FP_0 + P_0F' - P_0H'HP \tag{62}$$

which identifies $P_0(t,\tau)$ as the solution of a Riccati differential equation. A similar calculation will identify $\mathbf{O}_0(t,\tau)$ as an observability Gramian of the matrices $\{F-P_0H'H, H\}$

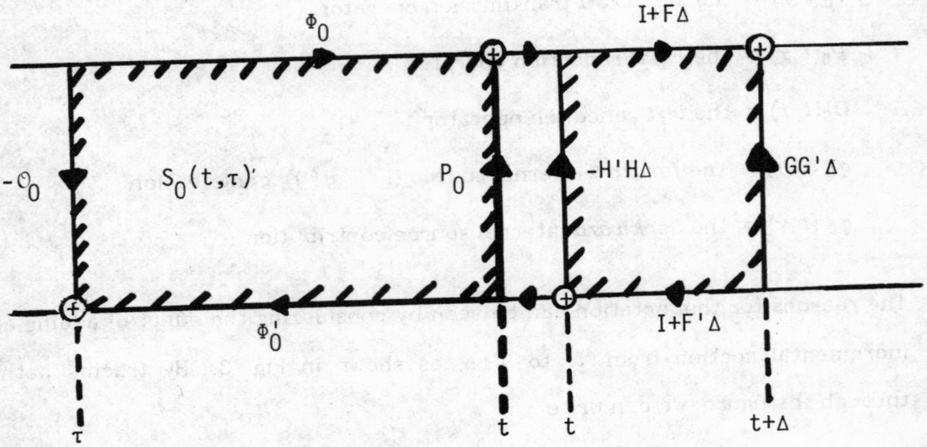

Figure 3. To determine the (forward) evolution equation of $S_0(t,\tau)$

We shall collect these operators in a so-called *scattering matrix*

$$S_0(t,\tau) = \begin{bmatrix} \Phi_0(t,\tau) & P_0(t,\tau) \\ -\mathbf{O}_0(t,\tau) & \Phi_0'(t,\tau) \end{bmatrix} \tag{63}$$

and our previous calculations show that

$$\frac{\partial}{\partial t} S_0(t,\tau) = \begin{bmatrix} (F-P_0H'H)\Phi_0 & FP_0+P_0F'+GG'-P_0H'HP_0 \\ \Phi_0'H'H\Phi_0 & \Phi_0'(F-P_0H'H)' \end{bmatrix} \tag{64a}$$

with initial conditions

$$S_0(\tau,\tau) = \begin{bmatrix} I & 0 \\ 0 & I \end{bmatrix} \tag{64b}$$

These initial conditions help to explain the subscript '0' on the various quantities above, and we can get some more insight into this by going back to the boundary conditions of the original Hamiltonian equations and trying to incorporate them into our scattering picture. The $\lambda_{t|t}=0$ condition means we have no backwards incoming wave, while the condition

$$\hat{x}_{\tau|t} = \Pi_0 \lambda_{\tau|t}$$

can be incorporated, as shown in Fig. 4, by adding a 'boundary' layer to the left of the section $S_0(t,\tau)$ of Fig. 2.

One immediate result from this figure is that we can identify $\{q_0^+, q_0^-\}$ as the *emerging waves from the medium when* $\Pi_0=0$. We shall denote these as

$$q_0^-(t,\tau) = \lambda_0(\tau|t), \quad q_0^+(t,\tau) = \hat{x}_0(t|t) \tag{65}$$

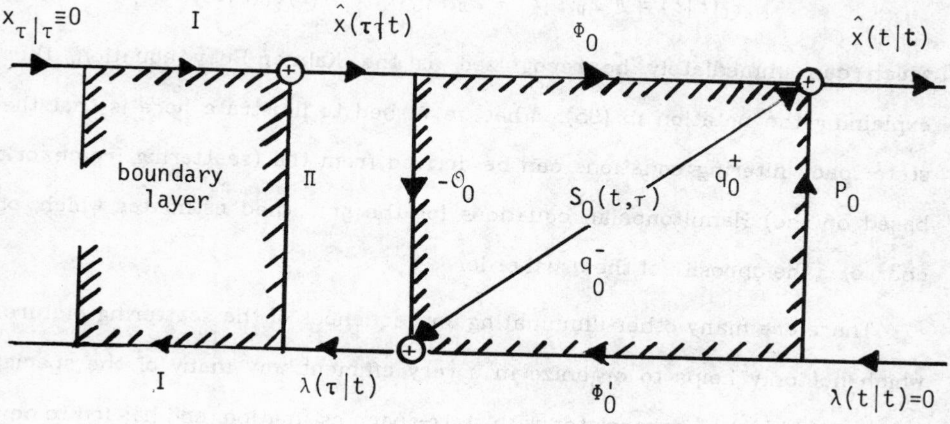

Figure 4. Incorporating the boundary conditions.

We shall now show how to derive forward differential equations for λ_0 and $\hat{x}_0$ and also backward equations as in (17)-(18). For forward equations, we start by adding an incremental section to the one in Fig. 4 (but with $\Pi=0=x_0$). Doing this gives the result shown in Fig. 5b.

Combining the signals in the two parts of Fig. 5a we obtain for the quantities in the combined section of Fig. 5b, the relations

$$\lambda_0(t\,|\,t+\Delta) = \lambda_0(\tau\,|\,t) + \Phi_0' H'(y - H\hat{x}_0(t\,|\,t+\Delta)\Delta + o(\Delta)$$

so that

$$\frac{\partial\lambda_0(\tau\,|\,t)}{\partial t} = \Phi_0'(t,\tau)H'(t)(y(t) - H(t)\hat{x}_0(t\,|\,t)) \tag{66}$$

So also, we can read off the relations

$$\hat{x}_0(t+\Delta\,|\,t+\Delta) = (I + F\Delta)\hat{x}_0(t\,|\,t+\Delta)$$
$$\hat{x}_0(t\,|\,t+\Delta) = \hat{x}_0(t\,|\,t) + P_0H'(y - H\hat{x}_0(t\,|\,t+\Delta)) + o(\Delta)$$

so that

$$\dot{\hat{x}}_0(t\,|\,t) = F\,\hat{x}_0(t\,|\,t) + P_0H'(y(t) - H(t)\hat{x}_0(t\,|\,t)) \tag{67}$$

which can immediately be recognized as the Kalman-Bucy equation, thus explaining the notation in (65). What we wished to illustrate here is that the state-space filtering equations can be derived from the (scattering framework based on the) Hamiltononian equations for the smoothed estimates which, of course, is the opposite of the usual order.

There are many other illuminating consequences of the scattering picture, which not only helps to organize in a very efficient way many of the special results and identities associated with state-space estimation, and has led to new results on backwards Markovian models, smoothing formulas, asymptotic properties, fast algorithms, decentralized estimation, etc. However, we shall content ourselves here with mention of the papers Kailath (1975), Ljung, Kailath and Friedlander (1976), Friedlander, Kailath and Ljung (1976), Verghese, Friedlander and Kailath (1980), Lévy (1981).

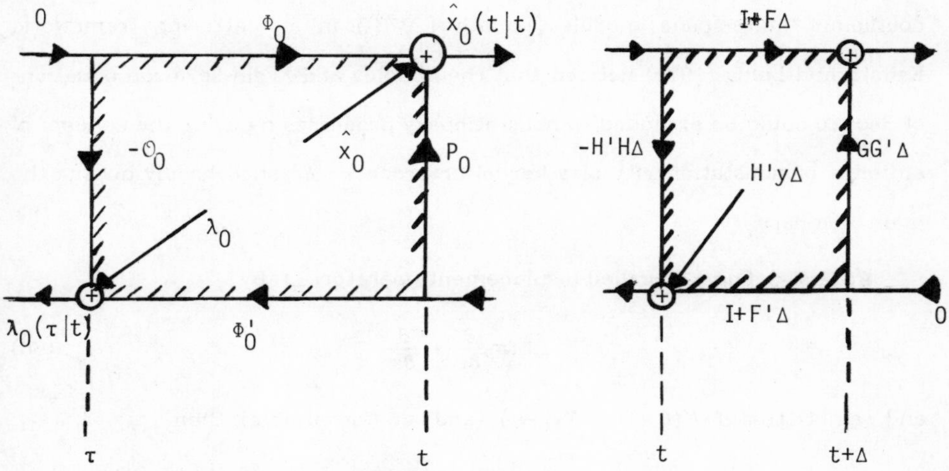

(a) Adding an incremental layer

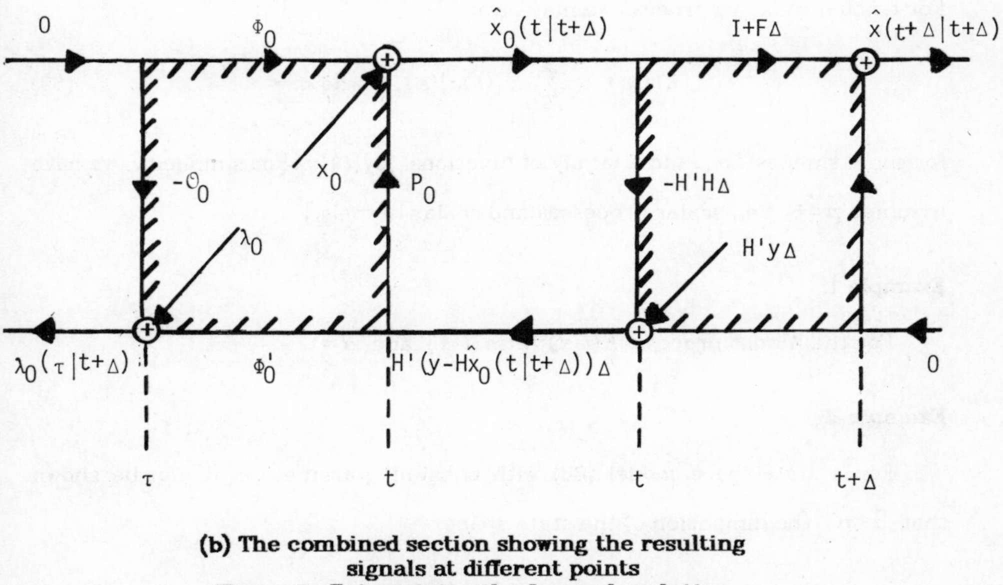

(b) The combined section showing the resulting signals at different points

Figure 5. Determining the forward evolution

3e. Nonstationary Processes Studied as Processes Close to Stationary

In studying the Ambartzumian-Chandrasekhar equations of Sec. 2b and especially their extension to nonstationary processes generated by constant coefficient state-space models (Kailath (1973)) in a scattering framework, Kailath and Ljung (1975) noticed that the Sobolev and Krein-Levinson equations of Sec. 2c could be extended to nonstationary processes by using the concept of an index of nonstationarity of a kernel or process. We shall briefly outline the main idea here.

We first define a so-called displacement operator $\lrcorner$ by

$$\lrcorner := \frac{\partial}{\partial t} + \frac{\partial}{\partial s} \tag{68}$$

and we note that if $K(t,s) = K(t-s)$ (and is differentiable), then

$$\lrcorner K = 0$$

For a non-stationary process we may have

$$\lrcorner K(t,s) = \sum_{t=1}^{\tilde{\alpha}} \varphi_i(t)\varphi_i'(s) \; , \quad \tilde{\alpha} < \infty \tag{69}$$

for some smallest $\tilde{\alpha}$, and a family of functions $\{\varphi_i(t)\}$. [For simplicity, we have assumed $p=1$, i.e., scalar processes and scalar kernels.]

Example 1.

For the Wiener process $K(t,s)=min(t,s)$ and $\tilde{\alpha}=1$.

Example 2.

For a state-space model (38) with constant parameters, it can be shown that $\tilde{\alpha} \le n$, the dimension of the state space.

Example 3.

For a composition of kernels

$$(K_1K_2)(t,s) = \int_0^T K_1(t,u)K_2(u,s)du$$

we obtain the rule

$$\underline{|}(K_1K_2) = (\underline{|}K_1)K_2 + K_1(\underline{|}K_2) + K_1\delta_0K_2 - K_1\delta_TK_2 \tag{70}$$

where

$$K_1\delta_gK_2 := \int_0^T K(t,u)\delta(u-g)K_2(u,s)\,du = K_1(t,g)K_2(g,s)\,.$$

The composition rule gives us an easy derivation of the Sobolev identity. For this, we apply the displacement operator to the equation $H+HK=K$ and use the rule (70) to obtain, after some rearrangement, the result

$$(\underline{|}H)(I+K) = (I-H)\underline{|}K + H(\delta_T - \delta_0)K$$

Noting that $(I+K)^{-1}=I-H$ and $K(I+K)^{-1}=H$ we then obtain

$$\underline{|}H = (I-JH)(\underline{|}K)(I-H) + H(\delta_T-\delta_0)H\,. \tag{71}$$

In the stationary case $\underline{|}K=0$ and we have

$$\underline{|}H = H\delta_TH - H\delta_0H$$

or written out

$$\underline{|}H(t,s;T) = H(t,T;T)H(T,s;T) - H(t,0;T)H(0,s;T)\,. \tag{72}$$

which is just the Sobolev identity (29) of Sec. 2c.

The interesting fact is that this identity can be extended to nonstationary processes by using a slight modification of the representation (69). Let us rewrite (69) in the form

$$\underline{|}K = K(t,0)K(0,s) + \sum_{i=1}^{\alpha} \lambda_j d_i(t)d_i'(s) = K\delta_0K + D_t'\Lambda D_s\,,\ \ say. \tag{73}$$

Substituting into (71) we will have

$$
\begin{aligned}
\_|H &= (I-H)(K\delta_0 K + D_t' \Lambda D_s)(I-H) + H\delta_T H - H\delta_0 H \\
&= H\delta_0 H + (I-H)D_t' \Lambda D_s (I-H) + H\delta_T H - H\delta_0 H \\
&= H\delta_T H + C_t' \Lambda C_s\,, \ say
\end{aligned}
\tag{74}
$$

where $C:=D(I-H)$, or equivalently, C satisfies the integral equation

$$C(1+K) = D$$

Note that C will be a $1\times\alpha$ vector of functions, with $\alpha=1$ in the stationary case. Thus α can serve as a measure of nonstationarity of the process, and (74) is the corresponding Sobolev identity for nonstationary processes. By simple further calculations we can also obtain generalized versions of the Krein-Levinson equations of Sec. 2c.

We refer to Kailath, Ljung, Morf (1976), (1978) for discussions of the computational aspects of these equations and of the role of the parameter α. We regret that reasons of space and time do not permit us to describe here results on some efficient, so-called ladder-form, implementations of these equations. These results allow us to carry over to processes without (finite-dimensional) state-space models, the basic computational advantages of the state-space assumption. These were briefly mentioned in the conference lectures; for more details we refer to the theses of D. T. L. Lee (1980) and H. Lev-Ari (1981).

4. A Concluding Remark

In the nonlinear filtering problem, the state-space assumption has by no means been as useful as in the linear case, since it leads to difficult nonlinear stochastic partial differential equations. It may be that return to an input-output formulation, perhaps based on the Wiener-Volterra reprsentation, can be combined with analysis along the lines of Secs. 2c and 3e to make some computational progress in the nonlinear filtering problem.

REFERENCES

V. A. Ambartsumian, "Diffuse Reflection of Light by a Foggy Medium", *Dokl. Akad. Sci., SSSR*, vol. 38, pp. 229-322, 1943.

R. Bellman and G. Wing, *An Introduction to Invariant Imbedding*, J. Wiley, N.Y., 1975.

A. Bryson and M. Frazier, "Smoothing for Linear and Nonlinear Dynamic Systems", Aeronaut. Syst. Div., Wright-Patterson AFB, Ohio, Tech. Rept. ASD-TDR-63-119, Feb. 1963.

J. L. Casti, R. E. Kalman and V. K. Murthy, "A New Initial-Value Method for On-Line Filtering and Estimation", *IEEE Trans. on Inform. Thy.*, vol. IT-18, pp. 515-518, July 1972.

S. Chandrasekhar, "On the Radiative Equilibrium of Stellar Atmosphere, Pt. XXI", *Astrophys. J.*, vol. 106, pp. 152-216, 1947; Pt. XXII, *ibid*, vol. 107, pp. 48-72, 1948.

S. Chandrasekhar, *Radiative Transfer*, Dover Publications, N.Y., 1960.

R. Courant and D. Hilbert, *Methods of Mathematical Physics*, Vol. I, Interscience, 1953.

M. Devinatz and M. Shinbrot, "General Wiener-Hopf Operators", *Trans. Amer. Math. Soc.*, vol. 145, pp. 47-494, Nov. 1967.

C. L. Dolph and H. A. Woodbury, "On the Relation Between Green's Functions and Covariances of Certain Stochastic Processes and is Applications to Unbiased Linear Prediction", *Trans. Amer. Math. Soc.*, vol. 72, pp. 519-550, 1952.

J. L. Doob, "The Elementary Gaussian Processes", *Ann. Math. Statit.*, vol. 15, pp. 229-282, 1944.

J. L. Doob, "Time Series and Harmonic Analysis", in *Proc. Berkeley Symp. Math., Stat., and Probability*, pp. 303-343, Berkeley, CA, Univ. California Press, 1949.

J. L. Doob, *Stochastic Processes*, J. Wiley, N.Y., 1953.

J. Durbin, "The Fitting of Time-Series Models", *Rev. Inst. Int. Statit.*, vol. 28, pp. 233-243, 1960.

P. Faurre, "Identification par minimisation d'une representation Markovienne de processus aleatoire", *Lecture Notes in Mathematics*, vol. 132, pp. 85-106, Springer-Verlag, N.Y., 1970.

P. Faurre, M. Clerget and F. Germain *Operateurs Rationnels Positifs*, Dunod, 1979.

B. Friedlander, T. Kailath and L. Ljung, "Scattering Theory and Linear Least Squares-Estimation, Pt. II: Discrete-Time Problems", *J. Franklin Inst.*, vol. 301, nos. 1 & 2, pp. 71-82, Jan. 1976.

B. Friedlander, T. Kailath and G. Verghese, "Scattering Theory and Linear Least-Squares Estimation, Pt. III: The Estimates, *Proc. 1977 IEEE Conf. on Decision and Contr.*, pp. 591-597, New Orleans. LA, Dec. 1977. Also, *IEEE Trans. Autom. Contr.*, vol. AC-25, no. 4, pp. 794-802, August 1980.

I. C. Gohberg and L. A. Fel'dman, "Convolution Equations and Projections Methods for their Solutions", *Trans. of Math. Monographs*, vol. 41, Amer. Math. Soc., 1974.

T. Kailath, "An Innovations Approach to Least-Squares Estimation, Pt. I: Linear Filtering in Additive White Noise", *IEEE Trans. on Autom. Contr.*, vol. AC-13, pp. 646-655, Dec. 1968.

T. Kailath, "Some Extensions of the Innovations Theorem", *Bell Syst. Tech. J.*, vol. 50, pp. 1487-1494, April 1971.

T. Kailath, "Some New Algorithms for Recursive Estimation in Constant Linear Systems", *IEEE Trans. on Inform. Thy.*, vol. IT-19, pp. 750-760, Nov. 1973.

T. Kailath, "A View of Three Decades of Linear Filtering Theory", *IEEE Trans. on Information Theory*, vol. IT-20, no. 2, pp. 145-181, March 1974.

T. Kailath, ed., *Benchmark Papers in Linear Least-Squares Estimation*, Academic Press, N.Y., 1977.

T. Kailath, "Some Alternatives in Recursive Estimation", *Int'l. J. of Control*, vol. 32, no. 2, pp. 311-328, August 1980.

T. Kailath, B. Lévy, L. Ljung and M. Morf, "Fast Time-Invariant Implementations of Gaussian Signal Detectors", *IEEE Trans. Inform. Thy.*, vol. IT-24, no. 4, pp. 469-477, July 1978.

T. Kailath and R. Geesey, "An Innovations Approach to Least-Squares Estmation, Pt. IV: Recursive Estimation Given Lumped Covariance Functions", *IEEE Trans. on Autom. Contr.*, vol. AC-16, pp. 720-72, Dec. 1971.

T. Kailath and L. Ljung, "A Scattering Theory Framework for Fast Least-Squares Algorithms", *Proc. Fourth Int'l. Multivariate Analysis Symposium*, Dayton, OH, June 1975; North Holland, Amsterdam, 1977.

T. Kailath and L. Ljung, "On the Stochastic Realization Problem", Inform. Systems Lab. Tech. Rept., Stanford Univ., 1981.

T. Kailath, L. Ljung and M. Morf, "Recursive Input-Output and State-Space Solutions for Continuous-Time Linear Estimation Problems", *Proc. 1976 IEEE Decision & Control Conference*, pp. 910-915, Florida, Dec. 1976.

T. Kailath, L. Ljung and M. Morf, "A New Approach to the Determination of Fredholm Resolvents of Nondisplacement Kernels", in *Topics in Functional Analysis*, ed. by I. Gohberg and M. Kac, Academic Press, 1978.

T. Kailath, A. Vieira and M. Morf, "Inverses of Toeplitz Operators, Innovations, and Orthogonal Polynomials", *SIAM Review*, vol. 20, no. 1, pp. 106-119, Jan. 1978.

R. E. Kalman, "A New Approach to Linear Filtering and Prediction Problems", *J. Basic Engg.*, vol. 82, pp. 34-45, March 1960.

R. E. Kalman and R. S. Bucy, "New Results in Linear Filtering and Prediction Theory", *Trans. ASME, Ser. D., J. Basic Engg.*, vol. 83, pp. 95-107, Dec. 1961.

M. G. Krein, "On a New Method of Solving Linear Integral Equations of the First and Second Kinds", *Dokl. Akad. Nauk SSSR*, vol. 100, pp. 413-416, 1955.

M. G. Krein, "Integral Equations on a Half-Axis with Kernel Depending on the Difference of the Arguments", *Usp. Math. Nauk.*, vol. 12, pp. 3-120, 1958.

H. Laning and R. Battin *Random Processes in Automatic Control*, McGraw-Hill Book Co., N.Y., 1958.

D. T. L. Lee, Ph.D Dissertation, Dept. of Elec. Engg., Stanford Univ., 1980.

H. Lev-Ari, "Ph.D. Dissertation, Dept. of Elec. Engg., Stanford, Univ., 1981.

N. Levinson, "The Wiener rms (Root-Mean-Square Error in Criterion in Filter Design and Prediction", *J. Math. Phys.*, vol. 25, pp. 261-278, Jan. 1947.

B. Lévy, "Scattering Theory and Some Problems of Decentralized Estimation", MIT Lab. for Inform. & Decision Sciences Tech. Rept., 1981.

B. Lévy, T. Kailath, L. Ljung and M. Morf, "Fast Time-Invariant Implementations for Linear Least-Squares Smoothing Filters", *IEEE Trans. Autom. Contr.*, vol. 24, no. 5, pp. 770-774, Oct. 1979.

A. Lindquist and G. Picci, "State Space Models for Gaussian Stochastic Processes", in *Stochastic Systems: The Mathematics of Filtering and Identification Applications*, eds. M. Hazewinkel and J. C. Willems, Reidel Publ. Co., 1981.

L. Ljung, T. Kailath and B. Friedlander, "Scattering Theory and Linear Least Squares Estimation, Pt. I: Continuous-Time Problems", *Proc. IEEE*, vol. 64, no. 1, pp. 131-138, Jan. 1976.

J. Meditch, *Stochastic Optimal Linear Estimation and Control*, McGraw-Hill, N.Y., 1969.

P. A. Meyer, "Sur un problème de filtration", Seminaire de probabilites, Pt. VII, *Lecture Notes in Math.*, vol. 321, pp. 223-247, Sprnger-Verlag, N.Y., 1973.

B. Sz. Nagy and C. Foias, *Harmonic Analysis of Operators on Hilbert Space*, Academic Press, N.Y., 1970.

F. Riesz and B. Sz. Nagy, *Functional Analysis*, F. Ungar, 1955.

G. Ruckebusch, "Theorie geometrique de la representation markovienne", These de doctorat d'etat, Univ. Paris VI, 1980.

M. Shinbrot, "A Generalization of a Method for the Solution of the Integral Equation Arising in Optmization of Time-Varying Linear Systems with Nonstationary Inputs", *IRE Trans. Inform. Thy.*, vol. IT-3, pp. 220-225, Dec. 1957.

V. V. Sobolev, *A Treatise on Radiative Transfer*, D. Van Nostrand, Princeton, N.J., 1963, (Russian edition 1956).

V. V. Sobolev, *A Treatise on Radiative Transfer, Appendices*, D. Van Nostrand Co., Princeton, N.J., 1963.

V. V. Solodovnikov, *An Introduction to the Statistical Dynamics of Automatic Control Systems*, Dover Publications, 1962.

R. L. Stratonovich, "On the Theory of Optimal Nonlinear Filtration of Random Functions", *Thy. Prob. & Its Appl.*, 4, pp. 223-225, 1959.

R. L. Stratonovich, "Application of the Theory of Markov Processes for Optimum Filtration of Signals", *Radio Engg. Elctron. Phys., (USSR)*, vol. 1, pp. 1-19, Nov. 1960.

R. L. Stratonovich, "Conditional Markov Processes", *Thy. Prob. & Its Appl.*, 5, pp. 156-178, 1960.

M. Wax, ed., *Selected Papers in Noise and Stochastic Processes*, Dover, 1954.

H. Weinert and V. B. Desai, "On Adjoint Models and Fixed-Interval Smoothing", Tech. Rept. JHU-EE No. 79-8, Johns Hopkins Univ., MD, 1979.

P. Whittle, *Prediction and Regulation*, Van Nostrand Reinhold, N.Y., 1963.

N. Wiener, *Extrapolation, Interpolation and Smoothing of Stationary* Time Series, with Engineering Applications, Technology Press and J. Wiley, N.Y., 1949, (originally issued in Feb. 1942, as a classified Nat. Defense Res. Council Rept.)

L. A. Zadeh and J. R. Ragazzini, "An Extension of Wiener's Theory of Prediction", *J. Appl. Phys.*, vol. 21, pp. 645-655, July 1950.

INVERSE SCATTERING AND LINEAR PREDICTION, THE TIME CONTINUOUS CASE

P. Dewilde, J.T. Fokkema and I. Widya

Delft University of Technology
Department of Electrical Engineering
Mekelweg 4
2628 CD Delft, The Netherlands

1. Introduction

Let be given a scalar, stationary stochastic process y(t) with covariance function

$$\delta(t) + k(|t|) \tag{1.1}$$

where k(t) = 0 for t < 0. We shall assume that k(t) is known as a function of time. It may have been estimated from measurements, e.g. form a datastream out of an A/D converter, and it may represent a Gaussian stochastic process, say a voice signal. In a later paragraph some of the applications of the theory presented here will be discussed.

The main goal of the paper will be to extract from k(t) a model filter consisting of a transmission line which, when inputed with white noise yields a process whose covariance function is approximately (1.1). This problem has been studied extensively in the past and has given rise, for the case where t is discrete, to the Levinson algorithm, and for the case where t is continuous to the theory of Krein orthogonal functions. An introduction to both subjects can be found in [1],[2]. Recently, e.g. in [3], the addition of scattering theory going back to Schur [4] has led to a substantial improvement in the quality of the algorithms, and has put the classical prediction theory in the context of inverse scattering theory. These ideas have been systematically studied in a recent paper [5] for the time discrete time case. In the time-continuous case the ideas used turn out to be even simpler and form the subject of the present paper.

M. Hazewinkel and J. C. Willems (eds.), Stochastic Systems: The Mathematics of Filtering and Identification and Applications, 351–382.

We start out by reviewing some salient features of both prediction and lossless scattering theory. Using the Fourier transform $F(j\omega)$ of a function $f(t)$ defined by:

$$F(j\omega) = \int_{-\infty}^{\infty} f(t)e^{-j\omega t}dt \tag{1.2}$$

it is well known [6] that the spectrum of the covariance funtion (1.1) is non-negative:

$$W(j\omega) = 1 + K(j\omega) + K(-j\omega) \geq 0 \tag{1.3}$$

We shall for simplicity assume that $k(t)$ is smooth and also that it is square integrable in the interval $[0,\infty)$. In that case $K(j\omega)$ is also an L^2 function. Based on (1.1) define the function:

$$z_0(t) = \delta(t) + 2k(t) \tag{1.4}$$

Using the Laplace transform $F(\lambda)$ of a function $f(t)$ defined by:

$$F(\lambda) = \int_0^{\infty} f(t)e^{-\lambda t}dt \tag{1.5}$$

it follows from the fact that $k(t) \in L^2$ and has support on $t \geq 0$, and from (1.3) that

$$Z(\lambda) = 1 + 2K(\lambda) \tag{1.6}$$

is positive real in the right half complex plane.

In prediction theory one tries to predict $y(t)$ by means of a linear prediction which is a weighted sum of the values of $y(t)$ in the interval $[t-T,t)$. In doing so a prediction error is incurred which is called an innovations. E.g. for $t=T$ the innovations is defined by:

$$\varepsilon(T) = y(T) - \int_0^T h(T,\tau)y(\tau)d\tau \tag{1.7}$$

where $h(T,\tau)$ is the predictor weight function of length T. According to the classical Wiener theorie (see [2]), $h(T,\tau)$ must be choosen so that $\varepsilon(T)$ is statistically orthogonal on $y(t)$ for t in $[0,T)$, so that $h(T,\tau)$ must satisfy:

$$h(T,\tau) + \int_0^T h(T,u)k(|u-\tau|)du = k(T-\tau) \quad \text{for } \tau \in [0,T) \tag{1.8}$$

A thorough discussion of the innovations theory leading to (1.8) may be found in [2].

The Laplace transform of functions related to $h(T,\tau)$ has been studied by Krein [7]. We define the Krein function of the first kind

$$P(T,\lambda) \triangleq e^{\lambda T}[1- \int_0^T h(T,T-t)e^{-\lambda t}dt] \tag{1.9}$$

For future use let us introduce the para-hermitian conjugate $F_*(\lambda)$ of a function $F(\lambda)$:

$$F_*(\lambda) = [F(-\bar{\lambda})]^- \tag{1.10}$$

and the reverse Krein function $P^*(T,\lambda)$:

$$\begin{aligned} P^*(T,\lambda) &= e^{\lambda T}P_*(T,\lambda) \\ &= 1 - \int_0^T \bar{h}(T,T-t)e^{\lambda t}dt \end{aligned} \tag{1.11}$$

It is not too hard to prove (see e.g. [8]) that $h(T,\tau)$ will be smooth in the triangle $0 \le \tau \le T$ (boundaries included) if $k(t)$ is. This means also that limits on the inside boundary may be defined. The function is taken zero outside the triangle. In our case the derivative $\partial_T h(T,T-\tau)$ exists and one has the Bellman-Siegert-Krein relation:

$$\partial_T h(T,T-\tau) = -\bar{h}(T,\tau)h(T,0) \tag{1.12}$$

It follows easily that:

$$\partial_\tau \begin{bmatrix} P(\tau,\lambda) \\ P^*(\tau,\lambda) \end{bmatrix} = \begin{bmatrix} \lambda & -h(\tau,0) \\ -\bar{h}(\tau,0) & 0 \end{bmatrix} \begin{bmatrix} P(\tau,\lambda) \\ P^*(\tau,\lambda) \end{bmatrix} \tag{1.13}$$

with initial conditions

$$\begin{bmatrix} P(0,\lambda) \\ P^*(0,\lambda) \end{bmatrix} = \begin{bmatrix} 1 \\ 1 \end{bmatrix} \tag{1.14}$$

Integration of (1.14) produces:

$$\begin{bmatrix} P(T,\lambda) \\ P\ (T,\lambda) \end{bmatrix} = \Theta(T,\lambda) \begin{bmatrix} 1 \\ 1 \end{bmatrix} \tag{1.15}$$

where

$$\Theta(T,\lambda) = \int_0^T \exp \begin{bmatrix} \lambda & -h(\tau,0) \\ -\bar{h}(\tau,0) & 0 \end{bmatrix} d\tau \tag{1.16}$$

where the product integral is to be defined as in [9] or [10]. In order to be able to express $\Theta(T,\lambda)$ in function of Krein functions we define the Krein functions of the second order kind by:

$$\partial_\tau \begin{bmatrix} Q(\tau,\lambda) \\ -Q^*(\tau,\lambda) \end{bmatrix} = \begin{bmatrix} \lambda & -h(\tau,0) \\ -\bar{h}(\tau,0) & 0 \end{bmatrix} \begin{bmatrix} Q(\tau,\lambda) \\ -Q^*(\tau,\lambda) \end{bmatrix} \tag{1.17}$$

with initial condition:

$$\begin{bmatrix} Q(0,\lambda) \\ -Q^*(0,\lambda) \end{bmatrix} = \begin{bmatrix} 1 \\ -1 \end{bmatrix}$$

$Q(\tau,\lambda)$ is exactly of the same type as $P(\tau,\lambda)$. There exists a smooth function $h_1(T,t)$ such that

$$Q(T,\lambda) = e^{\lambda T} \quad 1 - [\int_0^T h_1(T,T-t)e^{-\lambda t}dt] \tag{1.18}$$

Using the functions defined so far we have:

$$\Theta(T,\lambda) = \begin{bmatrix} \dfrac{P(T,\lambda) + Q(T,\lambda)}{2} & \dfrac{P(T,\lambda) - Q(T,\lambda)}{2} \\ \dfrac{P^*(T,\lambda) - Q^*(T,\lambda)}{2} & \dfrac{P^*(T,\lambda) + Q^*(T,\lambda)}{2} \end{bmatrix} \tag{1.19}$$

At this point we introduce scattering theory in a physical way. Consider the transmission situation pictured in fig. 1.

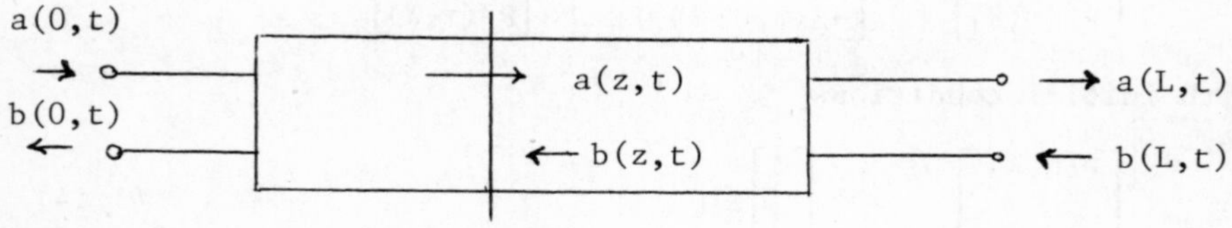

Fig.1 Picture of the transmission situation

In fig. 1, $a(0,t)$ is the incident wave at the left port and $b(L,t)$ the incident wave at the right port. Similarly $b(0,t)$ respect. $a(L,t)$ are reflected waves at the left respect. right ports. We shall suppose that the transmission line is characterized by a reflection function $\rho(z)$ along the line and the infinitesimal equation:

$$\partial_z \begin{bmatrix} A(z,\lambda) \\ B(z,\lambda) \end{bmatrix} = \begin{bmatrix} -\lambda & -\rho(z) \\ -\bar{\rho}(z) & 0 \end{bmatrix} \begin{bmatrix} A(z,\lambda) \\ B(z,\lambda) \end{bmatrix} \tag{1.20}$$

where A and B are Laplace transforms of a and b. Notice that the transmission line employed here is a non-reciprocal one with the left to right wave having infinite velocity and the right to left wave having velocity one. We shall show that this type of line enjoys the properties of a "normal" transmission line. In fact we would obtain (1.20) as characteristic equation if one would travel along a normal transmission line with the same velocity as the left to right wave ;(1.20) is used for convenience.

Let J be the matrix

$$J = \begin{bmatrix} 1 & 0 \\ 0 & -1 \end{bmatrix} \tag{1.21}$$

and define the "generator"

$$X(z,\lambda) = \begin{bmatrix} -\lambda & -\rho(z) \\ -\bar{\rho}(z) & 0 \end{bmatrix} \tag{1.22}$$

The generator has the property that it is -skew on the imaginary axis and -semi-negative in Reλ > 0 (denoting the hermitian conjugate with a ~):

$$JX(z,j\omega) + J\tilde{X}(z,j\omega) = 0 \tag{1.23}$$
$$JX(z,\lambda) + J\tilde{X}(z,\lambda) \leq 0 \text{ for } \mathrm{Re}\lambda > 0 \tag{1.24}$$

(1.20) may be integrated to yield:

$$\begin{bmatrix} A(L,\lambda) \\ B(L,\lambda) \end{bmatrix} = \theta(L,\lambda) \begin{bmatrix} A(0,\lambda) \\ B(0,\lambda) \end{bmatrix} \tag{1.25}$$

where

$$\theta(L,\lambda) = \overset{\curvearrowleft}{\int_0^L} \exp \begin{bmatrix} -\lambda & -\rho(z) \\ -\bar{\rho}(z) & 0 \end{bmatrix} dz \tag{1.26}$$

From (1.23)-(1.24) it follows immediately that $\theta(L,\lambda)$ is J-contractive for Reλ > 0 and J-unitary on the imaginary axis:

$$J - J\theta(L,j\omega)J\tilde{\theta}(L,j\omega) = 0 \tag{1.27}$$
$$J - J\theta(L,\lambda)\ J\tilde{\theta}(L,\lambda) \geq 0 \qquad \text{for } \mathrm{Re}\lambda > 0 \tag{1.28}$$

In the scattering theory litterature, $\theta(L,\lambda)$ is denoted as the "Chain Scattering Matrix" (CSM). Connected with it there is a scattering matrix $\Sigma(L,\lambda)$ which relates reflected waves to incident waves:

$$\begin{bmatrix} B(0,\lambda) \\ A(L,\lambda) \end{bmatrix} = \Sigma(L,\lambda) \begin{bmatrix} A(0,\lambda) \\ B(L,\lambda) \end{bmatrix} \tag{1.29}$$

With

$$P = \begin{bmatrix} 1 & 0 \\ 0 & 0 \end{bmatrix} ; \quad P^{\perp} = \begin{bmatrix} 0 & 0 \\ 0 & 1 \end{bmatrix} ; \quad E = \begin{bmatrix} 0 & 1 \\ 1 & 0 \end{bmatrix} \tag{1.30}$$

one easily expresses Σ in function of θ:

$$\Sigma(L,\lambda) = E(P^{\perp} + P\theta)(P + P^{\perp}\theta)^{-1} \tag{1.31}$$

$$= \begin{bmatrix} -\theta_{22}^{-1}\,\theta_{21} & \theta_{22}^{-1} \\ \theta_{11} - \theta_{12}\theta_{22}^{-1}\,\theta_{21} & \theta_{12}\theta_{22}^{-1} \end{bmatrix} \tag{1.32}$$

$\Sigma(L,\lambda)$ is a genuine lossless scattering matrix if θ is a J-contractive CSM which is J-unitary on the imaginary axis. In fact, $\Sigma(L,\lambda)$ is unitary on the imaginary axis and contractive for $\mathrm{Re}\lambda \geq 0$:

$$\Sigma(L,j\omega)\tilde{\Sigma}(L,j\omega) = 1_2 \tag{1.33}$$

$$1_2 - \Sigma(L,\lambda)\tilde{\Sigma}(L,\lambda) \geq 0 \quad \text{for } \mathrm{Re}\lambda > 0 \tag{1.34}$$

Additional properties of Σ and θ may be found in the litterature, see especially [5] where a summary of the relevant properties can be found. Important for our purpose will be the existence of θ_{22}^{-1} as a bounded analytic function in $\mathrm{Re}\lambda > 0$.

The identification of the scattering situation with the prediction problem is obtained by putting:

$$\rho(z) = h(z,0) \tag{1.35}$$

and hence an expression for $\theta(L,\lambda)$ is directly obtained in function of the Krein functions:

$$\theta(L,\lambda) = \begin{bmatrix} \dfrac{P(L,-\lambda) + Q(L,-\lambda)}{2} & \dfrac{P(L,-\lambda) - Q(L,-\lambda)}{2} \\ \dfrac{P^*(L,-\lambda) - Q^*(L,-\lambda)}{2} & \dfrac{P^*(L,-\lambda) + Q^*(L,-\lambda)}{2} \end{bmatrix} \tag{1.36}$$

From the properties of $\Sigma(L,\lambda)$ it follows that:

$$\Sigma_{12}(L,\lambda) = \theta_{22}^{-1}(L,\lambda) = \frac{2}{P^*(L,-\lambda)+Q^*(L,-\lambda)} \tag{1.37}$$

is a causal transfer function which is contractive in $\mathrm{Re}\lambda \geq 0$ and

$$\begin{aligned} \Sigma_{21}(L,\lambda) &= \theta_{11} - \theta_{12}\theta_{22}^{-1}\theta_{21} = \theta_{11*}^{-1} = e^{-\lambda L}(\theta_{11}^*)^{-1} \\ &= e^{-\lambda L}\Sigma_{12}(L,\lambda) \end{aligned} \tag{1.38}$$

It follows that the corresponding time function $\sigma_{21}(L,t) = 0$ for $t < L$.

The main goal of an inverse scattering problem is the computation of the reflection function $\rho(z)$ in function of some knowledge (obtained by measurements) of the functions $a(0,t)$, $b(0,t)$, $a(L,t)$ and $b(L,t)$. For instance, $a(0,t)$ may have been given as input wave and $b(0,t)$ may have been measured as the reflected wave. It may additionally be known that the transmission line is of infinite length or that the line is of finite length L and adapted at the right end. We shall show in the next paragraph that the stochastic problem defined earlier, i.e. the computation of $h(t,0)$ given $k(t)$ can be formulated as such an inverse scattering problem. To be more precise $\rho(z) = h(z,0)$ results as reflection function when

$$a(0,t) = \delta(t) + k(t)$$

and

$$b(0,t) = k(t)$$

In the sequel we shall first prove a number of theoretical results. Next we shall present the continuous-time equivalent of the Schur algorithm to solve the prediction problem. In a following paragraph we shall show that iterative inverse scattering algorithms may be translated into a prediction algorithm. A number of interesting relations and numerical checks may be obtained from conservation projection of the transmission situation. We shall deduce some of these properties.

2. The scattering formalism in prediction theory

In this paragraph we shall deduce the main theoretical results on which the subsequent algorithms are based. The results presented are contained in numerous previous publications see e.g. [11], [12]. However, we believe to present a shortcut to most of these. The first result is obvious from a physical point of view but its mathematical proof is somewhat more elaborate.

Proposition 2.1. [Causality principle]

In the transmission line defined in fig. 1 and equation (1.20) let $a(t,0) = 0$ for $t < 0$, and suppose that $b(L,t) = s(t)*a(L,t)$ where $s(t)$ is a passive scattering function and $*$ indicates convolution , then $a(z,t) = b(z,t) = 0$ for $t < z$ and any $z \leq L$.

Proof

It will be enough to show the property for z=L, for the assumptions of the proposition remain valid at any level z, due to the fact that a passive transmission line loaded in a passive scattering function is itself a passive scattering function. Using Laplace transforms we have at z=L:

$$B(L,\lambda) = \delta(\lambda)A(L,\lambda) \tag{2.1}$$

Let $\Sigma(L,\lambda)$ be the scattering matrix of the two-port defined in the previous paragraph by eqs. (1.29) and following. After eliminating $B(L,\lambda)$ and denoting the entries of $\Sigma(L,\lambda)$ by $\Sigma_{ij}(\lambda)$ we find:

$$[1-\Sigma_{22}(\lambda)S(\lambda)]A(L_1\lambda) = \Sigma_{21}(\lambda)\, A(0,\lambda) \tag{2.2}$$

Since $\Sigma_{22}(\lambda)$ and $S(\lambda)$ are both contractive in $\mathrm{Re}\lambda > 0$ the right hand side is invertible. Moreover,

$$[1 - \Sigma_{22}(\lambda)S(\lambda)]^{-1}\Sigma_{21}(\lambda) \tag{2.3}$$

is surely contractive in $\mathrm{Re}\lambda > 0$ because of energy conservation. From (1.38) it now follows that

$$A(L,\lambda) = e^{-\lambda L}F(\lambda)A(0,\lambda) \tag{2.4}$$

where $F(\lambda)$ is a causal and contractive transfer function. Translating back into the time domain we have

$$a(L,t) = 0 \qquad \text{for } t < L \tag{2.5}$$

and by (2.1) also

$$b(L,t) = 0 \qquad \text{for } t < L \tag{2.6}$$

□

Proposition 2.2

Suppose that the incident wave $a(0,t) = \delta(t) + k(t)$ and the reflected wave is $b(0,t) = k(t)$. Then the condition $a(z,t) = b(z,t) = 0$ for $t < z \le L$ forces $\rho(z) = h(z,0)$ for $z \le L$.

Remark: proposition 2.2 is a lemma preparing the main theorem. We do not assert the existence of a transmission line with the stated property but only the fact that "causality" in this situation forces the Fredholm equation to be satified.

Proof

Using the CSM formalism and using the definition of $z_0(t)$ (1.4) we have:

$$\begin{bmatrix} A(L,\lambda) \\ B(L,\lambda) \end{bmatrix} = \frac{1}{2}\theta[L,\lambda) \begin{bmatrix} Z_0(\lambda) + 1 \\ Z_0(\lambda) - 1 \end{bmatrix} \tag{2.7}$$

with $\theta(L,\lambda)$ given by (1.36) it follows:

$$\left\{\begin{aligned} 2A(L,\lambda) &= P(L,-\lambda)\, Z_0(\lambda) + Q(L,-\lambda) \\ 2B(L,\lambda) &= P^*(L,-\lambda)\, Z_0(\lambda) - Q^*(L,-\lambda) \end{aligned}\right\} \tag{2.8}$$

Let us define further in the time-domain:

$$\left.\begin{aligned} p(L,t) &\triangleq \delta(t+L) - h(L,-t) \\ q(L,t) &\triangleq \delta(L+L) - h_1(L,-t) \end{aligned}\right\} \tag{2.9}$$

where h and h_1 were defined in the introduction.
Transforming (2.8) back to the time domain we have:

$$2b(L,t) = 2k(t) - h(L,L-t)+h_1(L,L-t)-2h(L,L-t)*k(t) \tag{2.10}$$

$$2a(L,t) = 2\delta(L-t) - h(L,t)-h_1(L,t)+2k(t-L)-2h(L,t)*k(t) \tag{2.11}$$

Put $a(L,t) = b(L,t) = 0$ for $t < L$ and change arguments $t \to L-t$ in the second equation to obtain in $[0,L)$:

$$0 = -h(L,t) - 2\int_0^t h(L,u)k(t-u)du - h_1(L,t) \qquad (2.12)$$

$$0 = -h(L,t) + 2k(L-t) - 2\int_t^L h(L,u)k(u-t)du + h_1(L,t) \qquad (2.13)$$

Adding these two equations gives the original Fredholm equation:

$$h(L,t) + \int_0^L h(L,u)k(|t-u|)du = k(L-t) \qquad (2.14)$$

Because of the uniqueness of the solution of this equation [8] , it follows that $h(L,t)$ is uniquely determined and that, in particular $h(z,0) = \rho(z)$ for $z < L$.

□

The main theorem of this section asserts that the prediction filter, i.e. the determination of $h(z,0)$ from $k(z)$ may be solved by solving an inverse scattering problem. Proposition 2.2 indicates that if $a(0,t)=\delta(t) + k(t)$ and $b(0,t) = k(t)$ could be observed as left incident and reflected waves of a causal transmission line, then its reflection function $\rho(z)$ must be equal to $h(z,0)$. We must now show that for any $k(t)$ such a situation always exists. In the next paragrpahs we shall have to show how this knowledge is to be used to produce concrete algorithms.

The discussion of the main theorem starts by looking at the "adapted" situation, shown in fig. 2.

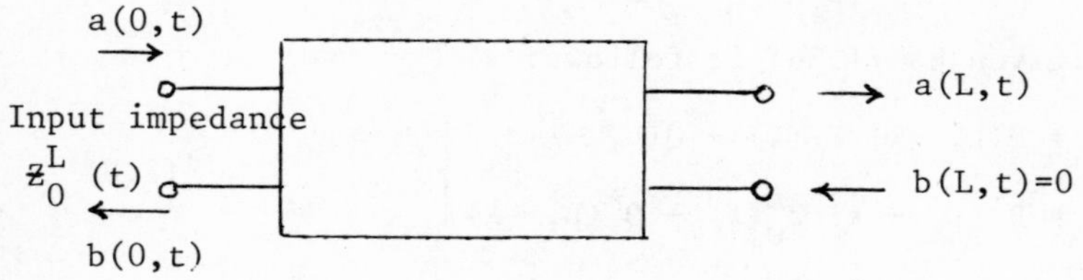

Let for the situation shown $s_0^L(t)$ denote the input scattering function i.e. the map $a(0,t) \to b(0,t) = s_0^L(t)*a(0,t)$.
In fact, its Fourier transform $S_0^L(\lambda)$ is the 11 entry in the scattering matrix Σ. Related to $s_0^L(t)$ and $S_0^L(\lambda)$ is the input impedance function $z_0^L(t)$ with Laplace transform $Z_0^L(\lambda)$ and which is the map

$$[a(0,t) + b(0,t)] \to [a(0,t) - b(0,t)]$$

so that

$$Z_0^L(\lambda) = \frac{1 + S_0^L(\lambda)}{1 - S_0^L(\lambda)} \qquad (2.13)$$

$Z_0^L(\lambda)$ exists in the open right half plane where it is a positive real function, because $S_0^L(\lambda)$ is contractive. $z_0^L(t)$ must have the form:

$$z_0^L(t) = \delta(t) + 2k^L(t) \tag{2.15}$$

for some smooth function $k^L(t)$. This follows from the expression deduced from (1.36):

$$Z_0^L(\lambda) = \frac{Q^*(L,-\lambda)}{P^*(L,-\lambda)} \tag{2.16}$$

and the properties of Q^* and P^*, see [11].

Proposition 2.3

$k^L(t) = k(t)$ for $0 \leq t < L$.

Proof

One way of producing $b(L,t) = 0$ in fig. 2 is to put $A(0,\lambda) = Z_0^L(\lambda) + 1$. From this one has automatically $B(0,\lambda) = Z_0^L(\lambda) - 1$. Hence the equations (2.10) and (2.11) remain valid with $k^L(t)$ replacing $k(t)$ and thus for $t \in [0,L)$ one has for all $t \in [0,L)$:

$$\int_0^t h(L,u)k(t-u)du = \int_0^t h(L,u)k^L(t-u)du$$

as well as

$$k(t) - \int_0^t h(L,L-u)k(t-u)du = k^L(t) - \int_0^t h(L,L-u)k^L(t-u)du \tag{2.17}$$

It follows that $k(t) - k^L(t)$ is a solution of the Volterra equation for $t \in [0,L)$:

$$y(t) - \int_0^t h(L,L-u)y(t-u)du = 0 \tag{2.18}$$

Hence $y(t)=0$ and $k(t)=k^L(t)$ in $[0,L)$

We are now ready for the main Theorem:

Theorem 2.4

Suppose that $h(L,t)$ is the solution of the Fredholm equation

$$h(L,t) + \int_0^L h(L,u)k(|t-u|)du = k(L-t) \tag{2.19}$$

then there exists a passive load $S_L(\lambda)$ such that the transmission line with chain scattering matrix $\theta(L,t)$ build on $h(L,t)$ and terminated in it, has input impedance $Z_0(\lambda)$.

Corollary 2.5

If $a(0,t) = \delta(t) + k(t)$ is the left incident wave to the transmission line with CSM $\theta(L,t)$, then $b(0,t) = k(t)$ is the left reflected wave. Furthermore $a(L,t)$ and $b(L,t)$ have the form:

$$a(L,t) = \delta(t-L) + \alpha(L,t) \qquad (2.20)$$

$$b(L,t) = \beta(L,t) \qquad (2.21)$$

where $\alpha(L,t) = \beta(L,t) = 0$ for $t < L$ are smooth functions such that

$$\hat{S}_L(\lambda) = \frac{B(L,\lambda)}{A(L,\lambda)} \qquad (2.21)$$

is a contractive scattering matrix.

Proof of the theorem

We give an indirect proof. For the time discrete case a direct proof is to be found in [5].

First remark that any finite length transmission line computed from k(t) by the Krein procedure described in par. 1 is uniquely determined by the function h(T,0) so that any two lines with length $L+L_1$, $L+L_2$ will coincide on [0,L). Take two such lines and enforce the boundary condition $b(L_1,t) = b(L_2,t) = 0$ (one says: the lines are right-end-matched), and separate out the first piece of length L. Call S'_{L1} and S''_{L2} the scattering functions which load the pieces of length L (see fig. 3).

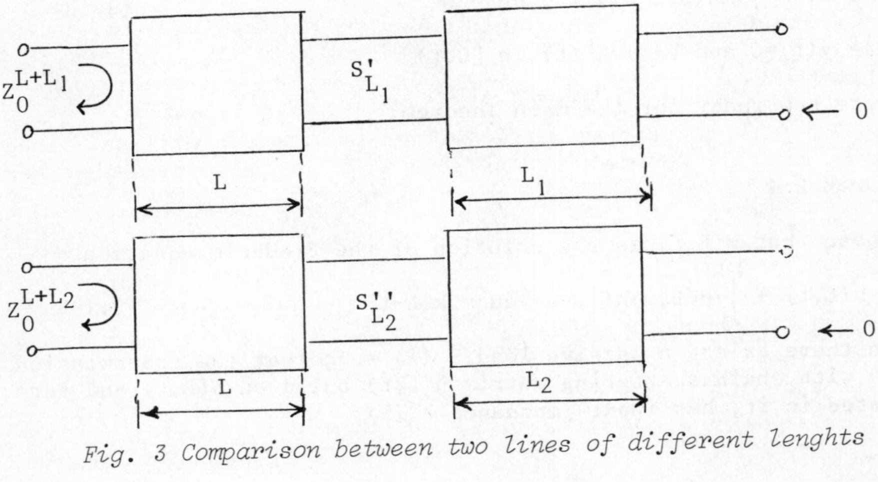

Fig. 3 Comparison between two lines of different lenghts

We study the relation between S'_{L_1} and S''_{L_2}. Suppose for that matter that $L_2 > L_1$. From proposition 2.1 specialised to the input of the lines of lengths L_1 and L_2 it follows that

$$S'_{L_1} - S''_{L_2} = e^{-L_1\lambda}\,\Delta(\lambda) \tag{2.22}$$

where $\Delta(\lambda)$ is analytic in the open right half plane. It follows that $S'_L(\lambda)$ is a uniformily bounded Cauchy net for any point λ in the open right half plane. Hence $S'_L(\lambda)$ converges uniformily on compacts to an analytic and contractive function $S_L(\lambda)$. Suppose further that $S_L(\lambda)$ is used as load for the piece of line of length L (Fig. 4).

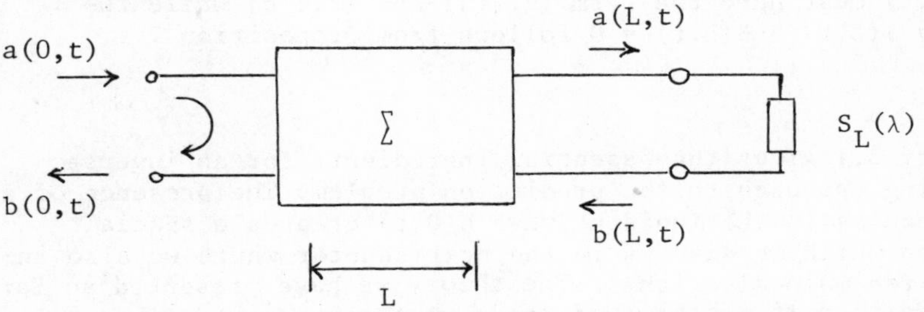

Fig. 4 The line of length L loaded in $S_L(\ )$.

I claim that the input impedance of the line in fig. 4 is precisely $Z_0(\lambda)$. From proposition 2.3 we know that

$$S_0(\lambda) - S_0^{L+L_1}(\lambda) = e^{-\lambda(L+L_1)}\,\Delta_1(\lambda) \tag{2.23}$$

with some $\Delta_1(\lambda)$ analytic in the open right halve plane. Just as before it follows from (2.23) that $S_0^{L+L1}(\lambda)$ converges to $S_0(\lambda)$ for $L_1 \to \infty$. But

$$S_0^{L+L_1}(\lambda) = \Sigma_{11} + \Sigma_{12}(1-\Sigma_{22}S'_{L1})^{-1}\Sigma_{21} \tag{2.24}$$

when Σ is the scattering matrix of the first piece of line. Since Σ_{22} is strictly contractive in the open right halve plane, we see that (2.24) converges to

$$\Sigma_{11} + \Sigma_{12}(1-\Sigma_{22}S_L)^{-1}\Sigma_{21} \tag{2.24}$$

which is precisely the input scattering function of fig. 4. Hence $S_0(\lambda)$ is obtained.

Proof of the Corollary

Given the situation of fig. 4, let $a(0,t) = \delta(t) + k(t)$. This means that

$$a(0,t) = \frac{1}{2}[\delta(t) + z_0(t)] \tag{2.25}$$

and from the definition of the input impedance we have

$$b(0,t) = \frac{1}{2}[z_0(t) - \delta(t)] = k(t) \tag{2.26}$$

The expression (2.22) follows immediately from the proof of the theorem. Finally, it follows from (2.12) and (2.13) that $a(L,t)$ and $b(L,t)$ must have the form (2.21a) and (2.21b) while the property $\alpha(L,t) = \beta(L,t) = 0$ follows from proposition 2.1.

□

Corollary 2.1 gives the essential ingredients for an inverse scattering approach to the prediction problem. The presence of a Dirac inpulse in the incident wave $a(0,t)$ creates a special situation which we discuss in the next chapter where we also introduce the main algorithms. The theory we have presented so far is an adaption to our present needs of theory due to Schur [4] and Krein [7].

3. Schur type recursive algorithm

In this section we shall derive a recursive algorithm to solve the inverse scattering problem for the stochastic prediction case. The algorithm will be called a "Schur-algorithm" in honor of Schur who gave the first version of this algorithm in a somewhat different context [4]. The algorithm is related to the Darlington synthesis as discussed in [3]. The derivation of the algorithm, as well as other iterative algorithms to be discussed later is done using a perturbation-type argument. Here, the occuring reflection terms are considered to be perturbations of an undisturbed wave equation. From par. 2 we know that the stochastic approximation problem is solved by an inverse scattering problem in which

$$\left.\begin{aligned} a(0,t) &= \delta(t) + k(t) \\ b(0,t) &= k(t) \end{aligned}\right\} \tag{3.1}$$

where $k(t)$ is the covariance function of the process and is supposedly a smooth function. The wave equation is:

$$\partial_z \begin{bmatrix} A(z,\lambda) \\ B(z,\lambda) \end{bmatrix} = \begin{bmatrix} -\lambda & 0 \\ 0 & 0 \end{bmatrix} \begin{bmatrix} A(z,\lambda) \\ B(z,\lambda) \end{bmatrix}$$
$$\begin{bmatrix} 0 & -\rho(z) \\ -\rho(z) & 0 \end{bmatrix} \begin{bmatrix} A(z,\lambda) \\ B(z,\lambda) \end{bmatrix} \qquad (3.2)$$

In (3.2) the reflection term has been separated off and is considered a "source" term. (3.2) is easily integrated, say in the interval [0,z]:

$$\left\{ \begin{aligned} A(z,\lambda) &= e^{-\lambda z} A(0,\lambda) - \int_0^z e^{-\lambda(z-\xi)} \rho(\xi) B(\xi,\lambda) d\xi \\ B(z,\lambda) &= B(0,\lambda) - \int_0^z \rho(\xi) A(\xi,\lambda) d\xi \end{aligned} \right\} \qquad (3.3)$$

(3.3) gives in the time domain:

$$\left\{ \begin{aligned} a(z,t) &= a(0,t-z) - \int_0^z \rho(\xi) b(\xi,t-z+\xi) d\xi \\ b(z,t) &= b(0,t) - \int_0^z (\xi) a(\xi,t) d\xi \end{aligned} \right\} \qquad (3.4)$$

In par. 2 we have shown that a(z,t) and b(z,t) have the form

$$\left\{ \begin{aligned} a(z,t) &= \delta(z-t) + \alpha(z,t) \\ b(z,t) &= \beta(z,t) \end{aligned} \right\} \qquad (3.5)$$

where $\alpha(z,t)$ and $\beta(z,t)$ are smooth functions. In terms of these new functions, and introducing a characteristic function

$$\chi_z(t) = \begin{cases} 1 & \text{for } t \in [0,z] \\ 0 & \text{otherwise} \end{cases} \qquad (3.6)$$

we obtain:

$$\left\{ \begin{aligned} \alpha(z,t) &= k(t-z) - \int_0^z \rho(\xi) \beta(\xi,t-z+\xi) d\xi \\ \beta(z,t) &= k(t) - \rho(t)\, \chi_z(t) - \int_0^z \rho(\xi) \alpha(\xi,t) d\xi \end{aligned} \right\} \qquad (3.7)$$

where we have used (3.1). As a word of caution one should remark that the $\delta(t)$ in (3.1) sits to the right of the point zero, i.e. is supported on $[0,\infty)$.

If $\rho(\xi)$ is to be determined recursively, it will be necessary to impose an additional constraint which will turn out to yield the

Schur recursion for this case. For that purpose we use the causality condition of proposition 2.1. The functions $\alpha(z,t)$ and $\beta(z,t)$ are both discontinuous at the points $z=t$. For $t < z$ causality requires $\alpha(z,t) = \beta(z,t) = 0$. We have:

(i) for $t=z^+$

$$\alpha(z,z^+) = k(0) - \int_0^z \rho(\xi)\beta(\xi,\xi^+)d\xi \tag{3.8}$$

$$\beta(z,z^+) = k(z) - \int_0^z \rho(\xi)\alpha(\xi,z)d\xi \tag{3.9}$$

(ii) for $t=z^-$ [causality]

$$0 = \alpha(z,z^-) = 0 \quad \text{trivially if } \beta(z,z^-) = 0$$

$$0 = \beta(z,z^-) = k(z) - \rho(z) - \int_0^z \rho(\xi)\alpha(\xi,z)d\xi \tag{3.10}$$

From (3.10) follows the boundary condition on the diagonal:

$$\rho(z) = k(z) - \int_0^z \rho(\xi)\alpha(\xi,z)d\xi \tag{3.11}$$

As an extra divident one obtains the equality

$$\alpha(z,z^+) = k(0) - \int_0^z |\rho(\xi)|^2 \, d\xi \tag{3.12}$$

from (3.8) - (3.10). We shall call (3.11) the Schur equation.

Finally, the set to be integrated recursively is given by:

(3.13a) $\alpha(z,t) = k(t-z) - \int_0^z \rho(\xi)\beta(\xi,t-z+\xi)d\xi \ , \ (t \geq z)$

(3.13b) $\beta(z,t) = k(t) - \int_0^z \rho(\xi)\alpha(\xi,t)d\xi, \quad (t \geq z)$

(3.13c) $\rho(z) = k(z) - \int_0^z \rho(\xi)\alpha(\xi,z)d\xi$

In fig. 3.1 we picture twice the (z,t) plane. In the first figure $\alpha(z,t)$ is thought to be represented along the vertical, while in the second $\beta(z,t)$ is thought to be represented. For a given point (z,t) the integration paths are given.

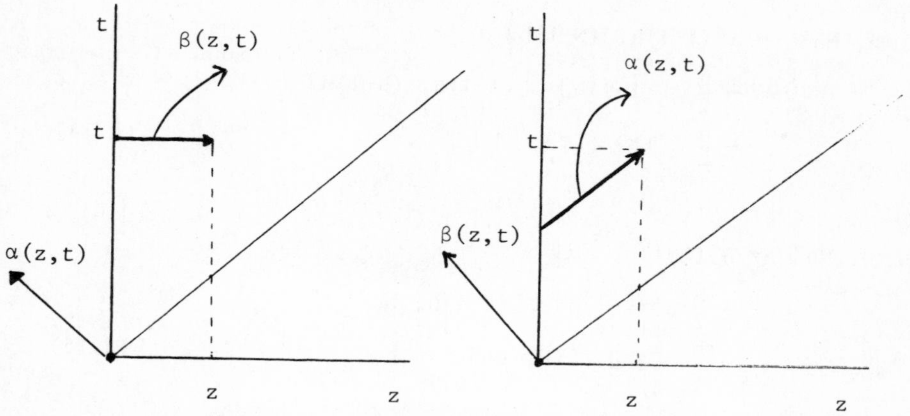

Fig. 3.1 Integration paths for α(z,t) and β(z,t)

In order to solve (3.13a-c) the simplest way is to use an integration rule to evaluate the integrals. In order to demonstrate the procedure let us use e.g. the simplest rule: a trapezoidal integration rule. Any other rule would yield similar results. We discretize the z-axis and the t-axis with steps of size h, and start out with the point t=z=0. The main recursion loop increments t=nh while the secondary loops increments z=ih. The order of integration is shown in fig. 3.2.

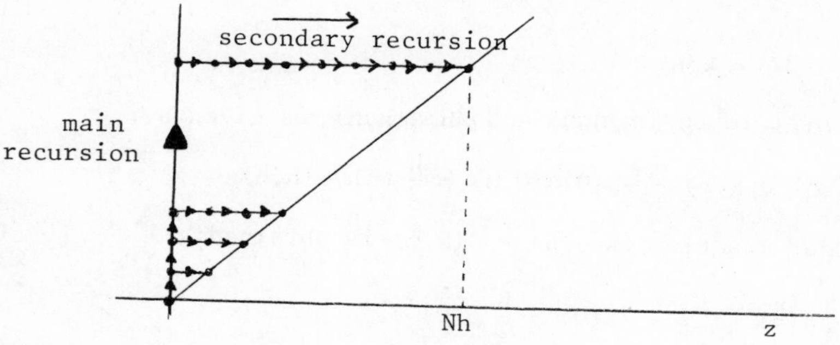

Fig. 3.2 Order of integration in the Schur algorithm

Typically one shall wish to computer recursively the quantities $\alpha[\ell h, Nh]$ and $\beta[\ell h, Nh]$, say with a Trapezoidal rule. It is easy to see that the integrals for $\alpha[\ell h, Nh]$ and $\alpha[(\ell-1)h, (N-1)h]$ differ only by two trapezoidal terms. The same is true for $\beta[\ell h, Nh]$ and $\beta[(\ell-1)h, Nh]$. Hence:

$$\alpha[\ell h, Nh] = \alpha[(\ell-1)h, (N-1)h] - \frac{h}{2}\rho[(\ell-1)h]\,\beta[(\ell-1)h, (N-1)h] - \frac{h}{2}\rho[\ell h]\,\beta[\ell h, Nh] \tag{3.14}$$

$$\beta[\ell h, Nh] = \beta[(\ell-1)h, Nh] - \frac{h}{2}\rho[(\ell-1)h]\,\alpha[(\ell-1)h, Nh] - \frac{h}{2}\rho[\ell h]\,\alpha[\ell h, Nh] \tag{3.15}$$

For the computation of $\alpha[\ell h, Nh]$, $\beta[\ell h, Nh]$ all quantities α and β with lower order indices are known. Thus (3.14) and (3.15) can be solved for them, provided the reflection function $\rho(z)$ is known. This will be the case so long $\ell < N$. For $\ell=N$, the extra Schur equation must be used. We exhibit a number of steps in the algorithm.

Step 0

$\alpha(0,0) = \beta(0,0) = k(0)$; $\rho(0) = k(0)$

Step 1

$\alpha(0,h) = \beta(0,h) = k(h)$

$\alpha(h,h) = k(0) - \frac{h}{2}\rho(0)\,\beta(0,0) - \frac{h}{2}\rho(h)\beta(h,h)$

$\beta(h,h) = \rho(h) = k(h) - \frac{h}{2}\rho(0)\alpha(0,h) - \frac{h}{2}\rho(h)\alpha(h,h)$

A third order equation in $\rho(h) = \beta(h,h)$ is obtained:

$$\frac{h^2}{4}\rho(h)^3 - \rho(h)\left[1 + \frac{h}{2}\left(k(0) - \frac{h}{2}\rho(0)^2\right)\right] + \left[k(h) - \frac{h}{2}\rho(0)k(h)\right] = 0 \tag{3.15}$$

which to an order h^2 solves as:

$$\rho(h) \cong \frac{k(h)\left[1 - \frac{h}{2}\rho(0)\right]}{\left[1 + \frac{h}{2}k(0)\right]} \tag{3.17}$$

This value can be used as a start for a Newton-Raphson iteration to solve (3.15). Because of the recursivity of the procedure one must computer $\rho(h)$ exact to the order h^2 in order to salvage the precision of the integration rule.

Step N

Substep 0

$$\alpha[0,Nh] = \beta[0,Nh] = k[Nh] \tag{3.18}$$

Substep ℓ with $0 < \ell < N$

Let $\alpha'[(\ell-1)h,\ (N-1)h] \triangleq \alpha[(\ell-1)h,\ (N-1)h] - \frac{h}{2}\rho[(\ell-1)h]$

$$\cdot\ \beta[(\ell-1)h,(N-1)h] \tag{3.19a}$$

$$\beta'[(\ell-1)h,\ Nh] \triangleq -\frac{h}{2}\rho[(\ell-1)h]\alpha[(\ell-1)h,\ Nh] + \beta[(\ell-1)h,\ Nh] \tag{3.19b}$$

Then
$$\alpha[\ell h,\ Nh] = \frac{1}{[1-\frac{h^2}{4}\rho[\ell h]^2]}\left\{\alpha'[(\ell-1)h,(N-1)h] - \frac{h}{2}\rho[\ell h]\,\beta'[(\ell-1)h,\ Nh]\right\} \tag{3.20a}$$

$$\beta[\ell h,\ Nh] = \frac{1}{[1-\frac{h^2}{4}\rho[\ell h]^2]}\left\{-\frac{h}{2}\rho[\ell h]\,\alpha'[(\ell-1)h,\ (N-1)h] + \beta'[(\ell-1)h,\ Nh]\right\} \tag{3.20b}$$

Notice that all quantities to the right can be computed.

Substep N

At this point the Schur condition is to be used:

$$\beta[Nh,\ Nh] = \rho(Nh)$$

This transforms (3.20b) into a third order equation:

$$[1-\frac{h^2}{4}\rho[Nh]^2]\rho[Nh] = -\frac{h}{2}\rho[Nh]\alpha'[(N-1)h,(N-1)h] + \beta'[(N-1)h,Nh] \tag{3.21}$$

where $\alpha'[(N-1)h,\ (N-1)h]$ and $\beta'[(N-1)h,\ Nh]$ are as in (3.19)

Neglecting terms of $O(h^2)$ one has the approximate solution:

$$\rho(Nh) \cong \frac{\beta'[(N-1)h,\ Nh]}{1+\frac{h}{2}\alpha'[(N-1)h,(N-1)h]} \tag{3.22}$$

To obtain an accurate solution, (3.22) has to be solved to $O(h^2)$, meaning that a couple of Newton-Raphson iterations msut be added to achieve the required precision.

The algorithm as presented is totally recursive and makes use only of quantities computed in the previous step as exhibited by fig. 3.3

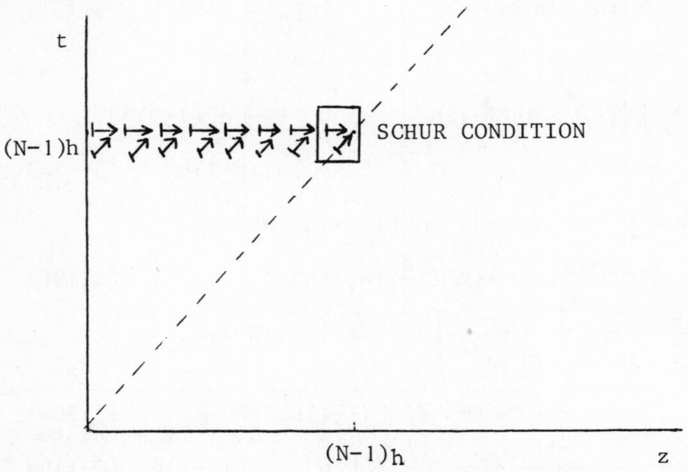

Fig. 3.3 Points used in the recursive computations.

The algorithm has been tried on a number of examples and works well. It can be used e.g. for reconstruction of a large number of points of a scattering medium e.g. a voice tract. The algorithm may be improved numerically by using a higher order integration rule. This however will not change the basic procedure and its recursive properties.

4. Iterative Algorithms

The Schur-algorithm presented in the previous paragraph is probably the best one can hope for in the case k(t) is given in an interval [0,T] and a Krein approximant of the same length is desired. In that case, a reflection function $\rho(z)$ is determined on the same interval [0,T] and a maximal use is made of the available data. Another situation, however, may arise when one desires a "best" $\rho(z)$ in an interval [0,L] which fits the data k(t) on a larger interval [0,T].In that case an estimate must be available of the line termination of z=L. A common assuption would be to take b(L,t) = 0, but here again one may be tempted to use a better estimate. Another, similar, case may arise when only a partial amount of data is given, e.g. k(t) in the interval [a,b]. That case may be treated by similar techniques as those we shall describe.

Hence suppose k(t) given in [0,T] and the problem is to determine

$\rho(z)$ in $[0,L]$ such that a "best" fit is obtained with a constant, passive boundary condition $\alpha(L,t) = s.\beta(L,t)$ where s is some coefficient.

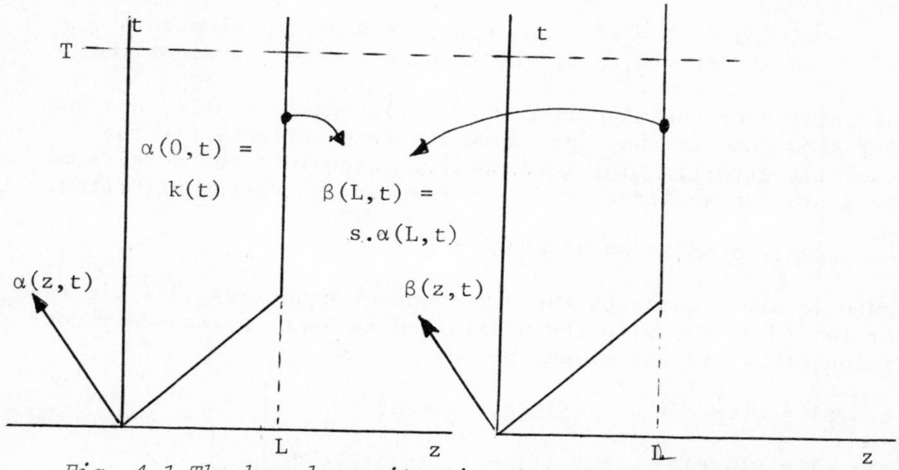

Fig. 4.1 The boundary situation for the iterative problem.

In the light of the previous theory one easily sees that the set of boundary conditions imposed is larger than the number of unknowns. In fact, the recursive algorithm presented in par. 3 allows one to compute an initial $k(z)$ for $z\in[0,L]$ on the basis of the information contained in $k(t)$ for $t \in [0,L]$ alone. Once this initial guess of $k(z)$ is known one can compute $\beta(0,t)$ for $t\in[0,T]$ from the knowledge of the boundary conditions $\alpha(0,t) = k(t)$ and $\beta(L,t) = s.\alpha(L,t)$ by a recursive algorithm which we shall derive immediately. If the $\rho(z)$ guess were correct and if the boundary condition at z=L is also exact, then the computed $\beta(0,t)$ should turn out to be equal to $k(t)$. This will not in general be the case and one shall wish to compute $\rho(z)$ such that an error measure like

$$\int_0^T [\beta(0,t) - k(t)]^2 \, dt \tag{4.1}$$

is minimal. This can be achieved by the following iterative algorithm:

Initial stage

Guess or compute recursively $\rho(z)$ in $[0,L]$
Set $\alpha(0,t) = k(t)$ and $\beta(L,t) = s\alpha(L,t)$ and compute (recursively) $\beta(0,t)$ by an inverse scattering algorithm.

Iterative stage

While error $\int_0^T [\beta(0,t)-k(t)]^2 \, dt$ is too large do:

- update ρ by solving

$$\rho(z) = k(z) - \int_0^z \rho(\xi)a(\xi,z)dz$$

- with $\alpha(0,t) = k(t)$ and $\beta(L,t) = s.\alpha(L,t)$ compute a new value for $\beta(0,t)$ by the inverse scattering algorithm.

One shall stop the algorithm when $\rho(z)$ becomes stable, and one must show that in that case some discrete estimate for the quadratic error is minimized. Before going into the details of the iteration we first given the inverse scattering algorithm.

The inverse scattering algorithm

Again we start out with the basic set of equations (3.4) in the version (3.7) but with the understanding that the boundary condition $\beta(0,t)$ is not given. Hence:

$$\left\{\begin{aligned} \alpha(z,t) &= k(t-z) - \int_0^z \rho(\xi)\beta(\xi,t-z+\xi)d\xi \\ \beta(z,t) &= \beta(0,t) - \rho(t)\chi_z(t) - \int_0^z \rho(\xi)\alpha(\xi,t)d\xi \end{aligned}\right\} \tag{4.2}$$

where $\beta(0,t)$ is unknown. Imposing the boundary condtion $\beta(L,t) = s.\alpha(L,t)$ for $t > L$ and the causality condition $\beta(t,t^-)=0$ and eliminating $\beta(0,t)$ we obtain:

$$\alpha(z,t) = k(t-z) - \int_0^z \rho(\xi)\beta(\xi,t-z+\xi)d\xi \tag{4.3}$$

for $z < t < L$:

$$\beta(z,t) = \rho(t) + \int_z^t \rho(\xi)\alpha(\xi,t)d\xi \tag{4.4}$$

for $z < L < t$:

$$\beta(z,t) = \int_z^L \rho(\xi)\alpha(\xi,t)d\xi + s.\alpha(L,t) \tag{4.5}$$

In general a discontinuity in $\beta(z,t)$ is obtained for t=L. One shall have:

$$\beta(z,L^-) = \beta(z,L^+) \tag{4.6}$$

iff:

$$\rho(L) = s.\alpha(L,L^+) \tag{4.7}$$

where, by (4.3) and (4.4)

$$\alpha(L,L^+) = k(0) - \int_0^L |\rho(\xi)|^2 d\xi \tag{4.8}$$

or:

$$s = \frac{\rho(L)}{k(0) - \int_0^L |\rho(\xi)|^2 d\xi} \tag{4.9}$$

With respect to solving the inverse problem one can either put s=0 and allow a discontinuity in β(0,t) or else use the value determined by the equations (4.9), if it is such that $|s| < 1$.

The equations (4.3)-(4.5) can be integrated recursively starting from the right boundary in steps as shown in fig. 4.2. A small distinction has to be made between $t \le L$ and $t \ge L$ where we have presumed continuity by the correct choice of s as discussed previously.

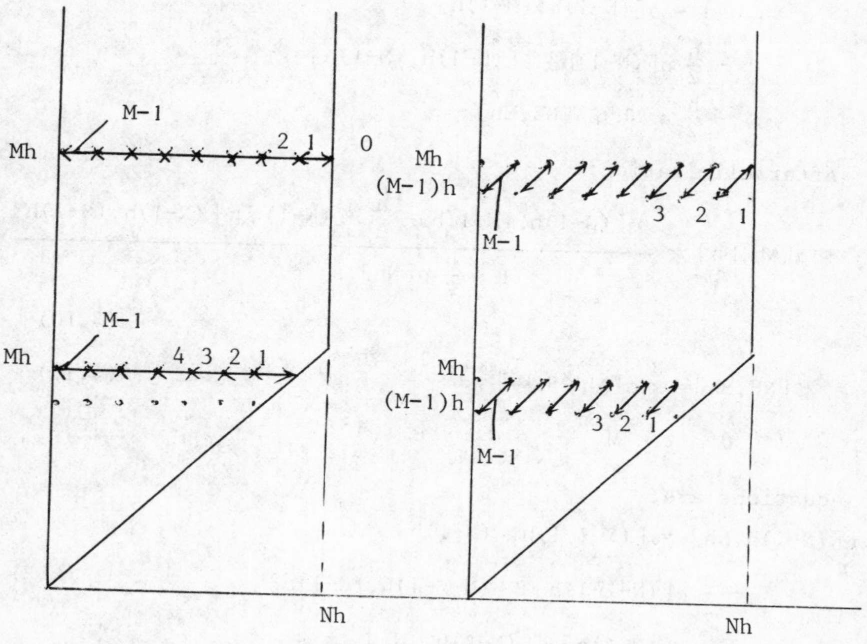

Fig. 4.2 Recursive iteration of the inverse scattering problem

Let L = Nh. We must distinguish between the triangular region and the strip region, but only in the first step of the algorithm.

Step 0

Triangular part $0 \leq M \leq N$

$$\alpha(Mh,Mh) = \alpha[(M-1)h,(M-1)h] - \frac{h}{2}\rho[(M-1)h]^2 - \frac{h}{2}\rho[Nh]^2$$

$$\beta(Mh,Mh) = \rho(Mh)$$

Strip part $N \leq M$

The equations are:

$$\beta[Nh,Mh] = s.\alpha[Nh,Mh]$$

$$\alpha[Nh,Mh] = \alpha[(N-1)h,(M-1)h] - \frac{h}{2}\rho[(N-1)h]\,\beta[(N-1)h,(M-1)h] - \frac{h}{2}\rho[Nh]\beta\,[Nh,Mh]$$

After elimination:

$$\alpha[Nh,Mh] = \frac{\alpha[(N-1)h,(M-1)h] - \frac{h}{2}\rho[(N-1)h]\beta[(N-1)h,(M-1)h]}{1 + \frac{h}{2}\rho[Nh].s} \tag{4.10}$$

$$\beta[Nh,Mh] = s.\alpha[Nh,Mh] \tag{4.11}$$

Step ℓ with $0 < \ell < M$

The equations are:

$$\alpha[(N-\ell)h,Nh] = \alpha[(N-\ell-1)h,\ (N-1)h] - \frac{h}{2}\rho[(N-\ell-1)h]\,\beta\,[(N-\ell-1)h,(N-1)h] - \frac{h}{2}\rho[(N-\ell)h]\,\beta\,[(N-\ell)h,\ Nh]$$

$$\beta[(N-\ell)h,Nh] = \beta[(N-\ell+1)h,\ Nh] + \frac{h}{2}\rho[(N-\ell+1)h]\,\alpha\,[(N-\ell+1)h,Nh] + \frac{h}{2}\rho[(N-\ell)h]\,\alpha\,[(N-\ell)h,Nh]$$

Elimination:

Define the intermediate quantities:

$\alpha'[(N-\ell-1)h,(N-1)h]$

$$= \alpha[(N-\ell-1)h,(N-1)h] - \frac{h}{2}\rho[(N-\ell-1)h]\beta[(N-\ell-1)h,(N-1)h]$$

$\beta'[(N-\ell+1)h,Nh]$

$$= \frac{h}{2}\rho[(N-\ell+1)h]\alpha[(N-\ell+1),Nh]+\beta[(N-\ell+1)h,Nh]$$

We have:

$$\alpha[(N-\ell)h,Nh] = \frac{\alpha'[(N-\ell-1)h,(N-1)h] - \frac{h}{2}\rho[(N-\ell)h]\beta'[(N-\ell+1)h,Nh]}{1+\frac{h^2}{4}\rho[(N-\ell)h]^2} \quad (4.12)$$

$$\beta[(N-\ell)h,Nh] = \frac{\frac{h}{2}\rho[(N-\ell)h]\ \alpha'[(N-\ell-1)h,(N-1)h] + \beta'[(N-\ell+1)h,Nh]}{1+\frac{h^2}{4}\rho[(N-\ell)h]^2} \quad (4.13)$$

The final step M

$\alpha(0,Mh) = k(Mh)$ [Boundary condition]

$$\beta(0,Mh) = \beta(1,Mh) + \frac{h}{2}\rho(h)\alpha(h,Mh) + \frac{h}{2}(0)\alpha(0,Mh)$$
$$= \beta(1,Mh) + \frac{h}{2}\rho(h)\alpha(h,Mh) + \frac{h}{2}\rho(0)k(Mh) \quad (4.14)$$

End of the algorithm

Update equation and iteration

The update equation is obtained by requesting $\beta(0,t) = k(t)$ for $0 \le t \le T$. We use the recently computed values for α and β in $\beta(0,t)$. The new value ρ' for the reflection function is then obtained from (4.4) by requesting (setting ρ^n for the new reflection coefficient)

for $0 \le t \le L$

$$k(t) = \rho^n(t) + \int_0^t \rho^n(\xi)\alpha(\xi,t)d\xi \quad (4.15)$$

for $L \le t \le T$

$$k(t) = s.\alpha(L,t) + \int_0^t \rho^n(\xi)\alpha(\xi,t)d\xi \quad (4.16)$$

Using the trapezium-rule with T=Mh and L=Nh this gives the following set of equations:

$$\begin{bmatrix} 1 & 0 & & 0 \\ \frac{h}{2}\alpha(0,h) & 1+\frac{h}{2}\alpha(h,h) & 0 \ldots & 0 \\ & & \ldots & \\ \frac{h}{2}\alpha(0,Nh) & \frac{h}{2}\alpha(h,Nh) & \ldots & 1+\frac{h}{2}\alpha(Nh,Nh) \\ \frac{h}{2}\alpha[0,(N+1)h] & \frac{h}{2}\alpha[h,(N+1)h] & \ldots. & (s+\frac{h}{2})\alpha[Nh,(N+1)h] \\ & \ldots & & \\ \frac{h}{2}\alpha[0,Mh] & \frac{h}{2}\alpha[h,Mh] & \ldots. & (s+\frac{h}{2})\alpha(Nh,Mh) \end{bmatrix}$$

$$\begin{bmatrix} \rho^n(0) \\ \rho^n(h) \\ \vdots \\ \rho^n(Nh) \end{bmatrix} = \begin{bmatrix} k(0) \\ k(1) \\ \vdots \\ k[(Nh)] \\ k[(N+1)h] \\ \\ k(Mh) \end{bmatrix} \tag{4.17}$$

An optimal solution for this set of equations is obtained using Householder transformations or the SVD [13].
This means that the newly computed "least distance" solution $\rho^n(\xi)$ produces a $\beta(0,t)$ given by the equations (4.4) and (4.5) for which the distance

$$\int_0^T [\beta(0,t) - k(t)]^2 \, dt \tag{4.18}$$

is minimal presuming the waves α and β are correctly computed. With the new values $\rho^n(\xi)$ one may compute a new set of waves α^n and β^n using the inverse scattering algorithm discussed earlier. Practice shows that convergence of the algorithm occurs very fast, although no analysis of the rate of convergence is available at present. Be it as it may, once a fixpoint solution $\rho(z)$, $\alpha(z,t)$

and $\beta(z,t)$ is obtained, we may from the previous resonably infer that the obtained $\beta(0,t)$ minimizes (4.18) although found proof is still lacking.

In the next paragraph we shall deduce a set of conservation relations which can serve as a check on the numerical procedures used.

5. Orthogonality Relations

In this paragraph we return to the original basic theory in order to prove a number of conservation relations which are useful in verifying the numerical computations as well as in deriving new results. We suppose that we have a line of length L with incident waves

$$\begin{aligned} &a(0,t) = \delta(t) + \alpha(0,t) \\ &\beta(L,t) \end{aligned} \qquad (5.1)$$

and reflected waves

$$\begin{aligned} &a(L,t) = \delta(t-L) + \alpha(L,t) \\ &\beta(0,t) \end{aligned} \qquad (5.2)$$

Moreover we shall only consider the (natural) case where all waves are zero for $t < 0$ (this does not necessarily imply causality, see par. 1).
Because of the presence of dirac impulses, energy relations must be derived in an indirect manner. Let $\Sigma(j\omega)$ be the scattering matrix of the given transmission line, as discussed in par. 1. Conservation of energy is equivalent to saying that Σ is unitary on the $j\omega$-axis, see (1.29) repeated here:

$$\Sigma(j\omega)\tilde{\Sigma}(j\omega) = 1_2 \qquad (5.3)$$

Let's define Fourier Transforms:

$$\begin{aligned} A(z,j\omega) &= \mathfrak{F}[\alpha(z,t)] \\ B(z,j\omega) &= \mathfrak{F}[\beta(z,t)] \end{aligned} \qquad (5.4)$$

and pre/post multiplying (5.3) with the waves one obtains (where ~ indicates Hermitian conjugation):

$$[1+\tilde{A}(0,j\omega),\tilde{B}(L,j\omega)]\tilde{\Sigma}\Sigma\begin{bmatrix}1+A(0,j\omega)\\ B(L,j\omega)\end{bmatrix}$$

$$=[1+\tilde{A}(0,j\omega),\tilde{B}(L,j\omega)]\begin{bmatrix}1+A(0,j\omega)\\ B(L,j\omega)\end{bmatrix} \tag{5.5}$$

or:

$$[\tilde{B}(0,j\omega),e^{jL\omega}+\tilde{A}(L,j\omega)]\begin{bmatrix}B(0,j\omega)\\ e^{-jL\omega}+A(L,j\omega)\end{bmatrix}$$

$$=[1+\tilde{A}(0,j\omega),\tilde{B}(L,j\omega)]\begin{bmatrix}1+A(0,j\omega)\\ B(L,j\omega)\end{bmatrix} \tag{5.6}$$

$$\tilde{B}(0,j\omega)B(0,j\omega)+\tilde{A}(L,j\omega)A(L,j\omega)+e^{jL\omega}A(L,j\omega)+e^{-jL\omega}\tilde{A}(L,j\omega)$$

$$=A(0,j\omega)+\tilde{A}(0,j\omega)+\tilde{A}(0,j\omega)A(0,j\omega)+\tilde{B}(L,j\omega)B(L,j\omega) \tag{5. 7}$$

Transforming back to the time-domain, using reality of the functions and restricting to $t \geq 0$ and symmetry, we obtain:

$$\int_t^{\infty}\{\beta(0,\tau)\beta(o,\tau-t)+\alpha(L,\tau)\alpha(L,\tau-t)$$
$$-\alpha(0,\tau)\alpha(0,\tau-t)-\beta(L,\tau)\beta(L,\tau-t)\}d\tau$$
$$+\alpha(L,t+L)+\alpha(L,L-t)-\alpha(0,t)-\alpha(0,-t)=0 \tag{5.8}$$

(5.8) is the general energy conservation relation in the given situation. The most common form is obtained by puting $t=0$ which gives:

$$\int_0^{\infty}\{[\beta(0,\tau)]^2+[\alpha(L,\tau)]^2$$
$$-[\alpha(0,\tau)]^2-[\beta(L,\tau)]^2\}d\tau$$
$$+\alpha(L,L^+)+\alpha(L,L^-)-\alpha(0,0^+)=0 \tag{5.9}$$

In the case where the finite line is a section of the infinite line realizing $a(0,t) = \delta(t) + k(t)$ and $b(0,t) = k(t)$, (5.9) simplifies to:

$$k(0)-a(L,L^+)=\int_L^{\infty}[\alpha(L,\tau)^2-\beta(L,\tau)^2]d\tau \tag{5.10}$$

Because of (3.7) we have (as already used and remarked in the previous paragraph):

$$\alpha(L,L^+) = k(0) - \int_0^L |\rho(\xi)|^2 d\xi$$

so that (5.10) reduces to:

$$\int_L^\infty [\alpha(L,\tau)^2 - \beta(L,\tau)^2]d\tau = \int_0^L |\rho(\xi)|^2 d\xi \qquad (5.11)$$

(5.11) is a major consistency equation between the three quantities computed in the Schur recursion and any deviation from it indicates computational errors.

In the case of the iterative algorithm presented in par. 4 one has:

$$\alpha(L,L^+) = k(0) - \int_0^L |\rho(\xi)|^2 d\xi$$

$$\alpha(L,L^-) = 0$$

$$\alpha(0,0^+) = k(0)$$

$$\beta(L,\tau) = s.\alpha(L,\tau)$$

so that:

$$\int_0^\infty [k(\tau)^2 - \beta(0,\tau)^2]d\tau$$
$$= -\int_0^L |\rho(\xi)|^2 d\xi + (1-s^2)\int_L^\infty \alpha(L,\tau)^2 d\tau \qquad (5.12)$$

Again (5.12) allows to check for the numerical correctness of the computations in the iterative algorithm.

Reciprocity relations

For the sake of completeness we mention that another set of invariants may be obtained usign symmetry relations of the Σ matrix. The Σ matrix is not exactly réciprocal but a slight modification of it is. In fact, using the Laplace formalism of part. 1 we have:

$$\Sigma_{12} e^{-\lambda L/2} = \Sigma_{21} e^{\lambda L/2} \qquad (5.13)$$

(5.13) may be expressed directly in terms of the Σ matrix by:

$$\begin{bmatrix} e^{-\lambda L/2} & 0 \\ 0 & e^{\lambda L/2} \end{bmatrix} \Sigma = \Sigma^T \begin{bmatrix} e^{-\lambda L/2} & 0 \\ 0 & e^{\lambda L/2} \end{bmatrix} \qquad (5.14)$$

The reciprocity invariant is obtained by applying two sets of incident waves

$$\begin{bmatrix} a^1(0,\lambda) \\ b^1(L,\lambda) \end{bmatrix} \quad \text{and} \quad \begin{bmatrix} a^2(0,\lambda) \\ b^2(L,\lambda) \end{bmatrix}$$

to (5.14) and observe two sets of reflected waves:

$$\begin{bmatrix} b^1(0,\lambda) \\ a^1(L,\lambda) \end{bmatrix} \quad \text{and} \quad \begin{bmatrix} b^2(0,\lambda) \\ a^2(L.\lambda) \end{bmatrix}$$

We obtain:

$$e^{-\lambda L/2}a^2(0,\lambda)b^1(0,\lambda) + e^{\lambda L/2}b^2(L,\lambda)a^1(L,\lambda)$$
$$= e^{-\lambda L/2}b^2(0,\lambda)a^1(0,\lambda) + e^{\lambda L/2}a^2(L,\lambda)b^1(L,\lambda) \quad (5.15)$$

(5.15) may then further be specialised to the α and β components used previously and to the concrete algorithms.

6. Ladder Filters

The recursions derived earlier (3.14), (3.15) and (4.10) following, define a ladder filter structure which is typical for the trapezium rule approximation used and which we now make explicit. The basic filter cell defined in those equations is shown in fig. 6.1.

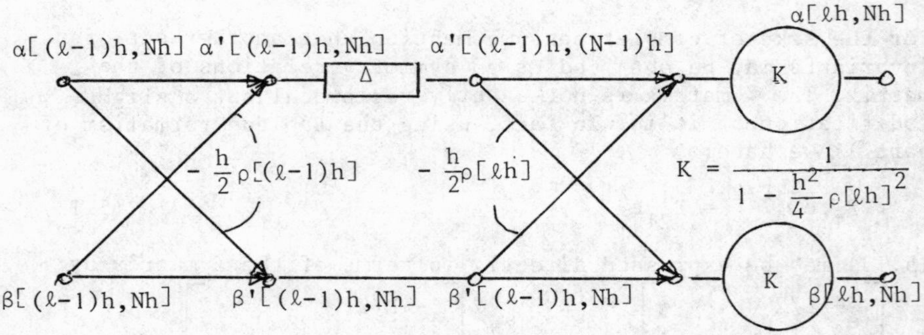

Fig. 6.1 The Basic Cell defined by the Trapezium Rule

The cell shows how $\alpha[\ell h,Nh]$, $\beta[\ell h,Nh]$ is computed from $\alpha[(\ell-1)h,Nh]$, $\beta[(\ell-1)h,Nh]$, i.e. from level $z=(\ell-1)h$ to level $z=\ell h$ at time Nh. The basic cell is $O(h^2)$ close to an orthogonal cell. The normalization factor is the right in fig. 6.1 is slightly off:

$$\frac{1}{1-\frac{h^2}{4}\rho(\ell h)^2} = \frac{1}{\sqrt{(1-\frac{h^2}{4}\rho[(\ell-1)h]^2)(1-\frac{h^2}{4}\rho(\ell h)^2)}} + 0(h^2) \tag{6.1}$$

The initial condition for the computation is $\alpha(0,Nh) = \beta(0,Nh) = k(Nh)$. The Schur conditions request $\beta(Nh,Nh) = \rho_N$. The Schur conditions request $\beta(Nh,Nh) = \rho_N$. This set of boundary conditions does not produce a causal situation but is otherwise consistent. It is not necessary to compute values of $\alpha(\ell h,Nh)$, $\beta(\ell h,Nh)$ for $N < \ell$. Also the very first reflection coefficient $\rho(0)$ is equal to $k(0)$ and does not obey the Schur condition. It seems that the computation of the discrete set of reflection coefficients $\rho(\ell h)$ does noet solve a discrete time inverse scattering problem for a discrete covariance function which is closely related to $k(\ell h)$.

Acknowledgement

The authors wish to thank Prof. H. Blok and Drs. A.G. Tijhuis for their suggestions. The method we have presented and followed has been very much inspired by their treatment of electromagnetic inverse scattering problems.

Bibliography

1. T. Kailath, "A view of three decades of linear filtering theory", IEEE Trans. on Information Theory, vol. IT-20, No.2, pp. 146-181, Mar. 1974.
2. T. Kailath, Lectures on Linear Least Squares Estimation, Springer Verlag, Wien, New York, 1976.
3. P. Dewilde, A. Vieira and T. Kailath, "On the generalized Szegö-Levinson realization algorithm for optimal linear predictors based on a network synthesis approach", IEEE Trans. on CAS, Vol. 25, No. 9, Sept. 1978.
4. J. Schur, "Ueber Potenzreihen, die im Innern des Einheitskreises beschränkt sind", J. für die reine und angewandte Mathematik, Vol. 147, (Berlin), pp. 205-232, 1917.
5. Dewilde, P. and H. Dym, "Schur Recursions, Error formulas for stationary stochastic sequences" IEEE Trans. on Information Theory (to appear).
6. Doob, J.L., Stochastic Processes, J. Wiley, New York, 1953.
7. Krein, M.G., "The continuous analogues of theorems on polynomials orthogonal on the unit circle", Dokl.Akad.Nank. SSSR, vol. 104, pp. 637-640, 1955.

8. Dunford, N., and J.T. Schwarz, Linear Operators, Interscience publishers Inc., New York.

9. Schlesinger, L., "Neue Grundlagen für einen Infenitesimalkalkul der Matrizen", Mathematische Zeitschrift 33.

10. Gantmacher, F.R., The Theory of Matrices, New York, Chelsa, 1959.

11. Dym, H. and H.P. McKean, Gaussian Processes, Function Theory and the Inverse Spectral Problem, Academic Press, New York, 1976.

12. Atkinson, F.B., Discrete and Continuous Boundary Problems, New York: Academic Press, 1964.

13. Wilkinson, J.H. and C. Reinsch, Handbook for Automatic Computation, Vol. II, Linear Algebra, Part 2, Springer Verlag, New York, 1971.

HARMONIC ANALYSIS AND THE MATHEMATICS BEHIND FILTERING AND PREDICTION(D'APRES CARTIER D'APRES WIENER AND KOLMOGOROV)

Michiel Hazewinkel
Dept. Math., Erasmus Univ. Rotterdam
P.O.Box 1738, 3000 DR Rotterdam, The Netherlands

There are fundamental connections between quantum theory (Heisenberg commutation relations, stochastic quantization) and (nonlinear) filtering. This is e.g. a main topic in S.K. Mitter's contribution to this volume, and these may well be just the surface manifestations of deeper and very fruitful links.

So it seemed a good idea to take another look at an old (1961) Seminaire Bourbaki exposé of Cartier in which he explains how Wiener- Kolmogorov prediction can be connected to a number of ideas in harmonic analysis (Mackey imprimitivity systems) having to do with the Stone-von Neumann uniqueness theorem for representations of the Heisenberg commutation relations.

During the actual talks I added a good many remarks, asides, results and speculations on how various other topics and concepts such as scattering and inverse scattering, theta functions, automorphic functions, the Weil- Segal- Shale representation, isospectral and isomonodromy deformations, path integrals, the Heisenberg group, and heat equation semigroups relate or might relate to these methods.

Beyond these remarks which at the moment do not seem to lend

M. Hazewinkel and J. C. Willems (eds.), Stochastic Systems: The Mathematics of Filtering and Identification and Applications, 383–384.

themselves readily to a coherent presentation of reasonable length (at least in my hands) I had little to add to Cartier's exposé which is available in the literature (Seminaire Bourbaki, 13e année, Exp. 218, (1960/1961), Benjamin, 1967). The remainder of this abstract is a translation of Cartier's introduction.

In 1941 Wiener and Kolmogorov simultaneously gave a complete answer for the prediction problem for stationary random functions. The problem is the following: given a random function $\{X_t\}$, (t real) and a value τ of the variable t, can one on the basis of the statistical data of the X_t for $t \leqq \tau$ give an estimate of the variable $X_{t+\nu}$ when $\nu > 0$? If one adopts the usual "least squares" methods this question translates as follows to a Hilbert space setting: given a one parameter group $\{U(t)\}$ of unitary operators in a Hilbert space H and an element $a \in H$, is $U(\nu)a$ in the closed subspace spanned by the $U(t)a$ for $t \leqq 0$. The Wiener-Kolmogorov theorem gives a very precise response in terms of the Fourier transform of the function $t \to \langle a|U(t)|a\rangle$. Instead of reporting on all the work which followed in the wake of the Wiener-Kolmogorov theorem we shall in this exposé show how a certain number of results from the spectral theory can be organized around the same idea, and how to deduce the Wiener-Kolmogorov theorem there from.

Approach your problems from the
right end and start with the
solutions. Then, perhaps, one
day you will find the final
question.

R.H. Van Gulik

Part 6

ADAPTIVE CONTROL

DETERMINISTIC AND STOCHASTIC MODEL REFERENCE ADAPTIVE CONTROL

(A unified presentation of model reference adaptive controllers and stochastic self-tuning regulators)

Y.D. Landau

Laboratoire d'Automatique de Grenoble (C.N.R.S.)
E.N.S.I.E.G. - B.P. 46
38402 ST MARTIN D'HERES

Abstract : Model Reference Adaptive Control Systems (MRAC) where the control objectives are specified in terms of a reference model and stochastic Self-Tuning Regulators (S-STURE) where the control objectives are specified in terms of an ARMA model feature a duality character. The duality between these two classes of adaptive control systems extends the duality existing in the linear case with known parameters between modal control (pole-placement) and minimum variance control. This allows a unified presentation of the two techniques, the S-STURE beeing interpreted as a stochastic M.R.A.C. The underlying concepts and the structure of these adaptive control systems are described. The available theoretical results concerning the analysis of these systems are briefly reviewed.

1. INTRODUCTION

The use of adaptive control techniques is motivated by the need of automatically adjusting the parameters of the controller when plant parameters and disturbances are unknown or change with time, in order to achieve (or to maintain) a certain index of performance for the controlled system. While this problem can be reformulated as a nonlinear stochastic control problem (the unknown parameters are considered as auxiliary states) the resulting solutions are extremely complicated and in order to obtain something useful, it is necessary to make approximations. Therefore adaptive control techniques can be viewed as approximations for nonlinear stochastic control problems. Model Reference Adaptive Control Systems (MRAC) and Stochastic Self-Tuning Regulators (S-STURE) can be considered as two approximations among others possible approximations.

M. Hazewinkel and J. C. Willems (eds.), Stochastic Systems: The Mathematics of Filtering and Identification and Applications, 387–419.

MRAC techniques have been initially developed for deterministic continuous time tracking problems and S-STURE techniques have been initially developed for stochastic discrete time regulation problems. Both techniques have led to a number of successful applications (for a detailed reference list, see [1] as well as [2] through [8] among other references) despite that not all the basic theoretical problems have been solved at that time. The two techniques have known in the past independent ways of development (partly explained by the difference in the formulation of the original problems to be solved) and the turn towards a mutual understanding and a more unified theoretical framework can reasonably be considered to have been done at the 6th IFAC Congress, Boston 1975 (Adaptive Control Round Table).

It is also important to recall that in the past a long standing controverse has been taken place between those arguing for direct adaptation of the controller parameters and those arguing for the use of an adaptive predictor of the plant output (which gives also estimates of the plant parameters) as an intermediate step for building an adaptive control. Progress has been made in the understanding of these two approaches [9], [10], [11], [12]. These two approaches can be equivalent if the output of the adaptive predictor behaves identically to that of a (deterministic or stochastic prediction) reference model which specifies the desired output [9] [14].

Connections between M.R.A.C. and Minimum Variance STURE have been investigated in the last few years [9], [10], [13], through [19]. Minimum variance STURE is in fact a particular case of S-STURE where the control objectives are specified by ARMA models (in the case of minimum variance STURE the control objective is specified in term of a moving average of order d-1 where d is the plant delay). The S-STURE where the objectives are specified in terms of an ARMA model can be interpreted as stochastic MRAC.

It is shown in [13], [17], [18] that the deterministic MRAC and stochastic MRAC are dual in the sense that they extend the known duality existing in the linear case between modal control (pole-placement) and minimum variance control or more generally between transfer function matching and ARMA model matching [18].

We will try in this paper to make a unified presentation of deterministic and stochastic MRAC (for minimum phase plants). The roots for this unified approach to MRAC and S-STURE are the presence of a reference model (explicit or implicit) in the deterministic environment and of a prediction reference model (implicit or explicit) in the stochastic environment which in both cases specifies the control objective as well as the similar structure of the parameter adaptation algorithms used in both techniques.

A brief review of the underlying concepts and configurations used in MRAC and S-STURE (stochastic MRAC) is given in Section II. The linear tracking and regulation problem in deterministic and stochastic environment which allows to define the structure of the controller used for both deterministic and stochastic MRAC is discussed in Section III. The structure and the objectives of deterministic and stochastic MRAC are discussed in Section IV. The parameter adaptation algorithms used for deterministic or stochastic MRAC are presented in Section V. A brief review of convergence analysis results for deterministic and stochastic MRAC is given in Section VI. The conclusions summarize the basic design assumptions and emphasize some of the open theoretical problems.

2. MODEL REFERENCE ADAPTIVE CONTROLLERS AND STOCHASTIC SELF-TUNING REGULATORS - BASIC PRINCIPLES

MRAC and S-STURE can be considered as possible approximations for the solutions of some nonlinear stochastic control problems. However, when making approximations, some hypothesis should be considered which can justify the approximations. The basic hypothesis for MRAC and S-STURE is of algebraic nature : for any possible values of the plant (and disturbance) parameters, there exists a linear controller with a fixed complexity such that the plant plus the controller has pre-specified characteristics.

The development of these two adaptive control techniques is based on the deep understanding of certain types of linear algebraic control problems and on an appropriate interpretation of the controller design strategy allowing an extension to adaptive control when plant (and disturbance) parameters are unknown or change in time. Fig. 2.1 illustrates the basic phylosophy for designing a linear controller where the desired performances are specified in terms of the characteristics of a dynamic system which is a "realization" of the desired input-output behaviour of the closed loop control system. The controller is designed such that the closed loop control system is characterized by the same parameters as those of the "desired" dynamic system.

Since the desired performance corresponds in fact to the output of the "desired" dynamic system which is pre-specified, the design problem can be recast as in Fig. 2.2. The objective is now to design a controller such that the error between the output of the plant and the output of the dynamic system which has the desired characteristics (it will be called the "reference model") is identically null for identical initial conditions and such that an eventual initial error will vanish with a certain dynamics.

These two interpretations of the linear control design in the case of plant with unknown or varying parameters lead to two adaptive control schemes shown in Fig. 2.3 and Fig. 2.4. Fig. 2.3 which is

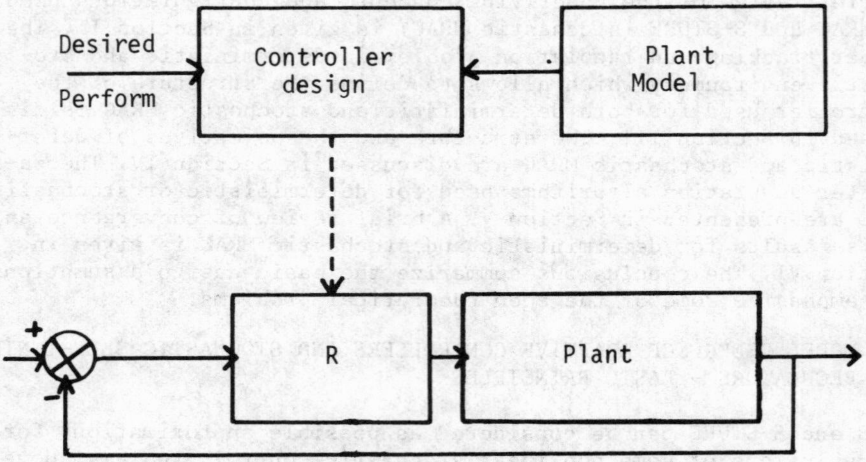

Fig. 2.1 Block diagram illustrating the design of linear controllers in a deterministic environment

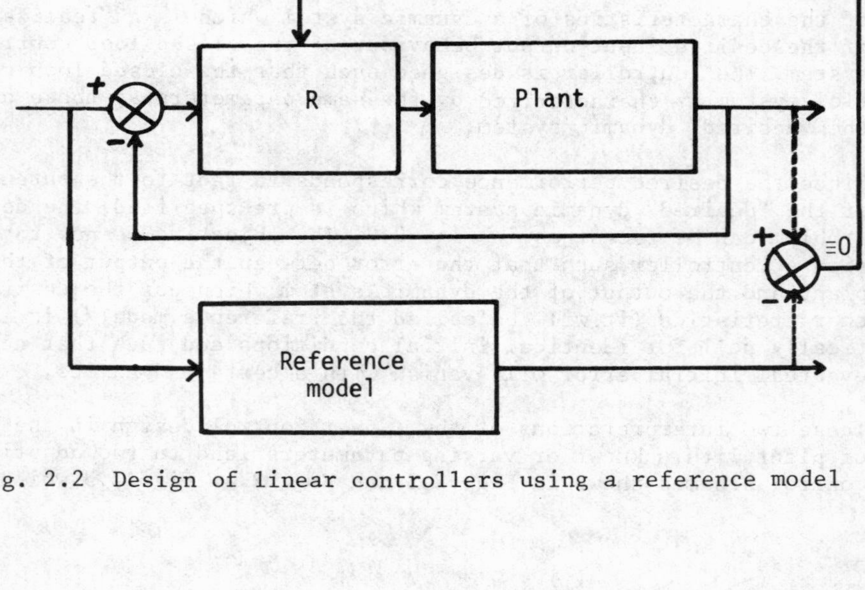

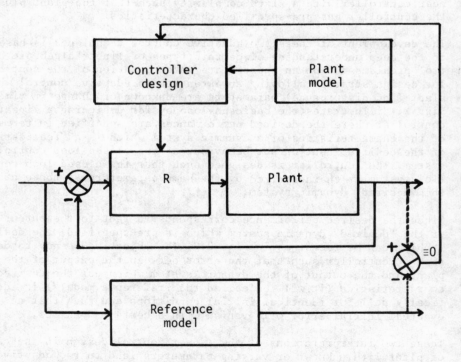

Fig. 2.2 Design of linear controllers using a reference model

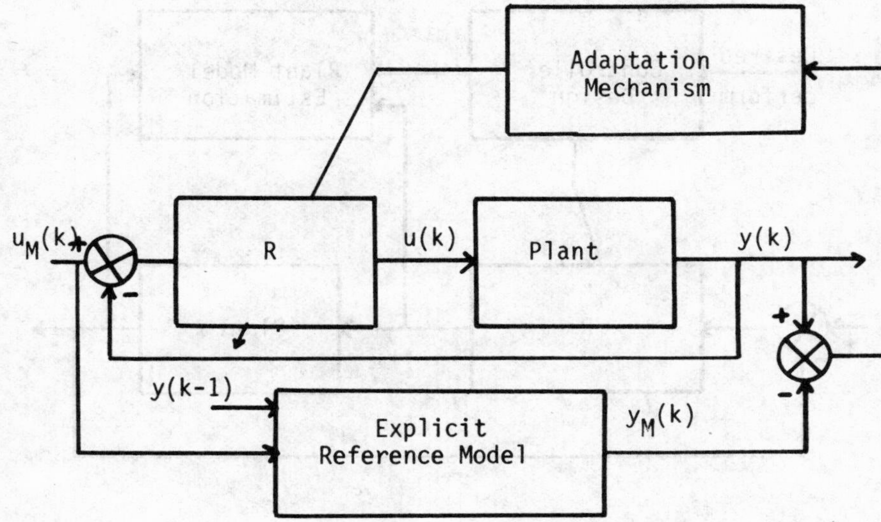

Fig. 2.3 Explicit Model Reference Adaptive Control

an extension of the scheme considered in Fig. 2.2 is called "Explicit Model Reference Adaptive Control". The difference between the output of the plant and the output of the reference model is a measure of the difference between the real performance and the desired one and this information is used through an "adaptation mechanism"(adaptation algorithm) to automatically adjust the parameters of the controller.

Fig. 2.4 a is an extension of the scheme considered in Fig. 2.1 in the sense that a correct design can be obtained if a plant model is estimated on-line based on input-output data available. Taking in account the structure of the on-line parameter estimation schemes, one obtains the more detailed scheme given in Fig. 2.4 b. The parameters of the adaptive predictor for the plant output are used for the design of the controller. This means that the controller plus the adaptive predictor will behave at each instant as the "desired" dynamic system, i.e. the output of the adaptive predictor will be equal to that of the explicit reference model in Fig. 2.3 (this is illustrated by the dotted line part in Fig. 2.4 b). Therefore the adaptive controller plus the adaptive predictor form an "implicit reference model". Note that this scheme is clearly inspired from the separation theorem in linear stochastic control. This type of structure was used for developing self tuning controllers and adaptive control systems using Model Reference Adaptive System techniques for designing the adaptive predictor (observer) [3], [11], [21], [23].

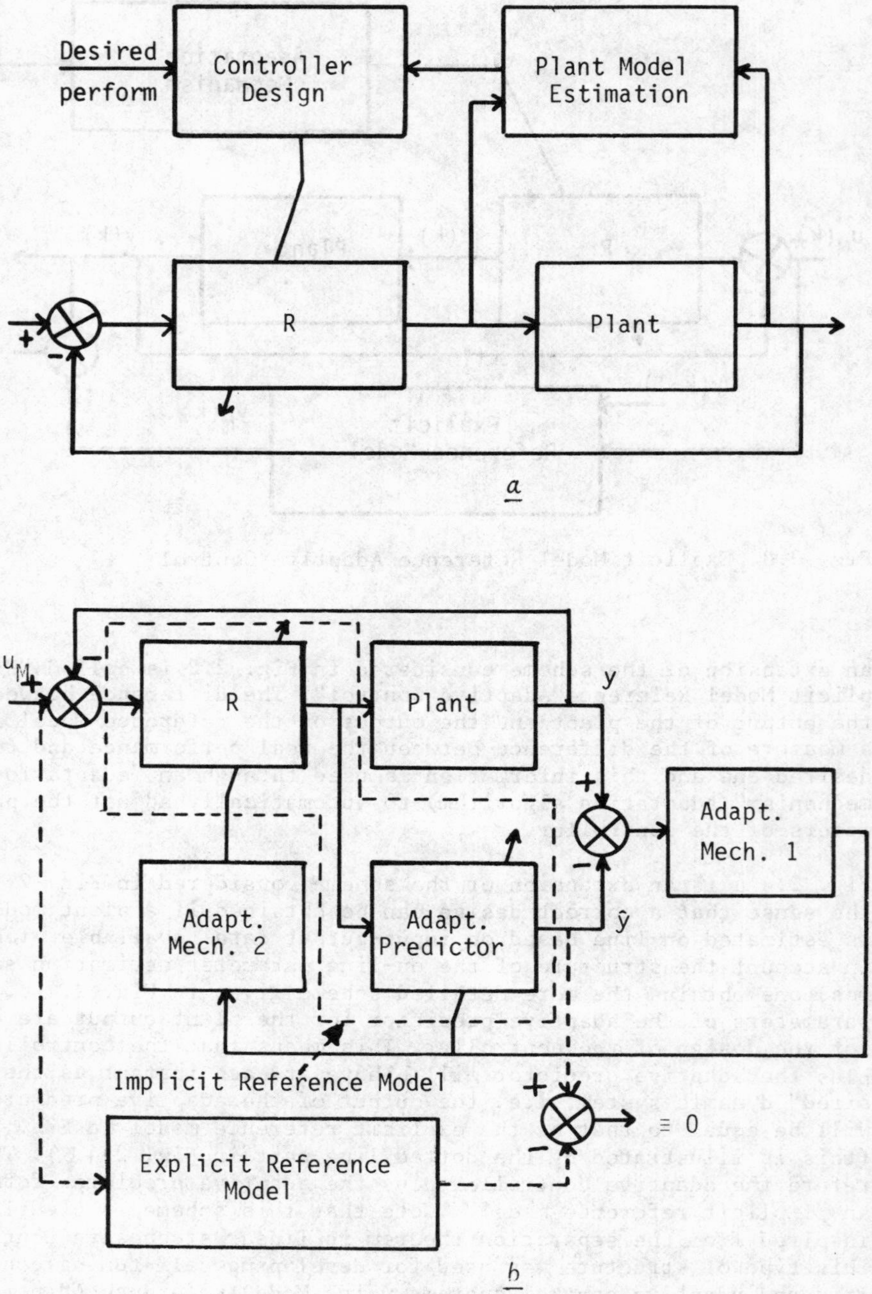

Fig. 2.4 Implicit Model Reference Adaptive Control : a) Basic scheme ; b) Detailed configuration

In the stochastic case, the problem can be approached in a similar way with the difference that in addition to the plant model one considers also a model for the disturbance (see Fig. 2.5).

Assuming that the disturbance model is an ARMA process the desired control objective will be specified in terms of an ARMA model (for the plant output) [18], [24].

However the design of a linear controller in the stochastic case can be also approached through the use of a prediction reference model which gives the desired predicted value for the plant output. (See Fig. 2.6). The design is done such that the difference between the plant output and the desired predicted output is an innovation sequence as illustrated in Fig. 2.6.

When the parameters of the plant and of the disturbance model are unknown or vary during operation, the schemes of Figs. 2.5 and 2.6 lead to the adaptive control schemes of the type presented in Figs. 2.3 and 2.4 where the deterministic reference model is replaced by a stochastic prediction reference model. For a more detailed discussion, see [13], [17], [18].

The objective of the adaptation mechanism is that after a certain time, the difference between the output of the reference model and the plant output becomes almost zero in the deterministic environment or becomes almost an innovation sequence (or a martingale difference sequence) in a stochastic environment. However the theoretical tools available for analysis suggest to consider asymptotic type objectives.

The adaptive control system is a nonlinear system which results clearly from Figs. 2.3 and 2.4 and it is possible to cast the analysis of the MRAC in a stability problem for a system disturbed from an equilibrium point (both in deterministic and stochastic case). Then, tools for stability analysis can be used either for analysis or for the design [21], [23], [25], [26].

Since the plant-model error depends on the misalignment of the parameters of the controller, the adaptation of the parameters of the controller can be approached as a recursive parameter estimation problem. This suggests the use of algorithms inspired by recursive parameter estimation techniques

A basic feature of these adaptive systems (at least for minimum phase plants) is that the adaptation algorithm for the parameters of the controllers has an explicit form in terms of the plant-model error and of the input and output of the plant. As a consequence the resulting adaptive control scheme is relatively simple and this explains in part the success of these schemes.

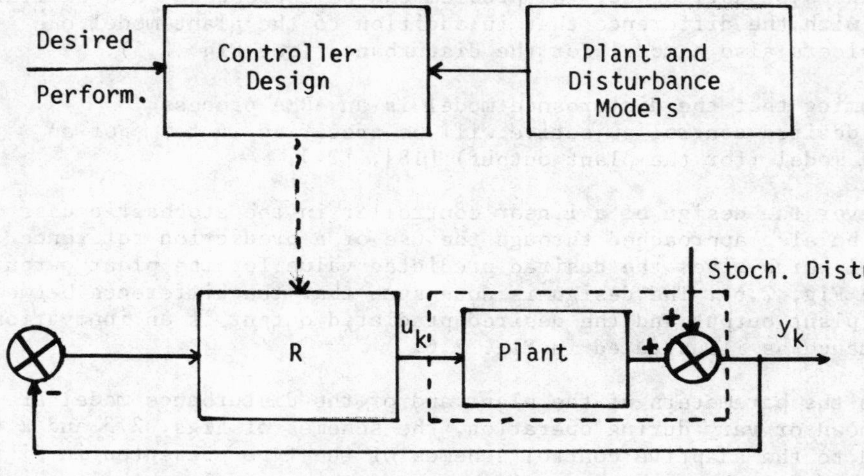

Fig. 2.5 Block diagram illustrating the design of linear controllers in a stochastic environment

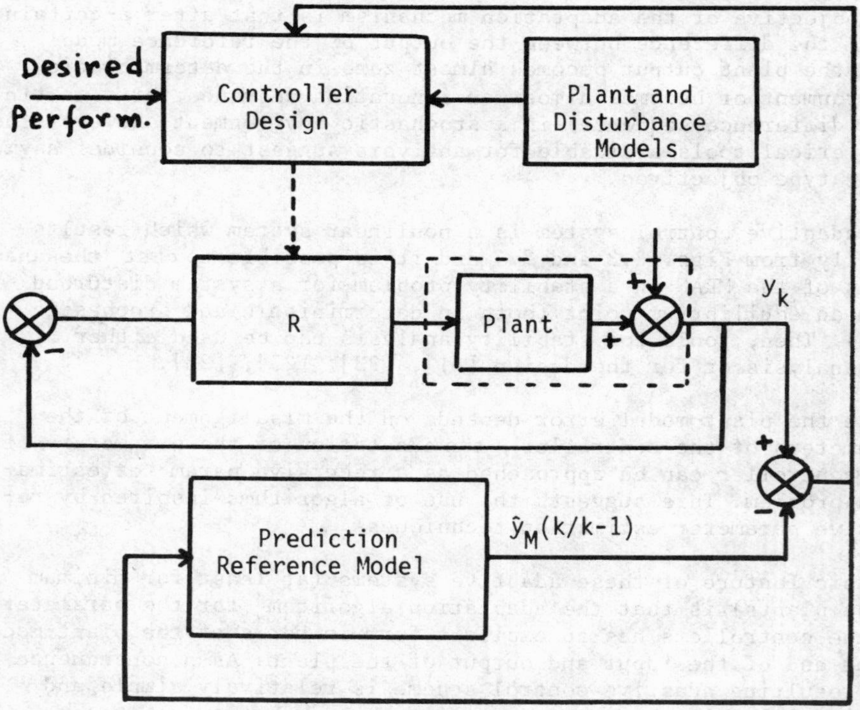

Fig. 2.6 Design of stochastic linear controllers using a prediction reference model

The implicit and explicit (deterministic or stochastic) MRAC are equivalent if the control strategy is such that the output of the adaptive predictor is equal to that of the explicit reference model and if the resulting adaptation algorithms for controller parameters are identical. This is discussed in detail in [11], [12], [13], [17], [18]. Moreover some types of adaptation algorithms can be used both for deterministic and stochastic MRAC. It was shown that deterministic and stochastic MRAC are dual in the sense that with the same adaptation algorithm and provided that the same positivity condition on a certain transfer function is verified (in order to assure the convergence), they asymptotically achieve dual objectives which can be obtained in the linear case with known parameters when using the same linear controller [13], [17], [18]. This has as a consequence the possibility of using a unique scheme for a combined environment.

In the following sections, we will try to present the basic algorithms and some of their properties. The presentation does not reflect how "hystorically" these algorithms have been developed and its aim is to introduce the algorithms through a series of successive logical steps by reducing to a minimum the feedback from the theoretical analysis which will be discussed last.

3. COMBINED LINEAR TRACKING AND REGULATION IN DETERMINISTIC AND STOCHASTIC ENVIRONMENT

We will briefly summarize in this section some results from the algebraic approach to deterministic and stochastic control of linear S.I.S.O. plants. This will allow to illustrate how the reference models (implicit and explicit) are introduced, to specify the complexity and structure of the controller and to emphasize the duality between deterministic and stochastic algebraic control. For more detailed results, see [14], [22], [24], [26], [27], [28].

Consider the S.I.S.O. (single input - single output) discrete linear time invariant plant described by :

a) deterministic environment :

$$A(q^{-1})\ y(k+d) = B(q^{-1})\ u(k),\ d > 0,\ y(0) \neq 0 \qquad (3.1)$$

b) stochastic environment :

$$A(q^{-1})\ y(k+d) = B(q^{-1})\ u(k) + C(q^{-1})\ \omega(k+d) \qquad (3.2)$$

where :

$$A(q^{-1}) = 1 + a_1 q^{-1} + \dots + a_n q^{-n}$$

$$B(q^{-1}) = b_o + b_1 q^{-1} + \dots + b_m q^{-m} \qquad b_o \neq 0 \qquad (3.3)$$

$$C(q^{-1}) = 1 + c_1 q^{-1} + \dots + c_n q^{-n}$$

q^{-1} is the backward shift operator, d represents the plant delay, u(k) and y(k) are the plant input and output respectively ω_k is a sequence of equally distributed independent normal (0, σ)' random variables which acts through a shaping filter as a stochastic disturbance upon the plant output (the effect of the disturbance upon the output is given by the ARMA process $\frac{C(q^{-1})}{A(q^{-1})}$). We assume that the zeroes of $B(z^{-1})$ are all in $|z| < 1$, therefore they can be cancelled without leading to an unbounded control (in practice they should be in fact enough damped). The degree of $A(q^{-1})$ and $C(q^{-1})$ are chosen equal to n without loss of generality.

3.1 Tracking and Regulation in Deterministic Environment

The objectives of the control which we are considering in the deterministic environment are :

Tracking : The control u_k should be such that in tracking, the output of the plant satisfies the difference equation :

$$C_1(q^{-1})\, y(k) = q^{-d}\, D(q^{-1})\, u^M(k) \tag{3.4}$$

where :

$$C_1(q^{-1}) = 1 + c_1^1 q^{-1} + \ldots + c_n^1 q^{-n} \tag{3.5}$$

is an asymptotically stable polynomial

$$D(q^{-1}) = d_0 + d_1 q^{-1} + \ldots + d_n q^{-n} \tag{3.6}$$

and $u^M(k)$ is a bounded reference sequence.

Regulation : The control should be such that in regulation ($u^M(k) \equiv 0$) an initial disturbance ($y(0) \neq 0$) is eliminated with the dynamics defined by :

$$C_2(q^{-1})\, y(k+d) = 0 \tag{3.7}$$

where : $C_2(q^{-1}) = 1 + c_1^2 q^{-1} + \ldots + c_n^2 q^{-n}$ (3.8)

is an asymptotically stable polynomial.

In order to design the controller, we will consider two strategies, one using an explicit reference model as part of the control system and the other using a d-step ahead predictor of the plant output which together with the controller will form an implicit reference model.

<u>Strategy 1</u> : Explicit Reference Model

One considers an explicit reference model given by :

$$C_1(q^{-1})\, y^M(k+d) = D(q^{-1})\, u^M(k) \tag{3.9}$$

where $y^M(k)$ is the output of the explicit reference model. The design objective is :

$$C_2(q^{-1})\ \varepsilon(k+d) = 0 \qquad k > 0 \tag{3.10}$$

where : $\varepsilon(k) = y(k) - y^M(k)$ (3.11)
is the plant model error. It is obvious that Eq. (3.10) includes the regulation objective specified by Eq. (3.7) (for $u_M(k) \equiv 0$, $\varepsilon(k) = y(k)$) as well as the tracking objective specified by Eq. (3.4).

Taking in account the plant and model equations and using in order to obtain a causal control the following polynomial identity :

$$C_2(q^{-1}) = A(q^{-1})\ S(q^{-1}) + q^{-d}\ R(q^{-1}) \tag{3.12}$$

which has a unique solution $S(q^{-1})$, $R(q^{-1})$ for deg. $S(q^{-1}) = d-1$, where :

$$S(q^{-1}) = 1+s_1 q^{-1} + \dots + s_{d-1} q^{-d+1}$$
$$R(q^{-1}) = r_1 + r_2 q^{-1} + \dots + r_n q^{-n+1} \tag{3.13}$$

one obtains :

$$C_2(q^{-1}) y(k+d) = R(q^{-1}) y(k) + b_o u(k) + B_s(q^{-1})\ u(k) \tag{3.14}$$

where :

$$B_s(q^{-1}) = B(q^{-1})\ S(q^{-1}) - b_o \tag{3.15}$$

and Eq. (3.10) becomes :

$$C_2(q^{-1})\varepsilon(k+d) = R(q^{-1})\ y(k) + b_o u(k) + B_s(q^{-1})\ u(k) - C_2(q^{-1})\ y_M(k+d) = 0 \tag{3.16}$$

which yelds the desired control :

$$u(k) = \frac{C_2(q^{-1})\ y^M(k+d) - R(q^{-1}) y(k) - B_s(q^{-1}) u(k)}{b_o} \tag{3.17}$$

Introducing the notation :

$$\phi_o^T(k) = [u(k-1)\ \dots\ u(k-d-m+1),\ y(k)\ \dots\ y(k-n+1)] \tag{3.18}$$

$$\theta_o^T = [b_{s_1}\ \dots\ b_{s(d+m-1)},\ r_1\ \dots\ r_n] \tag{3.19}$$

Eq. (3.17) can be written :

$$u(k) = \frac{C_2(q^{-1})\ y^M(k+d) - \theta_o^T\ \phi(k)}{b_o} \tag{3.20}$$

or in an equivalent form :

$$C_2(q^{-1})\ y^M(k+d) = \theta^T\ \phi(k) \tag{3.21}$$

where : $\phi(k)^T = [u(k), \phi_o^T(k)]$ (3.22)

$\theta^T = [b_o, \theta_o^T]$ (3.23)

The resulting control scheme is given in Fig. 3.1.

Strategy 2 : Implicit Reference Model. This strategy is directly inspired by the separation theorem : one first designs an appropriate predictor for the plant output and then a control will be computed such that the output of the predictor behaves as the desired output in tracking.

First step : (predictor design). The predictor will be designed such that the d-step ahead prediction error $\hat{\varepsilon}(k+d)$ defined by :

$$\hat{\varepsilon}(k+d) = y(k+d) - \hat{y}(k+d) \tag{3.24}$$

where $\hat{y}(k+d)$ is the predictor output will vanish according to :

$$C_2(q^{-1})\ \hat{\varepsilon}(k+d) = 0 \quad ; \quad k > 0 \tag{3.25}$$

Using the polynomial identity of Eq. (3.12), one obtains from Eq. (3.25) that the d-step ahead predictor is characterized by :

$$C_2(q^{-1})\hat{y}(k+d) = b_o u(k) + R(q^{-1})y(k) + B_s(q^{-1})u(k) = \theta^T\ \phi(k) \tag{3.26}$$

where $R(q^{-1})$, $B_s(q^{-1})$, θ, $\phi(k)$ are given by Eqs. (3.13), (3.15), (3.22), (3.23) respectively.

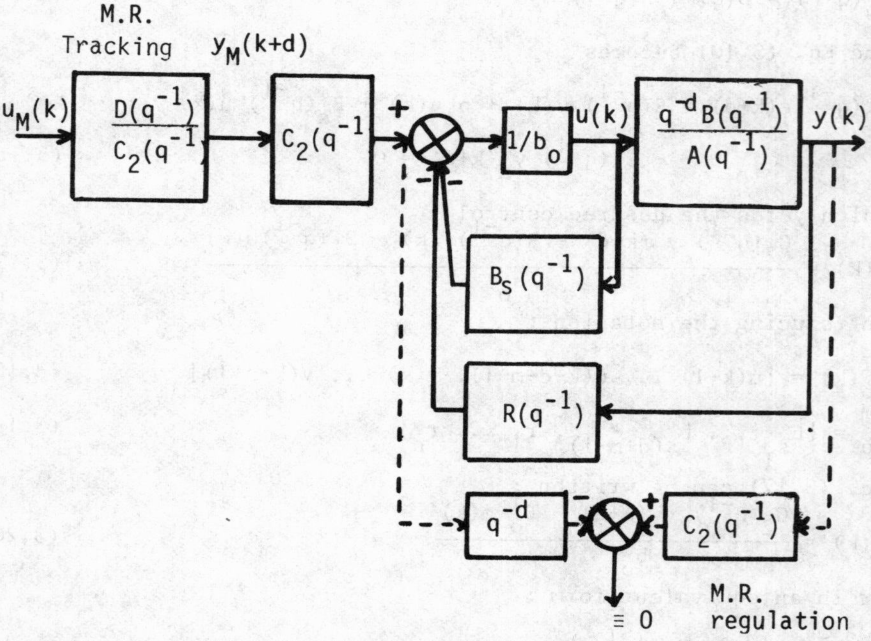

Fig. 3.1 Control scheme for independent tracking and regulation objectives

Second step : (computation of the control). The control is computed such that $\hat{y}(k+d) = y^M(k+d)$ where $y^M(k+d)$ is the desired output given by Eq. (3.9) and one finally obtains that :

$$C_2(q^{-1})\ \hat{y}(k+d) = C_2(q^{-1})\ y^M(k+d) = \theta^T\ \phi(k) \tag{3.27}$$

and the control is given by Eq. (3.21) as expected.

Because the output of the predictor is equal to the output of the explicit reference model, the predictor plus the control will form an "implicit reference model".

3.2 Tracking and Regulation in Stochastic Environment

The control objectives which we are considering in the stochastic environment are generalizations of the minimum variance control strategy and the solutions are obtained via an algebraic approach. For more details concerning the algebraic approach to linear stochastic control problems, see [24], [28].

Regulation : The plant and the disturbance beeing given by Eq. (3.2), the control u(k) should be such that :

$$C_2(q^{-1})\ y(k+d) = C(q^{-1})\ S(q^{-1})\ \omega(k+d) \tag{3.28}$$

where $C_2(q^{-1})$ defines in fact the poles of the closed loop system and $S(q^{-1})$ is a polynomial of degree d-1 solution of the identity (3.12). For $C_2(q^{-1}) = C(q^{-1})$, the control objective corresponds to the minimum variance control strategy [13], [22] and is the dual of the determinsitic modal control defined by the objective specified in Eq. (3.7). In addition for $d = 1$, $S(q^{-1}) = 1$ and $y(k+d) = \omega(k+d)$.

Tracking : Given a "reference" stochastic sequence x(k) modelized by :

$$C_3(q^{-1})\ x(k+d) = C_1(q^{-1})\ v(k+d) \tag{3.29}$$

where $C_1(q^{-1})$ and $C_3(q^{-1})$ are asymptotically stable polynomials of degree n and v(k) is a sequence of equally distributed independent (0, δ) random variables, the control should be such that for $\omega(k) \equiv 0$ the tracking error satisfies :

$$x(k+d) - y(k+d) = F(q^{-1})\ v(k+d) \tag{3.30}$$

where $F(q^{-1})$ is a polynomial of degree d-1 solution of the polynomial identity :

$$C_1(q^{-1}) = C_3(q^{-1})\ F(q^{-1}) + q^{-d}\ G(q^{-1}) \tag{3.31}$$

and this corresponds to a minimum variance tracking.

The appropriate control satisfying both objectives can be obtained by using an explicit or implicit prediction reference model.

Strategy 1 : Explicit prediction reference model. The explicit prediction reference model is the minimum variance d-step ahead predictor whose output will be denoted $y^M(k+d) = \hat{x}(k+d/k)$ and which is given by :

$$C_1(q^{-1})\, y^M(k+d) = G(q^{-1})\, x(k) \tag{3.32}$$

where $G(q^{-1})$ is a polynomial of degree n solution of Eq. (3.31) and one considers the design objective :

$$C_2(q^{-1})\varepsilon(k+d) = C_2(q^{-1})\, [y(k+d)-y^M(k+d)] = C(q^{-1})\, S(q^{-1})\omega(k+d) \tag{3.33}$$

The resulting control satisfying Eq. (3.33) as well as Eqs. (3.28) and (3.30) is the same as in deterministic environment (duality) and is given by Eq. (3.20).

Strategy 2 : Implicit Prediction Reference Model. As in the deterministic environment, one constructs first a predictor for the plant output such that the prediction error $\hat{\varepsilon}(k+d)$ satisfies the equation :

$$C_2(q^{-1})\, \hat{\varepsilon}(k+d) = C(q^{-1})\, S(q^{-1})\, \omega(k+d) \tag{3.34}$$

The predictor will be therefore given as in the deterministic environment by :

$$C_2(q^{-1})\, \hat{y}(k+d) = \theta^T\, \phi(k) \tag{3.35}$$

where $\phi(k)$ and θ are given by Eqs. (3.22) and (3.23). Then a control is computed such that the output of the predictor $\hat{y}(k+d)$ be equal to that of the explicit prediction model i.e. :

$$C_2(q^{-1})\hat{y}(k+d) = C_2(q^{-1})y^M(k+d) = \theta^T\, \phi(k) \tag{3.36}$$

and one obtains again the same control as in the deterministic case, given by Eq. (3.20).

3.3 Combined environment

An interesting situation from a practical point of view is the tracking of a deterministic trajectory in the presence of a stochastic disturbance. Combining the results given previously by using the control given in Eq. (3.17), one obtains that the output of the plant will be given by :

$$C_2(q^{-1})\, y(k+d) = C_2(q^{-1})y^M(k+d) + C(q^{-1})\, S(q^{-1})\,\omega\,(k+d) \tag{3.37}$$

For $C_2(q^{-1}) = C(q^{-1})$, one obtains a minimum variance tracking of the reference trajectory in the presence of a stochastic disturbance.

4. STRUCTURES AND OBJECTIVES OF DETERMINISTIC AND STOCHASTIC MODEL REFERENCE ADAPTIVE CONTROL

When the parameters of the plant (and those of the disturbance in the stochastic case) are unknown or change in time an adaptive approach should be considered with the hope that the design objectives specified in the linear case with known parameters will be achieved at least asymptotically.

When implementing an adaptive control scheme, one should answer several questions :

1) What will be the structure of the adjustable controller ?

2) What will be the reasonable objectives which can be assigned to the adaptive control system ?

3) What will be the parameter adaptation algorithm to be used in order to achieve the objectives ?

4) Does the resulting adaptive control system really achieve the assigned objectives ?

The question 2, 3 and 4 are clearly correlated since in order to make a complete design, one should be able to analyze analytically the resulting scheme. Since the resulting scheme will be nonlinear (the parameters will be up-dated using information extracted from the system itself) and the analysis which up to day have been done for these systems concern mainly asymptotic convergence, the objectives will be assigned according to this fact. However from a practical point of view to guarantee the global convergence and the boundedness of all the variables of the adaptive control system is a problem of prime importance and the interest granted to this problem is not at all "theoretical *speculation*".

Returning to the first question we recall that one of the basic assumption is that the structure of the plant and disturbance remains unchanged when parameters change (This assures the existence of the solution for the linear control problems described in section III for a fixed structure of the controller). Therefore one concludes that in the adaptive case the structure of the controller remains the same as in the linear case with known parameters (Eq. (3.20) or Eq. (3.21)) but the fixed parameter vector θ will be replaced by an adjustable parameter vector $\hat{\theta}(k)$.

$$\hat{\theta}^T(k) = [\hat{b}_o(k), \hat{\theta}_o^T(k)] \tag{4.1}$$

and the corresponding control law will be given (either in deterministic or stochastic environment) by :

$$u(k) = \frac{C_2(q^{-1})y^M(k+d) - \hat{\theta}_o^T(k)\phi_o(k)}{\hat{b}_o(k)} \tag{4.2}$$

or :

$$\hat{\theta}^T(k)\ \phi(k) = C_2(q^{-1})\ y^M(k+d) \tag{4.3}$$

Note that in the case of the schemes using an implicit (prediction) reference model the plant predictor will be replaced by an adaptive predictor governed by :

$$C_2(q^{-1})\ \hat{y}(k+d) = \hat{\theta}^T(k)\ \phi(k) \tag{4.4}$$

and the control will be computed according to the strategy in the linear case with known parameters which will lead to Eq. (4.3).

For the objectives of the adaptive control schemes in a deterministic environment, we will consider, instead of the objectives for the linear case with known parameters given in Eqs. (3.10) and (3.25) the following ones :

$$\lim_{k\to\infty} C_2(q^{-1})\ \varepsilon(k+d) = \lim_{k\to\infty} e^o(k+d) = 0 \tag{4.5}$$

in the case of Explicit Model Reference Adaptive Control and :

$$\lim_{k\to\infty} C_2(q^{-1})\hat{\varepsilon}(k+d) = \lim_{k\to\infty} e^o(k+d) = 0 \tag{4.6}$$

in the case of Implicit Model Reference Adaptive Control. We will also require that the input and output of the plant remain bounded for all k (i.e. $||\phi(k)|| < M < \infty, \forall k$).
The corresponding block diagrams of the two schemes are given in Fig. IV.1 a, b where (*) marks the points of the two schemes which follow the same trajectory.

In the stochastic case the structure of the adaptive control schemes will be the same as in deterministic case. It remains to specify the objectives.

We will consider the case of stochastic regulation (the situation is similar in tracking). From Eq. (3.28), one obtains that in the linear case with known parameters for d = 1 (the generalization for d > 1 is straightforward), one has :

$$\frac{C_2(q^{-1})}{C(q^{-1})}\ \varepsilon(k+1) = \omega(k+1) \tag{4.7}$$

where $\omega(k)$ is a sequence of equally distributed normal $(0, \sigma)$ random variables (i.e. the plant-model error, passed through a shaping filter is white). Note also that :

$$E[\omega(k+1)|\mathcal{F}_k] = 0 \tag{4.8}$$

where $\mathcal{F}_k$ is generated by all possible observations up to and including time k. This suggests that the disturbance can be modelled

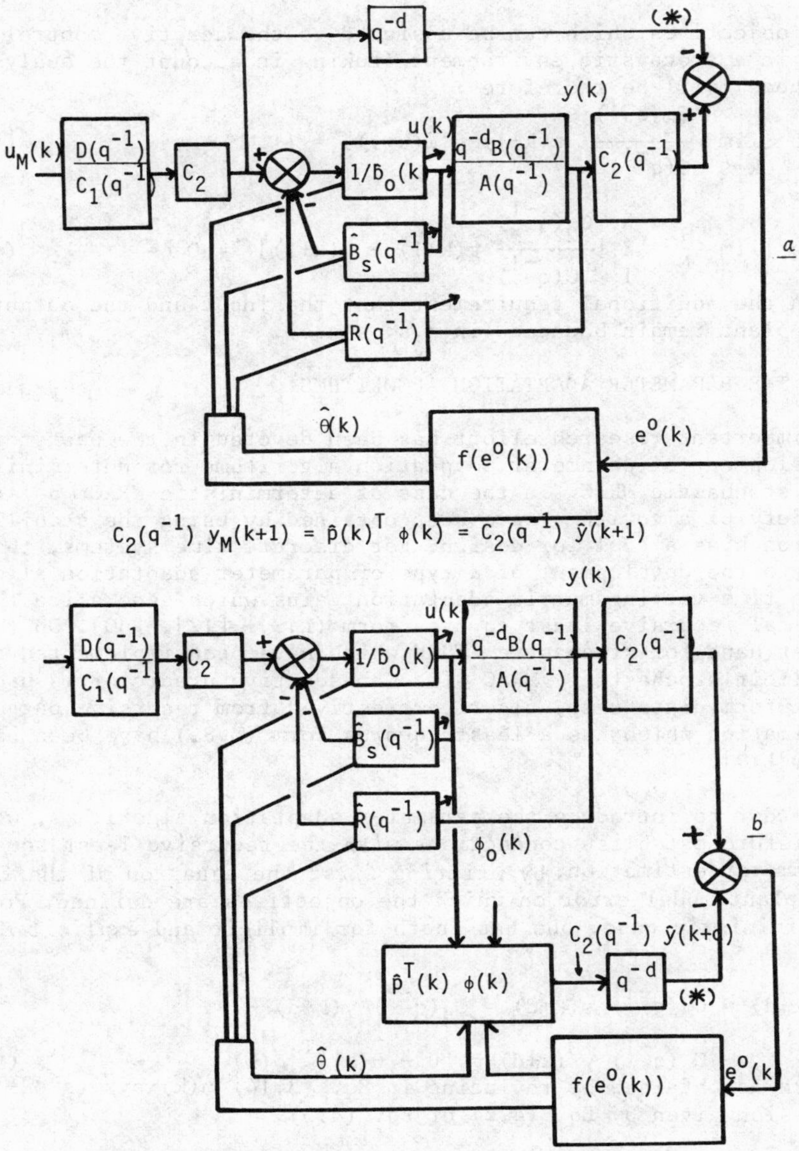

Fig. 4.1 Model Reference Adaptive Control for tracking and regulation : a) with explicit reference model
b) with implicit reference model

also by considering $\omega(k)$ as a martingale difference sequence [41], [42], [43].

The objectives which can be assigned to the adaptive control schemes in a stochastic environment (taking in account the analysis methods) will be therefore :

$$\text{Prob}\,\{\lim_{k\to\infty} \frac{C_2(q^{-1})}{C(q^{-1})}\, \varepsilon(k+1) = \omega(k+1)\} = 1 \qquad (4.9)$$

or :

$$\text{Prob}\,\{\lim_{N\to\infty} \frac{1}{N} \sum_{1}^{N} [\frac{C_2(q^{-1})}{C(q^{-1})}\, \varepsilon(k+1) - \omega(k+1)]^2 = 0\} = 1 \qquad (4.10)$$

with the additional requirement that the input and the output of the plant remain bounded (in m.s. sense).

5. THE PARAMETER ADAPTATION ALGORITHMS

An important research effort has been devoted in the past to the development of parameter adaptation algorithms for deterministic and stochastic MRAC. In the case of deterministic MRAC, a large variety of algorithms have been obtained by using the stability approach as a tool for design. For discrete time systems, this led to the development of a type of parameter adaptation algorithms with time-varying matrix adaptation gains which generalize the typical recursive least squares form (L.S.) [12], [30]. On the other hand for stochastic STURE which have been implemented as implicit stochastic MRAC since an adaptive predictor is used as an intermidiate step, algorithms derived from recursive parameter estimation which has a least squares form (L.S.) have been used [3], [19].

In order to introduce the parameter adaptation algorithms, we will therefore establish connections with the recursive least squares parameter estimation, by writting first the equation of the filtered plant-model error on which the objectives are defined. For the deterministic case, one has (both for implicit and explicit MRAC) :

$$e^o(k+d) = C_2(q^{-1})\,\varepsilon(k+d) = C_2(q^{-1})\,y(k+d) - C_2(q^{-1})\,y^M(k+d) = [\theta - \hat{\theta}(k)]^T\,\phi(k) \qquad (5.1)$$

which is obtained by replacing in Eq. (3.14) u(k) by its expression given by Eq. (4.2) or Eq. (4.3).

Our first objective is to find an adaptation algorithm :

$$\hat{\theta}(k+d) = \hat{\theta}(k+d-1) + f[e^o(k+d)] \qquad (5.2)$$

such that :

$$\lim_{k\to\infty} e^o(k+d) = 0 \qquad (5.3)$$

One can consider an associate deterministic least squares parameter estimation problem leading to an error equation similar to Eq. (5.1), and this suggests to use the following adaptation algorithm.

$$\hat{\theta}(k+d) = \hat{\theta}(k+d-1) + F_k\, \phi(k)\, \nu(k+d) \tag{5.4}$$

where F_k is the adaptation gain matrix given by :

$$F_{k+1}^{-1} = F_k^{-1} + \phi(k)\, \phi(k)^T \; ; \; F_0 > 0 \tag{5.5}$$

and $\nu(k+d)$ is the residual (or "a posteriori") adaptation error given by :

$$\nu(k+d) = [\theta - \hat{\theta}(k+d)]^T\, \phi(k) = \frac{\nu^o(k+d)}{1 + \phi(k)^T\, F_k\, \phi(k)} \tag{5.6}$$

The one step "a priori" adaptation error (prediction error) $\nu^o(k+d)$ is given by :

$$\nu^o(k+d) = \nu(k+d)/k+d-1) = [\theta - \hat{\theta}(k+d-1)]^T\, \phi(k) \tag{5.7}$$

$\nu(k+d)$ and $\nu^o(k+d)$ are not directly measurable but they can always be expressed in terms of $e^o(k+d)$ and their previous values as shown next. Eq. (5.6) can be written also :

$$\nu(k+d) = e^o(k+d) + [\hat{\theta}(k) - \hat{\theta}(k+d)]^T \phi(k) = e^o(k+d) + \bar{e}(k+d) \tag{5.8}$$

where $e^o(k+d)$ given by Eq. (5.1) is the d-steps "a priori" error (d-steps prediction error) and $\bar{e}(k+d)$ is an auxiliary error which can be computed using Eq. 5.4 :

$$\bar{e}(k+d) = -\phi^T(k)\, F_k\, \phi(k)\, \nu(k+d) - \phi(k)^T \sum_{i=1}^{d-1} F_{k-d+1}\, \phi(k-d+i)\, \nu(k+i) \tag{5.9}$$

Introducing Eq. (5.9) in Eq. (5.6), one obtains :

$$\nu(k+d) = \frac{e^o(k+d) - \phi(k)^T \sum_{i=1}^{d-1} F_{k-d+i}\, \phi(k-d+i)\, \nu(k+i)}{1 + \phi(k)^T F_k\, \phi(k)} \tag{5.10}$$

Because $\nu(k+d)$ can be put under the form of a sum of the measured error and an auxiliary error it was also called "augmented error" [25].

The important remark to be made is that the use of this algorithm will not guarantee that the objective specified by Eq. (5.3) will be achieved since the hypothesis that $\phi(k)$ is independent of $\hat{\theta}(k)$, $\theta(k-1)$ ... which is a basic one for least squares, is violated in the case of M.R.A.C. Therefore a special analysis should be carried on (see Section 6).

A second remark is that the adaptation gain F_k is time decreasing which implies that the algorithm should be systematically re-initialized if the parameters change in time. To overcome this, one can use either a "forgetting" factor in Eq. (5.5) which becomes :

$$F_{k+1}^{-1} = \lambda_1\, F_k^{-1} + \phi(k)\, \phi(k)^T \qquad 0 < \lambda_1 \leq 1 \tag{5.11}$$

or to use a constant adaptation matrix gain :

$$F_{k+1} = F_k = F_o \tag{5.12}$$

Both solutions have their own disadvantages. If $\phi(k)$ is null during a certain period, F_k will grow to much when using Eq. (5.11). If one uses Eq. (5.12) the disadvantage is that the variation of the parameter at each step is not done in general in the best direction (the convergence is slower).

A general form for up-dating the matrix adaptation gain which allows to overcome the various difficulties is [12], [26], [29] :

$$F_{k+1}^{-1} = \lambda_1(k)\, F_k^{-1} + \lambda_2(k)\, \phi(k)\, \phi(k)^T$$

$$0 < \lambda_1(k) \leq 1 \qquad 0 \leq \lambda_2(k) < 2 \tag{5.13}$$

Using the matrix inversion lemma, Eq. (5.13) becomes :

$$F_{k+1} = \frac{1}{\lambda_1(k)} \left[F_k - \frac{F_k\, \phi(k)\, \phi(k)^T\, F_k}{\frac{\lambda_1(k)}{\lambda_2(k)} + \phi(k)^T\, F_k\, \phi(k)} \right] \tag{5.14}$$

In the deterministic environment, the best results seem to be obtained using a "constant trace" algorithm, i.e. $\lambda_1(k)$ and $\lambda_2(k)$ are chosen such that $TrF_{k+1} = TrF_k = TrF_o$ [26].

Note that the use of the possible time varying coefficients of $\lambda_1(k)$ and $\lambda_2(k)$ in Eq. (5.14) allows also to avoid division by zero in Eq. (4.2) during adaptation transients [26].

In various designs, the Eq. (5.1) can have a more general form. For example in order to guarantee certain convergence conditions in the stochastic environment it is useful to filter the vector $\phi(k)$ through an asymptotically stable polynomial $L(q^{-1})$.

$$L(q^{-1})\, \phi^f(k) = \phi(k) \tag{5.15}$$

In that case, the structure of the controller given in Section IV is slightly modified. The filtered control is given by :

$$u^f(k) = \frac{C_2(q^{-1})\, y^{Mf}(k+d) - \hat{\theta}_o(k)^T\, \phi^f(k)}{\hat{b}_o(k)} \tag{5.16}$$

and the control applied to the plant will be :

$$u(k) = L(q^{-1})\, u^f(k) \tag{5.17}$$

Using this type of control, the equation for the filtered plant model error on which the objectives are defined becomes :

$$e^o(k+d) = C_2(q^{-1})\, \varepsilon(k+d) = L(q^{-1})\, [\theta - \hat{\theta}(k)]^T\, \phi^f(k) \tag{5.18}$$

In addition, for the same reasons mentionned above (convergence conditions for stochastic case) $e^o(k+d)$ is sometimes filtered prior to be used for adaptation i.e. :

$$H_2(q^{-1})\, v^o(k+d) = H_1(q^{-1})\, e^o(k+d) \tag{5.19}$$

where $H_1(q^{-1})$ and $H_2(q^{-1})$ are asymptotically stable polynomials and $v^o(k+d)$ is the measured adaptation error (or the d-step "a priori" adaptation error).

Therefore the equation of the "a priori" adaptation error will be:

$$v^o(k+d) = H(q^{-1}) \; [\theta - \hat{\theta}(k)]^T \; \phi^f(k) \tag{5.20}$$

where :

$$H(q^{-1}) = \frac{H_1(q^{-1}) \; L(q^{-1})}{H_2(q^{-1})} \tag{5.21}$$

It can be shown (through convergence analysis [29], [30]) that the appropriate adaptation algorithm has still the form of Eq. (5.4) where $\phi(k)$ is replaced by $\phi^f(k)$ and $\nu(k+d)$ is given by :

$$\nu(k+d) = H(q^{-1}) \; [\theta - \hat{\theta}(k+d)]^T \; \phi^f(k) = \frac{\nu^o_{k+d}}{1 + \phi^f(k)^T F_k \phi^F(k)} \tag{5.22}$$

where in the deterministic case :

$$\nu^o(k+d) = \hat{\nu}(k+d/k+d-1) \tag{5.23}$$

which is the prediction of $\nu(k+d)$ based on $\hat{\theta}(i)$ up to $\hat{\theta}(k+d-1)$. Note also that $\nu(k+d)$ can also in this case be expressed as an augmented error :

$$\nu(k+d) = \nu^o(k+d) + H(q^{-1}) \; [\hat{\theta}(k) - \hat{\theta}(k+d)]^T \; \phi^f(k) \tag{5.24}$$

Exactly the same types of parameter adaptation algorithms are used for stochastic model reference adaptive control (stochastic STURE) with the remark that convergence to a fixed controller will be obtained by setting in Eq. (5.14), for $k > k_o$, $\lambda_1(k) \equiv 1$ and $2 > \lambda_2(k) > 0$, which will give a decreasing adaptation gain.

In the presence of stochastic disturbances, Eq. (5.20) becomes :

$$v^o(k+d) = H(q^{-1}) \; [\theta - \hat{\theta}(k)]^T \phi^f(k) + w(k+d) \tag{5.25}$$

where $w(k+d)$ is the image of the stochastic disturbance acting on the plant and the signification of $\nu^o(k+d)$ in Eq. (5.30) is now :

$$\nu^o(k+d) = \hat{\nu}(k+d/k+d-1) + w(k+d) \tag{5.26}$$

5 . CONVERGENCE ANALYSIS OF DETERMINISTIC AND STOCHASTIC MODEL REFERENCE ADAPTIVE CONTROL

We will briefly review in this section the results of convergence analysis for deterministic and stochastic MRAC which give the conditions assuring that the objectives specified in Section IV are achieved.

We mention that if the implicit and explicit (prediction) reference model are the same (i.e. the output of the adaptive predictor used in implicit MRAC is equal to the output of the explicit reference model), the equations for the (filtered) plant-model error are the same. If in addition the eventual filters used for generating the "a priori" adaptation error are the same and the parameter adaptation algorithms are identical then the two schemes are equivalent [11], [13], [18]. Therefore the convergence analysis can be carried out for one or other configuration, the problem being the same.

We will discuss first the deterministic case and then the stochastic case.

6.1 Deterministic case

The equations characterizing the M.R.A.C. (either with implicit or explicit reference model) are the following :

- The measured (filtered) plant-model error :

$$e^o(k+d) = L(q^{-1})\, [\theta - \hat{\theta}(k)]^T\, \phi^f(k) \tag{6.1}$$

where $L(q^{-1})$ is an asymptotically stable polynomial , (or $e^o(k+d) = L(q^{-1})/M(q^{-1})[\theta - \hat{\theta}(k)]^T \phi^f(k)$ for some schemes with $M(q^{-1})$ an asymptotically stable polynomial which for convergence analysis can be included in the adaptation error equation).

- The d-step "a priori" adaptation error :

$$v^o(k+d) = H(q^{-1})[\theta - \hat{\theta}(k)]^T \phi^f(k) \tag{6.2}$$

where $H(q^{-1}) = L(q^{-1}).H_1(q^{-1})/H_2(q^{-1})$; $H_2(q^{-1})$ can include $M(q^{-1})$).

- The "a posteriori" adaptation error (the"residual" or the "augmented error") :

$$\nu(k+d) = H(q^{-1})[\theta - \hat{\theta}(k+d)]^T\, \phi^f(k) = v^o(k+d)+H(q^{-1})\, [\hat{\theta}(k)-\hat{\theta}(k+d)]\, \phi_f(k) \tag{6.3}$$

- The parameter adaptation algorithm :

$$\hat{\theta}(k+d) = \hat{\theta}(k+d-1) + F_k\, \phi^f(k)\, \nu(k+d) \tag{6.4}$$

- The adaptation gain updating equation :

$$F^{-1}_{k+1} = \lambda_1(k) F^{-1}_k + \lambda_2(k)\, \phi^f(k)\, \phi^{fT}(k) ; 0 < \lambda_1(k) \leqslant 1,\ 0 \leqslant \lambda_2(k) < 2 ;\quad F_o > 0 \tag{6.5}$$

- The measurement vector :

$$\phi^f(k) = G(q^{-1})\, [C_2(q^{-1})\, y^{Mf}(k+d) + [\theta - \hat{\theta}(k)]^T\, \phi^f(k)] \tag{6.6}$$

where $G(q^{-1})$ is an asymptotically stable vector transfer function which for the configurations considered throughout the paper has the form :

$$G^T(q^{-1}) = \left[\frac{A(q^{-1})}{C_2(q^{-1})B(q^{-1})}, \ldots \frac{A(q^{-1})q^{1-m-d}}{C_2(q^{-1})B(q^{-1})}, \frac{q^{-d}}{C_2(q^{-1})} \ldots \frac{q^{-n+1-d}}{C_2(q^{-1})}\right] \tag{6.7}$$

and $C_2(q^{-1})\, y^{Mf}(k+d)$ is bounded sequence.

The convergence analysis should show under what conditions the objectives of the M.R.A.C. specified by:

1) $\lim\limits_{k\to\infty} e^o(k+d) = 0 \quad \forall\, e^o(0), \forall\, [\theta - \hat{\theta}(0)]$ (6.8)

2) $||\phi^f|| < M < \infty \quad \forall\, k > 0$
are achieved. (6.9)

Condition (2) implies that the plant control input u(k) and output y(k) remain bounded ($\phi^f(k)$ contains y(k) and u(k) passed through an asymptotically stable filter).

The above equations can be represented as two interconnected feedback systems as shown in Fig. 6.1 (for d = 1). One distinguishes a main feedback loop with a linear invariant feedforward block characterized by the transfer function $H(q^{-1})$ and a non linear time varying feedback block essentially containing the adaptation algorithm where the vector $\phi(k)$ depends on the parameter error vector (which is the state vector of the feedback block in the main loop). $\phi(k)$ is generated in a secondary feedback loop (dashed lines) where the parameter error is a feedback gain.

In a first stage for analysis the second feedback loop can be neglected (dashed lines in Fig. 6.1) and the feedback block of the main loop can be considered to be only time varying (one neglects Eq. (6.6)). [26].

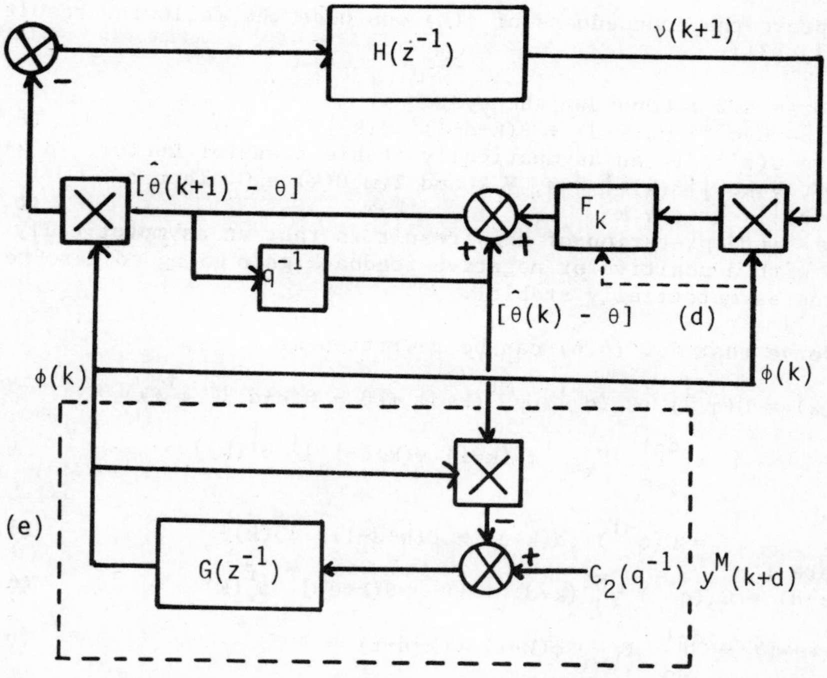

Fig. 6.1 Equivalent feedback representation of M.R.A.C.

For this configuration, one has the following basic results [26], [30].

Theorem 6.1 : Consider the adaptation algorithm of Eqs. (6.4) and (6.5). Assume that the relation between $\nu(k+d)$ and $\phi^f(k)$ is given by Eq. (6.3) where $\phi^f(k)$ is a bounded or unbounded vector sequence $H(z^{-1})$ is a discrete transfer function (ratio of monic polynomials) and θ is a constant vector. Then, if the transfer function

$$H'(z^{-1}) = H(z^{-1}) - \frac{\lambda}{2} \tag{6.10}$$

is strictly positive real, where :

$$2 > \lambda \geqslant \max_k \lambda_2(k) \tag{6.11}$$

one has for any $\nu(0)$ and $\hat{\theta}(0)$ bounded :

$$1)\ \lim_{k\to\infty} \nu(k+d) = 0 \tag{6.12}$$

$$2)\ \lim_{k\to\infty} F_k\ \phi^f(k)\ \nu(k+d) = 0 \tag{6.13}$$

$$3)\ \lim_{k\to\infty} [\theta - \hat{\theta}(k+d)]^T\ \phi^f(k) = 0 \tag{6.14}$$

$$4)\ [\theta(k+d-1) - \theta]^T\ F_k^{-1}\ [\hat{\theta}(k+d-1) - \theta] < \infty \qquad \forall\ k \tag{6.15}$$

To prove the boundedness of $\phi(k)$ one uses the following result [26], [31].

<u>Theorem 6.2</u> : Consider the system :

$$\phi(k) = G(q^{-1})\ [\alpha(k+d) + \beta(k+d-1)^T\ \phi(k)] \tag{6.16}$$

where $G(z^{-1})$ is an asympotically stable transfer vector, $|\alpha(k+d)| < \infty$, $\forall$ k, $||\beta(k)|| < \infty$, $\forall$ k and $\lim_{k\to\infty} \beta(k) = 0$. Then :

$$||\phi(k)|| < \infty\ ,\ \forall\ k \tag{6.17}$$

(the interpretation of this result is that an asymptotically system with a positive or negative feedback gain going to zero remains asymptotically stable).

Observe that Eq. (6.6) can be rewritten as :

$$\phi^f(k) = G(q^{-1})\ \{C_2(q^{-1})\ y^{Mf}(k+d) + [\theta - \hat{\theta}(k+d)]^T\ \phi^f(k) + [\sum_{i=0}^{d-1} F_{k-i}\ \phi^f(k-i)\ \nu(k+d-i)]^T\ \phi^f(k)\} =$$

$$= G(q^{-1})\ \{\alpha(k+d) + \beta(k+d-1)^T\ \phi^f(k)\} \tag{6.18}$$

where :

$$\alpha(k+d) = C_2(q^{-1})\ y^{Mf}(k+d) + [\theta - \hat{\theta}(k+d)]^T\ \phi^f(k) \tag{6.19}$$

$$\beta(k+d-1) = \sum_{i=0}^{d-1} F_{k-i}\ \phi^f(k-i)\ \nu(k+d-i) \tag{6.20}$$

If $F_k^{-1} > \varepsilon\ F_0^{-1}, \varepsilon > 0$, $F_0 > 0$, $\forall$ k > 0 (i.e. $\lambda_1(k)$, $\lambda_2(k)$ are chosen such that for $k > k_o$, F_k^{-1} is non decreasing) then from theorem 6.1 it results that the hypothesis of theorem (6.2) are satisfied which implies that $||\phi(k)|| < M < \infty \; \forall k$.

Note that an alternative proof for the boundedness of $\phi(k)$ can be done using the results given in [32]. For details, see also [26]. Since $\phi(k)$ is bounded, from Eq. (5.24), it results that : $\lim_{k\to\infty} H(q^{-1}) [\theta(k) - \theta(k+d)]^T \phi^f(k) = 0$ and one concludes from Eqs. (6.3), (6.2) and (6.1) that :

$$\lim_{k\to\infty} e^o(k+d) = 0 \tag{6.21}$$

The proof of boundnedness of $\phi(k)$ enlights the importance of the assumption that $B(z^{-1})$ is an asymptotically stable transfer function (this assures that $G(z^{-1})$ is asymptotically stable). For other stability proofs, see [33], [34], [35].

6.2 Stochastic M.R.A.C.

In the stochastic environment (when a disturbance is present) the Eq. (6.3) of the residual ("a posteriori" adaptation error) becomes :

$$\nu(k+d) = H(q^{-1}) [\theta - \hat{\theta}(k+d)]^T \phi^f(k) + w(k+d) \tag{6.22}$$

where $w(k+d)$ is the image of the disturbance. For the type of disturbance considered in Eq. (3.2), $w(k+d)$ has the form :

$$w(k) = \frac{H_1(q^{-1})}{H_2(q^{-1})} C(q^{-1}) S(q^{-1}) \omega(k+d) \tag{6.23}$$

The analysis of the convergence of stochastic M.R.A.C. can be carried on with two techniques.

1) Use of the O.D.E. method,[36] through [40]. It has the advantage of allowing a systematic analysis for various types of configurations using matrix type adaptation gains (time decreasing). However it has the disadvantage of assuming that $\hat{\theta}(k)$ belongs infinitely often to the domain for which stationary processes $\bar{w}(k, \hat{\theta})$ and $\bar{\phi}(k, \hat{\theta})$ can be defined when $\hat{\theta}(k) = \hat{\theta}$ = const. For this reason the results have a local validity around the possible convergence points of the algorithm and the problem of boundedness of $\phi^f(k)$ is not solved.

2) Use of the martingales convergence theorems. When applicable, it gives a complete answer to the convergence and boundedness problems. However up to now the proofs require important alteration of the adaptation algorithm (either a switch to stochastic approximation type algorithms or drastic reduction of the gain) [41], [42].

O.D.E. method : From the O.D.E. method, one can derive the following result, directly applicable to stochastic M.R.A.C. [37],[38], [40].

Theorem 6.3 : Consider the adaptation algorithm of Eqs. (6.4) and (6.5) with $\lambda_1(k) = 1$ and $\lambda_2(k) = \lambda_2$, $\forall k > k_0$. Assume that the stationary processes $\bar{\phi}^f(k, \hat{\theta})$ and $\bar{\nu}(k, \hat{\theta})$ can be defined for $\hat{\theta}(k) = \hat{\theta}$, and that $\hat{\theta}(k)$ belongs infinitely often to the domain for which these stationary processes can be defined. Assume that for $\hat{\theta}(k) = \hat{\theta}$.

$$\bar{\nu}(k+d, \hat{\theta}) = H(q^{-1})\, \bar{\phi}^T(k, \hat{\theta})\, [\theta^* - \hat{\theta}] + w(k+d) \tag{6.24}$$

where $\{w(k)\}$ is a white random sequence or uncorrelated with $\bar{\phi}^T(k, \theta)$. Then, if :

$$H'(z^{-1}) = H(z^{-1}) - \frac{\lambda}{2} \tag{6.25}$$

is strictly positive real, where :

$$\lambda \geq \lambda_2 \tag{6.26}$$

$$\text{Prob}\,\{\lim_{k\to\infty} \hat{\theta}(k) \in D_c\} = 1 \tag{6.27}$$

where :

$$D_c : \{\theta \mid (\theta^* - \theta)^T\, \phi(k) = 0\} \tag{6.28}$$

and :

$$\text{Prob}\,\{\lim_{k\to\infty} \nu(k+d) = w(k+d)\} = 1 \tag{6.29}$$

From Eq. (6.22) and (6.23), one obtains for $\hat{\theta}(k) = \hat{\theta}$, that :

$$\bar{\nu}(k+d, \hat{\theta}) = \frac{H_1(q^{-1})\, L(q^{-1})}{H_2(q^{-1})} [\theta - \hat{\theta}(k)]^T \phi^f(k, \hat{\theta}) + \frac{H_1}{H_2} C\, S(q^{-1})\, \omega(k+d) \tag{6.30}$$

and one sees that in order to have convergence $H_2(q^{-1}) = C$, $H_1(q^{-1}) = 1$ and $\frac{L(q^{-1})}{C(q^{-1})} - \frac{\lambda}{2}$ should be strictly positive real. $\nu(k+d)$ will converge to a moving average of order d-1 but one should know the model of the disturbance. To remove this restriction, one considers the regulation case, and using the polynomial identity of Eq.(3.12) in which C_2 has been replaced by $C(q^{-1})$, Eq. (6.30) can be written :

$$\bar{\nu}(k+d, \hat{\theta}) = \frac{H_1(q^{-1})\, C_2(q^{-1})\, L(q^{-1})}{H_2(q^{-1})\, C(q^{-1})} [\theta^* - \hat{\theta}]^T \phi(k, \hat{\theta}) + \frac{H_1(q^{-1})}{H_2(q^{-1})} C_2(q^{-1})\, S'(q^{-1})\, \omega(k+d) \tag{6.31}$$

(θ^* corresponds to the value assuring that the closed loop poles are defined by $C(q^{-1})$ and instead of $C_2(q^{-1})$.

Now applying theorem 6.3, one concludes that if $H_1(q^{-1}) = 1$, $H_2(q^{-1}) = C_2(q^{-1})$ and $L(q^{-1})$ is chosen such that $\frac{L(q^{-1})}{C(q^{-1})} - \frac{\lambda}{2}$ is strictly positive real,

$\nu(k+d)$ will converge again to a moving average of order d-1 (which corresponds to the minimum variance of the plant output) without knowing the disturbance model. The closed loop poles will move from their desired values in deterministic environment (defined by $C_2(q^{-1})$). Therefore in order to maintain the tracking objectives, the control law should take in account the drift of the closed loop poles. This can be achieved using the control law :

$$u^f(k) = \frac{C_2(q^{-1})\, y^{Mf}(k+d) - \hat{\theta}^T_{oe}(k)\, \phi^f_{oe}(k)}{\hat{b}_o(k)} \qquad (6.32)$$

where :

$$\hat{\theta}^T_{oe}(k) = [\hat{\theta}_o(k)^T, \Gamma^T(k)] \qquad (6.33)$$

$$\phi^T_{oe}(k) = [\phi_o(k)^T, y^{Mf}(k+d-1) \ldots y^{Mf}(k+d-n)] \qquad (6.34)$$

($\Gamma(k)$ is a vector formed by the estimates of the coefficients of the polynomial $[C_2(q^{-1}) - C(q^{-1})]$). The adaptation algorithm has the same form but for an extended parameter vector and the convergence conditions remain unchanged. Asymptotically (if the conditions for application of O.D.E. are satisfied), one has :
$y(k+d) = y^M(k+d) + S'(q^{-1})\, \omega(k+d)$
where $S'(q^{-1})$ is a polynomial of degree d-1, solution of the polynomial identity of Eq. (6.12) for $C_2(q^{-1})$ replaced by $C(q^{-1})$. One notes that the adaptive control system in the presence of stochastic disturbance will asymptotically assure (with probability one) the minimum variance tracking of the desired trajectory. One notes that in the absence of disturbance the poles of the closed loop will be defined by $C_2(q^{-1})$ while in the presence of the disturbance they tend to those defined by $C(q^{-1})$ or more exactly such that Eq. (6.28) holds.

In the absence of stochastic disturbance, the stability condition for this scheme is: $L(q^{-1})/C_2(q^{-1}) - \frac{\lambda}{2}$ be a strictly positive transfer function. According to [13], [18] this means that the corresponding deterministic and stochastic MRAC are dual since in addition to the fact that the adaptation algorithm are the same, the positive real condition is the same for $C_2(q^{-1}) = C(q^{-1})$ and under appropriate excitation they will converge to the same parameters. For other results, see [37].

Martingales : The reason for considering the modelization of the stochastic disturbance through the use of a martingale difference sequence instead of a white sequence can be easily explained by the fact that if $\nu(k+d)$ converges to martingale difference sequence the parameter adaptation algorithm will stop a.s. Therefore the convergence proofs try to establish the convergence of the residual $\nu(k+d)$ to a martingale difference sequence [43].

The adaptive control scheme considered just above for tracking in a stochastic environment has been analyzed through the use of a martingale convergence theorem for $L(q^{-1})$, $C_2(q^{-1}) = 1$, $d = 1$ and replacing $y^M(k)$ in $\phi_{eo}(k)$ (see Eq. (6.34)) by $\bar{y}_M(k)$ the "a posteriori" output of the prediction reference model (i.e. the "a posteriori" output of the adaptive predictor) defined by (for $C_2(q^{-1}) = 1$) $\bar{y}_M(k+1) = \hat{\theta}(k+1)^T \phi(k)$. [42]. The case $d > 1$ has been analyzed in [41] but for stochastic approximation type adaptation algorithm and using $y^M(k)$ in $\phi_{eo}(k)$.

The analysis in [42] has confirmed the results obtained through O.D.E. method (same positivity condition) and has proved the boundedness of the $\phi(k)$ (in mean square sense). However these results have required the introduction of restrictions on the adaptation algorithm. The algorithm of Eqs. (6.4) and Eq. (6.5) should either switch to a stochastic approximation type when the condition number of F_k^{-1} is too large or the adaptation gain in Eq. (6.4) should be reduced when $\phi(k)^T F_k \phi(k)$ is too large.

For the convergence analysis of stochastic M.R.A.C. through martingales, the following theorem was implicitely used.

Theorem 6.4 : 1) Assume that : $\exists \theta^* \rightarrow \nu(k+1)\Big|_{\hat{\theta}=\theta^*} = \omega(k+1)$

where $\omega(k+1)$ is a martingale difference sequence with respect to the increasing sequence of σ - algebras $\mathcal{F}_k$ generated by $\omega(i), i \leqslant k$ and characterized by :

$$E[\omega(k+1)\,|\,\mathcal{F}_k] = 0 \; ; \; E[\omega(k+1)^2\,|\,\mathcal{F}_k] = \sigma^2 < \infty \text{ a.s.} \tag{6.35}$$

2) Assume that :

$$\nu(k+1) = H(q^{-1})\,[\theta^* - \hat{\theta}(k+1)]^T \phi(k) + \omega(k+1) \tag{6.36}$$

3) Assume that in Eq. (6.5), $\lambda_1(k) \equiv 1$; $\lambda_2(k) > 0$.

4) Assume that :

$$\sum_0^{\infty} \frac{\phi(k)^T F_{k+1} \phi(k)}{r_{k+1}} < \infty \tag{6.37}$$

where

$$r(k+1) = r(k) + \lambda_2(k)\,\phi(k)^T \phi(k) \tag{6.38}$$

Then, if :

$$H'(z^{-1}) = H(z^{-1}) - \frac{\lambda}{2} \tag{6.39}$$

is strictly positive real, where $\lambda \geqslant \max_k \lambda_2(k)$, one has :

$$\limsup_{N\to\infty} \sum_0^N \frac{\left[[\theta^* - \hat{\theta}(k+1)]^T \phi(k)\right]^2}{r(k+1)} < \infty \text{ a.s.} \tag{6.40}$$

From Eq. (6.40) taking in account the specific properties of the adaptive control scheme, one obtains the desired results. It is hoped that more powerful results using martingale will be available in the near future.

What is perhaps relevant to the convergence analysis of deterministic and stochastic MRAC is that they have an equivalent representation as a feedback system with a linear feedforward part and a non-linear time-varying feedback part. This explains why in all the convergence results a positivity condition on the equivalent linear part enters. However the complete analysis requires a detailed analysis of the feedback part itself. For this reason, the convergence problems for deterministic and stochastic MRAC can not be considered as belonging to the class of problems solvable by now classical approaches to (deterministic or stochastic) stability. (They are not generalizations of Lurie-Postnikov problems but new problems in stability theory).

7 . CONCLUSIONS

An attempt has been made to present model reference adaptive controllers and stochastic self-tuning regulators from a unified point of view which is to consider them as adaptive control systems using for specifying the desired control objectives, a difference equation excited either by a deterministic signal (a deterministic model) or by a white stochastic sequence (an A.R.M.A. model) respectively. The two techniques correspond in fact to deterministic and stochastic MRAC respectively.

It was also shown that (at least for minimum phase plants) the two strategies, one using an explicit (prediction) reference model and the other using an adaptive predictor whose output is equal to that of the explicit (prediction) reference model are equivalent.

It was also pointed out that the duality in the linear case with known parameters between deterministic and stochastic control is extendable to the adaptive case between deterministic and stochastic MRAC. It is possible with a unique scheme to achieve desired performance in a deterministic environment and in a stochastic environment.

We will summarize next the basic assumptions for the design of deterministic and stochastic M.R.A.C.

1) Exact knowledge of the plant delay {d}
2) An upperbound for the degree of $A(q^{-1})$ ($n \geqslant \deg A(q^{-1})$) which is the denominator of the plant transfer function.
3) The zeroes of the plant transfer function must lie within the unit circle.
4) The sign of the leading coefficient b_o of the numerator plant transfer function is useful to be known (to avoid large adaptation transients).
5) The stochastic disturbances can be modelled by ARMA processes of given structure.

6) The design assures only infinite time convergence.

Despite many successfull applications of these techniques, important questions remain to be answered and these questions are motivated by the wish of improving the design and relaxing the assumption as well as by applications which can not be solved with the available techniques. Among the various open questions, we will mention [44].

1) Theoretical analysis of adaptation transients.
2) Analysis under various types of disturbances.
3) Convergence analysis of some known schemes (not yet completely analysed).
4) Adaptive control of non-minimum phase plants.
5) Parametrization studies and extension of MRAC techniques to M.I.M.O. systems.
6) Analysis of MRAC using reduced order models.
7) Design of adaptive control schemes for restricted domain of parameter variations.
8) Adaptive control of time-varying plants.
9) Adaptive control of certain classes of non-linear systems.
10) Adaptive control with constraints on the control and its variation.

The field of MRAC and Stochastic STURE has been since many years a rendez-vous point for practitioners, algorithms makers and theoricians. There are reasons to hope that this will continue in the future for the benefit of the development of the field of adaptive control in general.

Acknowledgments

The author would like to thank Prof. K.J. Aström, Prof. L. Ljung, Prof. R. Monopoly, Prof. K.S. Narendra, Dr. B. Egardt, Dr. E. Irving for useful discussions on the subject. Much insight to the problems discussed in the paper results from the collaboration with Dr. Silveira, Dr. Dugard and Mr. Lozano. The benefit derived from this is gratefully acknowledged.

REFERENCES

[1] P.C. Parks, W. Schaufelberger, Cr. Schmid, H. Unbehaunen"Applications of adaptive control systems", Proc. of International Conf. on Adaptive Systems, Bochum, March 1980
[2] B. Courtiol "Applying Model Reference Adaptive Techniques for the control of electromechanical systems", Proc. 6th IFAC Congress, Vol. 1.B, pp. 58.2-1 - 58.2-9, Boston 1975
[3] K.J. Aström, V. Borisson, L. Ljung, B. Wittenmark "Theory and applications of self-tuning regulators" Automatica, Vol.13,1977

[4] S. Negoesco, B. Courtiol, A. Françon "Application de la commande adaptative à la fabrication des guides d'ondes hélicoïdaux", Automatisme, n° 1, pp. 7-14, 1978

[5] C.G. Kallström, K.J. Aström, N.E. Thorell, J. Eriksson, L. Sten "Adaptive autopilots for tankers", Automatica, Vol. 15, n° 3, pp. 241-254, 1979

[6] J. Van Amerongen, A.J. Udink Ten Cate "Model reference adaptive autopilots for ships", Automatica, Vol. 10, pp. 125-131, 1975

[7] E. Irving, J.P. Barret, C. Charcossey, J.P. Monville "Improving power network stability and unit stress with adaptive control", Automatica, Vol. 15, n° 1, pp. 31-46, 1979

[8] G. Dumont, R. Belanger "Self-tuning control of a Titanium Dioxidekiln" I.E.E.E. Trans. on Aut. Control, Vol. AC-23, n° 4, pp. 532-537

[9] P.J. Gathrop "Some interpretations of the self-tuning controller", Proc. I.E.E.E., Vol. 124, pp. 889-894, 1977

[10] L. Ljung, I.D. Landau "Model Reference Adaptive Systems and Self-Tuning Regulators - Some connections", Proc. 7th IFAC Congress, Vol. 3, pp. 1973-1980, June 1978

[11] K.S. Narendra, L.S. Valavani "Direct and indirect adaptive control", Automatica, Vol. 15, n° 6, pp. 653-664, 1979

[12] H.M. Silveira "Contributions à la synthèse des systèmes adaptatifs avec modèle sans accès aux variables d'état", Thèse ès-Sciences Physiques, I.N.P.G., Grenoble, March 1978

[13] I.D. Landau "Adaptive controllers with explicit and implicit reference models and stochastic self-tuning regulators - Equivalence and duality aspects", Proc. 17th, I.E.E.E. - CDC Conference, San Diego, Jan. 1979

[14] B. Egardt "Unification of some discrete time adaptive control schemes", I.E.E.E. Trans. on Aut. Contr., to appear

[16] C.R. Johnson Jr. "Input matching error augmentation, self-tuning and output error identification : algorithmic similarities in discrete adaptive model following", I.E.E.E. Trans. on Aut. Contr., to be published

[17] I.D. Landau "Dualité asymptotique entre les systèmes de commande adaptative avec modèle et les régulateurs à variance minimale auto-ajustable", R.A.I.R.O., Série Jaune, Automatique, n° 2, pp. 189-204, 1980

[18] I.D. Landau "Model Reference Adaptive Control and Stochastic Self-Tuning Regulators - Towards cross fertilization", Proc. of AFO SR Workshop on Adaptive Control, Champain, Ill., May 1979 (revised form, repport LAG-79-13, June 1979, Grenoble)

[19] K.J. Aström "Self-Tuning Regulators - Design principles and applications", Yale Workshop on Applications of Adaptive Control, Yale Univ., Aug. 1979

[21] Y.D. Landau "Adaptive control - the model reference approach", Dekker, 1979

[22] K.J. Aström "Introduction to stochastic control theory", Academic Press, New York, 1970

[23] K.S. Narendra, B. Peterson "Recent developments in adaptive control", Proc. of Int. Conf. on Adaptive Systems, March 1980, Bochum
[24] V. Kucera "Algebraic approach to discrete stochastic control", Kybernetika, Vol. 11, n° 2, 1975
[25] T. Ionescu, R. Monopoli "Discrete model reference adaptive control with an augmented error signal", Automatica, Vol. 13, n° 5, pp. 507-517, Sept. 1977
[26] I.D. Landau, R. Lozano "Unification and evaluation of discrete time explicit Model Reference Adaptive Designs", Automatica, to be published
[27] R. Lozano, I.D. Landau "Redesign of explicit and implicit discrete time Model Reference Adaptive Control schemes", Int. J. of Control, to be published
[28] L. Dugard "Etude des systèmes adaptatifs avec modèle de référence dans un environnement stochastique", Thèse Docteur-Ingénieur, I.N.P.G. Grenoble, April 1980
[29] I.D. Landau, H.M. Silveira "A stability theorem with application to adaptive control", I.E.E.E. Trans. on Aut. Contr., Vol. AC-24, April 1979
[30] I.D. Landau "An extension of a stability theorem applicable to adaptive control", I.E.E.E. Trans. on Aut. Contr., Vol. AC-25, Aug. 1980
[31] K.S. Narendra, Y.H. Lin, L.S. Valavani "Stable adaptive controller design - Part II : Proof of stability", I.E.E.E. Trans. on Aut. Contr., Vol. AC-25, June 1980
[32] G.C. Goodwin, P.J. Ramadge, P.E. Caines "Discrete time multivariable adaptive control", I.E.E.E. Trans. on Aut. Contr., Vol. AC-25, June 1980
[33] A.S. Morse "Global stability of parameter adaptive control systems", I.E.E.E. Trans. on Aut. Contr., Vol. AC-25, June 1980
[34] B. Egardt "Stability analysis of discrete time adaptive control schemes", I.E.E.E. Trans. on Aut. Contr., to appear
[35] J.J. Fuchs "Adaptive control of single input, single output discrete time linear systems, a sufficient condition for stability", I.E.E.E. Trans. on Aut. Contr., to appear
[36] L. Ljung "On positive real transfer functions and the convergence of some recursive schemes", I.E.E.E. Trans. on Aut. Contr., Vol. AC-22, Aug. 1977
[37] L. Dugard, I.D. Landau "Stochastic model reference adaptive controllers", submitted to 19th I.E.E.E. - CDC, 1980
[38] L. Dugard, I.D. Landau, H.M. Silveira "Adaptive state estimation using M.R.A.S. techniques - Convergence analysis and evaluation", I.E.E.E. Trans. on Aut. Contr., Vol. AC-25, Aug. 1980
[39] L. Dugard "Utilisation de la méthode de l'équation différentielle moyenne pour l'étude des systèmes adaptatifs stochastiques", Proc. "Techniques probabilistes en Automatique et Télécommunications", CNRS, March. 1980, Paris

[40] L. Dugard, I.D. Landau "Recursive output error identification algorithms - Theory and evaluation", Automatica, Vol. Sept. 1980

[41] G.C. Goodwin, P.J. Ramadge, P.E. Caines "Discrete time stochastic adaptive control", S.I.A.M. Journal on Control and Optimization, to be published

[42] K.S. Sin, G.C. Goodwin "Stochastic adaptive control using modified least squares algorithms", Repport EE 7907, Dept. of El. Eng., Univ. of Newcastle, Febr. 1980

[43] V. Solo "The convergence of AML, I.E.E.E. Trans. on Aut. Contr. Vol. AC-24, Dec. 1979

[44] I.D. Landau "Report of the working group on Model Reference Adaptive Control and Stochastic Self-Tuning Regulators" Proc. of AFOSR-Workshop on Adaptive Control, (J.B. Cruz-editor), Urbana, May 1979

[45] P.E. Wellstead, J.H. Edmunds, D. Prager, P. Zanker "Self-tuning pole/zero assignement regulators", Int. J. of Control, Vol. 30, n° 1, pp. 1-26, July 1979

[46] P.E. Wellstead, P. Zanker "Servo self-tuners", Int. J. of Control, Vol. 30, n° 1, pp. 27-36, July 1979

[47] K.J. Aström, B. Wittenmark "Controllers based on pole zero placement", I.E.E.E. Proceedings, n° 3, pp. 120-130, May 1980

ADAPTIVE CONTROL OF SYSTEMS SUBJECT TO A CLASS OF RANDOM PARAMETER VARIATIONS AND DISTURBANCES

P.E. Caines* and D. Dorer †

*McGill University, Montreal, Quebec
†Harvard University, Cambridge, Massachusetts and
NASA-Ames Research Center, Moffett Field, California.

ABSTRACT: The adaptive control of linear discrete time parameter systems is studied for the case where both the (unobserved) disturbances and the (unknown) parameters are random. The class of disturbance processes considered is a generalization of the usual white noise process. The random parameters are permitted to be (convergent) martingale processes evolving within the set of parameters corresponding to (time varying) inverse stable systems whose moving average noise process satisfies a (time varying) positive real condition. The main result of this paper generalizes those found in [1-4] for constant parameter systems. Specifically, we show that a stochastic gradient parameter identification algorithm generating $\hat{\theta}_N$, N=1,2,.., combined with a minimum variance feedback controller (designed using $\hat{\theta}_N$), results in a closed loop system that is stable, and whose performance asymptotically approaches that of the system regulated by a controller designed knowing the true system parameter $\overset{o}{\theta}$.

1. INTRODUCTION

The past three decades have seen great progress within systems and control theory (i.e. theoretical control engineering) particularly in such central areas as (i) optimal deterministic and stochastic control (ii) system identification and parameter estimation and (iii) recursive filtering and recursive parameter estimation. One of the early objectives of theoretical control engineering was a theory of adaptive control, that is to say, an analysis of regulators which are intended to stabilize a given system when that system's parameters are unknown to the controller. Partly as a result of the advances in control theory, and partly

M. Hazewinkel and J. C. Willems (eds.), Stochastic Systems: The Mathematics of Filtering and Identification and Applications, 421–432.

as a result of the practical implementation (see e.g. [5]) of adaptive control techniques, there have recently been a sequence of advances (see e.g. [1-4,6-12] in the theory of adaptive control. The current situation is that classes of stabilizing adaptive control algorithms have been shown to exist for (i) certain scalar, deterministic, continuous time parameter and (ii) certain multivariable, stochastic and deterministic, discrete time parameter systems. All of these algorithms contain some form of parameter estimation algorithm although this is not necessarily self-evident in the analysis of the algorithms' behaviour. In general this 'learning' part of the control algorithm does not produce consistent (i.e. convergent) system parameter estimates.

In this paper we generalize the stochastic discrete time parameter adaptive control results of [1-4] to a class of systems that have random parameters. The parameters $\{\overset{o}{\theta}; N = 1,2,...\}$ are allowed to evolve in a random manner so that they converge to some random variable $\{\overset{o}{\theta}_\infty\}$. It will be seen that the nature of this random parameter variation enters the form of the (asymptotic) loss incurred by the control strategy. For constant parameters we recall [1-2] that this loss is the same as that for a minimum variance controller designed with the system parameters known. We believe these results may constitute the first steps towards a stochastic adaptive control theory for systems with non-convergent randomly varying parameters - a topic of theoretical and practical interest.

2. SYSTEM DESCRIPTION

We consider time varying autoregressive moving average system with exogenous control inputs (denoted (TV) ARMAX systems) of the form

$$y_k + a_1^k y_{k-1} + \ldots + a_n^k y_{k-n} = b_o^k u_{k-1} + \ldots + b_n^k u_{k-n-1}$$

$$+ c_o^k w_k + \ldots + c_n^k w_{k-n}, \; k \geqslant 0 \qquad (2.1)$$

where the superscript k on the system parameters indicates their time varying nature. The initial condition for (2.1) will be denoted x. In this paper we shall only consider (2.1) for the case of scalar inputs and outputs. As in [1-4] and [9] the techniques we describe generalize to the multivariable case when the multivariable versions of the minimum phase, positive real etc. conditions (see below) are employed.

We remark here that (TV) ARMAX systems of the form (2.1) are, of course, equivalent to time varying state space systems (see e.g. [13]).

We first make the following hypothesis on the parameters appearing on the RHS of (2.1). (This is a time variant version of the hypothesis used in the time invariant cases examined in [1-2].)

(H1) $\{b_0^k,\dots,b_n^k;\ k \geqslant 0\}$ is a <u>uniformly asymptotically stable</u> system in the sense that the collection of zeros of the set of polynominals $\{b_0^k + zb_1^k + \dots + z^n b_n^k\ ;\ z \in \mathbb{C}\ ;\ k \geqslant 0\}$ are bounded away from the unit circle. □

3. STOCHASTIC HYPOTHESES AND THE OPTIMUM PREDICTOR

The initial stochastic hypotheses on the system (2.1) are as follows:

(H2) The random variables $\{x_0, w_0, w_1, \dots\}$ are defined on a probability space $(\Omega, \mathcal{B}, P)$ with $x_0 \perp\!\!\!\perp w_0^k$, i.e. x_0 independent of w_0^k, for k = 0,1,... Further we assume $E\|x_0\|^2 < \infty$. □

(H3) Denote the sigma field generated by $\{x_0, w_0^k\}$ by F_k, that generated by $\{x, y_0^k\}$ by $F_x\{y_0^k\}$ and that by $\{y_0^k\}$ by $F\{y_0^k\}$. Then we assume w satisfies $E(w_k | F_{k-1}) = 0$ a.s., $E(w_k^2 | F_{k-1}) = \sigma^2$ a.s. and

$$\limsup_N \frac{1}{N} \sum_{k=1}^{N} w_k^2 < \infty \text{ a.s.}$$ □

(The first two conditions in (H3) imply the ergodic result $\lim_N \frac{1}{N} \sum_{k=1}^{N} w_k = 0$ a.s. This is a standard result in the theory of martingales [14].)

(H4) The control inputs u_k at the k-th instant will be chosen to be $F\{y_0^k\}$ measurable and for $k \leqslant 0$ the elements u_k in x are set to 0. □

(H5) Let Ψ_k denote the vector of random parameters appearing in (2.1) defined upon the probability space $(\Omega, \mathcal{B}, P)$. We shall assume $\Psi_j \perp\!\!\!\perp \{x_0, y_0^k\}$, k = 1,2...; $j > k$; conditioned on $\psi_0, \dots, \psi_k$ □

(It is clear from (2.1) that, in general, one must have y_k dependent upon Ψ_j for $k \geqslant j$, i.e. future outputs depend upon past parameters.)

Rissanen and Barbosa have derived the input-output version of the Kalman-Bucy filter and an exposition of their theory is given in [13]. The steady state version of this filter was independently derived by Åström [15] using the algebraic device of the division algorithm.

We wish to compute a recursion for the conditional expectation $\hat{y}_{k+1/k} \triangleq E\, y_{k+1} | F_x\{y_o^k, \psi_o^{k+1}\})$ i.e. we wish to give a recursion for the one step ahead predictor for the system (2.1) with the parameters entering the predictor as deterministic quantities. Applying the results of Rissanen-Barbosa we obtain

$$\hat{y}_{k+1/k} + \sum_{i=1}^{n} \bar{c}_i^{k+1} \hat{y}_{k+1-i/k-i} = \sum_{i=1}^{n} (\bar{c}_i^{k+1} - a_i^{k+1}) y_{k+1-i} + \sum_{i=0}^{n} b_i^{k+1} u_{k-i}, \quad k \geq 0 \qquad (3.1)$$

with the $\{a_i^k\}$ and $\{b_i^k\}$ coefficients just as in (2.1) and the $\{\bar{c}_i^k\}$ coefficients given via a Cholesky factorization of the symmetric matrix given by CC^T, where C is the infinite upper triangular matrix with c_i^k in the k-th row (counting upwards) and the (k-i)-th column (counting leftwards).

The initial conditions for (3.1) may be computed in terms of the given data x and the parameters $\{a_i^k\}$, $\{b_i^k\}$ and $\{\bar{c}_i^k\}$.

We now impose the following hypothesis on $\bar{c}^k(z)$:

(H6) $\{\bar{c}_o^k, \ldots, \bar{c}_n^k;\ k \geq 0\}$ is uniformly asymptotically stable in the sense described above and, further, $\bar{c}^k(z) - \frac{a}{2}$ is <u>uniformly strictly positive real</u> for some a in the sense that

(i) Each $\bar{c}^k(z) - \frac{a}{2}$ has a minimal state space realization, say $\{C_k, A_k, B_k, D_k\}$, of the same dimension for each $k \geq 0$.

(ii) There exists a sequence of matrices $\{P_k \geq 0, L_k, W_k\}$ s.t. for sufficiently small $\rho > 0$ and all k

$$A_{k-1}^T P_k A_{k-1} - P_{k-1} = -L_k L_k^T$$

$$A_{k-1}^T P_k B_{k-1} = C_{k-1}^T - L_k W_k$$

$$W_k^T W_k + I\rho/2 = D_{k-1} + D_{k-1}^T - B_{k-1}^T P_k B_{k-1}$$

□

This time varying version of the positive real lemma implies that if x_0 is the initial state and x_N the current state of the system $\{\bar{c}^k(z) - \frac{a}{2}\ ;\ k \geq 0\}$ with input $\{p_k, k \geq 0\}$ and output $\{q_k, k \geq 0\}$ then

$$x_N^T P_N x_N - x_0^T P_0 x_0 = -\sum_{s=1}^{N} \| L_s^T x_{s-1} + W_s p_{s-1} \|^2 + \sum_{s=1}^{N-1} p_s^T q_s$$

Further we require

(H7) $\{\bar{c}_0^k, \ldots, \bar{c}_n^k;\ k \geq 0\}$ is a <u>uniformly asymptotically stable system</u> in the sense described in (H1).

We remark that in the constant parameter case examined in [1-3] one has the freedom to assume that C(z) is asymptotically stable, in other words one parameterizes the system (2.1) with the moving average filter polynomial right from the start.

Equation (3.1) is of central importance and we denote the vector list of its coefficients by θ_k. Clearly there is a one-to-one relationship between θ_0^k and Ψ_0^k and hence $\mathcal{F}\{\theta_0^k\} = \mathcal{F}\{\Psi_0^k\}$ and $\mathcal{F}_x\{y_0^k, \theta_0^j\} = \mathcal{F}_x\{y_0^k, \Psi_0^j\}$ for $j \geq k$.

Our next hypothesis is

(H8) The θ parameter process is a martingale i.e. $E\ \theta_{k+1} | \mathcal{F}\{\theta_0^k\}) = \theta_k$ for $k \geq 0$ and a (second order) Doob's condition holds for θ i.e. $\sup_N E \| \theta_N \|^2 < \infty$ a.s.

□

(Again by a standard result in martingale theory [14] this implies $\theta_k \to \theta_\infty$ a.s. and $E||\Theta_\infty||^2 < \infty$).

We observe that by the implicit function theorem applied to the equations in (H6) there exists a non-empty open set U around any parameter corresponding to a strictly positive real system $\bar{c}(z) - \frac{\bar{a}}{2}$ such that a (possibly random) evolution of parameters satisfying (H6) to (H8) is possible within U. Hence non-constant random uniformly strictly positive real systems exist.

When the output demand sequence is denoted y* the equation (3.1) can be rewritten (in an obvious hybrid "z" notation) as

$$\bar{c}^{k+1}(z)(\hat{y}_{k+1|k} - y^*_{k+1})$$

$$= (\bar{c}^{k+1}(z) - a^{k+1}(z))y_k + b^{k+1}(z)u_k$$

$$- (\bar{c}_1^{k+1} y_k^* + \ldots + \bar{c}_n^{k+1} y^*_{k+1-n}) - y^*_{k+1} \qquad (3.2)$$

$$= \theta^T_{k+1}\, \phi_k - y^*_{k+1}\ ,$$

when $\theta^T_{k+1} \underline{\Delta} ((\bar{c}_1^{k+1} - \bar{a}_1^{k+1}), \ldots, (\bar{c}_n^{k+1} - \bar{a}_n^{k+1}),\ b_o^{k+1} \ldots, b_n^{k+1}\ ,$

$$\bar{c}_1^{k+1},\ \ldots,\ \bar{c}_n^{k+1})$$

and

$$\phi_k^T \underline{\Delta} (y_k, \ldots, y_{k-n},\ u_k, \ldots, u_{k-n},\ -y_k^*, \ldots, -y^*_{k+1-n}).$$

Writing $e_k = y_k - y_k^*$ and $\nu_k = y_k - Ey_k \,|\, F\{y_o^{k-1}, \theta_o^k\} = y_k - \hat{y}_{k/k-1}$

(3.2) may again be rewritten as

$$\bar{c}^{k+1}(z)\ (e_{k+1} - \nu_{k+1}) = \theta^T_{k+1}\, \phi_k - y^*_{k+1} \qquad (3.3)$$

4. CONTROL OBJECTIVE

As in [1-4] we take the control objective to be the minimization (in some average sense) of the difference between the output y_k and a demand level y_k^* at each instant k, y* is assumed to be deterministic and bounded.

The stationary ergodic minimum variance control case (with known parameters) which was treated by Åström [15] is such that the average loss $\lim_{N} \frac{1}{N} \sum_{k=1}^{N} (y_k - y_k^*)^2$ may be replaced by

$L_k = E(y_k - y_k^*)^2$. The resulting optimum control law is to choose

u_{k-1} so that $y_k^* = Ey_k(u_{k-1}) | \{y_{-\infty}^{k-1}\})$. The resulting minimum value

of the "risk" L_k is $L_o = L_1 = \ldots = L_k = E(y_o - Ey_o | \{y_{-\infty}^{-1}\}))^2$, the minimum one step ahead prediction error for (2.1), otherwise denoted σ^2.

The work in [1-4] shows that the adaptive control algorithms described there for constant parameter systems have the property that, in addition to stabilizing the system, the limit of the average loss function

$W_N = \frac{1}{N} \sum_{k=1}^{N} E(y_k - y_k^*)^2 | F(y_o^{k-1}\})$ is also σ^2. Now in the random

parameter case the quantity $\sigma_k^2 \underline{\Delta} E(y_k - Ey_k)^2 | F\{y_o^{k-1}\}$ is not meaningful if the θ process sample path has not been specified. However, $\sigma^2(\theta_k) = E(y_k - \hat{y}_{k/k-1})^2 | F\{y_o^{k-1}, \theta_o^k\})$ has meaning, as does the averaged version π_N defined via

$\pi_N \underline{\Delta} \frac{1}{N} \sum_{k=1}^{N} E\, \sigma^2 (\theta_k) | F\{\theta_o^{k-1}\})$. We can now make the hypothesis:

(H9) (a) $\sup_k E\, \sigma^2(\theta_k) | F\{\theta_o^{k-1}\}) < \infty$ a.s.

(b) $\pi_\infty \underline{\Delta} \lim_N \pi_N$ exists and is finite a.s. □

It will be convenient to denote the expected value in (a) above by $\hat{\sigma}^2_{k/k-1}$ on some occasions.

We observe that, like all the other "random parameter hypotheses" introduced above, (H9) is necessarily satisfied in the special case of constant parameters as treated in [1-3].

The main result of this paper will be that the appropriate version of W_N for the random parameter case has the limit π_∞ when our adaptive control algorithm is employed. This is interpretted to be a form of asymptotic optimality resulting from the adaptive control strategy.

We conclude this section by remarking that the standard decision theoretic risk, which is employed in all stochastic control theory except these adaptive studies, is given by taking a total expectation over <u>all</u> events.

5. THE ADAPTIVE CONTROL ALGORITHM

The adaptive control algorithm consists of two parts: first the application of a stochastic gradient system parameter identification algorithm and second the computation of a control action.

The parameter identification part takes the form of the recursive algorithm

$$\hat{\theta}_k = \hat{\theta}_{k-1} + \frac{a}{r_{k-1}} \cdot \phi_{k-1} \cdot [y_k - \phi_{k-1}^T \hat{\theta}_{k-1}], \tag{5.1}$$

where ϕ_k was defined in Section 3, $r_k \triangleq r_{k-1} + \phi_k^T \phi_k$, $r_o \triangleq 1$, and a is the quantity in (H6). Inspired by the control law used in the Åström minimum variance controller the adaptive control laws used in [1-4] took the form:

For each k u_k is the solution of the equation

$$y_{k+1}^* = \phi_k^T \hat{\theta}_k \equiv \phi_k^T(u_k)\hat{\theta}_k. \tag{5.2}$$

We do the same here, and so (5.1) takes the form

$$\hat{\theta}_k = \hat{\theta}_{k-1} + \frac{\bar{a}}{r_{k-1}} \phi_{k-1} e_k \tag{5.3}$$

A technicality that arises at this point is that solving (5.2) for u_k could, in principle, give rise to infinite values if the current estimate of b_o^k was 0. This may be avoided, in the sense of making it an event of zero probability, in several ways. We make two of these into the hypothesis:

(H10) Either (i) The distributions of all the underlying random variables (x, w_o^N, θ_o^N) are mutually absolutely continuous with respect to Lebesgue measures for each N, or
(ii) The value of $\bar{a}$ is randomly chosen at each N, independently of the underlying random variables and independently of all previous values of $\bar{a}$, over an interval $[\bar{a}_o - \varepsilon, \bar{a}_o + \varepsilon]$ where (H6) holds at each point in this interval.

□

We observe that since $\theta_k \to \theta_\infty$ a.s. for some random θ_∞ a direct application of the constant parameter adaptive control laws to stabilize (2.1) is a reasonable strategy.

6. STOCHASTIC LYAPUNOV FUNCTIONS FOR ADAPTIVE CONTROL

The basic proof technique of [1-3] was the use of the Lyapunov function $V_k = \| \mathring{\theta} - \hat{\theta}_k \|^2$, k = 1,2,... We call this a stochastic Lyapunov function since one attempts to show V_k is a super-martingale. In fact in [1-3] $V_k + \frac{1}{r_{k-1}} \cdot S_k$ is shown to be a "near-super-martingale" i.e. a positive super martingale less a negative quantity plus a positive quantity, the latter being a.s. summable. The term S_k is introduced in an apparently arbitrary manner in order to deal with a cross-term arising in the expansion of V_k via (5.3). S_k is positive by virtue of the positive real condition in (H6). Solo [16] showed how to exploit this property in parameter estimation and our technique is inspired by his ideas. Landau and his associates [17] have pointed out (for the constant parameter case) that a natural stochastic Lyapunov function for our adaptive control problem is

$$Z_k = \| \theta_k - \hat{\theta}_k \|^2 + \frac{1}{r_{k+1}} \| x_k \|^2_{P_k}, \tag{6.1}$$

where x_k is the state of the system $\bar{c}^k(z) - (\frac{a+\rho}{2})$ with input process $\{e_k - \nu_k;\ k \geqslant 0\}$, where ρ is a sufficiently small positive number that (H6) still holds, and where $\{P_k;\ k \geqslant 0\}$ is given by (H6) again.

It appears [17,18] that the entire closed loop adaptive control system may be viewed as a (suitably rearranged) interconnection of a device with state $\theta_k - \hat{\theta}_k$ and one with state x_k. Hence (6.1) is a suitable overall Lyapunov function.

By a tedious calculation, which involves as an intermediate step evaluating $E\, Z_k \mid \mathcal{F}_x\{y_o^{k-1}, \theta_o^k\}$, we obtain the near-super-martingale property

$$E\, Z_k \mid \mathcal{F}_x\{y_o^{k-1}, \theta_o^{k-1}\})$$

$$\leqslant Z_{k-1} + \frac{a^2}{r_{k-1}^2} \| \phi_{k-1} \|^2 . E\sigma^2(\Theta_k) \mid \mathcal{F}_x\{y_o^{k-1}, \Theta_o^{k-1}\})$$

$$- \rho \frac{a}{r_{k-1}} \cdot E\, z_{k-1}^2 \mid \mathcal{F}_x\{y_o^{k-1}, \theta_o^{k-1}\}) \tag{6.2}$$

$$+ E \| \theta_k - \theta_{k-1} \|^2 \mid \mathcal{F}_x\{y_o^{k-1}, \theta_o^{k-1}\})$$

when $z_{k-1} = e_k - \nu_k$.

The details of this calculation will be presented in a future paper.

7. MAIN RESULT

From equation (6.2) it follows by use of the techniques of [1-4] that

$$\frac{1}{N}\sum_{k=1}^{N} E\, z_{k-1}^2 \mid \mathcal{F}\{y_o^{k-1}, \theta_o^{k-1}\}) \to 0$$

a.s. as $N \to \infty$. Now it may be shown that the effect of the initial condition x decays geometrically and thence we obtain our main result which may be stated as follows:

Theorem

Consider the system (2.1) subject to the hypotheses (H1) - (H10). Let the adaptive control algorithm (5.2) be applied to the system. Then the system is stabilized, in the sense that

$$\text{(i)} \quad \lim_N \sup \frac{1}{N}\sum_{k=1}^{N} u_k^2 < \infty, \text{ a.s.}$$

$$\text{(ii)} \quad \lim_N \sup \frac{1}{N}\sum_{k=1}^{N} y_k^2 < \infty, \text{ a.s.,}$$

and is asymptotically optimized in the sense that

$$\lim_N \frac{1}{N}\sum_{k=1}^{N} E(y_k - y_k^*)^2 \mid \mathcal{F}\{y_o^{k-1}, \theta_o^{k-1}\})$$

$$= \lim_N \frac{1}{N}\sum_{k=1}^{N} E\, \sigma_k^2\, (\theta_o^k) \mid \mathcal{F}\, \{\theta_o^{k-1}\}) \text{ a.s.} \qquad (7.1)$$

□

We observe that under reasonable regularity conditions the right hand side of (7.1) takes a limiting value which, in an obvious notation, may be written $\sigma_\infty^2\,(\theta_\infty)$. In other words, the average control risk for (2.1) subject to (H1) - (H10) under the control law (5.2) is the steady state one step ahead prediction error variance $\sigma_\infty^2\,(\theta_\infty)$ evaluated at the limiting value of the random parameters.

We conclude by remarking, as we did in the introduction, that the result above may constitute a first step towards a theory of stochastic adaptive control for systems with non-convergent random parameters.

This work was supported in part by the Joint Services Electronics Program under Contract N00014-75-C-0648 and in part by the National Aeronautics and Space Administration under contract NCC2-39.

REFERENCES

[1] G.C. Goodwin, P.J. Ramadge, P.E. Caines. "Recent Results in Stochastic Adaptive Control." Conference on Information Sciences and Systems, The Johns Hopkins University, Baltimore MA, 1979.

[2] G.C. Goodwin, P.J. Ramadge, P.E. Caines. "Discrete Time Stochastic Adaptive Control." Conference on Decision and Control, Ft. Lauderdale,FLA, December 1979. To appear in SICOP.

[3] G.C. Goodwin, K.S. Sin and K.K. Saluja. "Stochastic Adaptive Control and Prediction: The General Delay - Coloured Noise Case." IEEE Trans. Auto. Cont. Vol. AC-25, No. 5, pp. 946-950, 1980.

[4] G.C. Goodwin and K.S. Sin. Adaptive Filtering Prediction and Control (To be published)

[5] K.J. Åström, U. Borisson, L. Ljung and B. Wittenmark, "Theory and Applications of Self-tuning Regulators." Automatica, 1977, Vol. 13, pp. 457-476.

[6] G.C. Goodwin and K.S. Sin, "Stochastic Adaptive Control using Modified Least Squares Algorithms." University of Newcastle Technical Report Number 7907, June 1979.

[7] A. Feuer and S. Morse, "Adaptive Control of Single-input Single-output Linear Systems." IEEE Trans. Auto. Control, AC-23, N. 4, Aug. 1978, pp. 557-570.

[8] K. Narendra and L. Valavani, "Stable Adaptive Controller Design-Direct Control." IEEE Trans. Automatic Control, Vol. AC-19, No. 4, Aug. 1978, pp. 570-583.

[9] G.C. Goodwin, P.J. Ramadge, P.E. Caines, "Discrete Time Multivariable Adaptive Control." IEEE Trans. Automatic Control, Vol. AC-25, No. 3, June 1980, pp. 449-456.

[10] B. Egardt, "A Unified Approach to Model Reference Adaptive Systems and Self-tuning Regulators." Dept. of Automatic

Control, Lund Institute of Technology, Sweden TFRT-7134/1/ 67/1978.

[11] K.J. Åström and B. Wittenmark, "Self-tuning controllers based on pole-zero placement." IEE Proceedings, Vol. 127, Pt. D., No. 3, May 1980.

[12] L. Ljung, "The ODE Approach to the Analysis of Adaptive Control Systems - Possibilities and Limitations." Joint Automatic Control Conference, San Francisco, CA, August, 1980.

[13] P.E. Caines, Linear Discrete Time Parameter Stochastic Systems, John Wiley, N.Y.C. (To be published.)

[14] J. Neveu, Mathematical Foundations of the Calculus of Probability, Holden-Day, San Francisco, 1965.

[15] K.J. Åström, Introduction to Stochastic Control Theory, Academic Press, 1970.

[16] V. Solo, Time Series Recursions and Stochastic Approximation. Ph.D. Dissertation, The Australian National University, Sept. 1978.

[17] Y. Landau, Private Communication.

[18] J.B. Moore and G. Ledwich, "Multivariable Adaptive Parameter and State Estimators with Convergence Analysis." J. Austral. Math. Soc. 21 (Series B), pp. 176-197, 1979.

ON STOCHASTIC SELF-TUNING METHODS

J.-J. J. Fuchs

IRISA - Laboratoire d'Automatique
Campus de Beaulieu
35042 RENNES Cédex - France

A general scheme which allows the analysis of self-tuning controllers is outlined. The emphasis is on indirect adaptive control methods applied to a given class of single input-single output, discrete time stochastic systems. These methods consist of an identification algorithm and a control law calculation algorithm. Sufficient conditions for both algorithms are given, which if fulfilled, guarantee the overall system to be stable, under an additional assumption on the estimated model of the system.

1. INTRODUCTION

Stochastic adaptive control algorithms have attracted considerable interest in recent years. Convergence has been established for quite general systems, i.e. the general delay-coloured noise case [1]-[3]. It should be noted however that these results require the system to be minimum-phase and to our knowledge no complete results have been obtained for non-minimum phase systems. We shall propose an approach allowing to control both type of systems, but do not claim it to be the ultimate solution since we are left with a condition which appears to be difficult to verify.

All self-tuning methods can be seen as consisting of two parts: an identification algorithm and a control law calculation algorithm. In direct adaptive control schemes [1]-[3], the second part is trivial since the control law parameters are directly identified. If in the first part, the parameters of the system itself are identified, the approach is termed indirect adaptive control and relies upon the following fairly natural idea : since

M. Hazewinkel and J. C. Willems (eds.), Stochastic Systems: The Mathematics of Filtering and Identification and Applications, 433–438.

the true system's parameters are unknown, identify them -recursively in time- and use the estimates as if they were correct in the control law calculation. This approach is taken in this paper. We present a theorem giving sufficient conditions for both parts, which, if fulfilled guarantee some kind of satisfactory stochastic stability. The basic assumptions about the system to be controlled are given in chapter 2. The feasibility of the conditions required for the identification part is established in chapter 3 where an algorithm satisfying them is presented. The allowable controller structure is outlined in chapter 4. The theorem and some comments are to be found in chapters 5 and 6. Due to limited space no proofs will be given, these can be found in the clearly referenced papers.

2. STATEMENT OF THE PROBLEM

We consider single input -single output linear time invariant systems which admit a representation of the form :

$$A(q^{-1})\, y_t \;=\; q^{-1}\, B(q^{-1}) u_t + e_t \tag{1}$$

A and B are polynomials in the backward shift operator q^{-1}, $\{y_t\}$, $\{u_t\}$ and $\{e_t\}$ denote the scalar output, input and disturbance respectively. We do not assume $b_0, b_1 \ldots$ to be nonzero - i.e. the exact delay to be known. The system is assumed to be stabilizable and upperbounds $\bar{n}$ and $\bar{m}$ of the true degrees of A and B are known.

The scalar sequence $\{e_t\}$ is a real stochastic process defined on a probability space $(\Omega, \mathcal{A}, P)$ on which we define the sequence of increasing sigma-algebras $(\mathcal{F}_t,\ t \in \mathbb{N})$, where $\mathcal{F}_t$ is generated by the observations up to and including time t. We assume $\{e_t\}$ to be white gaussian noise with zero mean and variance σ^2, this assumption could be slightly weakened [4].

3. THE IDENTIFICATION ALGORITHM

The system (1) can be rewritten

$$y_t \;=\; \phi^T_{t-1}\, \theta + e_t \tag{2}$$

with:

$$\phi^T_{t-1} \;=\; [y_{t-1} \ldots\ldots y_{t-\bar{n}} \qquad u_{t-1} \ldots\ldots u_{t-\bar{m}}] \tag{3}$$

$$\theta^T \;=\; [-a_1 \ldots -a_n \;\; 0 \ldots 0 \qquad b_1 \ldots b_m \;\; 0 \ldots 0]$$

Let us consider the following identification algorithm:

$$\theta_{t+1} \;=\; \theta_t + \gamma_t\, G_t\, \phi_t\, \varepsilon_{t+1} \tag{4a}$$

$$\varepsilon_{t+1} = y_{t+1} - \phi_t^T \theta_t \tag{4b}$$

$$G_t^{-1} = G_{t-1}^{-1} + \phi_t \phi_t^T \; ; \; G_o = I \tag{4c}$$

$$\gamma_t = \frac{\phi_t^T G_{t-1} \phi_t}{\phi_t^T \phi_t} \tag{4d}$$

If (4d) is replaced by $\gamma_t = 1$ for all t, we obtain the recursive least square algorithm [5].

Proposition [6]: Algorithm (4a)-(4d) has the following properties, with probability one:

P1: $\|\tilde{\theta}_t\| < M(\omega) \;\; \forall t$ where $\tilde{\theta}_t = \theta_t - \theta$

P2: $\|\theta_t - \theta_{t-1}\| \to 0$ as $t \to \infty$

P3: $\sum_{t=1}^{\infty} \frac{(\phi_t^T \theta_t)^2}{r_t} < \infty$ where $r_t = 1 + \sum_{j=1}^{t} \phi_j^T \phi_j$

Other algorithms satisfying these properties can be found in [1]-[4], [7].

4. THE CONTROLLER STRUCTURE

We shall not consider any specific control strategy but just assume that the control law arrived to can be written in the following form:

$$u_t = d_t^T \phi_{t-1} \tag{5}$$

where d_t is a column vector which contains the adaptive controller parameters at sample instant t. The analysis which follows readily extends to control laws [8] including an additional input and/or depending upon y_t the current output. The only knowledge at hand about the system being the estimates θ_j, the controller parameter d_t in (5) will be a function of $\theta_t, \theta_{t-1} \ldots$, in most cases [1]-[4] d_t depends only upon the current estimate θ_t, however we definitly believe that, from a practical point of view, control calculation methods taking into account "all" previous estimates are preferable.

From the parameter estimates θ_t, we deduce the polynomials $A_t(q^{-1})$ and $B_t(q^{-1})$, the "stability assumption" -and lack in our proof!- we shall need in the sequel is then:

Stability assumption: the common roots of A_t and B_t are strictly and uniformly inside the unit disc as t goes to infinity, with probability one.

This simply means that the time varying estimated model of the system must be uniformly "stabilizable", to establish this property is not a trivial task and it does not seem to be -to our knowledge- and intrinsic property of prediction-error identification methods [5], [7].

5. THEOREM

Let us close the loop and analyse the overall adaptive system in order to get a feeling of what kind of conditions the control law should verify. We shall write the evolution equation of the vector ϕ_t (3). Using relations (2) and (5) we obtain:

$$\phi_t = F_{t-1}\,\phi_{t-1} + B(-\phi_{t-1}^T\,\tilde{\theta}_{t-1} + e_t)$$

where F_{t-1} is a $(\overline{n}+\overline{m})$ square matrix and B a $(\overline{n}+\overline{m})$ column vector equal to:

$$F_{t-1} = \left[\begin{array}{c|c} \multicolumn{2}{c}{\theta_{t-1}^T} \\ \hline I_{\overline{n}-1} & \\ \hline \multicolumn{2}{c}{d_t^T} \\ \hline & I_{\overline{m}-1} \end{array}\right] \qquad B = \begin{bmatrix} 1 \\ 0 \\ \vdots \\ 0 \end{bmatrix}$$

The eigenvalues of F_t are nothing but the poles of the closed-loop transfer function of the controlled system's model, it seems reasonable to expect control laws to lead to d_t-vectors guaranteeing: $\rho(F_t) \leq \delta < 1$ asymptotically ; the "stability assumption" enters here.

Theorem [8]: For the considered class of systems and under the "stability assumption", any identification algorithm satisfying P1-P3 and any control vector sequence $\{d_t\}$ verifying w.p.1.:

C1: $\|d_t\| < M(\omega) \quad \forall t$

C2: $\|d_t - d_{t-1}\| \to 0$ as $t \to \infty$

C3: $\exists$ finite $T(\omega) \ni \rho(F_t) \leq \delta < 1 \quad \forall\ t > T$

lead to a system which is stable in the sense that, w.p.1.:

$$\sup_N \frac{1}{N} \sum_1^N \phi_t{}^T \phi_t < K(\omega) \tag{6}$$

$$\lim_N \frac{1}{N} \sum_1^N E((y_{t+1} - \phi_t{}^T \theta_t)^2 / F_t) = \sigma^2 \tag{7}$$

The proof is given in [8]. Properties (6) and (7) are quite satisfactory features for stochastic self-tuning methods, and it seems that in each specific situation they can be further strenghtened.

6. CONCLUSION

For minimum variance type control laws the present analysis can be easily modified in order to remove the need for the "stability assumption" and interesting new results can be obtained [4], [9].

The "stability assumption" is obviously a lack in the proof and one could argue about it. Let us say that the approach is applicable to non-minimum phase system, and in this context this assumption can be considered as weaker as those usually required. Moreover one can expect that if closed loop identifiability conditions are satisfied the true system's parameters may eventually be obtained. Work in this direction is under progress.

REFERENCES

1. Goodwin,G.C., Ramadge,P.J., Caines,P.E.: "Discrete time stochastic adaptive control", 1979, to appear SIAM Jnl. on Control and Optimization.

2. Goodwin,G.C., Sin,K.S.: "Stochastic adaptive control: the general delay-coloured noise case", Tech. rep. n° 7904, Dept. of Electr. Eng., Newcastle Univ., 1979.

3. Goodwin,G.C., Sin,K.S.: "Stochastic adaptive control using a modified least squares algorithm" ibid., n° 7907, 1979.

4. Fuchs,J.-J.,J.: "Indirect stochastic adaptive control: the general delay-white noise case", submitted to IEEE-AC, June 1980.

5. Ljung,L.:
"Convergence analysis of parametric identification methods", IEEE-AC-23, pp. 770-783, 1978.

6. Fuchs,J.-J.,J.:
"The recursive least squares algorithm revisited", submitted to IEEE-AC, May 1980.

7. Ljung,L.:
"Analysis of recursive stochastic algorithms", IEEE-AC-22, pp. 551-575, 1977.

8. Fuchs,J.-J.,J.:
"Explicit self-tuning methods", Proc. IEE-CTA, Nov. 1980.

9. Fuchs,J.-J.,J.:
"Commande adaptative explicite: un exemple", 4th Int. Conf. on Anal. and Optimization of systems, Springer-Verlag, Dec. 1980.

The full area of ignorance is
not mapped: we are at present
only exploring its fringes.

J.D. Bernal

Part 7

NONLINEAR FILTERING

NONLINEAR SYSTEMS AND NONLINEAR ESTIMATION THEORY

R.W. Brockett

Harvard University
Division of Applied Sciences
Cambridge, Massachusetts 02138
U.S.A.

This research was supported in part by the Army Research Office under Grant DAAG29-76-C-0139, the U.S. Office of Naval Research under the Joint Services Electronics Program Contract N00014-75-C-0648 and the National Science Foundation under Grant ENG-79-09459. The final manuscript was prepared while the author was visiting the Mathematical Center in Amsterdam using a set of lecture notes compiled and reworked by J. van Gelderen, A. van der Schaft, and A. van Swieten. The hospitality of the Mathematical Center in Amsterdam, and, in particular, that of Jan van Schuppen is greatfully acknowledged.

M. Hazewinkel and J. C. Willems (eds.), Stochastic Systems: The Mathematics of Filtering and Identification and Applications, 441–477.

I. NONLINEAR SYSTEMS AND LIE ALGEBRAS

1.1 Lie Algebras

Definition: A Lie algebra over a field $\mathscr{F}$ ($\mathbb{R}$ or $\mathbb{C}$ for these lectures) is a triple $(V,+,[\cdot,\cdot])$ where $(V,+)$ is a vector space over $\mathscr{F}$ and where $[\cdot,\cdot]$ is a bilinear map from $V \times V$ into V such that

i) $[v_1,v_2] = -[v_2,v_1]$ (antisymmetry)

ii) $[v_1,[v_2,v_3]]+[v_2,[v_3,v_1]]+[v_3,[v_1,v_2]] = 0$
(Jacobi identity)

Every Lie algebra we will encounter here can be thought of as a Lie algebra of linear operators with the bracket being $[v_1,v_2] = v_1v_2-v_2v_1$.

Example 1: Let $C^\infty(M)$ be the vector space of all infinitely differentiable functions defined on a differentiable manifold M. The vector space of all differential operators $A : C^\infty(M) \to C^\infty(M)$ becomes a Lie algebra if the Lie bracket $[A,B]$ is defined as $[A,B] = AB-BA$ where AB denotes the ordinary composition of the operators.

Example 2: A subclass of all differential operators on $C^\infty(M)$ is the set of all vector fields $\sum_i f_i(x) \frac{\partial}{\partial x_i}$. The Lie bracket of two vector fields turns out to be not a second order partial differential operator but another first order partial differential operator. So the set of all vector fields is a Lie algebra with addition and bracketing defined as in example 1.

Example 3: The vector space of all n×n matrices over a field with $[A,B] = AB-BA$ is a Lie algebra.

Example 4: A subclass of all vector fields on $\mathbb{R}^n$ is the set of all vector fields of the form $\sum_{i,j} a_{ij}x_j \frac{\partial}{\partial x_i}$. We call these linear vector fields. The Lie bracket of two linear vector fields is another linear vector field and the set of all linear vector fields is a Lie algebra under addition and bracketing as defined in example 2.

Example 5: A common way in which Lie algebras arise is the following. Given a smooth function $f : R^n \to R^n$, by identifying the tangent space of R^n with R^n we can think of f as defining a vector field on R^n; $\Sigma f_i \frac{\partial}{\partial x_i}$. We are, in this way, led to the following definition of the Lie bracket:

$[g,f] = \frac{\partial f}{\partial x} g - \frac{\partial g}{\partial x} f$ where $\frac{\partial f}{\partial x}$ and $\frac{\partial g}{\partial x}$ are the Jacobian matrices of f and g respectively.

1.2 The Exponential

We will use the symbol "exp" in several ways. For a linear equation $\dot{x} = Ax$ in a linear space the unique solution which satisfies $x(0) = x_0$ can be written as $x(t) = (\exp At)x_0$ where exp At is a linear map defined by $\exp At = 1 + At + \frac{1}{2!} A^2 t^2 + \ldots$. For a nonlinear equation $\dot{x} = f(x)$ we cannot in general explicitly calculate a solution. However we can still denote points on the trajectory that passes through x_0 at $t = 0$ by $(\exp ft)x_0$. For a third way of using the expression "exp", suppose we have m vector fields $f_1, \ldots f_m$ defined on some open subset of R^n. Then we define $\exp\{f_1, \ldots, f_m\}x_0$ to be the set of all points reachable by following all possible integral curves of the vector fields f_i one after another in a piecewise fashion.

1.3 Controllability

In order to see why we care about Lie brackets in control theory consider the differential equation:

$$\dot{x} = u_1(t)g_1(x) + u_2(t)g_2(x), \quad x(0) = x_0 \tag{1}$$

If we apply the following control:

$$\begin{array}{ll} u_1 = 1, \; u_2 = 0 & \text{from } 0 \text{ to } \varepsilon \text{ units of time} \\ u_1 = 0, \; u_2 = 1 & \text{from } \varepsilon \text{ to } 2\varepsilon \text{ units of time} \\ u_1 = -1, \; u_2 = 0 & \text{from } 2\varepsilon \text{ to } 3\varepsilon \text{ units of time} \\ u_1 = 0, \; u_2 = -1 & \text{from } 3\varepsilon \text{ to } 4\varepsilon \text{ units of time} \end{array} \tag{2}$$

then we reach at time 4ε the point

$$x(4\varepsilon) = (\exp -g_2\varepsilon)(\exp -g_1\varepsilon)(\exp g_2\varepsilon)(\exp g_1\varepsilon)x_0$$

Using the expansion

$$\begin{aligned}(\exp g_i\varepsilon)x_0 &= x(\varepsilon) = x_0 + \varepsilon\dot{x}(0) + \frac{1}{2}\varepsilon^2\ddot{x}(0) + 0(\varepsilon^3) \\ &= x_0 + \varepsilon g_i(x_0) + \frac{1}{2}\varepsilon^2\left(\frac{\partial g_i}{\partial x} g_i\right)(x_0) + 0(\varepsilon^3) \\ &\qquad i = 1,2\end{aligned}$$

one finds

$$x(4\varepsilon) = x_0 + \varepsilon^2[g_1,g_2](x_0) + 0(\varepsilon^3) \approx (\exp[g_1,g_2]\varepsilon^2)x_0$$

<u>Definition</u>: The real Lie algebra generated by a pair of vector fields g_1 and g_2 is the real linear span of the following set

of vector fields

$$\{g_1,g_2,[g_1,g_2], [g_1,[g_1,g_2]],[g_2,[g_1,g_2]],\ldots\}$$

This association of a Lie algebra with a pair of vector fields is preserved under a change of coordinates in the manifold; if one brackets two vector fields and then changes coordinates the result is the same as if one were to change coordinates first and then compute the bracket in the new coordinate system. Given a control system

$$\dot{x} = f(x) + \sum_{i=1}^{m} u_i g_i(x) \qquad (3)$$

then we can form the Lie algebra $\{f,g_1,\ldots,g_m\}_{LA}$ generated by the vector fields f, $g_1,\ldots, g_m$. It is called the <u>controllability Lie algebra</u>. A theorem of Frobenius states that under a mild hypothesis there exists for each x_o in the manifold X a manifold $\exp\{f,g_1,\ldots,g_m\}_{LA}x_o$. It contains the reachable set for (3) starting at x_o but unless special assumptions are made it does not equal the reachable set.

<u>Remark 1</u>: The circle of ideas under discussion here stems in part from Hermann [1] who, to my knowledge, was the first person to point out the connection between Lie algebras and controllability, after some work in the 60's by Hermes and Haynes. These ideas were developed extensively in the period 1970-1973 by Krener, Lobry, Sussmann, and other people [2-6].

We are also interested, for reasons that have to do with the assymetry of f as compared with the g_i, in an algebra which we shall write $\bar{\mathscr{L}}$.

<u>Definition</u>: $\bar{\mathscr{L}}$ is the smallest subalgebra which contains $\{g_i\}_{i=1}^{m}$ and is closed under bracketing with f.

<u>Remark 2</u>: I am using here $\bar{\mathscr{L}}$ for something which is often written $\mathscr{L}_o$ but the reason is that $\mathscr{L}_o$ is used in other ways in a stochastic setting.

<u>Example</u>:

$$\dot{x}_1 = x_2$$
$$\dot{x}_2 = -x_1+u$$
$$\dot{x}_3 = 1$$

We see that x_3 is just the time. This system is controllable in the sense that as time goes on we can reach any point in the half-space (x_1,x_2,x_3); $x_3 > 0$. Unfortunately if we plot x_3 versus time we obtain a straight line so we cannot hit a desired point at an arbitrary point in time.

Let's compute $\mathscr{L}$ and $\overline{\mathscr{L}}$. We have $F = x_2 \frac{\partial}{\partial x_1} - x_1 \frac{\partial}{\partial x_2} + \frac{\partial}{\partial x_3}$ and $G = \frac{\partial}{\partial x_2}$ respectively. Now $\mathscr{L}$ contains F, G and $[F,G] = -\frac{\partial}{\partial x_1}$ and $[F,[F,G]] = -\frac{\partial}{\partial x_2}$. Note that $[G,[F,G]] = 0$. Thus $\mathscr{L}$ spans the tangent space of $\mathbb{R}^3$ and $\overline{\mathscr{L}}$, the two-dimensional subalgebra spanned by $\frac{\partial}{\partial x_1}$ and $\frac{\partial}{\partial x_2}$, does not. This is a manifestation of the following phenomenon. There are some systems for which one can reach every point but the time at which a certain point is reached may not be adjustable. If this is true (as in the example) then $\overline{\mathscr{L}}$ is a proper subalgebra of $\mathscr{L}$. If one wants what is sometimes called "exact time controllability" it is appropriate to focus attention on $\overline{\mathscr{L}}$. To fix ideas let's consider systems for which the control enters linearly as, for example, a deterministic version of the conditional density equation.

Example 1

$$\dot{x} = Ax + \sum_i u_i B_i x \qquad x \in R^n \tag{4}$$

Here $f = Ax$ and $g_i = B_i x$ so that $[f,g_i] = [Ax,B_i x] = -(AB_i - B_i A)x$. Each vector field which occurs in $\mathscr{L}$ is of the form Mx for some matrix M. Because the set of all n by n matrices is an n^2-dimensional vector space we have $\dim \mathscr{L} \leqslant n^2$.

Example 2

$$\dot{x} = ux + vx^3; \qquad x \in R^1 \tag{5}$$

Here $g_1(x) = x$, $g_2(x) = x^3$, so that $[g_1,g_2] = [x,x^3] = 2x^3 = 2g_2$. In this case the Lie algebra is two-dimensional; it is the real linear span of x and x^3.

Example 3

$$\dot{x} = u + vx^3 \qquad x \in R^1 \tag{6}$$

Now $g_1(x) = 1$ and $g_2(x) = x^3$

$$[1,x^3] = 3x^2 \qquad [x^2,x^3] = x^4 \quad \text{and so on.}$$

We generate an infinite-dimensional Lie algebra on a 1-dimensional manifold (the real line).

Example 4: Let B_n be the set of all nxn intensity matrices. An intensity matrix is defined to be a square matrix $A = (a_{ij})$ for which $\sum_{i=1}^{n} a_{ij} = 0$ for each j and $a_{ij} \geqslant 0$ for all $i \neq j$. Intensity matrices play a role as generators of finite state stochastic processes. B_n is not a vector space because the difference of two intensity matrices need not be an intensity matrix. In order to form the appropriate Lie algebra form the linear closure of B_n, that is the smallest linear space that contains B_n, and then take the Lie algebra generated by that linear span. This Lie algebra turns out to be isomorphic to the Lie algebra of the group of affine transformations on an (n-1)-dimensional vector space.

1.4 Observability

A second Lie algebra associated with a control problem arises in connection with questions of observability (or indistinguishability, etc.). Consider the control problem

$$\dot{x} = f(x) + \sum_i u_i g_i(x)$$

where x takes on values in a differentiable manifold X and suppose we observe $y = h(x(t))$ where y takes values in a differentiable manifold Y and h is a mapping from X into Y. We want to deduce information about x from the observation of y. Assuming enough smoothness we can differentiate y. If, for example, $u(\cdot) = 0$, then

$$\begin{aligned} \dot{y} &= \sum_{i=1}^{n} \frac{\partial h}{\partial x_i} f_i = h_1(x) \\ \ddot{y} &= \sum_{i,j}^{n} \frac{\partial}{\partial x_j}\left(\Sigma \frac{\partial h}{\partial x_i} f_i\right) f_j = \sum_{j}^{n} \frac{\partial h_1}{\partial x_j} f_j = h_2(x) \end{aligned} \tag{7}$$

etc. We may think of this in the following way. The vector $(h, h_1, h_2, \ldots, h_n)$ maps X into $T^{n-1}Y$, the (n-1)st jet bundle over Y.

Are two or more initial states in the manifold X compatible with these observations? If

$$\det \begin{bmatrix} \frac{\partial h}{\partial x_1} & \frac{\partial h}{\partial x_2} & \frac{\partial h}{\partial x_n} \\ \cdot & \cdot \;\cdot\;\cdot\;\cdot\;\cdot & \cdot \\ \frac{\partial h_{n-1}}{\partial x_1} & \frac{\partial h_{n-1}}{\partial x_2} & \frac{\partial h_{n-1}}{\partial x_n} \end{bmatrix} \neq 0 ,$$

then in view of the inverse function theorem we can assert that in some neighborhood of the true initial state there are no other points which give rise to the same response. However that does not preclude the possibility that there are some other points some distance away in the manifold X which give rise to exactly the same y's.

Example: Consider Newton's law for rotational motion

$$\ddot{\theta} = u \qquad\qquad (8)$$
$$y = \sin\theta$$

where the moment of inertia is one and u is the applied torque. A natural state space for this system is the cylinder $S^1 \times R^1$. With this state space representation the system is observable. On the other hand the equations

$$\ddot{x} = u$$
$$y = \sin x$$

often appear in the literature with the interpretation that x is a real number. This corresponds to cutting the cylinder along the side, flattening it out, laying it down on R^2 and covering R^2 with a countable number of copies of it. As far as observability is concerned the countable number of points x, $x \pm 2\pi$, $x \pm 4\pi$... can be regarded as the same. In terms of this model for the state space the system is only locally observable and this local observability is not enough to determine the initial state uniquely.

We want to now code the information about observability in a different way, one that is compatible with the way we will be looking at the conditional density equation. The vector field associated with the free motion is $F = \sum_i f_i \frac{\partial}{\partial x_i}$. The formal adjoint of this linear operator is the operator $F^* = -\sum_i \frac{\partial}{\partial x_i} f_i$, not a differential operator. F^* can be thought of as operating on the space $C^\infty(X)$ of all infinitely differentiable functions defined on the manifold X.

A function $h(.) \in C^\infty(X)$ also defines a linear operator on $C^\infty(X)$, namely "multiplication by h(.)". This maps $\phi \in C^\infty(X)$ to $h\phi \in C^\infty(X)$.
Thus we can form the Lie algebra of operators generated by the two operators $-\sum_i \frac{\partial}{\partial x_i} f_i$ and $h(\cdot)$. We propose to call this the <u>little observability algebra</u>. This algebra contains the commutator $[h(\cdot), -\sum_i \frac{\partial}{\partial x_i} f_i] = -\sum_i f_i \frac{\partial h}{\partial x_i} = -h_1$. It also contains $h_2, h_3, \ldots$ A sufficient condition for local observability around the free motion is that the little observability algebra contains n functions whose Jacobian is nonsingular.

We also have the <u>big observability algebra</u> associated with the controlled motion ($u \neq 0$). It is defined as
$$\{\sum_i \frac{\partial}{\partial x_i} f_i,\ \sum_i \frac{\partial}{\partial x_i} g_{1i}, \ldots h(x)\}_{LA}.$$

To build some intuition we apply the foregoing ideas to a linear system:

$$\dot{x} = Ax + bu \qquad\qquad (9)$$
$$y = cx$$

The controllability algebra is $\mathscr{L} = \{Ax,b\}_{LA}$ = the linear span of $\{b,Ab,\ldots A^{n-1}b,Ax\}$. In $\bar{\mathscr{L}}$ the Ax term is missing but the other terms remain. The little observability algebra is

$$\{\sum_{i,j} \frac{\partial}{\partial x_i} a_{ij}x_j, cx\}_{LA} = \text{linear span of } \{cx, cAx, \ldots cA^{n-1}x, \sum_{i,j} \frac{\partial}{\partial x_i} a_{ij}x_j\}$$

Consider now a second class of examples which is a little closer to the conditional density equation. By <u>a real Lie group</u> we understand a real Hausdorff manifold $\mathscr{G}$ with a multiplication $\cdot : \mathscr{G} \times \mathscr{G} \to \mathscr{G}$ under which $\mathscr{G}$ is a group and such that $\cdot : \mathscr{G} \times \mathscr{G} \to \mathscr{G}$ is continuous (and hence analytic by virtue of the solution of Hilberts 5th problem). The most pedestrian type of Lie group is a matrix Lie group. These include the general linear group $GL(n,r)$ the orthogonal group with det = 1, SO(n), etc. Lie groups are assumed to be finite-dimensional. This is in constrast with Lie algebras which can be finite or infinite dimensional.

A class of systems somewhat analogous to linear systems is the class defined by

$$\dot{X} = AX + \Sigma u_i B_i X$$
$$y = h(X) \tag{10}$$

where X takes values in a matrix Lie group and AX and B_iX are vector fields on that Lie group. In this case the controllability Lie algebra is necessarily finite dimensional and is given by

$$\{AX, B_iX\}_{LA} = \{AX, B_iX, [A,B_i]X, \ldots\}$$

An example of a system defined on a Lie group is that of controlled rigid body motion. The state space is the tangent bundle to SO(3); a manifold which admits the structure of a 6 dimensional Lie group.

1.5 System Isomorphism

Suppose we have two different linear control systems with zero initial conditions

$$\dot{x} = Ax+Bu; \quad y = Cx; \quad x(0) = 0; \quad x(t) \in R^n$$
$$\dot{z} = Fz+Gu; \quad y = Hz; \quad z(0) = 0; \quad z(t) \in R^{n'} \tag{11}$$

Assume that these systems, as models for a real system with input u and output y, possess exactly the same input-output behavior. We then have the following result.

Theorem: If (A,B,C) is controllable and observable then there exists a linear map $P : R^n \to R^n$ such that P preserves trajectories i.e. $Pz(t) = x(t)$ for every input $u(\cdot)$.

A theorem with the same hypothesis and conclusion holds for bilinear systems:

$$\begin{aligned} \dot{x} &= Ax+uBx \qquad y = Cx \qquad x(0) = 0 \\ \dot{z} &= Fz+uGz \qquad y = Hz \qquad z(0) = 0 \end{aligned} \tag{12}$$

The following result is very much in spirit of results in the literature [6, 7, 8] but probably does not appear exactly this way. Consider the control systems

$$\begin{aligned} \dot{x} &= f(x)+ug(x); \quad y = h(x); \quad x(0) = x_o; \quad x \in X \\ \dot{z} &= a(z)+ub(z); \quad y = c(z); \quad z(0) = z_o; \quad z \in Z \end{aligned}$$

where X and Z are analytic manifolds, f and g are analytic vector fields on X, and a and b are analytic vector fields on Z. We assume that f, g, a and b are complete; i.e. the integral curves of the vector fields can be continued for all time from $-\infty$ to $+\infty$. Finally, assume that the system on X is controllable and observable. Observable in this case means that we can distinguish between any two points on X provided we use the right input. Now we have the following theorem.

Theorem: Under the above hypothesis, if both systems generate the same input-output map then there exists an analytic mapping $\Phi : Z \to X$ that preserves trajectories i.e. $\phi(z(t)) = x(t)$ for every input $u(\cdot)$.

Example: Consider Newton's law for rotational motion. Let $X = S^1 \times R^1$ and $Z = R^2$ exactly as was done above.

We have a mapping $\phi : \begin{cases} x+2\pi k \ \text{(k integer)} \longmapsto \theta = x \\ \dot{x} \longmapsto \dot{\theta} = \dot{x} \end{cases}$

The mapping $\Phi : Z \to X$ induces a mapping Φ_* from the tangent bundle of Z into the tangent bundle of X. Because of the preservation of trajectories under Φ we must have $\Phi_*: a \longmapsto f$ and $\Phi_*: b \longmapsto g$. Now we assert that Φ_* extends to a homomorphism of the Lie algebra generated by a and b into the Lie algebra generated by f and g: $\Phi_* : \{a,b\}_{LA} \to \{f,g\}_{LA}$. In general Φ_* is not an isomorphism, it may have a nontrivial kernel because $\dim\{f,g\}_{LA}$ may be smaller than $\dim\{a,b\}_{LA}$.

Without going into detail, we mention one important construction. Suppose we start the systems on X and Z at x_o and z_o, respectively, and suppose we apply some input in both systems so that we arrive at x_1 and z_1 respectively. At that point we begin experimenting with the input. For instance if we have systems with controls u_1 and u_2 and choose them as indicated by equation (2) then we take both systems around in a loop. From this kind of construction we get a relationship between the Lie algebras for the X-system and that of the Z-system. In more detail; if we have two systems without a drift term

$$\begin{aligned} \dot{x} &= vf + ug \\ \dot{z} &= va + ub \end{aligned} \tag{13}$$

then, choosing v and u as before (2), we get

$$\Phi((\exp -a\varepsilon)(\exp -b\varepsilon)(\exp a\varepsilon)(\exp b\varepsilon)z_o) = (\exp -f\varepsilon)(\exp -g\varepsilon)(\exp f\varepsilon)(\exp g\varepsilon)x_o$$

$$\Phi((\exp[a,b]\varepsilon^2)z_o) = (\exp[f,g]\varepsilon^2)x_o$$

1.6 The Wei-Norman Equations

There exist representations of solutions of differential equations that will let us establish a connection between the unnormalized conditional density equation and a certain Lie algebra. This material is most explicit in Wei-Norman [9], an earlier paper by Chen [10] covers similar ground and the basic ideas could probably be traced back at least to Lie and Cartan.

To begin with, consider the finite dimensional linear equation:

$$\dot{x} = (uA+vB)x \tag{14}$$

with u and v functions from R^1 to R^1 and A and B constant n×n matrices. Naively one might expect to find that the fundamental solution $\Phi(\cdot)$ is

$$\begin{aligned} \Phi(t) &= e^{(\int_0^t u(\sigma)d\sigma)A + (\int_0^t v(\sigma)d\sigma)B} \\ &= I + (\int_0^t u(\sigma)d\sigma)A + (\int_0^t v(\sigma)d\sigma)B + \\ &\qquad \frac{1}{2}((\int_0^t u(\sigma)d\sigma)A + (\int_0^t v(\sigma)d\sigma)B)^2 + \ldots \end{aligned} \tag{15}$$

But this would imply that

$$\dot{\Phi}(t) = (uA+vB) + \frac{1}{2}(uA+vB)\cdot \int_0^t (uA+vB)d\sigma + \frac{1}{2}(\int_0^t (uA+vB)d\ )(uA+vB) + \ldots \tag{16}$$

It is not possible to factor out (uA+uB) from this expression because in general A and B do not commute and thus the above expression for Φ does not work. However, for n×n matrices A and B we can use the identity

$$\begin{aligned} e^{-A}Be^{A} &= (1+A+\tfrac{1}{2!}A^2+\dots)B(1-A+\tfrac{1}{2}A^2\dots) \\ &= B+AB-BA+\tfrac{1}{2}(A^2B-2ABA+BA^2) \\ &= B+[A,B]+\tfrac{1}{2}[A,[A,B]]+\dots \end{aligned} \tag{17}$$

This is sometimes called the Baker-Campbell-Hausdorff formula. Introduce the notation

$$\mathrm{ad}_A^k B = \underbrace{[A,[A,[A....[A,B]]...]}_{k \text{ times}}; \quad k \geqslant 1$$

$$\mathrm{ad}_A^0 B = B \tag{18}$$

($\mathrm{ad}_A^k B$ is an operator taking a pair of matrices into a single matrix) and define

$$\begin{aligned} \exp \mathrm{ad}_A B &= \mathrm{ad}_A^0 B + \mathrm{ad}_A^1 B + \tfrac{1}{2!}\mathrm{ad}_A^2 B + \dots \\ &= B + [A,B] + \tfrac{1}{2!}[A,[A,B]] + \dots \end{aligned} \tag{19}$$

and write

$$e^{A}Be^{-A} = \exp \mathrm{ad}_A B \tag{20}$$

Wei and Norman investigated the differential equation

$$\dot{x} = (\Sigma u_i A_i)x \tag{21}$$

by looking for a solution $\Phi(t)x_o$ which can be represented as a product of exponentials

$$\Phi(t)x_o = e^{g_1A_1}e^{g_2A_2}\dots e^{g_mA_m}x_o \tag{22}$$

in which $g_1,.,g_m$ are real valued functions of time. Differentiating gives:

$$\begin{aligned} \frac{d}{dt}(e^{g_1A_1}\dots e^{g_mA_m}) &= \dot{g}_1A_1e^{g_1A_1}\dots e^{g_mA_m}+e^{g_1A_1}\dot{g}_2A_2e^{g_2A_2}\dots e^{g_mA_m} \\ &\quad + \dots e^{g_1A_1}\dots e^{g_{m-1}A_{m-1}}\dot{g}_mA_me^{g_mA_m} \end{aligned}$$

Inserting exponentials and their inverses we can transform this into an expression in which all terms have a common factor $e^{g_1A_1}\dots e^{g_mA_m}$ on the right. Apart from this common factor, we get the expression:

$$\dot{g}_1A_1+e^{g_1A_1}\dot{g}_2A_2e^{-g_1A_1}+\dots \tag{23}$$

Applying the Baker-Campbell-Hausdorff-formula (23) can be written as:

$$\dot{g}_1A_1+\dot{g}_2(A_2+g_1[A_1,A_2]+\frac{1}{2}g_1^2[A_1,[A_1,A_2]]+\ldots)+\ldots$$

$$\ldots+\dot{g}_m \text{ (an expression containing matrices in } \{A_i\}_{LA} \text{ and } g_i\text{'s)} \quad (24)$$

Suppose now that the A_i's in (22) are not the matrices A_i from the differential equation (21) but suppose that we first construct the Lie algebra generated by the A_i's in the differential equation (21), then pick a basis from that Lie algebra, and use that basis in (22). Under these circumstances the coefficients of the $\dot{g}_i$ in (24) are linear combinations of the A_i's.

Assuming that the A_i are a basis of the Lie algebra associated with (21) we have

$$[A_i,A_j] = \sum_k \gamma_{ijk}A_k \quad (25)$$

with certain coefficients γ_{ijk}, the so-called structure constants of the Lie algebra. In order that $\Phi(t)x_0$ satisfies the differential equation (21) we must have

$$\dot{g}_1A_1+\dot{g}_2(A_2+g_1[A_1,A_2]+\frac{1}{2}g_1^2[A_1,[A_1,A_2]]+\ldots)+\ldots$$

$$+\ldots\dot{g}_m(\ldots) = u_1A_1+\ldots u_mA_m \quad (26)$$

and because the A_i are independent as vectors in $R^{n\times n}$ we get, on equating coefficients a set of equations of the form

$$\begin{aligned}\dot{g}_1 &= f_1(g_1,\ldots g_m, \dot{g}_2\ldots\dot{g}_m) + u_1\\ \dot{g}_2 &= f_2(g_1,\ldots g_m, \dot{g}_2\ldots\dot{g}_m) + u_2\\ &\cdots\cdots\\ \dot{g}_m &= f_m(g_1,\ldots g_m, \dot{g}_2\ldots\dot{g}_m) + u_m\end{aligned} \quad (27)$$

We will refer to these as the Wei-Norman equations associated with the differential equation (21) and a particular ordering of the exponential factors. Because $\Phi(0) = I$ we have initial conditions $g_1(0) = g_2(0) = \ldots = g_m(0)$. An analysis shows that the Wei-Norman equations can always be solved on some interval $|t| \leqslant \varepsilon$ however in most cases the solution cannot be continued for all time. A significant point is that the functions $f_1,\ldots,f_m$ only depend on the structure constants γ_{ijk}. That is, regardless of the representation of the Lie algebra we get the same Wei-Norman equations. We have here a situation such that by solving one set of nonlinear differential equations we simultaneously solve a whole family of linear evolution equations.

Example

$$\begin{bmatrix} \dot{x}_1 \\ \dot{x}_2 \end{bmatrix} = \begin{bmatrix} a(t) & c(t) \\ 0 & b(t) \end{bmatrix} \begin{bmatrix} x_1 \\ x_2 \end{bmatrix}$$

$$= \left[a(t) \begin{bmatrix} 1 & 0 \\ 0 & 0 \end{bmatrix} + b(t) \begin{bmatrix} 0 & 0 \\ 0 & 1 \end{bmatrix} + c(t) \begin{bmatrix} 0 & 1 \\ 0 & 0 \end{bmatrix} \right] \begin{bmatrix} x_1 \\ x_2 \end{bmatrix} \tag{28}$$

Let $\mathscr{L}$ be the Lie algebra generated by

$$\begin{bmatrix} 1 & 0 \\ 0 & 0 \end{bmatrix}, \quad \begin{bmatrix} 0 & 0 \\ 0 & 1 \end{bmatrix} \text{ and } \begin{bmatrix} 0 & 1 \\ 0 & 0 \end{bmatrix}$$

We choose an ordered basis for $\mathscr{L}$:

$$A_1 = \begin{bmatrix} 1 & 0 \\ 0 & 0 \end{bmatrix}, \quad A_2 = \begin{bmatrix} 0 & 1 \\ 0 & 0 \end{bmatrix} \text{ and } A_3 = \begin{bmatrix} 1 & 0 \\ 0 & 1 \end{bmatrix}$$

The differential equations can now be written

$$\begin{bmatrix} \dot{x}_1 \\ \dot{x}_2 \end{bmatrix} = (A_1\eta_1 + A_2\eta_2 + A_3\eta_3) \begin{bmatrix} x_1 \\ x_2 \end{bmatrix} \tag{29}$$

with $\eta_1 = a-b$, $\eta_2 = c$, $\eta_3 = b$. If we look for a fundamental solution having the form

$$\phi(t) = e^{A_1 g_1} e^{A_2 g_2} e^{I g_3}$$

Then

$$\dot{\Phi} = (A_1\dot{g}_1 + e^{A_1 g_1} A_2 \dot{g}_2 e^{-A_1 g_1} + \dot{g}_3 I) e^{A_1 g_1} e^{A_2 g_2} e^{I g_3}$$

Because we have a basis with $A_3 = I$ the expression for $\dot{\Phi}$ contains the term $\dot{g}_3\Phi$ instead of a term

$$e^{A_1 g_1} e^{A_2 g_2} \dot{g}_3 A_3 e^{-g_2 A_2} e^{-g_1 A_1}$$

that is

$$\dot{\Phi} = (A_1\dot{g}_1 + \dot{g}_2(A_2 + g_1[A_1,A_2] + \tfrac{1}{2} g_1^2 [A_1,[A_1,A_2]] + \ldots) + \dot{g}_3 I) e^{A_1 g_1} e^{A_2 g_2} e^{I g_3}$$

Now $[A_1,A_2] = A_2$ so that

$$\dot{\Phi} = (A_1\dot{g}_1 + \dot{g}_2(A_2 + g_1 A_2 + \frac{1}{2} g_1^2 A_2 + \ldots) + \dot{g}_3 I) e^{g_1 A_1} e^{g_2 A_2} e^{g_3 I}$$

$$= (A_1\dot{g}_1 + \dot{g}_2 e^{g_1} A_2 + \dot{g}_3 I) e^{g_1 A_1} e^{g_2 A_2} e^{g_3 I}$$

So

$$A_1\dot{g}_1+\dot{g}_2e^{g_1}A_2+\dot{g}_3I = \eta_1A+\eta_2A_2+\eta_3I \tag{31}$$

The Wei-Norman equations become

$$\begin{aligned} \dot{g}_1 &= \eta_1 = a-b \\ \dot{g}_2 &= e^{-g_1}\eta_2 = ce^{-g_1} \\ \dot{g}_3 &= \eta_3 = b \end{aligned} \tag{32}$$

The differential equations can be solved directly. By solving them we do not only find a fundamental solution of the particular set of equations (28) but also a fundamental solution of any family of operators that commute according to the same commutation relations.

1.7 The Covering Group

We can think about the Wei-Norman equation in another way. Consider the pair of linear equations

$$\dot{x} = Ax + \sum_i u_iB_ix\ , \qquad x \in \mathbb{R}^n \tag{33}$$

$$\dot{z} = Fz + \sum_i u_iG_iz, \qquad z \in \mathbb{R}^m \tag{34}$$

and suppose that there exists a map ϕ which maps A into F, B_i into G_i, and which extends to a Lie algebra isomorphism from $\{A,B_i\}_{LA}$ to $\{F,G_i\}_{LA}$. Can we determine the fundamental solution of (34) from the fundamental solution of (33)?

It is clear that since (33) and (34) are related by a Lie algebra isomorphism, we can pick an ordered basis for the Lie algebras such that the Wei-Norman equations of (34) are the same as those of (33). However, this only tells us that the fundamental solutions are related for g near to zero. The global picture may be radically different as becomes clear from an example.

Example:

$$\begin{bmatrix}\dot{x}_1\\ \dot{x}_2\end{bmatrix} = u\begin{bmatrix}0 & 1\\ -1 & 0\end{bmatrix}\begin{bmatrix}x_1\\ x_2\end{bmatrix}$$

$$\begin{bmatrix}\dot{z}_1\\ \dot{z}_2\end{bmatrix} = u\begin{bmatrix}0 & 2\\ -2 & 0\end{bmatrix}\begin{bmatrix}z_1\\ z_2\end{bmatrix}$$

The Lie algebra's of these two equations are isomorphic, i.e.

$$\begin{bmatrix} 0 & 1 \\ -1 & 0 \end{bmatrix} \simeq \begin{bmatrix} 0 & 2 \\ -2 & 0 \end{bmatrix}$$

but the solution of the first is

$$\Phi_x = \begin{bmatrix} \cos \int udt & \sin \int udt \\ -\sin \int udt & \cos \int udt \end{bmatrix}$$

while the solution of the second is

$$\Phi_z = \begin{bmatrix} \cos \int 2udt & \sin \int 2udt \\ -\sin \int 2udt & \cos \int 2udt \end{bmatrix}$$

and so only for $\int$ udt small can the fundamental solutions Φ_x and Φ_z be mapped onto each other. The z-system is a "double covering" of the x-system in the sense that when the x-system makes one rotation the z-system rotates two times.

In this example the group that we obtained by exponentiating the Lie algebra was S^1 which is not simply connected. To arrive at an x-system that reveals as much as possible about exponentiating the Lie algebra we must arrange the x-system in such a way that the corresponding group is simply connected. Such an x-system would cover all systems with the same Lie algebra.

If we assume that the set of matrices $\exp\{A_i x\}$ is a simply connected group and if

$$\phi : A_i \to F_i$$

is extendable to a Lie algebra homomorphism, then there exists a group homomorphism

$$\psi : \exp\{A_i x\} \to \exp\{F_i z\}$$

such that for

$$\dot{x} = (\Sigma u_i A_i)x \text{ and } \dot{z} = (\Sigma u_i F_i)z$$

we have $\psi(X) = Z$. Here X and Z are the fundamental solutions of the respective systems. Thus X solves all systems generated this way.

In our example we can take as the covering system

$$\dot{x} = u, \qquad x \in \mathbb{R} \tag{35}$$

Exponentiating this Lie algebra gives $\mathbb{R}$, which is the simply connected covering group of S^1 (the circle), and clearly from (35) we can obtain the solution of (33) and (34). This idea has a limitation: there may be no matrix group which is "big enough", i.e. the simply connected covering group may not have a matrix representation. A well known example due to Birkhoff shows that this is the case for $S\ell(2)$ (the 2×2 matrix of det +1). The study of global equivalence is important for recursive filters. If we don't take care the resulting filter will work only for a finite time and questions of a steady-state behavior will be completely inaccessible.

A less ambitious approach to global equivalence is the following. Let us consider $x = (x_1,\ldots,x_n)$. Introduce the monomials homogeneous of degree p in x_i

$$x_\alpha^{[p]} := (x_1^p, \alpha_{p-1,1} x_1^{p-1} x_2, \ldots, x_n^p)$$

This is a vector with $\binom{n+p-1}{p}$ components. The constants α denote a certain normalization. In fact we can define a set of constants α^q such that for the ℓ_q-norm $||x|| = (\Sigma|x_i|^q)^{1/q}$ equality: $||x||^p = ||x^{[p]}||$. For each normalization we can define a mapping

$$\text{sub}[p]: \ n\times n \text{ matrices } \binom{n+p-1}{p} \times \binom{n+p-1}{p} \text{ matrices},$$

$A_i \mapsto A_{i[p]}$, such that if

$$\dot{x} = (\sum_{i=1}^{m} u_i A_i)x \tag{36}$$

then

$$\frac{d}{dt} x^{[p]} = (\sum_{i=1}^{m} u_i A_{i[p]})x^{[p]} \tag{37}$$

and if we map $x(0) \longmapsto x^{[p]}(0)$ as above then solving (36) solves (37). Since we can find a linearly independent set in $\mathbb{R}^\beta$, $\beta = \binom{n+p-1}{p}$ consisting of vectors of the form $x^{[p]}$, this means that we can find the fundamental solution of (37) based on (36). Thus, again, (36) is a kind of universal simulator for all equations of the form (37). Note that (37) has the same Lie algebra as (36); this follows from the fact (37) can be regarded as being just (36) in a different coordinate system (locally, for $x \neq 0$).

<u>Example</u>

$$\frac{d}{dt}\begin{bmatrix} x_1 \\ x_2 \end{bmatrix} = \begin{bmatrix} u & v \\ w & -u \end{bmatrix}\begin{bmatrix} x_1 \\ x_2 \end{bmatrix} \qquad u,\ v,\ w : \mathbb{R} \to \mathbb{R}$$

then

$$\frac{d}{dt}\begin{bmatrix} x_1^2 \\ 2x_1x_2 \\ x_2^2 \end{bmatrix} = \begin{bmatrix} 2u & v & 0 \\ 2w & 0 & 2v \\ 0 & w & -2u \end{bmatrix}\begin{bmatrix} x_1^2 \\ 2x_1x_2 \\ x_2^2 \end{bmatrix}$$

(We have chosen the normalization α corresponding to the ℓ^1-norm.) One might think that under the mapping $x(0) \to x^{[p]}(0)$ we only reach a linear subspace of $\mathbb{R}^\beta$, because the image of the map in $\mathbb{R}^\beta$ is very thin. It is a basic fact from tensor theory that the image of this map contains a basis. For instance in our example:

$$\begin{bmatrix}1\\0\end{bmatrix} \to \begin{bmatrix}1\\0\\0\end{bmatrix}, \quad \begin{bmatrix}1\\1\end{bmatrix} \to \begin{bmatrix}1\\2\\1\end{bmatrix}, \quad \begin{bmatrix}0\\1\end{bmatrix} \to \begin{bmatrix}0\\0\\1\end{bmatrix}$$

The advantage of the present approach over the Wei-Norman g-equations is that here we have a construction which works for all u and all t. The limitation is that there are isomorphisms between matrix Lie algebras which do not arise in this way. Thus this is not a completely general tool.

II. FINITE STATE ESTIMATION PROBLEMS

2.1 Stochastic Differential Equations

Consider the stochastic differential equation (in Itô form) on the manifold X

$$dx = f(x)dt + \sum_{i=1}^{m} g_i(x)dw_i \qquad x \in X \tag{1}$$

where w_i are independent standard Wiener processes. This class includes, for instance, linear equations:

$$dx = Axdt + \sum_{i=1}^{m} b_i dw_i \tag{2}$$

and bilinear equations:

$$dx = Axdt + \sum_{i=1}^{m} B_i x dw_i \tag{3}$$

We recall the Itô rule for differentiating a function

$$d\phi = \langle \frac{\partial\phi}{\partial x}, f(x)\rangle dt + \langle \frac{\partial\phi}{\partial x}, \sum_{i=1}^{m} g_i(x)dw_i\rangle + \frac{1}{2}\sum_{i=1}^{m} \mathrm{tr}\, \frac{\partial^2\phi}{\partial x\partial x^T}(g_i,g_i)dt,$$

where $(c,d) = cd^T$.

The adjoint of the operator which appears on the right hand side before dt is the Fokker-Planck operator.

One of the early results exemplifying the interplay between stochastic equations and geometry is the connection between the controllability of certain control systems related to the stochastic differential equation and the smoothness of the solutions of the Fokker-Planck equation. Consider an equation $L\phi = \psi$ where L is a differential operator and ψ and ϕ are distributions in sense of L. Schwartz. Both ϕ and ψ are defined on a manifold M and may have singular parts (e.g. delta functions).

<u>Definition</u>: L is called <u>hypoelliptic</u> if every solution of $L\phi = \psi$ is C^∞ off the support of the singular part of ψ after a suitable modification of ϕ on a set of measure zero. For instance elliptic operators with C^∞ coefficients are hypoelliptic.

<u>Theorem</u>: (Hörmander [11]) Let L be a differential operator of the form $L = L_o + \sum_i (L_i)^2$ where $L_o = \sum_j a_j(x)\frac{\partial}{\partial x_j}$ and $L_i = \sum_j b_{ij}(x)\frac{\partial}{\partial x_j}$ in which the functions $a_j(\cdot)$ and $b_{ij}(\cdot)$ are smooth functions on a manifold of M. If at every point $x \in M$ the Lie algebra $\{L_o, L_i\}_{LA}$ spans the tangent space T_xM then L is hypoelleptic.

<u>Example</u>: Let's look at the Green's function of the diffusion equation:

$$-\frac{\partial p}{\partial t} + \frac{1}{2}\frac{\partial^2 p}{\partial x^2} = -\delta(t,x) \tag{4}$$

In this case the manifold is $M = \mathbb{R} \times \mathbb{R}$, $L_o = \frac{\partial}{\partial t}$ and $L_1 = \frac{\partial}{\partial x}$. Clearly $\frac{\partial}{\partial t}$ and $\frac{\partial}{\partial x}$ commute and $\{\frac{\partial}{\partial t}, \frac{\partial}{\partial x}\}_{LA}$ spans, at every point, the tangent space to $\mathbb{R} \times \mathbb{R}$ so that $(-\frac{\partial}{\partial t} + \frac{1}{2}\frac{\partial^2}{\partial x^2})$ is hypoelliptic. The impulse response is

$$p(x,t) = \frac{\theta(t)}{\sqrt{2\pi t}}\, e^{-\frac{x^2}{2t}} \tag{5}$$

where $\theta(t)$ is Heaviside's step function.

Consider again the stochastic differential equation

$$dx = f(x)dt + \sum_{i=1}^{m} g_i(x)dw_i \tag{6}$$

The associated Fokker-Planck equation is of the form:

$$(-\frac{\partial}{\partial t} + L + \sum_i (L_i)^2)\rho = 0 \tag{7}$$

The differential operator at the left hand side is again of the form of a first order part plus a second order part which is a sum of squares. We have now the following result.

<u>Theorem</u>: The Fokker-Planck operator is hypoelliptic if and only if the associated system

$$\dot{x} = f(x) - \frac{1}{2} \sum_{i=1}^{m} \frac{\partial g_i}{\partial x} g_i + \sum_{i=1}^{m} u_i g_i(x) \tag{8}$$

is controllable in the sense that the Lie algebra $\mathscr{L}$ spans the tangent space of X at each point.

It we write the stochastic differential equation as a Stratonovich equation

$$dx = a(x)dt + \sum_{i=1}^{m} b_i(x)dw_i \tag{9}$$

then the Fokker-Planck operator is hypoelliptic if and only if the associated control system

$$\dot{x} = a(x) + \sum_{i=1}^{m} u_i b_i(x) \tag{10}$$

has the exact time reachability property.

2.2 The Conditional Probability Equation

The deterministic system

$$\dot{x} = Ax+uBx \tag{11}$$

may be thought of as being a Fisk-Stratonövich equation

$$\bar{d}x = Axdt + Bx\bar{d}w \tag{12}$$

($\bar{d}$ denotes a Stratonovich differential as opposed to Itô notation) with the noise replaced by a control. It is sometimes useful to notice that the substitution

$$z = e^{-\int uBdt} x$$

which implies

$$\dot{z}(t) = e^{-\int_0^t Bud\sigma} A e^{\int_0^t Bud\sigma} z(t)$$

gives an equivalent equation in which u only appears as an integrand. Thus only the integrated form of u enters the equation. The same is true of (12)

$$x = e^{\int dwB} z \quad \Rightarrow \dot{z} = e^{-Bw} A e^{Bw}$$

Stated differently, for the input-output system

$$\dot{x} = Ax+\dot{u}Bx, \quad y = x, \quad \dot{u} = \frac{du}{dt}$$

we can find z so that

$$\dot{z} = C(u)z, \quad y = f(z,u)$$

Moreover, the same is true for

$$\dot{x} = Ax+(\dot{u}_1B_1+\dot{u}_2B_2+\ldots+\dot{u}_mB_m)x$$

if the B_i's commute. These ideas have been explored by Freedman and Willems [12], and by Doss [14] and Sussmann [13]. Under such a hypothesis the description of what the system does to continuous functions tells us what it will do to white noise "$\frac{dw}{dt}$" even though differentiable functions are a set of (Wiener) measure zero. In a certain sense this class of systems is more robust than others.

We want to separate the <u>algebraic</u> complexity in nonlinear filtering theory from the analytical problems. For this purpose we will look at finite state processes. We follow the ideas in Brockett and Clark [15]. Let $x(t)$ be a stochastic process taking on values in a finite set $S \subset \mathbb{R}$, say

$$S = \{b_1,\ldots,b_n\}$$

Suppose $p_i(t)$ is the probability that $x(t)$ is b_i and suppose

$$\dot{p} = Ap$$

with $p = (p_1,\ldots,p_n)^T$ and $A = (a_{ij})$ a generator (intensity matrix). Suppose we observe

$$dy = xdt+dw$$

with w a Brownian motion and want to propagate the conditional probability $p(t)$ given $y(s)$, $0 \leqslant s \leqslant t$. In the analogous discrete time situation the equation of the conditional probability follows easily from Bayes' rule. Actually it's far more convenient to deal with an unnormalized version of the conditional probability which satisfies

$$\tilde{p}(k+1) = (e^B)^{y_k} e^{-B^2/2} e^A \tilde{p}(k)$$

where B is the matrix

$$\begin{bmatrix} b_1 & & 0 \\ & \ddots & \\ 0 & & b_n \end{bmatrix}$$

and $p(k) = \tilde{p}/N$ where N is the normalization, $N = \tilde{p}_1+\tilde{p}_2+\ldots+\tilde{p}_n$. For the continuous time version Wonham [16] derived the conditional probability equation as an Itô equation

$$d\rho = A\rho dt + (B-\langle b,\rho\rangle I)\rho.(dy-\langle b,\rho\rangle dt)$$

where $b := (b_1,\ldots,b_n)^T$. By converting this to Stratonovich form we obtain

$$\bar{d}\rho = [A\rho - \frac{1}{2}(B-\langle b,\rho\rangle I)^2\rho + \frac{1}{2}\langle b,(B-\langle b,\rho\rangle I)\rho\rangle\rho]dt + (B-\langle b,\rho\rangle I)\rho(\bar{d}y-\langle b,\rho\rangle dt) \quad (13)$$

In unnormalized form the equations take the much simpler form

$$d\rho = A\rho dt + B\rho dy \quad (14)$$

and the Stratonovich version is

$$\bar{d}\rho = (A- \frac{1}{2} B^2)\rho dt + B\rho\bar{d}y \quad (15)$$

where ρ now denotes the unnormalized conditional probability (so $\Sigma\rho_i$ need not be equal to 1).

We remark that for the Itô equation

$$dx = Axdt + \Sigma B_i xdw_i$$

it is possible to study the moments by passing to the Stratonovich version

$$\bar{d}x = (A-\Sigma \frac{1}{2} B_i^2)xdt + \sum_{i=1}^{m} B_i xdw_i$$

and then using the map $x \to x^{[p]}$ to get a Stratonovich equation for $x^{[p]}$. This, in turn can be used to get a differential equation for the moments of x, e.g.

$$\frac{d}{dt} \mathcal{E} x^{[p]} = ((A- \frac{1}{2} \Sigma_i B_i^2)_{[p]} + \Sigma_i (B_{i[p]})^2)\mathcal{E} x^{[p]}$$

where $\mathcal{E}$ denotes expectation. (see [17])

2.3 The UCP Equation as a Nonlinear System

Our point of view is that the conditional probability equation together with any functional of the conditional probability defines an input-output system. We may use the results from nonlinear system theory, i.e. the controllability, observability, and minimality properties, to study the unnormalized conditional probability equation. Specifically, consider the equation

$$\dot{\rho} = (A- \frac{1}{2} B^2)\rho + uB\rho$$

with a scalar output map defined by the conditional mean

$$y = \frac{\Sigma b_i\rho_i}{\Sigma\rho_i}$$

This defines an input-output system but as a realization of that input-output system it may not be controllable or observable. For efficient implementation of the filter we are interested in knowing if we can make the conditional density equation simpler. The questions of controllability and observability are very natural in this context.

The controllability algebra of the unnormalized conditional probability equation (UCP) is the matrix algebra

$$\{A - \frac{1}{2} B^2, B\}_{LA}$$

and the little observability algebra, for output $h(\rho)$, is

$$\{\sum_{ij} \frac{\partial}{\partial x_i} \hat{a}_{ij} x_j, h(\rho)\}_{LA}$$

with $(\hat{a}_{ij}) = A - \frac{1}{2}B^2$. We make the following remarks.

1. The UCP equation is never minimal, because the equation is unnormalized. The normalized version which evolves on a manifold of dimension n-1 also generates the same input-output map.

2. The controllability algebra of the unnormalized equation is closely related to the controllability algebra of the normalized equation.

3. One way to make sure the recursive filter is low dimensional is to force $\{A - \frac{1}{2} B^2, B\}_{LA}$ to be low dimensional. The dimension of this Lie algebra is an upper bound on the number of sufficient statistics.

We are lead by this approach to the following conceptual view of the nonlinear filtering problem. The UCP is to be thought of as an input-output system with, say, the conditional mean as the output

A recursive filter is a realization of this input-output map

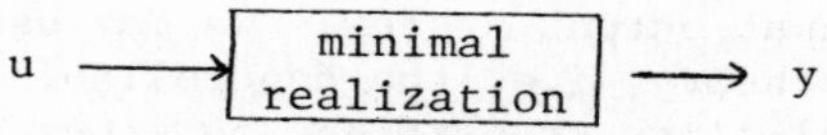

If we have a <u>finite</u> state process the question of <u>finite dimensionality</u> of the filter is not relevant because the UCP equation is automatically finite dimensional, however there can be a significant difference in the dimensionality of the UCP equation and its minimal realization.

We now focus on the following question. When does the UCP system have a simpler realization as an input-output system, and for what initial values of ρ is it the simplest? Recall that an intensity matrix (generator) A is called <u>irreducible</u> if there is no permutation matrix P such that

$$PAP^{-1} = \begin{bmatrix} A_1 & A_3 \\ 0 & A_2 \end{bmatrix} \quad \text{with } A_1, A_2 \text{ square}$$

A matrix representation of a Lie algebra is said to be <u>irreducible</u> if there is no choice of a basis such that all elements of the representation simultaneously appear as

$$\begin{bmatrix} A_1 & A_3 \\ 0 & A_2 \end{bmatrix} \quad \text{with } A_1, A_2 \text{ square}$$

<u>Fact:</u> If A is irreducible as an intensity matrix and the $\{b_i\}_{i=1}^n$ are distinct then $\{A-\frac{1}{2}B^2, B\}$ is irreducible as a representation of a Lie algebra.

<u>Proof</u>: Since B is diagonal and has distinct eigenvalues its invariant subspaces are all spanned by collections of the basis vectors. A leaves no such space invariant by hypothesis.

2.4 A Class of Examples

Consider the Lie algebra $s\ell(2,\mathbb{R}) \simeq \text{span}\ \left\{\begin{bmatrix} 0 & 0 \\ 1 & 0 \end{bmatrix}, \begin{bmatrix} 1 & 0 \\ 0 & -1 \end{bmatrix}, \begin{bmatrix} 0 & 1 \\ 0 & 0 \end{bmatrix}\right\}$ with the objective of finding A and B such that controllability algebra $\{A-\frac{1}{2}B^2, B\}_{LA}$ is equal to $s\ell(2,\mathbb{R})$. Now take the operators

$$I = x\frac{\partial}{\partial y}, \quad II = y\frac{\partial}{\partial x}, \quad III = x\frac{\partial}{\partial x} - y\frac{\partial}{\partial y}$$

An easy calculation gives

(a) $[I, II] = III$
(b) $[I, III] = -2I$
(c) $[II, III] = 2II$

These commutation relations are the same as for $\left\{\begin{bmatrix} 0 & 1 \\ 0 & 0 \end{bmatrix}, \begin{bmatrix} 0 & 0 \\ 1 & 0 \end{bmatrix}, \begin{bmatrix} 1 & 0 \\ 0 & -1 \end{bmatrix}\right\}$, so the operators I, II, III form a representation of $s\ell(2,\mathbb{R})$. Notice that the vector space of forms $\phi(x,y) = \Sigma a_i x^i y^{p-i}$ homogeneous of degree p are mapped into itself by these operators. So we can construct a matrix representation of $s\ell(2,\mathbb{R})$ by matrices of <u>any</u> given dimension in the following way:

$$I \simeq \begin{bmatrix} 0 & p & 0 & & \\ 0 & 0 & p-1 & \ddots & \\ & 0 & \ddots & \ddots & 0 \\ & & \ddots & 0 & 1 \\ & & & 0 & 0 \end{bmatrix}; \quad II \simeq \begin{bmatrix} 0 & 0 & & & \\ 1 & 0 & & & \\ & 2 & \ddots & & \\ & & \ddots & 0 & 0 \\ & & & p & 0 \end{bmatrix}$$

$$\text{III} \simeq \begin{bmatrix} p & 0 & & & \\ 0 & p-2 & & & \\ & & \ddots & & \\ & & & -p+2 & 0 \\ & & & 0 & -p \end{bmatrix}$$

A second point of view is the following. Consider the mapping

$$\begin{bmatrix} x \\ y \end{bmatrix} \to \begin{bmatrix} x \\ y \end{bmatrix}^{[p]}$$

as applied to

$$\begin{bmatrix} \dot{x} \\ \dot{y} \end{bmatrix} = \begin{bmatrix} a & b \\ c & -a \end{bmatrix}\begin{bmatrix} x \\ y \end{bmatrix} = \left(a\begin{bmatrix} 1 & 0 \\ 0 & -1 \end{bmatrix} + b\begin{bmatrix} 0 & 1 \\ 0 & 0 \end{bmatrix} + c\begin{bmatrix} 0 & 0 \\ 1 & 0 \end{bmatrix}\right)\begin{bmatrix} x \\ y \end{bmatrix}$$

This gives

$$\frac{d}{dt}\begin{bmatrix} x^p y^0 \\ \binom{p}{1} x^{p-1} y^1 \\ \vdots \\ x^0 y^p \end{bmatrix} = \begin{bmatrix} pa & b & & & \\ pc & (p-2)a & 2b & & \\ & (p-1)c & \ddots & \ddots & \\ & & \ddots & \ddots & pb \\ & & & c & -pa \end{bmatrix}\begin{bmatrix} x^p y^0 \\ \binom{p}{1} x^{p-1} y^1 \\ \vdots \\ x^0 y^p \end{bmatrix}$$

where we have chosen the ℓ^1-normalization. If we want the map sub p to take intensity matrices into intensity matrices, then we have to take this normalization. Following [18] we show how to modify matrices of this form by a transformation R defined by a diagonal matrix with positive elements

$$A \to RAR^{-1}$$

such that $\{A - \frac{1}{2} B^2, B\}_{LA}$ is of a suitable form. Here p is taken equal to n-1, and B is taken equal to III. Because B is diagonal and R is diagonal, $RBR^{-1} = B$. Now we want to select A in the Lie algebra together with R and β such that for given α

$$A = R\bar{A}R^{-1} + \frac{1}{2}\alpha B^2 + \beta I$$

is an intensity matrix (note that we always can add a multiple of the identity βI, because the identity commutes with everything in the algebra and does not change the Lie algebra except in a trivial way). We can enforce the condition that A be an intensity matrix by asking that cA = 0 where c = (1,...,1). Thus we require

$$cR\bar{A}R^{-1} + \frac{1}{2}\alpha cB^2 + \beta cI = 0$$

or

$$cR\bar{A} + \frac{1}{2}\alpha cB^2 R + \beta cIR = 0$$

which can be written as

$$cR(\bar{A} + \frac{1}{2}\alpha B^2 + \beta I) = 0$$

It is a consequence of the Perron-Frobenius theorem on positive matrices that this equation always has a solution R. Notice that because R is diagonal and positive it induces only a change of measure.

To summarize, given a positive integer n, we can construct a finite state process with n states, which has a four-dimensional estimation algebra. (Four because we had to add the identity to the three dimensional Lie algebra $s\ell(2,\mathbb{R})$.) In fact we can write down explicitly a filter for this problem [18]. Moreover we can actually simulate the nonlinear filtering equation with a two-dimensional equation because this x-equation covers (globally!) the $x^{[p]}$-equation. What plays the role of the Gaussian initial conditions as in the Kalman filter? The analogue in our set-up comes from the binomial distributions transformed by the change of measure defined by R. Additional details are found in [18]. We see that for a binomial initial condition the conditional density is propagated in a manifold of dimension three.

III. ESTIMATION OF DIFFUSION PROCESSES

3.1 A Class of Models

Consider a stochastic process h(x) where x satisfies the Itô equation

$$dx = f(x)dt + \sum_{i=1}^{m} g_i(x)dw_i \qquad (1)$$

with $w_1, w_2, \ldots w_m$ independent, standard Wiener processes. If h(x) is observed in the presence of white noise, i.e. if there is available for processing

$$dy = h(x)dt + d\nu \qquad (2)$$

where ν is an additional independent Wiener process, then under suitable assumptions, the conditional density for x at time t given dy on the interval [0,t] satisfies a stochastic, nonlinear, partial differential equation, directly analogous to the situation we encountered in the finite state case. This is the class of problems which is of interest here.

As in the finite state case, it is more convenient to work with an unnormalized version of the conditional density equation. In the present context this equation, written as an Itô equation, takes the form

$$d\rho(t,x) = L_0\rho(t,x)dt + h(x)\rho(t,x)dy \qquad (3)$$

where L_o is the Fokker-Planck operator. As was pointed out in [15] the Lie theoretic point of view suggests that this be rewritten as a Stratonovich equation

$$\eth\rho(t,x) = (L_o - \frac{1}{2}h^2)\rho(t,x)+h(x)\rho\eth y \qquad (4)$$

thus bringing into prominence the Lie algebra generated by the operator $L_o - \frac{1}{2}h^2$, and h. Because this algebra will be important in determining the complexity of the estimation problem we call it the estimation algebra. We refer to (4) as the UCD equation.

We also point out that if one inserts a parameter to adjust the magnitude of the noise terms

$$dx = f(x)dt+\alpha\Sigma g_i(x)dw_i$$
$$dy = h(x)dt+d\nu$$

then the estimation algebra is of the form

$$\{\alpha\hat{L}_o+F^*-\frac{1}{2}h^2,h\}_{LA}$$

Setting α to zero gives an algebra which is isomorphic to the little observability algebra.

3.2 Representation Theory for the Estimation Algebra

Writing the UCD equation in Stratonovich form makes it clear that our understanding of nonlinear filtering would be enhanced if we knew what kind of Lie algebras can be generated by $L_o - \frac{1}{2}h^2$ and h. This is a question about the representation of Lie algebras in terms of linear operators on an infinite dimensional space. However the sort of representation which is of direct concern has a number of special features which set it apart from the standard theory which was developed by physicists and mathematicians to fill the needs of quantum mechanics. Briefly, the situation is this:

(a) We cannot assume that the elements of the Lie algebras exponentiate to give a one parameter group. Typically there is a cone in the Lie algebra which exponentiates to generate a semigroup of bounded operators which cannot be extended to a group.

(b) The representations of interest act on real L_1 spaces not complex L_2 spaces.

(c) The Lie algebra contains a cone whose exponentials map the nonnegative functions in L_1 into themselves.

The most basic example consists of a four-dimensional Lie algebra represented by the operators

$$L_0 = \frac{1}{2}\frac{\partial^2}{\partial x^2} - \frac{1}{2}x^2 \quad ,$$

$$L_1 = x, \quad L_2 = \frac{\partial}{\partial x}, \quad L_3 = 1$$

which are to be thought of as acting on $S \subset L_1(\mathbb{R})$ where S is a suitable dense subset. Of course, exp $L_0 t$ makes sense only for $t \geq 0$. (One easily sees that L_0 does generate a semigroup of bounded operators on $L_1(\mathbb{R})$.) Likewise, L_2 is a generator, in fact it generates a group. However, e^{xt} is not bounded for t positive or negative. We define the cone

$$K = \{\alpha L_0 + \beta L_1 + \gamma L_2 + \delta L_3 | \alpha,\beta,\gamma,\delta \in \mathbb{R},\ \alpha > 0\}$$

Each element of K is a generator of a semigroup. Since the cone has nonempty interior in the four-dimensional Lie algebra and since the operators formally satisfy the desired commutation relations, we will call $\{L_0,L_1,L_2,L_3\}$ a representation of the given Lie algebra.

Because the operator L_0 is always second order the Lie algebra representation $\{L_0 - \frac{1}{2}h^2, h\}_{LA}$ must contain at least one 2nd order operator. With a view toward understanding as much as possible about representations involving 2nd order operators we begin by describing one such family. This family is of interest because it includes, as a subalgebra, the estimation algebra associated with linear systems and it suggests several possibilities for nonlinear filtering. There is also a great deal of interest in this representation from pure mathematicians [19].

The set of real 2n by 2n matrices of the form

$$M = \begin{bmatrix} A & B \\ C & -A' \end{bmatrix} ; \quad B = B'; \quad C = C'$$

form a n(2n+1)-dimensional vector space under ordinary matrix addition. It is easliy verified that if M_1 and M_2 are of this form, then $M_1M_2 - M_2M_1$ is also and thus they form a Lie algebra with respect to the standard commutator product. Matrices of this form are called <u>Hamiltonian</u> or <u>infinitesimally symplectic</u>. The Lie algebra is called the <u>symplectic Lie algebra</u> and is denoted here by Sp(n).

We are also interested in an extension of this algebra consisting of real matrices of dimension 2n+1 by 2n+1, and having the form

$$M = \begin{bmatrix} A & B & b \\ C & -A' & c \\ 0 & 0 & 0 \end{bmatrix}; \quad B = B', \quad C = C'$$

Again, it is easy to verify that this set of matrices forms a Lie algebra. We will call it the extended symplectic algebra.

Turning now to Lie algebras of operators, consider the n(2n+1)-dimensional vector space consisting of sums of the form

$$L = \sum_{i,j=1}^{n} b_{ij} \frac{\partial^2}{\partial x_i \partial x_j} + a_{ij}\left(\frac{\partial}{\partial x_i} x_j - \frac{1}{2}\delta_{ij}\right) + c_{ij} x_i x_j$$

where δ_{ij} is one if i = j and zero otherwise. For example, if n = 1, we are looking at real linear combinations of the type

$$L = b\frac{\partial^2}{\partial x^2} + a\left(\frac{\partial}{\partial x} x - \frac{1}{2}\right) + cx^2$$

The identities

$$\left[\frac{\partial^2}{\partial x^2}, x^2\right] = \frac{\partial^2}{\partial x^2} x^2 - x^2 \frac{\partial}{\partial x^2} = 4\frac{\partial}{\partial x} x - 2$$

$$\left[\frac{\partial^2}{\partial x^2}, \frac{\partial}{\partial x} x - \frac{1}{2}\right] = \frac{\partial^3}{\partial x^3} x - \frac{\partial}{\partial x} x \frac{\partial^2}{\partial x^2} = 2\frac{\partial^2}{\partial x^2}$$

$$\left[\frac{\partial}{\partial x} x - \frac{1}{2}, x^2\right] = \frac{\partial}{\partial x} x^3 - x^2 \frac{\partial}{\partial x} x = 2x^2$$

verify that in the case n = 1, the above set of operators are closed under commutation and hence form a Lie algebra. It is straightforward to verify that the same is true for n > 1. For example, for i,j and b distinct

$$\left[\frac{\partial^2}{\partial x_i \partial x_j}, x_k x_j\right] = \frac{\partial}{\partial x_i} x_k$$

$$\left[\frac{\partial^2}{\partial x_i \partial x_j}, \frac{\partial}{\partial x_k} x_j\right] = \frac{\partial}{\partial x_i}\frac{\partial}{\partial x_k}$$

$$\left[\frac{\partial}{\partial x_i} x_j, x_k x_i\right] = x_j x_k$$

and the other cases follow similarly.

In order to understand the structure of this Lie algebra we establish an isomorphism between it and the symplectic algebra introduced above. The isomorphism works as follows:

$$\Sigma a_{ij}\left(\frac{\partial}{\partial x_i} x_j - \frac{1}{2}\delta_{ij}\right) \longmapsto \begin{bmatrix} (a_{ij}) & 0 \\ 0 & -(a_{ji}) \end{bmatrix}$$

$$\frac{1}{2}\Sigma b_{ij}\frac{\partial^2}{\partial x_i \partial x_j} \longmapsto \begin{bmatrix} 0 & (b_{ij}) \\ 0 & 0 \end{bmatrix}$$

$$\frac{1}{2}\Sigma c_{ij} x_i x_j \longmapsto \begin{bmatrix} 0 & 0 \\ (c_{ij}) & 0 \end{bmatrix}$$

This then allows one to understand the symplectic algebra a different way, i.e., as the Lie algebra of 2nd order partial differential operations of the given form with commutation of operators being the Lie bracket.

We leave the verification of the fact that this is a Lie algebra isomorphism to the reader. However, the identities we have given do most of the work and a study of the case n = 2 involving

$$\frac{1}{2}\left(b_{11}\frac{\partial}{\partial x^2} + 2b_{12}\frac{\partial}{\partial x}\frac{\partial}{\partial y} + b_{22}\frac{\partial^2}{\partial y^2}\right) \longmapsto \begin{bmatrix} 0 & 0 & b_{11} & b_{12} \\ 0 & 0 & b_{12} & b_{22} \\ 0 & 0 & 0 & 0 \\ 0 & 0 & 0 & 0 \end{bmatrix}$$

$$a_{11}\frac{\partial}{\partial x}x + a_{12}\frac{\partial}{\partial x}y + a_{21}\frac{\partial}{\partial y}x + a_{22}\frac{\partial}{\partial y}y - \frac{1}{2}(a_{11}+a_{22})$$

$$\longmapsto \begin{bmatrix} a_{11} & a_{12} & 0 & 0 \\ a_{21} & a_{22} & 0 & 0 \\ 0 & 0 & -a_{11} & -a_{21} \\ 0 & 0 & -a_{12} & -a_{22} \end{bmatrix}$$

and

$$\frac{1}{2}(c_{11}x^2 + 2c_{12}xy + c_{22}y^2) \longmapsto \begin{bmatrix} 0 & 0 & 0 & 0 \\ 0 & 0 & 0 & 0 \\ c_{11} & c_{12} & 0 & 0 \\ c_{12} & c_{22} & 0 & 0 \end{bmatrix}$$

should be convincing.

With respect to the remarks on the generation of semigroups, if we regard the operators defined above as acting on a suitable dense subset of $L_1(\mathbb{R}^n)$, then the theory of diffusion processes tells us that

$$L = \Sigma b_{ij} \frac{\partial^2}{\partial x_i \partial x_j} + a_{ij} \frac{\partial}{\partial x_i} x_j$$

generates a semigroup provided (b_{ij}) is a positive definite matrix. Moreover, if (c_{ij}) is positive definite along with (b_{ij}), then

$$\hat{L} = \sum_{i,j} b_{ij} \frac{\partial^2}{\partial x_i \partial x_j} + a_{ij} \frac{\partial}{\partial x_i} x_j - c_{ij} x_i x_j$$

will generate a subgroup. Of course, without some kind of definiteness assumptions, no such conclusions are valid. Since the conditions on b_{ij} and c_{ij} give a cone in the Lie algebra having a nonempty interior, we have here a representation of the symplectic algebra which meets the conditions set forth at the beginning of this section.

There is a generalization of this construction which enables one to capture other important examples. Consider appending to the family of operators discussed above the first order constant coefficient operators $\partial/\partial x_i$, multiplication by linear functions x_i and multiplication by constants. We notice that

$$\left[\frac{\partial^2}{\partial x_k^2} , b \frac{\partial}{\partial x_i} + c_j x_i \right] = 2 \frac{\partial}{\partial x_k} c_j \delta_{ik}$$

$$\left[\frac{\partial}{\partial x_i} , x_j \right] = \delta_{ij}$$

$$\left[\frac{\partial}{\partial x_i} x_j , x_i \right] = x_j \qquad i \neq j$$

$$\left[x_i x_j , \frac{\partial}{\partial x_i} \right] = -x_j \qquad i \neq j$$

which together with other obvious identities show that the $(2n^2+3n+1)$-dimensional vector space of operators of the form

$$L = \Sigma b_{ij} \frac{\partial^2}{\partial x_i \partial x_j} + a_{ij} \frac{\partial}{\partial x_i} - \frac{1}{2} \delta_{ij} +$$

$$c_{ij}x_ix_j + b_i\frac{\partial}{\partial x_i} + c_ix_i + d$$

is a Lie algebra.

The mapping which sends this operator into the element of the extended symplectic algebra is given by

$$L \longmapsto 2\begin{bmatrix} A & B & b \\ C & -A' & c \\ 0 & 0 & 0 \end{bmatrix}$$

defines a Lie algebra homomorphism. The constants ("d") lie in the kernel but otherwise this homomorphism is faithful. In Section 2 we discussed a 4-dimensional algebra which, under this homomorphism, goes into the set of all 3 by 3 matrices of the form

$$\begin{bmatrix} 0 & b & \alpha \\ -b & 0 & \beta \\ 0 & 0 & 0 \end{bmatrix}$$

There is one last comment on the structure of this Lie algebra. The subalgebra consisting of terms of the form

$$\hat{L} = \Sigma b_i\frac{\partial}{\partial x_i} + c_ix_i+d$$

forms an ideal in the whole algebra having dimension 2n+1. The terms of the form

$$\tilde{L} = \Sigma b_{ij}\frac{\partial^2}{\partial x_i\partial x_j} + a_{ij}(\frac{\partial}{\partial x_i}x_j - \frac{1}{2}\delta_{ij}) + c_{ij}x_ix_j$$

form a complementary subalgebra that is simple.

3.3 Linear Estimation Theory

A general linear model for recursive estimation theory is

$$dx = Axdt + \sum_{i=1}^{m} b_idw_i$$

$$dy_j = \langle c_j,x\rangle dt+dv_j; \quad j = 1,2,\ldots,p$$

with $w_1,w_2,\ldots,w_m,v_1,v_2,\ldots v_p$ being independent brownian motions. The unnormalized conditional density equations for such systems take the form (as a Stratonovich equation)

$$\frac{\partial p}{\partial t} = (\Sigma\hat{b}_{ij} \frac{\partial^2}{\partial x_i \partial x_j} + \frac{\partial}{\partial x_i} a_{ij}x_j - \frac{1}{2} <c_j,x>^2 + \dot{y}_i <c_i,x>)\rho$$

or

$$\frac{\partial p}{\partial t} = (L_o + \dot{y}L_1)\rho$$

Both the operator L_o and the operator L_1 belong to the $(2n^2+3n+1)$-dimensional Lie algebra defined in the previous section and so the Lie algebra which they generate is necessarily a subalgebra of the algebra discussed above. There is more to be said. Because L_1 belongs to the ideal L mentioned above, it is clear that the Lie algebra generated by L_o and L_1, which is the relevant algebra for studying the solution of the unnormalized conditional density equation, cannot have dimension higher than 2n+2. In fact, only L_o intersects the "symplectic part" of the algebra. Typically the Lie algebra generated by L_o and L_1 is of dimension 2n+2 in the case of single-input/single-output systems. In fact, if one assumes controllability and observability then only the presence of a degeneracy in the form of all-pass factors stands in the way of this conclusion (see [20]). Based on the homomorphism into the extended symplectic group given here, we see that a same general conclusion holds in the multivariable case.

We now consider a single-input/single-output situation and write down side-by-side the UCD realization

$$\partial\rho = L_o\rho dt + L_1\rho dy$$

$$\bar{y} = \int\rho<c,x>dx/\int\rho dx$$

and the usual Kalman-Bucy filter for the state estimation and the error variance

$$dz = (A-Pcc^T)zdt+Pcdy; \qquad \bar{y} = <c,z>$$
$$\dot{P} = AP + PA^T + bb^T - Pcc^TP \quad .$$

From the latter we obtain vector fields corresponding to the coefficient of dt and the coefficient of dy, respectively,

$$\begin{bmatrix} (A-Pcc^T)z \\ AP+PA^T+bb^T-Pcc^TP \end{bmatrix} , \quad \begin{bmatrix} Pc \\ 0 \end{bmatrix}$$

This pair is to be compared with the pair coming from UCD

$$(L_o - \frac{1}{2} L_1^2) , \quad L_1$$

Computing the successive Lie brackets of the vector fields which appear in the recursive filter gives vector fields of the form

$$\begin{bmatrix} P(A^T)^k c+d_k \\ 0 \end{bmatrix}$$

where d_k does not depend on P and $P(A^T)^k c$ is linear in P. There exist 2n vector fields of this form if c is cyclic for A^T. In fact there is a Lie algebra homomorphism which maps the unnormalized conditional density operators to the Lie algebra so obtained. The homomorphism has a one dimensional kernel corresponding to multiplication by a constant. (The reason for this kernel is that the UCD equation is not normalized.)

Finally, we show how to pass from the unnormalized conditional density equation to the recursive estimation equation in a logical way. Consider the case where x is a scalar and

$$dx = dw$$

$$dy = xdt+dv$$

The UCD equation is given by

$$d\rho = \frac{1}{2}\frac{\partial^2}{\partial x^2}\rho - \frac{1}{2}x^2\rho)dt + dyx\rho$$

The estimate is

$$\hat{x} = \frac{\int x\rho dx}{\int \rho dx}$$

Suppose x(0) has a gaussian distribution and look for a solution of the form

$$\rho(t,x) = e^{a(t)x^2+b(t)x+c(t)}$$

where a(t), b(t) and c(t) are not yet known. By differentiating this expression with respect to time and space we get three equations which make it possible to determine the unknowns a(t), b(t) and c(t).

$$\frac{\partial}{\partial t}(\rho(t,x)) = (\dot{a}x^2+\dot{b}x+\dot{c})\rho(t,x)$$

$$\frac{\partial}{\partial x}(\rho(t,x)) = [2ax+b]\rho(t,x)$$

$$\frac{\partial^2}{\partial x^2}(\rho(t,x)) = 2a\rho(t,x)+(2ax+b)^2\rho(t,x)$$

Substitution into the UCD equation and changing to Itô form gives

$$\dot{a}x^2+\dot{b}x+c = (2a^2-\tfrac{1}{2})x^2+(2ab+\dot{y})x+2a+b$$

Equating the respective coefficients of x results in

$\dot{a} = 2a^2 - \frac{1}{2}$ This plays the role of the Riccati equation

$\dot{b} = 2ab+\dot{y}$ This equation propagates the mean

$\dot{c} = 2a+b$ Normalization equation

We get three sufficient statistics, one of them is a function which does not depend on the sample path. (That is the reason why it is possible to compute the solution of the Riccati equation off-line.)

3.4 A Nonexistence Result

In section 3.2 we discussed a class of representations involving 2nd order differential operators, a simple example of which is the set spanned by

$$\{ \frac{d^2}{dx^2} , \ x\frac{d}{dx} , \ x^2 , \ 1\}$$

One direct way to generate different representations from a given one is to pick a nonvanishing function ψ and make the substitutions

$$\frac{d^2}{dx^2} \longmapsto \psi^{-1} \frac{d^2}{dx^2}\psi \ ; \qquad x^2 \to x^2$$

$$x\frac{d}{dx} \longmapsto \psi^{-1} x \frac{d}{dx}\psi \ ; \qquad 1 \to 1$$

This kind of substitution has been investigated by Mitter [21] in an estimation theory context and can be used to explain how the example of Beneš [22] can be obtained from a linear problem.

There is a second group of transformations which can be used to generate new representations of Lie algebras from old ones in the present context. Consider the estimation algebra generated by

$$dx = f(x)dt+g(x)dw; \quad dy = h(x)dt+d\nu$$

For any given diffeomorphism of the real line into itself say $x \to \phi(x) = z$ we obtain from the Itô rule a new equation

$$dz = \tilde{f}(z)dt + \tilde{g}(z)dw; \quad dy = \tilde{h}(z)dt+d\nu$$

having an estimation algebra which is isomorphic to the original one. Thus this is a second way to generate new representations of Lie algebras in terms of 2nd order operations (see [18]).

It is obvious that as it stands $\{\frac{d^2}{dx^2} , \ x\frac{d}{dx} , \ x^2 , \ 1\}$

does not contain operators of the form required in an estimation algebra. In view of the many alternative forms a representation of these same algebras may take, however, it makes sense to ask if there is any pair L_o, h such that L_o is a Fokker-Planck operator on $\mathbb{R}^1$, $h : \mathbb{R}^1 \to \mathbb{R}^1$, and $\{L_o, h\}_{LA}$ is isomorphic to this algebra. Correcting a claim to the contrary in [18] the answer is "no", provided the Fokker-Planck operator is sufficiently well behaved to avoid explosions.

The full proof is long and will not be given here. It can be put together out of the following remarks.

(a) If we have a representation of the Lie algebra in terms of 2nd order operators we can find a basis for the first derived algebra which takes the form

$$m \frac{d^2}{dx^2} + n \frac{d}{dx} + r, \quad p \frac{d}{dx} + q, \quad h$$

(b) The commutation relations then imply certain differential equations relating the coefficients m,n,r,p,q,n.

(c) There is an explicit criterion for explosions, see McKean [23] page 65, which is incompatible with the conditions on m etc. implied by the differential equations.

Some related work, but apparently not this particular result occurs in Occone's thesis [24].

REFERENCES

1. R. Hermann, "On the Accessibility Problem in Control Theory," in Nonlinear Differential Equations and Nonlinear Mechanics, (J.P. LaSalle and S. Lefschetz eds.) Academic Press, N.Y., 1963.

2. H. Hermes and G.W. Haynes, "On the Nonlinear Control Problem with Control Appearing Linearly," SIAM J. on Control, Vol. I, (1963) pp. 85-108.

3. C. Lobry, Quelques Aspects Qualitatifs de la theories de la Commande, L'Universite Scientifique et Medicak de Grenoble, pour obtenir le titre de Docteur es Sciences Mathematiques, May 19, 1972.

4. H. Sussmann and V. Jurdjevic, "Controllability of Nonlinear Systems," J. of Diff. Eqs. Vol. 12, (1972), pp. 95-116.

5. A. Krener, "A Generalization of Chow's Theorem and the Bang-Bang Theorem to Nonlinear Control Problems," SIAM J. on Control, Vol. 12, No. 1 (1974), pp. 43-52.

6. R.W. Brockett, "System Theory on Group Manifolds and Coset Spaces," SIAM J. on Control, Vol. 10, No. 2, (1972) pp. 265-284.

7. H.J. Sussmann, "Existence and Uniqueness of Minimal Realizations of Nonlinear Systems," Math. Systems Theory, Vol. 10, No. 3, (1975), pp. 1476-147

8. A.J. Krener and R. Hermann, "Nonlinear Controllability and Observability," IEEE Trans. on Automatic Control, Vol. 22, No. 5, (1977) pp. 728-740.

9. J. Wei and E. Norman, "On the Global Representation of the Solutions of Linear Differential Equations as a Product of Exponentials," Proc. Am. Math. Soc., April 1964.

10. K.T. Chen, "Decomposition of Differential Equations," Math. Ann., Vol. 146, pp. 263-278, 1962.

11. L. Hörmander, "Hypoelliptic Second Order Differential Equations," Acta Math., Vol. 119, (1967), pp. 147-171.

12. M.I. Freedman and J.C. Willems, "Smooth Representation of Systems with Differential Inputs," IEEE Trans. on Aut. Control, Vol. 23, (1978), pp. 16-21.

13. H.J. Sussmann, "On the Gap Between Deterministic and Stochastic Ordinary Differential Equations," Ann. of Probability, Vol. 6, (1978), pp. 19-41.

14. H. Doss, "Liens entre équations différentielles stochastiques et ordinaires," Ann. Inst. H. Poincaré, 13, pp. 99-125.

15. R.W. Brockett and J.M.C. Clark, "The Geometry of the Conditional Density Equations," Proc. of the Oxford Conf. on Stochastic Systems, Oxford, England, 1978.

16. M.W. Wonham, "Some Applications of Stochastic Differential Equations to Optimal Nonlinear Filtering," SIAM J. on Control, Vol. 2, (1965).

17. R.W. Brockett, "Lie Theory and Control Systems Defined on Spheres," SIAM J. on Applied Mathematics, Vol. 25, No. 2, Sept. 1973, pp. 213-225.

18. R.W. Brockett, "Classification and Equivalence in Estimation Theory," 1979 Conf. on Decision and Control, IEEE, N.Y., 1979.

19. R. Howe, "On the Role of the Heisenberg Group in Harmonic Analysis," Bulletin of the AMS, Vol. 3, (1980), pp. 821-843.

20. R.W. Brockett, "Remarks on Finite Dimensional Nonlinear Estimation," in Analyse des Systèmes, Astérisque, Vol. 75-76, 1980, pp. 47-55.

21. S.K. Mitter, "On the Analogy Between Mathematical Problems of Nonlinear Filtering and Quantum Physics," Ricerche di Automatica (to appear).

22. V. Benes, "Exact Finite Dimensional Filters for Certain Diffusions with Nonlinear Drift," Stochastics (to appear).

23. H.P. McKean Jr., Stochastic Integrals, Academic Press, N.Y. 1969.

24. D. Ocone, Topics in Nonlinear Filtering Theory, Sc.D. Thesis, Massachusetts Institute of Technology, 1980.

NON-LINEAR FILTERING AND STOCHASTIC MECHANICS (1)

Sanjoy K. Mitter

Department of Electrical Engineering and Computer Science
and Laboratory for Information and Decision Systems
Massachusetts Institute of Technology
Cambridge, Mass. 02139

Table of Contents

---

(1) This research has been supported by the Air Force Office of Scientific Research under Grant AFOSR 77-3281B and by the U.S. Dept. of energy, Contract ET-76-C-01-2295.

M. Hazewinkel and J. C. Willems (eds.), Stochastic Systems: The Mathematics of Filtering and Identification and Applications, 479–503.

1. INTRODUCTION

Many of the basic ideas of non-linear filtering for diffusion processes were developed in the early and mid sixties by Stratanovich, Kushner, Wonham and others [for references, see the paper by Davis and Marcus, this volume]. Indeed this line of development could be considered as an extension of linear filtering as viewed by Kalman and Stratanovich in the sense that non-linear filtering is considered as a theory of conditional Markov processes. In the late sixties, the innovations approach to non-linear filtering was emphasized by Kailath (and Frost in his Stanford doctoral dissertation). The idea of using the innovations in an essential way in the Gaussian case dates back to Wold and Kolmogoroff in a discrete-time situation and to Bode-Shannon in the continuous situation. In this approach the observation process is first whitened in a causal and causally-invertible fashion and the form of the filter then becomes transparent. The contribution of Kailath was to see that this approach extended to a non-Gaussian situation and to conjecture that the whitening of the observation process in a causal and causally invertible manner could also be carried out in the non-linear case.

This conjecture has recently been proved by D. Allinger and the author [1] and leads to the most transparent derivation of the non-linear filtering equations (at least, when the signal and noise are independent).

It was realized by Fujisaki and Kunita that something weaker than observation-innovations equivalence would suffice to prove the basic representation results of non-linear filtering, namely that all square-integrable martingales adapted to the observations could be represented as a stochastic integral on the innovations, the integrand being square integrable and adapted to the observations. The celebrated paper of Fujisaki, Kallianpur and Kunita [2] brings to a culmination the innovations approach to non-linear filtering of diffusion processes.

This approach emphasizes the semi-martingale representation of the filter and is rooted in the intrinsic form of the Ito-differential rule due to Kunita and Watanabe. It has the disadvantage that it uses the innovations process which is an "invariant" but derived object. In most non-linear situations of interest this is not explicitly computable.

An alternative approach to the non-linear filtering problem can be traced back to the pioneering doctoral dissertations of Mortensen [3] and Duncan [4] and the important paper of Zakai [5], and is striking in its similarities to the path-space approach to Quantum Mechanics due to Feynman and its rigorization using

Wiener space ideas by Kac and Ray. In this development one works directly with the observations process and computes an unnormalized conditional density using path space integration. The computation of the actual estimate requires a further integration. The recent developments of non-linear filtering have focussed on this approach. In a previous paper [6] the author has given a systematic expository account of this point of view and many of the ideas presented in that paper has been further developed [see, for example the papers of Davis, and of Marcus and Hazewinkel in this volume]. It is also this view that led Beneš [7] to discover finite-dimensional filters for a class of non-linear filtering problems.

This viewpoint is completely consistent if not identical to the viewpoint of quantum mechanics as stochastic mechanics and quantum field theory as euchidean (stochastic) field theory. For a lucid account of these ideas see F. Guerra [8]. To make the identification, it is necessary to admit open quantum systems, that is, admit both self interactions and external interactions. Indeed, it is correct to think of the observation of a stochastic process as producing an external interaction. Just as non-trivial Markov (euclidean) fields are constructed from the free field using a Multiplicative functional transformation (see for example, Nelson [9]), similarly non-trivial filtering problems arise out of a time-dependent multiplicative functional transformation (see the article of Davis, this volume). In this framework the Kalman Filter occupies the same role as the harmonic oscillator does in quantum mechanics or the free field does in quantum field theory.

It is therefore hardly surprising that the Heisenberg algebra, the Oscillator algebra and other Lie algebras and their infinite-dimensional representations have a central role to play in this theory. Indeed in field theory, the Nelson-Feynman-Kac formula provides such a representation. The infinite-dimensional representations one seeks in filtering theory are however semi-group representations which are positivity prpeserving (leaves a certain cone invariant), and these are obtained from the Bayes formula due to Kallianpur and Striebel:

The final topic in this line of thought is the question of variational principles for non-linear filtering and the duality between filtering and control. These ideas date back to Bryson-Frazier [10] and Mortensen [11] and more recently to Hijab [12] but has never been satifactorily resolved. We show in this paper that indeed there exists a variational principle for non-linear filtering and that the equation for the unnormalized conditional density (Duncan-Mortensen-Zakai equation) is closely connected to a Bellmann- Hamilton-Jacobi equation can be replaced by a deterministic Hamilton-Jacobi equation and this is the fundamental idea behind the duality between filtering and control in the Gaussian case. These questions were touched upon in the author's previous paper [loc. cit.] but receive for the first time a complete resolution in this paper. As a by-product we obtain a stochastic variational principle for the Guerra-Nelson Stochastic Mechanics (at least in the ground state) and also stochastic control analogues of Benes' filtering problems. These ideas also allow us to study the behaviour of the unnormalized conditional density in the presence of small process and observation noise, and it is clear that there is an analogue of quasi-classical approximations of quantum mechanics in filtering theory.

The most important avenue of generalization of these ideas is in the context of diffusion processes on manifolds. This generalization seems to be necessary both for treating new filtering problems as well as stochastic mechanics.

2. FORMULATION OF NON-LINEAR FILTERING PROBLEM

The filtering problems we consider are consistent with the general model considered by Davis and Marcus in this volume, excepting we specialize the model for the signal.

Let $(\Omega, \mathcal{F}, P)$ be a complete probability space equipped with an increasing family of σ-fields. We shall generally be considering stochastic processes on a fixed time interval $[0,T]$, except in the section dealing with stochastic mechanics.

We consider the following stochastic differential system describing the model of the signal and observation processes:

$$dy_t = z_t dt + d\eta_t \qquad \text{(Observation)} \tag{2.1}$$

$$z_t = h(x_t) \qquad \text{(Signal)} \tag{2.2}$$

$$dx_t = b(x_t)dt + dw_t \tag{2.3}$$

We make the following assumptions (for simplicity)

(H.1) y_t and x_t are real-valued processes

(H.2) $(w_t, \mathcal{F}_t)$ and $(\eta_t, \mathcal{F}_t)$ are independent standard Brownian motions

(H.3) $E \int_0^T |h(x_t)|^2 dt < \infty$

(H.4) x_t and η_t are independent

(H.5) $b(x_t) = f_x(x_t)$, where f_x denotes the derivative with respect to x and equation (2.3) has a unique strong solution. Further the process x is assumed to have a density p(t,x).

Remark: For much of what we do in the sequal, it is enough to assume that (2.3) has a unique weak solution and the assumption $b = f_x$ can be taken in the sense of distributions.

For simplicity we have assumed that the processes involved are scalar-valued. There is no difficulty in generalizing what follows to vector-valued processes.

Then from Theorem 6 and Example 4 of Davis and Marcus, this volume, the unnormalized conditional density $q(t,x,\omega,y_0^t)$ (where the arguments ω and y_0^t will be omitted) satisfies the stochastic partial differential equation:

$$dq(t,x) = L_0^* q(t,x)dt + L_1 q(t,x) \cdot dy_t \ , \tag{2.4}$$

where

$$\begin{cases} (L_0^*\phi)(x) = \dfrac{1}{2}\dfrac{\partial^2\phi}{\partial x^2} - \dfrac{\partial}{\partial x}(b(x)\phi(x)) - \dfrac{1}{2}h^2(x)\phi(x) \\ \\ (L_1\phi)(x) = h(x)\phi(x) \end{cases} \tag{2.5}$$

and $\cdot$ denotes Stratonovich differential.

We assume

$$q(0,x) = q_0(x) > 0. \tag{2.6}$$

Understanding the invariance properties of equation (2.4) and its explicit solution is the fundamental problem of non-linear filtering. We however mention that computing an estimate $E[\phi(x_t)|F_t^y] \triangleq \hat{\phi}_t$, where $E\int_0^T \phi(x_t)^2 dt < \infty$ requires a further integration

$$\hat{\phi}_t = \int_{\mathbb{R}} \phi(x)q(t,x)\,dx \tag{2.7}$$

and a normalization.

The remainder of the paper is devoted to an understanding of equation (2.4).

3. THE FEYNMAN-KAC FORMULA, GIRSANOV FORMULA AND GUERRA-NELSON STOCHASTIC MECHANICS

We first try to understand the autonomous (no external inputs) system:

$$\frac{\partial\rho(t,x)}{\partial t} = L_0^*\rho(t,x) \qquad (3.1)$$

Let $\psi(x) = e^{\int_0^x b(z)dz}$ and write $\rho(t,x) = \psi(x)\hat{\rho}(t,x)$. Then $\hat{\rho}$ satisfies the equation

$$\frac{\partial\hat{\rho}}{\partial t} = \left(\frac{1}{2}\frac{\partial^2}{\partial x^2} - V(x)\right)\hat{\rho}(t,x), \quad \text{where} \qquad (3.1)$$

$$V(x) = \frac{1}{2}h^2 + \frac{1}{2}[b_x(x) + b^2(x)] = \qquad (3.2)$$

$$\frac{1}{2}[f_{xx}(x) + f_x^2(x)] + \frac{1}{2}h^2(x)$$

(since b is assumed to be a gradient vector field f_x).

We remark that the operator $H = -\frac{1}{2}\frac{d^2}{dx^2} + V(x)$ is a Schrodinger operator and equation (3.2) has an important role to play in Euclidean (Quantum) mechanics.

Let us make the assumption:

(H6) $V \in L^2_{loc}(\mathbb{R})$, positive, and $\lim_{|x|\to\infty} V(x) = \infty$

Theorem 3.1: [13, Theorem XIII.47, p.207]

The operator $H = -\frac{1}{2}\frac{d^2}{dx^2} + V$ considered as an operator on $L^2(\mathbb{R}, dx)$ has an eigenvalue as the lowest point in its spectrum and the corresponding eigenfunction $\psi_0(x)$ is strictly positive.

The lowest eigenfunction is referred to as the ground state. Let us normalize the lowest eigenvalue to be 0 and hence the corresponding eigenfunction $\psi_0(x) > 0$ satisfies

$$-\frac{1}{2}\frac{d^2\psi_0}{dx^2} + V\psi_0 = 0$$

Then by the Feyman-Kac formula [14, Theorem x.68, p. 279]

$$\psi_0(x) = \int_{\mathcal{W}} \psi_0(x_t)\exp(-\int_0^t V(x_s)ds)d\mu_w^x$$

where $\mathcal{W} = C(\mathbb{R}_+;\mathbb{R})$ equipped with its family of Borel sets $\mathcal{B}$ corresponding to the topology of uniform convergence on compacts and μ_w^x is Wiener measure starting at x.

Our objective is to construct a stochastic process associated with the operator $H = -\frac{1}{2}\frac{d^2}{dx^2} + V$. Let us normalize ψ_0 such that $\int_{\mathbb{R}} |\psi_0(x)|^2 dx = 1$. Define the probability measure $d\mu = |\psi_0(x)|^2 dx$ and consider the space $L^2(R;\mu)$. Now, the spaces $L^2(\mathbb{R};dx)$ and $L^2(\mu)$ are unitarily equivalent under the unitary operator $U: f \to \psi_0^{-1}f: L^2(\mathbb{R};dx) \to L^2(\mathbb{R};\mu)$. On $L^2(\mathbb{R};d\mu)$ the operator H is equivalent to $H' = UHU^{-1}$ (assuming the lowest eigenvalue = 0) and $H'\phi = -\frac{1}{2}\frac{d^2\phi}{dx^2} + f_x\frac{d}{dx}$, where $f = -\ln\psi_0$.

H' is a contraction semigroup on $L^2(\mathbb{R};d\mu)$.

We now construct the stochastic differential equation defining the Markov process corresponding to the operator H' by exploiting the relationship between the Feynman-Kac Formula and the Girsanov formula and the generalized Ito-Differential rule due to Krylov and others (for the generalized Ito differential rule, see [15]).

Now the function $\psi_0 \in \mathcal{D}(H)$, we get since $f = -\ell n\psi_0$, that (i) f is continuous (ii) f_x in the sense of distributions belongs to $L^2_{loc}(\mathbb{R};dx)$ and (iii) f_{xx} in the sense of distributions belongs to $L^1_{loc}(\mathbb{R};dx)$.

By direct calculation using $H' = UHU^{-1}$ and $f = -\ln\psi_0$, we get

$$-\frac{1}{2} f_{xx} + \frac{1}{2}(f_x)^2 = V(x) \text{ almost all } x \in \mathbb{R} \tag{3.4}$$

Now

$$\text{(3.5)} \qquad \exp(-\int_0^t V(x_s)ds) = \exp(\frac{1}{2}\int_0^t f_{xx}(x_s)ds - \frac{1}{2}\int_0^t f_x^2(x_s)ds$$

Applying the generalized Ito Differential Rule to $f(x_s)$, we get

$$df(x_s) = f_x(x_s)dx_s + \frac{1}{2} f_{xx}(x_s)ds \quad . \tag{3.6}$$

Hence from (3.5) and (3.6)

$$\exp(-\int_0^t V(x_s)ds) = \exp\,[f(x_t)-f(x_0) - \int_0^t f_x(x_s)dx_s - \frac{1}{2}\int_0^t f_x^2(x_s)ds]$$

and hence

$$\exp[-\int_0^t f_x(x_s)dx_s - \frac{1}{2}\int_0^t f_x^2(x_s)dx]$$

$$= \exp(-f(x_t))\exp(f(x_0)\exp(-\int_0^t V(x_s)ds)$$

$$= \psi(x_0)^{-1}\psi(x_t)\exp(-\int_0^t V(x_s)ds) \quad .$$

Define

$$L_t = \psi(x_0)^{-1}\psi(x_t)\exp(-\int_0^t V(x_s)ds). \tag{3.7}$$

Now L_t is (i) a well defined random variable μ_w^x - a.s. for all $x \in \mathbb{R}$ (ii) positive and from (3.3) (iii) $\int L_t d\mu_w^x = 1$.

Therefore from the properties of Wiener Process (as a Markov process), L_t is a $(\mathcal{W}, \mathcal{B}_t, \mu_w^x)$ - martingale, where $\mathcal{B}_t$ is the sigma field generated by the coordinate functions of x after time t: Hence we may define a probability measure on $(\Omega,\mathcal{B})$ by

$$\left.\frac{d\nu^x}{d\mu_w^x}\right|_{\mathcal{F}_t} = L_t \quad .$$

Therefore by the Girsanov theorem the process w_t, $t \geq 0$ defined by

$$w_t = x_t - x_0 + \int_0^t f_x(x_s)ds$$

is a $(\mathcal{B}_t, \nu^x)$ - Brownian motion and therefore the process x_t, $t \geq 0$, considered as a stochastic process on $(\mathcal{W}, \mathcal{B}_t, \nu^x)$ is a weak solution of the stochastic differential equation

$$dx_t = -f_x(x_t)dt + dw_t \quad . \tag{3.8}$$

By construction, this equation has the unique invariant measure μ.

From our constructions, we see that

(a) $\int_{\mathcal{W}} \phi(x_t)d\nu^x = \int_{\mathcal{W}} \phi(x_t)L_t d\mu_w^x$, for all bounded continuous functions ϕ.

(b) the process x_t has a transition density $q(t,y;0,x)$ given by

$$q(t,y;0,x) = \psi^{-1}(x)\psi(y)E_{\mu_w^x}[\exp(-\int_0^t V(x_s)ds)\,|\,x_t = y] \times \times p(t,y;0,x) \quad . \tag{3.9}$$

where $E_{\mu^x}[\cdot|\cdot]$ denotes conditional expection with respect to the measure μ_w^x conditioned on $\{x_t = y\}$ and $p(t,y;0,x)$ is the transition density of the Wiener process. It follows from the work of Carmona [16] that q is a continuous functionof x and y.

We can summarize what we have done in:

<u>Theorem 3.2</u>: Under hypothesis (H6), there exists a unique family of probability measures $(\nu_x | x \in \mathbb{R})$ on the canonical probability space $(\mathcal{W},\mathcal{B})$ such that

(i) $(\mathcal{W},\mathcal{B},\mathcal{B}_t,x_t,\nu_x)$ is a strong Markov process;

(ii) the martingale problem corresponding to (3.8) has a unique solution;

(iii) the Markov process x_t has a unique invariant measure μ;

(iv) the process x_t is symmetric (i.e., the reverse Markov process is itself).

3.1 The case where V(x) is quadratic

In (2.2) and (2.3) let us assume

$$h(x) = x \tag{3.10}$$

$$b_x + b^2 = Q(x),\ \text{where } Q(x) \text{ is a positive quadratic function} \tag{3.11}$$

In this case V(x) = a positive quadratic function. To simplify matters, let us assume that $V(x) = x^2$. This case corresponds to the (euclidean) Harmonic oscillator. The ground state $\psi_0(x)$ can be explicitly computed to be $(\pi)^{-1/4} \exp\left(-\frac{x^2}{2}\right)$ and $H' = -\frac{1}{2}\frac{d^2}{dx^2} + x\frac{d}{dx}$ and is a contraction on $L^2(\mathbb{R};\pi^{-1/2}\exp(-x^2)dx)$.

We note that the corresponding Markov process is the Ornstein-Uhlenbeck process

$$dx_t = -x_t dt + dw_t \tag{3.12}$$

The ideas presented in this example have played an essential role in the original discovery by Benes of explicit finite-dimensional filters for a class of non-linear problems.

Furthermore the Ornstein-Uhlenbeck operator is an example of a self-adjoint operator which generates a hypercontractive semi-group on $L^2(\mathbb{R}; d\mu)$ [14, Theorem X.56, p.260].

3.2 Discussion

Our previous development is at the heart of Nelson's stochastic mechanics. Let us first note that the operator $H = -\frac{1}{2}\frac{d^2}{dx^2} + V(x)$ corresponds to the generator of a multiplicative functional of the Wiener process. What we have shown in Section 3 is that this operator is unitarily equivalent to a self-adjoint Markov semi-group on a suitable $L^2(\mathbb{R};\mu)$ space and we have explicitly constructed that Markov process. This Markov process is symmetric in the sense that the reversed Markov process is itself. This follows from the fact that the operator unitarily equivalent to

H on $L^2(\mathbb{R};\mu)$ is self-adjoint (μ is the unique invariant measure of the Markov process). It is in this sense that the reversibility of quantum mechanics is preserved in stochastic mechanics. Moreover the expectation values of all quantum mechanical observables in the ground state that can be computed using the quantum mechanical formalism can also be computed in terms of the measure on the path space of the corresponding stochastic process. Finally, the field operators can be constructed from the stochastic formalism. We emphasize that what we have really done is shown the relationship between the Feynman-Kac formula and the Girsanov formula.

3.3 An Associated Stochastic Control Problem

Our main objective in this section is to give a variational interpretation of Nelson's stochastic mechanics. The key to this is the remark that equation (3.4) is a stationary Bellman equation arising out of a stochastic control problem.

We first consider the non-stationary situation. In equation (3.1) let us make the transformation

$$\hat{\rho}(t,x) = \exp\left(-S(t,x)\right) \tag{3.13}$$

Then $S(t,x)$ satisfies the Bellman equation:

$$\begin{aligned}\frac{\partial S}{\partial t} &= \frac{1}{2}\frac{\partial^2 S}{\partial x^2} - \frac{1}{2}\left(\frac{\partial S}{\partial x}\right)^2 + V(x); \quad S(0,x) = S_0(x) \\ &\qquad\qquad\qquad\qquad\qquad\qquad = -\ln \rho_0(x)\end{aligned} \tag{3.14}$$

It is also worth observing that $\frac{\partial S}{\partial x}$ satisfies the "Navier-Stokes-like" equation:

$$\begin{cases}\dfrac{\partial}{\partial t}\left(\dfrac{\partial S}{\partial x}\right) = \dfrac{1}{2}\dfrac{\partial^2}{\partial x^2}\left(\dfrac{\partial S}{\partial x}\right) - \dfrac{1}{2}\dfrac{\partial S}{\partial x}\dfrac{\partial}{\partial x}\left(\dfrac{\partial S}{\partial x}\right) + \dfrac{\partial V}{\partial x} \\[2ex] \dfrac{\partial S}{\partial x}(0,x) = -\dfrac{1}{\rho_0(x)}\dfrac{\partial \rho_0}{\partial x}\end{cases} \tag{3.15}$$

We now make the assumption that the potential V in addition to satisfying the previous hypotheses is convex and of class C^1 (the case of most interest is where V is an even positive polynomial).

The development that follows is essentially due to Fleming [17] and Karatzas [18] and we explicitly follow Karatzas.

Consider the stochastic control problem:

$$\begin{cases} dx_t = u_t dt + dw_t \\ x_s = x_0 \end{cases} \qquad s \le t \le T \tag{3.16}$$

As before, let $\mathcal{W} = C(0,T)$ equipped with its Borel σ-algebra and let $\mathcal{B}_t$ denote the σ-algebra generated by the coordinate functions. Consider also the σ-field $\mathcal{D}$ of subsets of $[0,T] \times \mathcal{W}$ with the property that each t-section belongs to $\mathcal{B}_t$ and each x-section is Lebesgue measurable.

An admissible control function

$$\bar{u}: ([0,T] \times \mathcal{W}, \mathcal{D}) \to \big(\mathbb{R}, \mathcal{B}(\mathbb{R})\big)$$

is a measurable map such that $\bar{u}(x,t) = u_t$ and the stochastic differential equation

$$dx_t = \bar{u}(x,t)dt + dw_t \tag{3.17}$$

has a unique weak solution $(x_t, w_t | 0 \le t \le T)$ on some probability space $(\Omega, \mathcal{F}_T, P_x^u, \mathcal{F}_t)$ for any $x \in \mathbb{R}$, with $(w_t, \mathcal{F}_t)$ a Wiener process and

$$E_x^u \int_0^T |\bar{u}(x,t)|^p dt < \infty$$

$$\sup_{0 \le t \le T} E_x^u |x_t|^p < \infty \quad .$$

Let U denote this class of controls.

Consider the problem of minimizing

$$J(x,t;\bar{u}) = E_x^{\bar{u}} \int_t^T \left(\frac{1}{2} u_s^2 + V(x_s) \right) dS + E_x^{\bar{u}} (S_0(x_T))$$

$$0 \le t \le T$$

(To be consistent with equation (3.14) we should reverse time; S_0 is as defined in (3.14)).

Let

$$S(x,t) = \inf_{\bar{u} \in U} J(x,t,\bar{u})$$

Then S satisfies the Bellman equation

$$\begin{cases} \dfrac{\partial S}{\partial t} = \dfrac{1}{2}\dfrac{\partial^2 S}{\partial x^2} + \min\limits_{u \in \mathbb{R}} \left(u \dfrac{\partial S}{\partial x} + \dfrac{1}{2} u^2 \right) + V(x), \quad (x,t) \in \mathbb{R} \times [0,T] \\ S(x,T) = S_0(x) \quad . \end{cases} \tag{3.18}$$

It is shown by Karatzas [loc. cit.] that the optimal control law

$$u^*(x) = -\frac{\partial S}{\partial x}(x_t^*, t)$$

is admissible and the corresponding stochastic differential equation (3.17) has a unique strong solution.

Consider now the infinite-time problem with the cost function:

$$J(x;\bar{u}) = \lim_{T \to \infty} \frac{1}{T} E_x^{\bar{u}} \int_0^T [\frac{1}{2} u_t^2 + V(x_t^{\bar{u}})]dt \tag{3.19}$$

The minimization is now carried out over all admissible control laws which are Markovian and which given rise to an ergodic x_t process. This class is characterized by control laws $u_t = \bar{u}(x_t)$, such that $F^u(\infty) < \infty$ where

$$F^u(x) \triangleq \int_{-\infty}^{\infty} \exp\{2 \int_0^y \bar{u}(z)dz\}dy, \quad \text{and} \tag{3.20}$$

$$\int_{-\infty}^{\infty} \{\bar{u}^2(x) + V(x)\}dF^u(x) < \infty \tag{3.21}$$

Again it is shown by Karatzas that an optimal control law u* exists such that the limit in (3.19) is independent of the starting point x. Moreover, the optimal control is given by

$$u^*(x_t^*) = -\frac{\partial S}{\partial x}(x_t^*), \quad \text{where } S \tag{3.22}$$

is the solution of the stationary Hamilton-Jacobi equation

$$\frac{1}{2}\frac{\partial^2 S}{\partial x^2} - \frac{1}{2}\left(\frac{\partial S}{\partial x}\right)^2 + V(x) = 0 \quad . \tag{3.23}$$

It is interesting to note that from our previous considerations, (see equation (3.4) and the definition of f) we know the optimal control explicitly,

$$u^*(x) = (\psi_0(x))^{-1} \frac{\partial \psi_0}{\partial x}, \tag{3.24}$$

where ψ_0 is the non-degenerate ground state of the operator $-\frac{1}{2}\frac{d^2}{dx^2} + V(x)$.

In the case $V(x) = \frac{1}{2}x^2$, we get $u^*(x) = -x$, a well known result in stochastic control theory.

Our final observation is that to this class of finite-time stochastic control problems we can rigorously apply Bismut's duality theory of stochastic control [19]. According to this theory, there exists an adapted right continuous process p_t and and an adapted measurable process π_t, $E\int_0^T \pi_t^2 dt < \infty$ such that

$$\begin{cases} dx_t = u_t^* dt + dw_t \\ x_0 = x \end{cases} \tag{3.25}$$

$$\begin{cases} dp_t = -\frac{\partial V}{\partial x}(x_t)dt + \pi_t dw_t \\ p_T = \frac{\partial S_0}{\partial x}(x_T) \end{cases} \tag{3.26}$$

$$u^* = \arg\min_{u \in \mathbb{R}} \left(\frac{1}{2}u^2 + u.p\right) \tag{3.27}$$

Equations (3.25) - (3.27) are the "stochastic" bi-characteristics of the Bellman equation.

What we have shown in this section is that certain multiplicative functionals of Brownian motion (and indeed more general Markov processes) have associated stochastic canonical equations of motion, and in this sense stochastic mechanics is exactly like classical mechanics.

4. A NON-LINEAR STOCHASTIC CONTROL PROBLEM WITH AN EXPLICIT SOLUTION

We now consider a class of stochastic control problems which are the analogues of non-linear filtering problems first considered by Benes which have an explicit solution. This solution is obtained by exploiting the ideas of Section 3. A prototype example of this class is:

$$dx_t = f(x_t)dt + u_t dt + dw_t \tag{4.1}$$

where f satisfies the Riccati equation

$$\frac{df}{dx} + f^2 = x^2 \quad . \tag{4.2}$$

The cost function is

$$J(T,x;u) = E_x^u\left(\frac{1}{2}x_T^2\right) + E_x^u\int_0^T\left(\frac{1}{2}u_s^2 + \frac{1}{2}x_s^2\right)ds$$

We shall place ourselves under the hypotheses of the previous section for the control laws.

If $S(t,x) = \operatorname{Inf}_u J(t,x;u)$, then the Bellman equation for S is

$$\begin{cases} \dfrac{\partial S}{\partial t} = \dfrac{1}{2}\dfrac{\partial^2 S}{\partial x^2} + f\dfrac{\partial S}{\partial x} - \dfrac{1}{2}\left(\dfrac{\partial S}{\partial x}\right)^2 + \dfrac{1}{2}x^2 \\ S(T,x) = \dfrac{1}{2}x^2 \end{cases} \tag{4.3}$$

Let us introduce the transformation

$$S(t,x) = -\ln \rho(t,x)$$

Then ρ satisfies the equation

$$\begin{cases} \dfrac{\partial \rho}{\partial t} = \dfrac{1}{2}\dfrac{\partial^2 \rho}{\partial x^2} + f\dfrac{\partial \rho}{\partial x} - \dfrac{1}{2}x^2\rho \\ \rho(T,x) = \exp\left(-\dfrac{x^2}{2}\right). \end{cases} \tag{4.4}$$

Now we remove the drift term in the above equation by introducing the Gauge transformation

$$\rho(t,x) = \psi(x)\hat{\rho}(t,x), \quad \text{where}$$

$\psi \in C^{\infty}(\mathbb{R})$ invertible is to be chosen. A direct computation shows that $\hat{\rho}$ satisfies:

$$\frac{\partial\hat{\rho}}{\partial t} = \frac{1}{2}\frac{\partial^2\hat{\rho}}{\partial x^2} + \left(\psi^{-1}\frac{\partial\psi}{\partial x} + f\right)\frac{\partial\hat{\rho}}{\partial x} + \left(\frac{1}{2}\psi^{-1}\frac{\partial^2\psi}{\partial x^2} + \psi^{-1}\frac{\partial\psi}{\partial x}f - \frac{1}{2}x^2\right)\hat{\rho}$$

$$\hat{\rho}(T,x) = \psi^{-1}\rho(T,x) \tag{4.5}$$

Now choose ψ to satisfy

$$\frac{\partial\psi}{\partial x} + f\psi = 0$$

Then $\hat{\rho}$ satisfies

$$\frac{\partial\hat{\rho}}{\partial t} = \frac{1}{2}\frac{\partial^2\hat{\rho}}{\partial x^2} + \frac{1}{2}\left(\frac{\partial f}{\partial x} + f^2 - x^2\right)\hat{\rho}$$

$$= \frac{1}{2}\frac{\partial^2\hat{\rho}}{\partial x^2} - \frac{1}{2}x^2\hat{\rho} \tag{4.6}$$

This backward equation has a unique solution given by the Feynman-Kac formula:

$$\hat{\rho}(t,x) = E_{tx}[\hat{\rho}(x_T)\exp(-\int_t^T \frac{1}{2}x_s^2 ds)], \text{ where} \tag{4.7}$$

E_{tx} denotes expectation with respect to Wiener measure conditioned on $x_t = x$.

The integration in (4.7) can be carried out by using Gaussian integrals or by using the method of bicharacteristics introduced by the author in [6] (see section 3.2). By transforming back we get an explicit solution for S.

The developments in this section show that the fundamental solution to (4.2) can be written down in terms of a Riccati equation that arises in the gain computation in optimal control and Kalman filtering (not to be confused with the Riccati equation (4.2)).

5. LIE ALGEBRAIC CONSIDERATIONS

The fundamental Hamiltonian in the previous considerations is the Hamiltonian $H = -\frac{1}{2}\frac{d^2}{dx^2} + \frac{1}{2}x^2$ acting on $L^2(\mathbb{R}; dx)$ or the unitarily equivalent operator $H = -\frac{1}{2}\frac{d^2}{dx^2} + x\frac{d}{dx}$ acting on $L^2(\mathbb{R}; dg)$ where g is Gauss measure. This Hamiltonian corresponds to the Harmonic Oscillator. Underlying the Harmonic oscillator is the solvable Lie algebra with basis $\{r,p,q,i\}$ the oscillator algebra whose commutation relations are

$$\begin{cases} [r, p] = q \\ [p, q] = i \\ [r, q] = p. \end{cases} \tag{5.1}$$

A representation of this Lie Algebra by a Lie Algebra of unbounded operators is obtained by the correspondence

$$r \to L_0 = -\frac{1}{2}\frac{d^2}{dx^2} + \frac{1}{2}x^2$$

$$q \to L_1 = x$$

$$p \to L_2 = \frac{d}{dx}$$

$$i \to L_3 = I$$

Let T denote this representation. Let G denote the connected Lie group whose Lie algebra is the oscillator algebra. We are interested in a representation π of G by bounded operators on $L^2(\mathbb{R}; g)$, such that $\pi(g)$, $g \in G$ is a C_0-semi-group and $\pi(e^{tr}) = e^{-tT(r)}$ is a positivity preserving contraction semi-group.

The semi-group e^{-tL_0} can be constructed via the Feynman-Kac formula. The interest of the Lie algebraic viewpoint is that a finite dimensional sufficient statistic can be obtained for evaluating

$(e^{-tL_0}\rho_0)(x)$, by considering the basis

$\{\frac{1}{2}\frac{d^2}{dx^2}, \frac{1}{2}x^2, x\frac{d}{dx}, I\}$ for the oscillator algebra and writing

in analogy with the Wei-Norman theory (as first suggested by Brockett)

$$(e^{-tL_0}\rho_0)(x) = [e^{g_1(t)\frac{d^2}{dx^2}} e^{g_2(t)x^2} e^{g_3(t)x\frac{d}{dx}} e^{g_4(t)} \rho_0](x)$$

and obtaining differential equations for g_1, g_2, g_3 and g_4. Indeed g_1 is t and required to be nonnegative. For the oscillator' algebra this has been rigorously proved in the doctoral dissertation of D. Ocone [20].

Consider now the Lie algebra of operators with generator

$-\frac{1}{2}\frac{d^2}{dx^2} + V(x)$ and x , with V an even positive polynomial (but not quadratic). The work of Avez and Heslot [21] suitably modified shows that the corresponding Lie algebra is infinite-dimensional and simple. Hence it is unlikely that there are other examples of multiplicative functions of Brownian motion whose semigroups can be constructed using a finite dimensional sufficient statistic.

Indeed, it is not difficult to show that the only perturbations one can allow in the operators L_0 and L_1 so that the Lie algebra remains finite dimensional are $a\frac{d}{dx}$, x, x^2 (or their linear combinations; the only variable coefficient first order linear differential operator that we can allow must satisfy the condition $\frac{da}{dx} + a^2$ = quadratic).

6. NON-LINEAR FILTERING

It is a pleasant fact that the ideas expressed in the previous sections generalize in a very natural way to non-linear filtering theory. The main observation to make in that non-linear filtering theory corresponds to the (euchidean) quantum mechanical situations when we allow time-dependent random external interactions (in addition to self-interactions).

6.1 Pathwise Non-Linear Filtering

To give a variational interpreation of the non-linear problem it is necessary to consider the pathwise solution to non-linear filtering problem originally initiated by Clark (c.f. the paper of Davis this volume). For our purpose we could think , that in this approach, the stochastic integral in equation(2.4) is removed by a time-dependent Gauge transformation.

Define

$$\tilde{q}(t,x) = \exp(-h(x)y_t)q(t,x) \qquad (6.1)$$

Then $\tilde{q}(t,x)$ satisfies

$$\begin{cases} \dfrac{d\tilde{q}(t,x)}{dt} = \exp(-h(x)y_t)[L_0^* - \frac{1}{2}L_1^2](\exp(h(x)y_t\tilde{q}(t,x) \\ \tilde{q}(0,x) = \tilde{q}_0(x) = q_0(x) > 0 \end{cases} \quad (6.2)$$

We consider equation (6.2) for a fixed $y_{(\cdot)} \in C(0,T;\mathbb{R})$ and indeed if necessary we could approximate $y_{(\cdot)}$ by smoother functions.

It is convenient to write equation (6.2) in two other equivalent forms

$$\frac{d\tilde{q}(t,x)}{dt} = (L_0^* - \frac{1}{2}L_1^2)\tilde{q}(t,x) + y_t L_2\tilde{q}(t,x) - y_t^2 L_3\tilde{q}(t,x) \quad (6.3)$$

where

$$L_2 = [\frac{1}{2}L_0^* - \frac{1}{2}L_1^2, L_1] = \frac{dh}{dx}\frac{d}{dx} + \left(\frac{1}{2}\frac{d^2h}{dx^2} - b\frac{dh}{dx}\right)$$

$$L_3 = [L_1, L_2] = -\left(\frac{dh}{dx}\right)^2$$

$$\tilde{q}(0,x) = \tilde{q}_0(x) = q_0(x) > 0 .$$

and

$$\begin{cases} \dfrac{\partial\tilde{q}}{\partial t} = \dfrac{1}{2}\dfrac{\partial^2\tilde{q}}{\partial x^2} - \tilde{b}(t,x)\dfrac{\partial\tilde{q}}{\partial x} - V(t,x)\tilde{q}(t,x) \\ \tilde{q}(0,x) = q_0(x) > 0 , \end{cases} \quad (6.4)$$

where

$$\begin{cases} \tilde{b}(t,x) = b(x) - y_t\dfrac{dh}{dx} \\ V(t,x) = -\dfrac{1}{2}y_t^2\left(\dfrac{dh}{dx}\right)^2 + y_t b(x)\dfrac{dh}{dx} + \dfrac{db}{dx} + \dfrac{1}{2}h^2(x) . \end{cases} \quad (6.5)$$

Equation (6.3) exhibits the role of the commutators of $L_0^* - \frac{1}{2}L_1^2$ and L_1 and equation (6.4) shows that basically we are dealing with a situation not unlike that considered in Section 3.

6.2 Transformation into a Stochastic Control Problem

As in Section (3.3), introduce the transformation

$$\tilde{q}(t,x) = \exp(-S(t,x)) \qquad (6.6)$$

Then $S(t,x)$ satisfies the Hamilton-Jacobi-Bellman equation

$$\begin{cases} \dfrac{\partial S}{\partial t} = \dfrac{1}{2}\dfrac{\partial^2 S}{\partial x^2} - \tilde{b}(t,x)\dfrac{\partial S}{\partial x} - \dfrac{1}{2}\left(\dfrac{\partial S}{\partial x}\right)^2 + V(t,x) \\ S(0,x) = S_0(x) = -\ln q_0(x) \end{cases} \qquad (6.7)$$

Equation (6.7) corresponds to the stochastic optimal control problem

$$\begin{cases} dx_t = u_t dt + dw_t \\ x_s = x_0 \end{cases} \qquad s \leq t \leq T \qquad (6.8)$$

with cost function

$$J(s,x_0;\bar{u}) = E[\int_0^T L(t,x_t,u_t)dt + S_0(x_T)] \quad ,$$

where

$$L(t,x,u) = \frac{1}{2}[u + \tilde{b}(t,x)]^2 + V(t,x)$$

and the minimization is to be carried out over the class of Markov feedback controls

$$u_t = \bar{u}(t,x_t).$$

satisfying the conditions in section 3.3. To be consistent we should reverse time, as remarked in Section 3.

If h is a polynomial and if the drift b satisfies some **mild** conditions, then using the work of Fleming [23], it is possible to show that equation (6.7) has a solution with appropriate regularity. In this way we can prove that equation (6.4) has a solution in the strong (classical) sense. To prove uniqueness of (6.4) we can invoke a maximum principle argument directly on equation (6.4) (cf. [24] for maximum principles for equations with unbounded coefficients). It is worthwhile making the remark that transforming the Bellman equation into an equivalent Zakai equation and analyzing it may also be a useful tool for stochastic control problems. The details of these ideas will be presented in a joint paper with Wendell Fleming.

The relation between the pathwise equations of non-linear filtering and stochastic control introduced in this section explains in the clearest possible way the "duality" that exists between filtering and control. Previous difficulties in defining a likelihood functional because of the non-differentiability of Wiener paths are completely avoided using the pathwise equation (6.2) for fixed $y \in C(0,T;\mathbb{R})$.

6.3 Various Examples

Example 1 (Kalman Filtering)

$$\begin{cases} x_t = w_t \\ dy_t = x_t dt + d\eta_t \end{cases} \tag{6.10}$$

Here w_t and η_t are standard Brownian motions which are independent. Then from (6.5)

$$\tilde{b}(t,x) = -y_t$$

$$V(t,x) = -\frac{1}{2} y_t^2 + \frac{1}{2} x^2, \text{ and hence from (6.9)}$$

$$L(t,x_t,u_t) = \frac{1}{2} [u_t - y_t] - \frac{1}{2} y_t^2 + \frac{1}{2} x_t^2 = \frac{1}{2} u_t^2 + \frac{1}{2} x_t^2 - u_t y_t$$

and we have a stochastic control problem with a quadratic cost criterion. The theory of this is essentially the same as the theory of deterministic linear optimal control with a quadratic cost criterion.

Example 2 (Bilinear Filtering)

$$\begin{aligned} dx_t &= x_t\, dw_t \\ dy_t &= x_t dt + d\eta_t, \end{aligned} \tag{6.11}$$

with the same hypothesis as in Example 1.

For this problem the pathwise equation of non-linear filtering is:

$$\frac{\partial \tilde{q}}{\partial t} = \frac{1}{2} x^2 \frac{\partial^2 \tilde{q}}{\partial x^2} + (2x-y)\frac{\partial \tilde{q}}{\partial x} + \frac{1}{2} y^2 - \frac{1}{2} x^2 + 1 \tag{6.12}$$

and the corresponding Bellman equation is:

$$\frac{\partial S}{\partial t} = \frac{1}{2} x^2 \frac{\partial^2 S}{\partial x^2} + (2x-y_t) \frac{\partial S}{\partial x} - \frac{1}{2} x^2 \frac{\partial^2 S}{\partial x^2} + \frac{1}{2} x^2 - \frac{1}{2} y^2 - 1 \tag{6.13}$$

and the stochastic control problem is:

$$dx_t = x_t(u_t + dw_t) \tag{6.14}$$

$$L(t,x_t,u_t) = \frac{1}{2} (u_t + 2x_t - y_t)^2 + \frac{1}{2} x_t^2 - \frac{1}{2} y_t^2 - 1$$

These calculations also show that although the perfectly observable LQG-stochastic control problem with state dependent noise has a linear solution, this problem does not have a corresponding "dual" filtering problem.

Remarks

We may apply the Davis-Varaiya theory or the Bismut theory to obtain necessary conditions of optimality for problem (6.8) - (6.9) in the form of a maximum principle. This would give rise to stochastic bi-characteristics for equation (6.7) and in general these are necessary to compute the solution of (6.7) or equivalently (6.4). It appears that only for Kalman filtering or Benes problems is it possible to reduce these to the characteristics of an ordinary Hamilton-Jacobi equation parametrized by dy_t. This was done in the author's earlier paper cited in the introduction. The reason for this is the fact that the theorqy of deterministic LQ-control and perfectly observable LQG-control is essentially the same.

6.4 Remarks on Approximations and Perturbation Theory

The relationship between non-linear filtering and stochastic control appears to clarify various approximation schemes currently used and provides guidelines for their systematic analysis. For example, if in (6.7) $\tilde{b}$ is approximated by a linear function in x and V is approximated by a quadratic function in x, locally in t, then we have a Kalman filtering problem for a small time interval $[0,\tau]$, say. Having obtained $S(t,x)$, $t \in [0,\tau]$ the above approximation and iteration procedure can be continued. Thus the extended Kalman Filtering algorithm is the analogue of Newton's method.

Furthermore, the study of filtering problems of the form

$$dx_t = b(x_t)dt + \sqrt{\varepsilon}\, dw_t$$

$$dy_t = h(x_t)dt + \sqrt{\varepsilon}\, d\eta_t$$

can be reduced to the study of the corresponding stochastic control problems with small parameter ε and the asymptotic behaviour of $S(t,x;\varepsilon)$ as a function of ε can be studied using methods developed by Fleming.

6.5 Lie Algebraic Considerations

It is clear from the previous development that the Lie algebraic approach to the study of non-linear filtering problems is entirely analogous to our considerations in section 5 for the autonomous system. The reason for this is that the noise enters the Zakai equation in a "finite-dimensional" way and only has the function of parametrizing the manifold in which the Zakai equation is evolving. Therefore the Lie algebrda of the Kalman filter is the oscillator algebra and the class of examples considered by Benes give rise to Lie algebras which are gauge equivalent to the oscillator algebra.

7. FINAL REMARKS

We have shown in this paper that a close relationship exists between non-linear filtering theory and stochastic Hamilton-Jacobi theory. This work requires generalization in the direction of the study of filtering and control problems on Riemannian manifolds. A beginning in this direction has already been made by Duncan ([25], [26]). The most intriguing possibility however is to discover the analogues of completely integrable Hamiltonian systems in non-filtering and stochastic control.

Acknowledgment

This paper owes much to conversations with Wendell Fleming and portions of this paper are joint work with him.

References

1. D. Allinger and S.K. Mitter: New Results on the Innovations Problem for Non-Linear Filtering, to appear Stochastics, 1981.

2. M. Fujisaki, G. Kallianpur and H. Kunita: Stochastic Differential Equations for the Non-Linear Filtering Problem, Osaka J. of Math., Vol. 9, 1972, 19-40.

3. R. Mortensen: Doctoral Dissertation, Department of Electrical Engineering, University of California, Berkeley, California, 1966.

4. T.E. Duncan: Doctoral Dissertation, Department of Electrical Engineering, Stanford University, Stanford, California 1967.

5. M. Zakai: On the Optimal Filtering of Diffusion Processes, Z. Wahr. Verw. Gebiete, 11, 1969, 230-243.

6. S.K. Mitter: On the Analogy Between Mathematical Problems of Non-Linear Filtering and Quantum Physics, to appear Ricerche di Automatica, 1981.

7. V.E. Benes: Exact Finite Dimensional Filters for Certain Diffusions with Non-Linear Drift, to appear Stochastics, 1981.

8. F. Guerra: Structural Aspects of Stochastic Mechanics and Stochastic Field Theory, to appear in Proceedings of the 1980 Les Houches School on "Stochastic Differential Equations in Physics".

9. E. Nelson: Construction of Quantum Fields from Markoff Fields, J. Functional Analysis, 12, 1973, 97-112.

10. A.E. Bryson and M. Frazier: Smoothing for Linear and Non-Linear Dynamic Systems Proc. Optimum System Synthesis Comprence, Wright Patt. AFB, September 1962.

11. R.E. Mortensen: Maximum Likelihood Recursive Non-Linear Filtering, J. Optimization Theory and App., 2, 1968, 386-394.

12. O. Hijab: Minimum Energy Estimation, Doctoral Dissertation, University of california, Berkeley, California, 1980.

13. M. Reed and B. Simon: Methods of Modern Mathematical Physics, Vol. IV, Academic Press, New York, 1978.

14. M. Reed and B. Simon: Methods of Modern Mathematical Physics, Vol. II, Academic Press, New York, 1975.

15. P. Meyer: La Formule de Ito pur le Mouvement Brownian d'apres G. Brossmaler, Sem. Prob. Strasbourg 1976-77, Springer Lecture Notes in Math. 649, 1978, 763-769.

16. R. Carmona: Regularity Properties of Schrodinger and Dirichlet Semigroups, J. Functional Analysis 33, 1979, 259-296.

17. W. Fleming: Stochastically Perturbed Dynamical Systems, Rocky Mountain J. of Math. 4, 1974, 407-433.

18. I. Karatzas: On a Stochastic Representation for the Principal Eigenvalue of a Second Order Differential Equations, Stochastics, Vol. 3, 1980, 305-321.

19. J.M. Bismut: Analyse Convexe et Probabilité , Thèse, Faculté des Sciences de Paris, Paris, 1973.

20. D. Ocone: Topics in Non-Linear Filtering, Doctoral Dissertation, Mathematics Department, M.I.T., June 1980.

22. A. Avez and A. Heslot: L'algebre de Lie des Polynomes en les Coordonnee's Canonique muni de Crochet de Poisson, C.R. Acad. Science Paris, t.288, Serie A, Mai 1979, 831-833.

23. W. Fleming: Controlled Diffusions under Polynomial Growth Conditions in Control Theory and the Calculus of Variations, ed. A.V. Balakrishnan , Academic Press, New York, 1969.

24. D.G. Aronson and A. Besala: Uniqueness of Positive Solutions of Parabolic Equations with Unbounded Coefficients, Colloquium Mathematicum, Vol. XVIII, 1967, 125-135.

25. T.E. Duncan: Stochastic Systems in Riemannian Manifolds: J. Opt. Th. and Applns., 27, 1979, 399-426.

26. T.E. Duncan: Some Filtering Results in Riemann Manifolds, Information and Control, 35, 1977, 182-195.

Note added in proofs. The fact that the process corresponding to the optimal value function is a martingale has an important implication in this theory. Indeed a martingale is the analogue of a 'conserved quantity' and the interpretation says that the optimal conditional energy process is analogous to an integral of motion.

PATHWISE NON-LINEAR FILTERING

M.H.A. Davis

Department of Electrical Engineering,
Imperial College, London SW7 2BT, England.

CONTENTS

I. INTRODUCTION

This paper concerns the nonlinear filtering problem of calculating recursively estimates $E[f(x_t)|y_s, 0 \leq s \leq t]$ where x_t is a Markov process and y_t is a real-valued "observation process" given by

$$dy_t = h(x_t)dt + dw^{\circ}_t \tag{1.1}$$

Here h is a bounded function (additional smoothness assumptions

M. Hazewinkel and J. C. Willems (eds.), Stochastic Systems: The Mathematics of Filtering and Identification and Applications, 505–528.

will be imposed later) and $\overset{\circ}{w}_t$ is a standard Brownian motion. The introductory article [1] in this volume can be consulted for general background and most of the standard results in filtering theory used below.

Let us denote $\mathcal{Y}_t = \sigma\{y_s, s \leq t\}$ and

$$\pi_t(f) = E[f(x_t)|\mathcal{Y}_t] \tag{1.2}$$

π_t should be thought of as the conditional distribution of x_t given $\mathcal{Y}_t$, so that

$$\pi_t(f) = \int_M f(x)\pi_t(dx)$$

(Here M is the state space for x_t). It is convenient to calculate an unnormalized form σ_t of this, π_t then being given by

$$\pi_t(f) = \sigma_t(f)/\sigma_t(1).$$

If x_t and w_t are independent, an appropriate unnormalized distribution can be obtained in two alternative forms :

(i) the *Kallianpur-Striebel formula*, giving σ_t non-recursively as a function-space integral :

$$\sigma_t(f) = \int_\Xi f(x_t)\exp\left(\int_o^t h(x_s)dy_s - \tfrac{1}{2}\int_o^t h^2(x_s)ds\right)\nu(dx) \quad .. \tag{1.3}$$

(Here (Ξ,ν) is the sample path probability space.)

(ii) the *Zakai equation*, giving σ_t in recursive form as the solution of a measure-valued stochastic differential equation:

$$d\sigma_t(f) = \sigma_t(Af)dt + \sigma_t(hf)dy_t \tag{1.4}$$

$$\sigma_o(f) = \pi(f)$$

(A is the differential generator of the x_t process, $hf(x) = h(x)f(x)$ and π is the distribution of x_o).

Recently several authors (e.g. [2] - [5]) have shown how the solution of (1.3) and (1.4) can be obtained *separately for each sample path* of y_t. The reason why this is important is explained clearly by J.M.C. Clark in [2]. It has to do with the question of stochastic modelling. First, recall some facts about the conditional expectation (1.2). Since $\mathcal{Y}_t$ is generated by $\{y_s, o \leq s \leq t\}$, $\pi_t(f)$ is a functional of the continuous process y_t, i.e. there is a function $\phi : C[0,t] \to R$ such that

$$E[f(x_t)|\mathcal{Y}_t] = \phi(y) \qquad \text{a.s.} \tag{1.5}$$

ϕ is not uniquely defined, in that any other function ϕ' such that $\phi'(y) = \phi(y)$ a.s. would be an equally good "version" of the conditional expectation. Here "a.s." refers to the distribution of y on C[0,t], and this distribution has the same null sets as Wiener measure, In particular *the set of functions with bounded variation is a null set.* Now in the observation equation (1.1), y_t is a mathematical model for
$\tilde{y}_t = \int_0^t z_s ds$ where z_t is the physical observation

$$z_t = h(x_t) + n_t$$

and n_t is (physical) "wide band" noise. As an estimate for $f(x_t)$ we then plan to take $\phi(\tilde{y})$. But since $\tilde{y}_t$ has bounded variation, ϕ is undefined for $\tilde{y}$, and indeed on the whole set of physical sample paths. Thus nonlinear filtering theory cannot be applied in practice unless we are able to choose a particular version of the conditional expectation which has "nice" properties. Specifically, what is really required is a function $\phi : C[0,t] \to R$ such that

(i) (1.5) holds

(1.6)

(ii) ϕ is continuous (with respect to the supremum norm on C[0,t])

Then $\phi(\tilde{y})$ is a "sensible" estimator, in that the mean square error $E[f(x_t) - \phi(\tilde{y})]^2$ is close to the predicted value $E[f(x_t) - \phi(y)]^2$ as long as the distributions of $\tilde{y}$ are close to those of y in the sense of weak convergence (and this certainly includes all the usual bounded-variation approximations to Brownian motion).

In [3], [4] it was shown that such "robust" filtering algorithms could be produced for a very wide class of Markov signal processes x_t, when the signal and observation noise w_t^o are independent. The main purpose of this paper is to extend the results to certain cases where there is correlation between signal and observation noise. This cannot be done at the same level of generality as in [3], and the signals we consider are diffusions on finite-dimensional manifolds.Such signals appear in important potential areas of application of nonlinear filtering theory, for example in alignment problems in inertial navigation where the signal is an orientiation, represented by a quaternion vector. Also, the coordinate-free signal description (introduced in §2 below) adds insight even for R^d-valued diffusions.

The basis of our approach is that the unnormalized conditional distribution σ_t can be expressed in the form

$$\sigma_t(f) = <T^y_{o,t} B_{y_t} f, \pi>$$

where $T^y_{s,t}$ is the y-dependent semigroup associated with a certain multiplicative functional transformation, B_t is a group of operators and π is the distribution of the initial state x_o. A recursive form of estimator can then be obtained by considering the forward equation corresponding to $T^y_{s,t}$ (see §3.1). Our main concern is therefore to calculate the generator of $T^y_{s,t}$ and this is most readily done by factoring the relevant multiplicative functional (§3.2). The relation with "pathwise solutions" of the Zakai equation is explored in §3.4.

In the case of independent signal and noise, B_t is the operator of "multiplication by exp[th(x)]". Our main result is that, for the type of noise correlation considered in §4, the situation is formally analogous to the independence case, but with B_t now being the flow corresponding to a certain differential operator (see (4.12)). Showing this involves decomposing the signal equation in the way described by Kunita [7] in order to elucidate precisely the dependence of x_t on the observations y.

It must, regretfully, be pointed out that the results for correlated noise cannot, unlike those for the independence case, be extended to vector observations. This is because the corresponding operators B^i_t do not in general commute whereas with no noise correlation they are just multiplication operators which automatically commute.

2. THE SIGNAL PROCESS

The formulation here follows very closely that of Kunita in [7]. The signal process x_t envolves on a σ-compact, connected C^∞ manifold M of dimension d. Suppose $X_0,\ldots,X_r$ are C^∞ vector fields on M and $w^1_t \ldots w^r$ are independent scalar Brownian motions which are independent of w^o_t of equation (1.1). Then x_t is the solution of the stochastic differential equation

$$dx_t = X_o(x_t)dt + X_j(x_t) \circ dw^j_i \qquad (2.1)$$

This means that x_t is the unique M-valued process satisfying

$$f(x_t) = f(x_o) + \int_o^t X_o f(x_s)ds + \int_o^t X_j f(x_s) \circ dw_s^j, \; 0 \le t \le \tau \qquad \ldots. \; (2.1)'$$

for any real valued C^∞ function f. In these equations the $\circ$ denotes the Stratonovich stochastic integral, and the convention of implied summation from j = 1 to r is used †. The initial point x_o is supposed to be a random variable with a given distribution π, independent of all other r.v.'s. In (2.1)', τ is the *life-time* and we assume conditions are such that $\tau = \infty$ a.s. (automatically true if M is compact). The relation of (2.1) with the corresponding Ito equation is the following : for any k, $X_k f \in C^\infty(M)$ and hence from (2.1)

$$dX_k f(x_t) = X_o X_k f(x_t)dt + X_j X_k f(x_t) \circ dw_t^j$$

Thus the joint quadratic variation of the semimartingales $X_k f(x_t)$ and w_t^k is

$$d\langle X_k f, w^k\rangle_t = X_k^2 f(x_t)dt$$

and the Ito version of (2.1) is therefore

$$df(x_t) = (X_o + \tfrac{1}{2} \Sigma X_j^2) f(x_t)dt + X_j f(x_t)dw_t^j \qquad (2.2)$$

x_t is a Markov diffusion process; its *generator* A is an operator acting on C^∞ functions such that the Dynkin formula††

$$E_x f(x_t) - f(x) = E_x \int_o^t Af(x_s)ds$$

is satisfied for $f \in C_o^\infty(M)$ (= C^∞ functions of compact support). In view of the time-homogeneity this is equivalent to saying that the process

$$C_t^f := f(x_t) - f(x) - \int_o^t Af(x_s)ds$$

is a martingale. Now in (2.2) the stochastic integral is a martingale for $f \in C_o^\infty(M)$ since $X_j f$ is then bounded, and it follows that

$$A = X_o + \tfrac{1}{2} \Sigma X_j^2$$

A is sometimes called the *extended generator* of x_t: it is an

---

† All sums are over this range unless otherwise specified

†† E_x is the expectation starting at $x_o = x$ w.p.1.

extension of the infinitestimal generator of the semigroup of operators on C(M)

$$T_t f(x) := E_x[f(x_t)] \tag{2.3}$$

associated with the process x_t.

For the sequel, we shall need to compute some Lie brackets. We suppose that the observation function h of (1.1) is in $C_b^\infty(M)$ (i.e. is bounded and C^∞) h will also denote the zeroth-order operator of "multiplication by h", i.e. for $f \in C^\infty(M)$

$$hf(x) = h(x)f(x)$$

and similarly for other functions below). If D is any differential operator, $ad_h D$ denotes the Lie derivative

$$(ad_h D)f(x) := [h,D]f(x) = h(x)Df(x) - D(hf)(x)$$

and $ad_h^2 D = ad_h(ad_h D)$, etc. If X is a vector field then we find using the Leibnitz rule that

$$ad_h X = -Xh \qquad ad_h X^2 = -2Xh\,X - X^2 h$$

$$ad_h^2 X = 0 \qquad ad_h^2 X^2 = 2(Xh)^2$$

$$ad_h^3 X^2 = 0$$

Thus in particular

$$\begin{aligned} ad_h A &= -\Sigma X_j h\, X_j - Ah \\ ad_h^2 A &= \Sigma (X_j h)^2 \\ ad_h^k A &= 0 \qquad k > 2 \end{aligned} \tag{2.4}$$

3. INDEPENDENT SIGNAL AND NOISE

3.1 The KS formula as a multiplicative functional transformation

Let us return to the Kallianpur-Striebel formula, (1.3). In view of (2.1) the real-valued process $h(x_t)$ is certainly a semimartingale, and we can write

$$\int_0^t h(x_s)dy_s = h(x_t)y_t - \int_0^t y_s dh(x_s)$$

Note that the right hand side of this equality involves no stochastic integration with respect to dy and makes sense for any function $y(.) \in C[0,t]$. Thus (1.3) can be written in the form

$$\sigma_t(f) = E[f(x_t)e^{y(t)h(x_t)} \alpha_t^o(y)] \tag{3.1}$$

where $\alpha_t^s(y)$ is defined for $s \leq t$ by

$$\alpha_t^s(y) = \exp[- \int_s^t y(u)dh(x_u) - \tfrac{1}{2} \int_s^t h^2(x_u)du] \tag{3.2}$$

In these expressions $y \in C[0,t]$ is regarded as a parameter and the expectation is taken over the distribution of (x_t). If we now define $\phi(y) = \sigma_t(f)/\sigma_t(1)$ where $\sigma_t(f)$ is given by (3.1), then this is the desired version of the conditional expectation, in that (1.6) holds (property (ii) was shown by Clark [2] and Kushner [8]).

It remains to show how $\sigma_t(f)$ can be computed recursively.

Associated with the process x_t is a semigroup $(T_t)_{t\geq o}$ of operators on C(M) defined by (2.3) above. Now the process $\alpha_t^s(y)$ of (3.2) is a *multiplicative functional* of x_t, i.e. an adapted process satisfying

$$\alpha_t^r = \alpha_s^r \alpha_t^s \quad \text{for } r \leq s \leq t$$

It is easily checked that, if we define

$$T_{s,t}^y f(x) = E_{x,s}[f(x_t)\alpha_t^s] \quad , \tag{3.3}$$

then $T_{s,t}^y$ is another (two-parameter) semigroup of operators on C(M) (Note, however, that it is not Markovian, i.e. does not satisfy $T_{s,t}^y 1 = 1$). For $f \in C(M)$ and μ a measure on M, denote

$$\langle f,\mu\rangle = \int_M f(x)\mu(dx)$$

Then from (3.1) - (3.3) we see that

$$\sigma_t(f) = \langle T_{o,t}^y(e^{y(t)h}f), \pi\rangle$$

This provides us, in principle, with a recursive way of computing σ_t. Let $U^y_{t,s}$ be the adjoint semigroup to $T^y_{s,t}$, defined by

$$\langle T^y_{s,t} f, \mu\rangle = \langle f, U^y_{t,s}\mu\rangle$$

and define

$$\pi^y_t = U^y_{t,o}\pi$$

Then, formally, π^y_t is the solution of the forward (Fokker-Planck) equation

$$\frac{d}{dt}\pi^y_t = (A^y_t)^* \pi^y_t , \qquad \pi^y_o = \pi \tag{3.4}$$

Here A^y_t is the differential generator of $T^y_{s,t}$. (3.4) is a recursive equation for π^y_t, and σ_t is then given by

$$\sigma_t(f) = \langle e^{y(t)h} f, \pi^y_t\rangle \tag{3.5}$$

Notice that (3.4), (3.5) constitute a recursive filter in a form in which *no stochastic integration is involved.* The forward equation (3.4) has been investigated in detail for the case $M = R^d$ by Pardoux [5], [6].

The remainder of this section is devoted to explicit calculation of the generator A^y_t of the semigroup $T^y_{s,t}$.

3.2 Factorization of multiplicative functionals

This section follows up some ideas contained in a paper of S.K. Mitter [9]. We introduce three simple types of multiplicative functional, all relative to the Markov process x_t - we call them the Gauge, Feynman-Kac and Girsanov types - and explore the relations between them. Further general information on m.f.s. can be found in the books of Blumenthal and Getoor [10] and Dynkin [11].

The m.f. $\alpha^s_t(y)$ of (3.2) is of course time-varying in that it depends on the sample path y, but here we shall consider time-invariant m.f.s. A m.f. β^s_t is time invariant if for any r, and $s \leq t$,

$$\beta^{s+r}_{t+r} = \beta^s_t \circ \theta_r$$

where θ_r is the shift operator $(\theta_r x)_s = x_{r+s}$. In particular this implies that $\beta_t^s = \beta_{t-s}^o \circ \theta_s$, so that β_t^s is really a one-parameter functional; indeed, denoting $\beta_t = \beta_t^o$ we can write the multiplicative property as

$$\beta_{t+s} = \beta_t \beta_s \circ \theta_t$$

Let T_t^β be the semigroup corresponding to β, defined by

$$T_t^\beta f(x) = E_x[f(x_t)\beta_t] \tag{3.6}$$

We wish to consider the generator of T_t^β. If β satisfies

$$E_x[\beta_t] \leq 1 \tag{3.7}$$

(or, equivalently, $T_t^\beta 1 \leq 1$) then, as shown in [10], one can construct (possibly on an enlarged state space) the β-*sub-process* of x_t, which is a Markov process x_t^β satisfying

$$T_t^\beta f(x) = E_x[f(x_t^\beta)]$$

As in §2 above, the extended generator of x^β is an operator A^β such that

$$f(x_t^\beta) - f(x) - \int_o^t A^\beta f(x_s^\beta)ds$$

is a martingale for $f \in C_o^\infty(M)$, and this is equivalent (from(3.6)) to saying that

$$C_t^{\beta f} := \beta_t f(x_t) - f(x) - \int_o^t \beta_s A^\beta f(x_s)ds \tag{3.8}$$

is a martingale. The latter formulation has however the advantage that it does not involve the β-sub process or condition (3.7) (which is not satisfied in any of the applications we have in mind). We thus define the extended generator of T_t^β as an operator A^β such that $C_t^{\beta f}$ given by (3.8) is a martingale for all $f \in C_o^\infty(M)$.

Here then are the types of multiplicative functional.

(β) <u>Gauge transformation type</u>. Suppose $a \in C_b^\infty(M)$ and $a(x) > 0$ for all $x \in M$. Define

$$\beta_t = a(x_t)/a(x_o) \tag{3.9}$$

This is clearly an m.f., and from (3.6),

$$T_t^\beta g(x) = \frac{1}{a(x)} \; T_t(ag)(x)$$

Using the signal equation (2.2) with $f = ag$ we see that $C_t^{\beta g}$ is a martingale if

$$A^{\beta}g(x) = \frac{1}{a(x)} A(ag)(x)$$

and this is therefore the generator of T_t^{β}.

(γ) Feynman-Kac type. For $v \in C_b^{\infty}(M)$ define

$$\gamma_t = \exp[-\int_o^t v(x_s)ds] \tag{3.10}$$

Computing the product of the semimartingales $f(x_t)$ and γ_t using (2.2) and the Ito formula shows immediately that

$$A^{\gamma}f(x) = Af(x) - v(x)f(x)$$

(δ) Girsanov type. The above transformations can be applied to any Markov-process but this one is more specifically tied to the model (2.1). Fix $g \in C_b^{\infty}(M)$ and define

$$\delta_t = \exp(-\int_o^t X_j g(x_u)dw_u^j - \tfrac{1}{2}\int_o^t \Sigma(X_j g(x_u))^2 du) \tag{3.11}$$

A standard application of the Girsanov theorem shows that we can define a new measure P^{δ} by taking $dP^{\delta}/dP = \delta_t$, and that under P^{δ}

$$d\tilde{w}_u^j := dw_u^j + X_j g(x_u)du, \qquad u \leq t$$

is a standard Brownian motion, $j = 1,2...r$. Thus (2.1) becomes

$$df(x_u) = (X_o f(x_u) - X_j g(x_u)X_j f(x_u))du + X_j f(x_u)\circ d\tilde{w}_u^j$$

Now

$$E_x[f(x_t)\delta_t] = E_x^{\delta}[f(x_t)]$$

and it follows that

$$\text{(3.10)} \qquad A^{\delta} = A - \Sigma(X_j g)X_j \tag{3.12}$$

These three transformations are related by the x_t equation written in Ito form (2.2); indeed, from (2.2)

$$\int_s^t X_j g(x_u)dw_u^j = g(x_t) - g(x_s) - \int_s^t Ag(x_u)du$$

and inserting this in (3.11) we see that δ_t factors in the form

$$\delta_t = \beta_t \gamma_t$$

where β, γ are given by (3.9),(3.10) respectively with

$$a(x) = e^{-g(x)}$$

$$v(x) = -Ag(x) + \tfrac{1}{2} \Sigma(X_j g(x))^2$$

Applying β and γ successively (the order is immaterial) we conclude that

$$A^{\delta} f = e^{g} A(e^{-g} f) - (Ag - \tfrac{1}{2}\Sigma(X_j g)^2) f \qquad (3.13)$$

But this is just a disguised form of the Baker-Campbell-Hausdorff formula : using the expression (3.12) for A^{δ} and the relations (2.4), (3.13) becomes

$$e^{g} A e^{-g} = A - \Sigma(X_j g) X_j - Ag + \tfrac{1}{2}\Sigma(X_i g)^2 = A + ad_g A + \tfrac{1}{2} ad_g^2 A \qquad \ldots\ (3.14)$$

(Recall that $ad_g^k A = 0$ for $k > 2$)

3.3 The generator of $T^y_{s,t}$

Recall from §3.1 that the m.f. appearing in the KS formula is

$$\alpha_t^s(y) = \exp[-\int_s^t y(u)\, dh(x_u) - \tfrac{1}{2} \int_s^t h^2(x_u) du]$$

Using (2.2) we can factor this into the product of a Girsanov m.f. and a Feynman-Kac m.f. as follows:

$$\alpha_t^s(y) = \exp[-\int_s^t y(u) Ah(x_u) du - \int_s^t y(u) X_j h(x_u) dw_u^j - \tfrac{1}{2} \int_s^t h^2(x_u) du]$$

$$= \exp[-\int_s^t y(u) X_j h(x_u) dw_u^j - \tfrac{1}{2} \int_s^t y^2(u) \Sigma(X_j h(x_u))^2 du]$$

$$\times \exp[\int_s^t (\tfrac{1}{2} y^2(u) \Sigma(X_j h(x_u))^2 - y(u) Ah(x_u) - \tfrac{1}{2} h^2(x_u)) du]$$

It follows immediately that the corresponding generator is

$$A_s^y f = (Af - y(s)\ \Sigma(X_j h) X_j f) + [\tfrac{1}{2} y^2(s) \Sigma(X_j h)^2 - y(s) Ah - \tfrac{1}{2} h^2] f \qquad \ldots\ (3.15)$$

$$= Af + y(s)(ad_h A)f + \tfrac{1}{2}y^2(s)(ad_h^2 A)f - \tfrac{1}{2}h^2 f$$

$$= e^{y(s)h}\, A(e^{-y(s)h} f) - \tfrac{1}{2}h^2 f, \tag{3.16}$$

the last equality being immediate from (3.14) with $g = y(s)h$.

It is clear *a priori* that (3.16) must be the right formula: in (3.15) the calculation is done for an arbitrary function $y(.)$ but the result depends only on $y(s)$. Therefore $A_s^y = A_s^{\bar{y}}$ where $\bar{y}$ is the constant function

$$\bar{y}(u) = y(s) \qquad \text{for } u \geq s$$

But

$$\alpha_t^s(\bar{y}) = \exp[-y(s)h(x_t) + y(s)h(x_s)]\exp[-\tfrac{1}{2}\int_s^t h^2(x_u)du]$$

This factors $\alpha_t^s(\bar{y})$ into the product of a gauge m.f. and a Feynman-Kac m.f. , and (3.16) is immediate for $\bar{y}(.)$. But of-course some extra work has to be done to show that the same for mula works for non-constant $y(.)$.

Note from (3.15) that A_s^y is of the form

$$A_s^y f(x) = \tfrac{1}{2}\Sigma X_j^2(x) + Y_o(y(s))f(x) + \psi(x,y(s))f(x)$$

i.e. the second-order part of A_s^y is the same as that of A, and the effect of the m.f. transformation is only to add y-dependent "drift" and "potential" terms. Thus essentially the same conditions that ensure smooth solutions of the Fokker-Planck equation of the signal process also ensure smooth solutions of (3.4). The general conclusion is that computing the *conditional* distribution of x_t given $\mathcal{Y}_t$ is not in any essential way more complicated than computing the *unconditional* distribution. See Pardoux [5], [6] for the case $M = R^d$.

3.4 Dossing the Zakai equation

There is another way of looking at the basic formula

$$\sigma_t(f) = \langle T_{o,t}^y (e^{y(t)h} f), \pi \rangle \tag{3.17}$$

and that is as a Doss-Sussmann "pathwise solution" [12], [13] of the Zakai equation (1.4). This was indeed how (3.17) was originally arrived at, and although the m.f. approach turns out to be more fundamental, the pathwise solution idea is of value in understanding the picture and particularly in unravelling the complexities of the correlated noise case (see §4.4 below).

Let us recall the Doss-Sussmann construction for the simplest type of scalar equation

$$dx_t = f(x_t)dt + g(x_t) \circ dM_t, \qquad x_o = x \tag{3.18}$$

Here M_t is a real-valued continuous semimartingale and f,g are smooth functions. (The same basic idea is used with considerably more elaboration in §4.2 below). Let G(t,x) be the flow of g, i.e. the solution of the ordinary differential equation

$$\frac{\partial}{\partial t} G(t,x) = g(G(t,x))$$

$$G(0,x) = x$$

Then the solution of (3.18) is of the form

$$x_t = G(M_t,\eta_t) \tag{3.19}$$

where η_t is the solution of another o.d.e., parametrized by the sample path (M_t). Indeed, defining x_t by (3.19) we have

$$dx_t = g(x_t) \circ dM_t + G_x(M_t,\eta_t)\dot{\eta}_t\, dt \tag{3.20}$$

But

$$G_x(t,x) = \exp\left(\int_o^t g_x(G(s,x))ds\right) \tag{3.21}$$

so that (3.18) and (3.20) agree as long as

$$\dot{\eta}_t = \exp\left(-\int_o^{M(t)} g_x(G(s,\eta_t))ds\right)f(G(M_t,\eta_t)), \qquad \eta_o = x \tag{3.22}$$

This is an ordinary differential equation for η_t, parametrized by the sample path (M_t), and shows that the solution of (3.18) can be calculated separately for each sample path of M_t : first solve (3.22) and then evaluate (3.19). The same construction works for $x_t \in R^n$ (except for the explicit expression(3.21)), but not, in general, for vector M_t (see below). Things are particularly simple in the bilinear case: f(x) = Ax, g(x) = Hx. Then (3.22) and (3.19) become respectively

$$\dot{\eta}_t = e^{-HM(t)} Ae^{HM(t)} \eta_t \quad , \quad \eta_0 = x$$

$$x_t = e^{HM(t)}\eta_t$$

Let us now apply the same argument to the bilinear measure-valued Zakai equation (1.4). In Stratonovich form this is

$$d\sigma_t(f) = \sigma_t((A-\tfrac{1}{2}h^2)f)dt + \sigma_t(hf) \circ dy_t \qquad (3.23)$$

If the "drift" term in (3.23) were absent then the solution would be, as is easily checked,

$$\sigma_t(f) = \langle e^{y(t)h}f, \pi\rangle$$

An argument exactly analogous to the above shows that the solution *with* the drift term is given by (3.17), if $T^y_{o,t}$ is a semigroup with generator

$$A^y_t f = e^{y(t)h}A(e^{-y(t)h}f) - \tfrac{1}{2}h^2 f \qquad (3.24)$$

But we saw in (3.13) above that this precisely *is* the generator of the semigroup given by the Kallianpur-Striebel formula. Thus the two approaches lead to the same result. If, however, one starts with the Zakai equation, one has somehow to show that there exists a semigroup whose generator is (3.24). The only way to do this that I know of is through a probabilistic argument [14] which leads straight back to the KS formula. This is why I describe the m.f. approach as "more fundamental".

Finally, let us note that all of the above results extend without difficulty to the case of vector observations

$$dy^i_t = h^i(x_t)dt + dw^{oi}_t \qquad i = 1,\dots,m$$

if x_t and w^{oi}_t are independent for all i. The Zakai equation is

$$d\sigma_t(f) = \sigma_t(Af)dt + \sum_1^m \sigma_t(h^i f)dy^i$$

and a pathwise solution is constructed from this (or from the KS formula) as before. The reason this "works" is that the operators of "multiplication of h^i", which appear in the diffusion term of the Zakai equation, commute : $h^i h^j f(x) = h^j h^i f(x) = h^i(x)h^j(x)f(x)$. Recall that the condition under which the Doss-Sussmann construction for (3.18) can be extended to multiple inputs $\Sigma g^i(x_t) \circ dM^i_t$ is precisely that the vector fields g^i commute.

4. THE CORRELATED NOISE CASE

4.1 The signal and observation equations

We now wish to consider the filtering problem given by (1.1) and (2.1) as before, but allowing for possible correlation between the signal noise $(w^1,\ldots,w^r)$ and the observation noise w^o. We assume the simplest form of correlation; it will be obvious how to extend the results to more general cases. Specifically, we suppose

(i) w_t^i is a standard Brownian motion (i.e. $w_o^i = 0$ and $\langle w^i\rangle_t = t$), $i = 0,1\ldots,r$.

(ii) w^i, w^j are independent for $i \neq j \neq 0$

(iii) $\langle w^i,w^o\rangle_t = \alpha_i t$ for some constant α_i, $|\alpha_i| < 1$.

The first two of these are the same as before, and the third implies in particular that

$$Ew_t^i w_s^o = \alpha_i \; t \wedge s$$

The Kallianpur-Striebel formula is no longer valid in the form (1.3); we shall derive the correct form in §4.3 below. As regards the Zakai equation, it follows directly from the general filtering equation of Fujisaki-Kallianpur-Kunita [9] that (1.4) should be amended to

$$d\sigma_t(f) = \sigma_t(Af)dt + \sigma_t(Df)dy_t \tag{4.1}$$

where D is an operator defined as follows :
Let Z be the vector field

$$Z = \Sigma\, \alpha_j X_j . \tag{4.2}$$

Then

$$D = Z + h . \tag{4.3}$$

To get the appropriate form of the KS formula, introduce a measure P_o via the Girsanov transformation

$$\frac{dP_o}{dP} = \exp\Big(-\int_o^T h(x_s)dw_s^o - \tfrac{1}{2}\int_o^T h^2(x_s)ds\Big)$$

and for $i = 1,2\ldots,r$ define

$$dv^i := dw^i + \alpha_i h(x_t)dt \tag{4.4}$$

Then, under P_o,

(i) y_t and v_t^i, $i = 1,\ldots,r$ are standard Brownian motions

(ii) v^i, v^j are independent for $i \neq j$

(iii) $\langle v^i, y\rangle_t = \alpha_i t$

Now project the v^i onto y, i.e. define

$$\tilde{b}_t^i := v_t^i - \alpha_i y_t \tag{4.5}$$

Then each $\tilde{b}^i$ is an unnormalized Brownian motion, which is uncorrelated with, and hence independent of, y. Denote $\tilde{b}'_t = (\tilde{b}_t^1,\ldots,\tilde{b}_t^r)$, $\alpha' = (\alpha_1,\ldots\alpha_r)$ and I = rxr identity matrix. Then

$$\langle\tilde{b}\rangle_t = (I - \alpha\alpha')t$$

Now $I - \alpha\alpha'$ is positive definite and can be factored into a product $\Delta\Delta'$ of positive definite matrices. Defining

$$b_t := \Delta^{-1}\tilde{b}_t \tag{4.6}$$

we find that

$$\langle b\rangle_t = It$$

i.e. $b^1,\ldots,b^r$ are (under measure P_o) independent standard Brownian motions independent of y. Using (4.4) - (4.6), the signal equation (2.1) becomes

$$df(x_t) = Y_o f(x_t)dt + Zf(x_t) \circ dy_t + Y_j f(x_t) \circ db_t^j \tag{4.7}$$

where Z is given by (4.2),

$$Y_o := X_o - hZ$$

and, in an obvious notation, the Y_j are given by

$$Y := \Delta' X$$

(4.7) is the key formula for the filtering problem, as it expresses x_t in the form of an equation driven by the observation process y_t and the other "inputs" $b^1,\ldots,b^r$ which are independent of y. The next task is to decompose (4.7) in such a way that the dependence of x_t on y is explicitly brought out.

4.2 Decomposition of the signal equation

This section follows the approach of Kunita [7] very closely. Essentially, the idea is to use a tra formation of the Doss-Sussmann [12], [13] type to express the solution of (4.7) sample-path-wise in y. First we need some ideas from differential geometry.

The tangent space $T_p(M)$ at $p \in M$ consists of the set of derivations, i.e. linear functionals W_p on $C^\infty(U_p)$ (U_p is a neighborhood of p) such that the Leibnitz rule

$$W_p(fg) = f\,W_p g + g\,W_p f$$

is satisfied. Now let $\phi: M \to M$ be a diffeomorphism and denote $q = \phi(p)$. Then ϕ defines a map $\phi^* : C^\infty(U_q) \to C^\infty(U_p)$ by composition :

$$\phi^* f := f \circ \phi\,, \qquad f \in C^\infty(U_q)$$

and also a map $\phi_* : T_p(M) \to T_q(M)$ as follows.

$$(\phi_* W_p) f := W_p(f \circ \phi) \qquad f \in C^\infty(U_q)$$

$$= W_p(\phi^* f)$$

Since ϕ^{-1} is also a diffeomorphism, $\phi_*^{-1} : T_q(M) \to T_p(M)$ is given likewise by

$$(\phi_*^{-1} W_q) g = W_q(g \circ \phi^{-1}) \qquad g \in C^\infty(U_p)$$

If W is a vector field and W_p denotes its restriction to $p \in M$ then this relation defines a mapping, also denoted ϕ_*^{-1}, between vector fields, which, since $q = \phi(p)$, we can write

$$(\phi_*^{-1} W) g(p) = W(g \circ \phi^{-1})(\phi(p))$$

Let $\zeta_t(x) = \zeta(t,x)$ denote the flow of the vector field Z (see (4.2), (4.7)), i.e. the unique solution of the equation

$$\frac{d}{dt}f(\zeta_t(x)) = Zf(\zeta_t(x)), \qquad f \in C^\infty(M)$$

$$\zeta_o(x) = x$$

This is a diffeomorphism for each $t \geq 0$. Define

$$\xi_t(x) = \zeta_{y(t)}(x)$$

As is easily checked, $\xi_t = \xi_t(x)$ is the solution of

$$d\xi_t = Z(\xi_t) \circ dy_t$$

and, obviously, $\xi_t(.)$ is almost surely a diffeomorphism for each $t > 0$. Now consider the equation

$$df(\eta_t) = \xi_{t*}^{-1} Y_o f(\eta_t)dt + \xi_{t*}^{-1} Y_j f(\eta_t) \circ db_t^j \qquad (4.8)$$

This equation has a unique solution and it follows by applying the Ito formula that

$$x_t(x) = \xi_t \circ \eta_t(x)$$

$$= \zeta(y_t, \eta_t) \qquad (4.9)$$

The representation (4.8), (4.9) describes the behavior of x_t conditioned on y under measure P_o. Recall that the map ξ_{t*}^{-1} is parametrized by y and that y,b are independent. Thus, conditioned on y, η_t is a diffusion process whose differential generator is

$$A_t^* = \xi_{t*}^{-1} Y_o + \Sigma(\xi_{t*}^{-1} Y_j)^2$$

and, for each $t > 0$, x_t is diffeomorphically related to η_t by equation (4.9).

4.3 The KS formula and associated multiplicative functional

It follows from a standard formula of conditional expectations that $\pi_t(f)$ of (1.2) is given in terms of the measure P_o by

$$\pi_t(f) = \sigma_t(f)/\sigma_t(1)$$

where

$$\sigma_t(f) := E_0[f(x_t)\exp(\int_0^t h(x_s)dy_s - \tfrac{1}{2}\int_0^t h^2(x_s)ds)\,|\,\mathcal{Y}_t]$$

It is immediate from (4.7) that

$$d\langle h(x_\cdot),y\rangle_t = Zh(x_t)dt$$

and hence that the Stratonovich version of this is

$$\sigma_t(f) = E_0[f(x_t)\exp(\int_0^t h(x_s)\circ dy_s - \tfrac{1}{2}\int_0^t Dh(x_s)ds)\,|\,\mathcal{Y}_t] \qquad \ldots. \quad (4.10)$$

where D is given by (4.3). Now use (4.9): $x_t = \zeta(y_t,\eta_t)$ and η_t is a functional of the ide endent processes y_t and $b'_t = (b^1_t,\ldots,b^r_t)$. Thus (4.10) can be expressed in the form

$$\sigma_t(f) = E^b[\xi^*_t f(\eta_t)\exp(\int_0^t \xi^*_s h(\eta_s)\circ dy_s - \int_0^t \xi^*_s Dh(\eta_s)ds)] \qquad \ldots. \quad (KS)$$

where E^b means integration over the sample space measure for b_t (= Wiener measure on $C([0,T];R^r)$). This is the "Kallianpur-Striebel" formula for the correlated-noise problem. In order to get it in "robust" form we need to calculate the stochastic integral in (KS) as an explicit functional of y. Introduce the function

$$H_t(x) = H(t,x) := \int_0^t \xi^*_s h(x)ds$$

and calculate $H(y_t, \eta_t)$ using the Ito formula and (4.9). This gives

$$H(y_t,\eta_t) = \int_0^t h(x_s)\circ dy_s + \int_0^t (\xi^{-1}_{s*}Y_0)H_{y_s}(\eta_s)ds + \int_0^t (\xi^{-1}_{s*}Y_j)H_{y_s}(\eta_s)\circ db^j_s \qquad (4.11)$$

The stochastic integral with respect to b^j in (4.11) can be re-expressed in Ito form in the standard way using (4.8). Do this and introduce the notation

$$g_s(x) := H_{y_s}(x)$$

$$Y_j^* := \xi_{s*}^{-1} Y_j$$

$$B_s f(x) := \exp\left(\int_0^s \zeta_u^* h(x)\,du\right)\zeta_s^* f(x) \tag{4.12}$$

Then using (4.11) in (4.10) gives

$$\sigma_t(f) = E[B_{y_t} f(\eta_t)\alpha_t^o(y)] \tag{4.13}$$

where

$$\alpha_t^s(y) := \exp\Big[-\int_s^t Y_j^* g_u(\eta_u)\,db_u^j - \tfrac{1}{2}\int_0^t (Y_j^*)^2 g_u(\eta_u)\,du$$

$$- \int_s^t Y_o^* g_u(\eta_u)\,du - \tfrac{1}{2}\int_s^t \xi_u^* Dh(\eta_u)\,du\Big] \tag{4.14}$$

Equation (4.13) is the desired multiplicative functional formula. For each sample path of y, η_t is a diffusion process governed by vector fields Y_j^* as in (4.8), and $\alpha_t^s(y)$ given by (4.14) is a m.f. of η. The expectation in (4.13) is taken over the distributions of η for fixed y. It is possible to show that $\sigma_t(f)$ given by (4.13) is continuous in $y \in C[0,t]$, i.e. that this is a "robust" version in the sense of (1.6), but the details are complicated and space limitations preclude their inclusion here. In outline, one starts with functions $f \in C_o^\infty(M)$ whose support is contained within a single chart of M. Then, working in local coordinates, equation (4.8) satisfies the standard Ito conditions and continuous dependence (in the mean square sense) of $f(\eta_t)$ on $y \in C[0,t]$ follows from known results on parametric dependence of solutions of stochastic differential equations; see Theorem 2, §2.7, of Gihman and Skorohod [15]. One completes the argument for $f \in C_b^\infty(M)$ by considering a decomposition of the form

$$\sigma_t(f) = \sum_i \sigma_t(\rho_i f)$$

where (ρ_i) is a partition of unity : $\rho_i \in C_o^\infty(M)$ for all i and $\Sigma_i \rho_i(x) \equiv 1$.

As in §3.3 above, we can compute the generator A_t^y corresponding to the m.f. $\alpha_t^s(y)$ by factorization. Indeed

$$\alpha_t^s(y) = \exp[-\int_0^t Y_j^* g_u(\eta_u) db_u^j - \tfrac{1}{2}\int_s^t (Y_j^* g_u(\eta_u))^2 du]$$

$$\times \exp[\int_s^t (\tfrac{1}{2}\Sigma(Y_j^* g_u(\eta_u))^2 - \tfrac{1}{2}\Sigma(Y_j^*)^2 g_u(\eta_u) - Y_o^* g_u(\eta_u) - \tfrac{1}{2}\xi_u^* Dh(\eta_u))du]$$

It now follows as before that

$$A_t^y f = A_t^* f - \Sigma Y_j^* g_t Y_j^* f + (\tfrac{1}{2}\Sigma(Y_j^* g_t)^2 - A^* g_t - \tfrac{1}{2}\xi_t^* Dh) f \quad (4.15)$$

where A^* is the generator for η_t, i.e.

$$A_t^* = Y_o^* + \tfrac{1}{2}\Sigma(Y_j^*)^2 \quad (4.16)$$

A_t^y can be expressed in somewhat more explicit form by noting that

$$Y_j^* g_t(x) = \xi_{t*}^{-1} H_{y_t}(x)$$

$$= \int_0^{y_t} \xi_{t*}^{-1} Y_j \zeta_u^* h(x) du$$

$$= \int_0^{y_t} Y_j \zeta_{-u}^* h(\zeta_{y_t}(x)) du$$

The similarity of (4.15) to (3.15) is obvious (of course, (4.15) reduces to (3.15) if $\alpha = 0$) and similar remarks are pertinent: A_t^y differs from A^* only in the "drift" and "potential" terms and therefore the complexity of computing the conditional distribution is essentially that of computing the distribution of the decomposition η_t of the signal process x_t.

4.4 Solution of the Zakai equation

The Zakai equation for the correlated noise problem was given in (4.1); in Stratonovich form it is

$$d\sigma_t(f) = \sigma_t((A-\tfrac{1}{2}D^2)f)dt + \sigma_t(Df) \circ dy_t \quad (4.17)$$

Now

$$A - \tfrac{1}{2}D^2 = \tfrac{1}{2}(\Sigma X_j^2 - (\Sigma\alpha_j X_j + h)^2) + X_o$$

and a completion-of-squares calculation shows that

$$A - \tfrac{1}{2}D^2 = Y_o + \Sigma Y_j^2 - \tfrac{1}{2}Dh \tag{4.18}$$

where Y_o, Y_j are as in (4.7). The operator $D = Z + h$ is the generator of the group B_t on $C^\infty(M)$ given by (4.12). An argument analogous to that of §3.4 shows that the Doss-Sussmann solution of (4.15) is

$$\sigma_t(f) = \langle T^y_{o,t}(B_{y_t} f), \pi\rangle$$

where $T_{o,t}$ is the semigroup whose generator is

$$\tilde{A}^y_t = B_{y_t}(A - \tfrac{1}{2}D^2)B_{-y_t}. \tag{4.19}$$

We propose to show, using the Baker-Campbell-Hausdorff formula, that this coincides with A^y_t given by (4.15) above. Denote by C_t the multiplication operator

$$C_t f(x) = f(x)\exp[\int_o^{y_t} \zeta^*_s h(x)ds]$$

so that

$$B_{y_t} = C_t \xi^*_t \tag{4.20}$$

and

$$B_{-y_t} = B^{-1}_{y_t} = \xi^*_{-t} C^{-1}_t \tag{4.21}$$

(Note that $C^{-1}_t \neq C_{-t}$). Using (4.18) - (4.21) we can see that

$$\tilde{A}^y_t = C_t \xi^*_t (Y_o + \tfrac{1}{2}\Sigma Y_j^2 - \tfrac{1}{2} Dh)\xi^*_{-t} C^{-1}_t$$

Now

$$\xi^*_t Dh\ \xi^*_{-t} f(x) = \xi^*_t Dh(x) f(x)$$

and

$$\xi^*_t (Y_o + \tfrac{1}{2}\Sigma Y_j^2)\xi^*_{-t} = A^*_t$$

where A^*_t is given by (4.16). Thus

$$\tilde{A}_t^y = C_t A_t^* C_{-t} - \tfrac{1}{2}\xi_t^* Dh \tag{4.22}$$

Since C_t is a multiplication operator, we can expand the right-hand side using the Baker-Campbell-Hausdorff formula (3.14). Indeed, from (3.14)

$$C_t A^* C_{-t} = A^* - \Sigma(Y_j^* g_t) Y_t^* - A^* g_t + \tfrac{1}{2}\Sigma(Y_j^* g_t)^2$$

Using this expression in (4.22) we see that $\tilde{A}_t^y$ coincides with A_t^y given by (4.15). Thus, as claimed in §1, the results are formally analogous to the independence case with the operator B_t replacing the operator of multiplication by exp[th(x)]. However, while it is (with a bit of hindsight) in the independence case fairly obvious from the KS formula that the appropriate generator is (3.16), the interpretation of (4.15) in the correlated case is by no means obvious and it seems essential to look at the Zakai equation to get the full picture.

Finally, if there are vector observations then the last term in (4.17) will be of the form

$$\sum_i \sigma_t(D^i f) \circ dy_t^i$$

where

$$D^i = Z^i + h^i$$

for some vector fields Z^i. There cannot be a pathwise solution of (4.17) unless the D^i commute, but this only happens under extremely artificial conditions. If the D^i do not commute a decomposition of the type (4.9) is still possible, where ξ_t is almost surely a diffeomorphism (see [7]), but no continuous dependence of ξ_t on y can be expected. Thus the present results are essentially limited to the scalar case.

Acknowledgements: This paper owes a lot to lectures [7] of H. Kunita at the London Mathematical Society Symposium on Stochastic Integration, to conversations with S.K. Mitter at Les Arcs and a preview of his paper [9], and to a little patient instruction in differential geometry from A.J. Krener.

5. REFERENCES

1. M.H.A. Davis and S.I. Marcus, An introduction to nonlinear filtering, this volume.
2. J.M.C. CLark, The design of robust approximations to the stochastic differentials equations of nonlinear filtering, in Communication Systems and Random Process Theory, ed. J.K. Skwirzynski, NATO Advanced Study Institute Series, Sijthoff and Noorshoff, Alphen aan den Rijn, 1978
3. M.H.A. Davis, On a multiplicative functional transformation arising in nonlinear filtering theory, Z. Wahrscheinlichkeitstheorie verw. Geb., (to appear)
4. M.H.A. Davis, A pathwise solution of the equations of nonlinear filtering, Teoria Veroyatnostei i ee Prim., (to appear)
5. E. Pardoux, Backward and forward stochastic partial differential equations associated with a nonlinear filtering problems, Proc. 18th IEEE Conference on Decision and Control, Ft. Lauderdale, Florida, 1979
6. E. Pardoux, this volume
7. H. Kunita, On the decomposition of solutions of stochastic differential equations, London Mathematical Society Symposium on Stochastic Integrals, Durham, 1980
8. H.J. Kushner, A robust discrete state approximation to the optimal nonlinear filter for a diffusion, Stochastics 3 (1979) 75-83
9. S.K. Mitter, On the analogy between mathematical problems of nonlinear filtering and quantum physics, Ricerche di Automatica, to appear [also: LIDS Report P-1006, MIT]
10. R.M. Blumenthal and R.K. Getoor, Markov Processes and Potential Theory, Academic Press, New York, 1968
11. E.B. Dynkin, Markov Processes, Springer-Verlag, Berlin, 1965
12. H. Doss, Liens entre équations differéntielles stochastiques et ordinaires, Ann Inst. H. Poincaré, 13 (1977), 99-125
13. H.J. Sussmann, On the gap between deterministic and stochastic ordinary differential equations, Ann. Prob. 6 (1978), 19-41
14. M.H.A. Davis, Pathwise solutions and multiplicative functionals in nonlinear filtering, Proc. 18th IEEE Conference on Decision and Control, Ft. Lauderdale, 1979
15. I.I. Gihman and A.V. Skorohod, Stochastic differential equations, Springer-Verlag, Berlin, 1972

NON-LINEAR FILTERING, PREDICTION AND SMOOTHING.

E. PARDOUX

UER de Mathématiques
Université de Provence
13331 Marseille Cedex 3

CONTENTS

M. Hazewinkel and J. C. Willems (eds.), Stochastic Systems: The Mathematics of Filtering and Identification and Applications, 529–557.

I. INTRODUCTION

Let X_t be an unobserved Markov diffusion process, and suppose we observe the process :

$$Y_t = \int_0^t h(X_s)ds + W_t \qquad (0.1)$$

where W_t is a brownian motion - called the noise - possibly correlated with X_t. Denote by $\mathcal{F}_t$ the σ- field generated by $\{Y_s, s \leq t\}$. We want to study the three following problems :

a. the filtering problem : compute the conditional law of X_t, given $\mathcal{F}_t$.
b. the prediction problem : compute the conditional law of X_t, given $\mathcal{F}_s$, for $s < t$.
c. the smoothing problem : compute the conditional law of X_s, given $\mathcal{F}_t$, for $s < t$.

Under the hypotheses we will make, these conditional laws have densities with respect to Lebesque measure. Our aim is to establish equations - driven by the observed process Y_t - such that the densities are expressible in terms of the solutions to these equations.

The filtering problem has been extensively studied in the literature - see other articles in this volume, especially DAVIS-MARCUS [7], and the bibliographies therein. As we will see, the solution of the prediction problem is an easy corollary of that of the filtering problem. But-as for as we know - the solution of the smoothing problem, in the non linear case, had been solved until now only when X_t is a finite - state Markov process - see LIPTSER-SHIRYAYEV [10].

We will solve the three problems by an original method, based on the use of a pair of Stochastic Partial Differential Equations (S.P.D.E.), one backward and one forward, the forward one being the Zakai equation. The backward equation will serve as an intermediate tool for the solution of the filtering problem, and will be essential for the smoothing problem.
In the case without observation, our S.P.D.Es. reduce to the well-known backward and forward Kolmogorov equations.

In §2, we first apply our method to the derivation of the Kolmogorov forward equation, in order to make the idea clear. In §3, we consider our general filtering, prediction and smoothing problems. In §4, we restrict to the case where noise and signal are independent, and establish directly the "robust" form of the equations. In §5, we consider the case where Y_t, instead of being given by (0.1), is a Poisson point process, whose intensity is a given function of the unobserved process X_t.

Those proofs in §3 and 4, which are not given in detail, can be found in [13] or [15].

2. KOLMOGOROV BACKWARD AND FORWARD EQUATIONS

2.1 The X_t - process

Let X_t be a Markov diffusion process in R^N, satisfying the following stochastic differential equation :

$$dX_t = b(t,X_t)dt + \sigma(t,X_t)dW_t \tag{2.1}$$

Here W_t is an R^N-valued standard Wiener process, and the differential is taken in the Ito sense. We make the following hypotheses on the coefficients :

b and σ are bounded and measurable functions, from $R_+ \times R^N$, into R^N and R^{N^2} respectively. (2.2)

σ is continuous in x, uniformly on each compact subset of $R_+ \times R^N$. (2.3)

$$\left.\begin{array}{l}\exists\, \alpha > o \quad \text{s.t.} \\ a(t,x) = \sigma\sigma^*(t,x) \geq \alpha I, \qquad \forall (t,x).\end{array}\right\} \tag{2.4}$$

$$\frac{\partial \sigma_{ij}}{\partial x_j} \text{ is a bounded function of } (t,x). \quad i,j = 1 \ldots N \tag{2.5}$$

If the coefficients b an σ were lipschitz in x, then by Ito theory - see CURTAIN [5], this volume - to each initial condition at time s, would correspond a unique continuous process $(X_t, t \geq s)$, solution of (2.1).

But $(X_t, t \geq s)$ can be considered as a random variable, with values in the path space $C([s,+\infty[;R^N)$. We then can define its probability law, as a measure on $C([s,+\infty[;R^N)$, endowed with its Borel σ-field.

We then denote by P_{sx} the law of $(X_t, t \geq s)$, with the initial condition :

$$X_s = x,$$

and by P the law of $(X_t, t \geq o)$, with the initial condition X_o = given random variable, whose law possesses a density $p_o(x)$ with respect to Lebesque measure in R^N.
We suppose moreover :

$$p_o \in L^2(R^N) \tag{2.6}$$

It follows from the theory of STROOCK-VARADHAN [16]that, under the hypotheses (2.2), (2.3) and (2.4), the laws P_{sx} and P are still uniquely defined by the requirements :

(i) $P_{sx}(X_s=x)=1$ [resp. $P(X_o \in A)=\int_A p_o(x)\,dx$, $\forall A$ Borel subset of R^N]

(ii) there exists an R^N-valued standard Wiener $(W_t, t \geqslant o)$ such that (2.1) is satisfied P_{sx}-a.s. for $t \geqslant s$ [resp. P-a.s. for $t \geqslant o$].

here $(X_t,\ t \geqslant s)$ is the generic point in $C(R_+\ ;\ R^N)$. In the next subsections, E_{sx} [resp. E] will denote the expectation with respect to P_{sx} [resp. P].

It is easily verified that for any bounded and measurable functional Φ on $C(R_+;R^N)$,

$$E\ [\ \Phi(X.)] = \int E_{ox}[\Phi(X.)]\,p_o(x)\,dx \tag{2.7}$$

We define finally the infinitesimal generator of the Markov process X_t :

$$L = \frac{1}{2}\sum_{i,j=1}^{N} a_{ij}(t,x)\frac{\partial^2}{\partial x_i \partial x_j} + \sum_{i=1}^{N} b_i(t,x)\frac{\partial}{\partial x_i}$$

2.2 Kolmogorov equations

Let $u,v \in C_o^2(R^N)^{(1)}$. From (2.5), we can make the following integration by parts :

$$\int_{R^N} Lu(x)v(x)\,dx = -\frac{1}{2}\sum_{i,j}\int_{R^N} a_{ij}\frac{\partial u}{\partial x_i}\frac{\partial v}{\partial x_j}\,dx + \sum_i \int_{R^N} \overline{b}_i \frac{\partial u}{\partial x_i} v\,dx \tag{2.8}$$

where $\overline{b}_i = b_i - \frac{1}{2}\sum_{j=1}^{N}\frac{\partial a_{ij}}{\partial x_j}$

The right - hand side of (2.8) is defined as soon as u and v belong to the Sobolev space H^1, where :

$$H^1 \triangleq H^1(R^N) = \{u \in L^2(R^N);\ \frac{\partial u}{\partial x_i} \in L^2(R^N)\ i = 1\ldots N\}$$

We define L and L* as (families indexed by t of)elements of $\mathcal{L}(H^1,(H^1)')$ by :

$$< L\,u,v > = < u,L^*v >$$

$$= -\frac{1}{2}\sum_{i,j}\int a_{ij}\frac{\partial u}{\partial x_i}\frac{\partial v}{\partial x_j}\,dx + \sum_i\int b_i\frac{\partial u}{\partial x_i}\,v dx$$

where <.,.> denotes the pairing between H^1 and its dual.

Thanks to (2.4), -L and -L* are coercive operators, in the following sense :

$\exists\lambda \in R$, s.t. $\forall u \in H^1$,

$$- < L u ,u > + \lambda \|u\|^2_{L^2} \geq \frac{\alpha}{2}\|u\|^2_{H^1}$$

Consider now the Kolmogorov backward and forward equations associated with the X_t process :

$$\left.\begin{array}{l}\frac{dv}{ds}(s) + L\,v(s) = 0, \quad s \leq t \\ \qquad v(t) = f\end{array}\right\} \tag{2.9}$$

where $f \in C_o(R^N)$ is given

$$\left.\begin{array}{l}\frac{dp}{ds}(s) = L^*\,p(s) \quad , s \geq o \\ \qquad p(o)=p_o\end{array}\right\} \tag{2.10}$$

We have the following result, from the theory of P.D.Es (see e.g. BENSOUSSAN-LIONS [1]):

Theorem 2.1 Equations (2.9) and (2.10) have unique solutions :

$$v,p\in L^2(\,o,t;H^1)\cap C(\,[o,t];\; L^2(R^N))$$

Remark 2.2 The variational theory of PDEs we are refering to in theorem 2.1, is particularly adequate for our purpose, since it permits us to treat equation (2.10), after having differentiated only once the a_{ij} coefficients.

2.3 Relation with the X_t-process

Itis well known that if the law of X_t has a sufficiently smooth density, then this density obeys equation (2.10). We will now indicate a procedure-which we will follow in §3 to establish the Zakaï equation-which permits to prove that under the hypotheses of §2.1, p(t,x)-given by Theorem 2.1 - is the density of the law of X_t.

Theorem 2.3 $\forall(s,x) \in [o,t]\times R^N$,

$$v(s,x) = E_{sx}[f(X_t)] \tag{2.11}$$

Hint of proof : It suffices to prove the result with regular b,σ and f, since both v and P_{sx} depend continuously on these ; the result is then obtained as a limit in the results with regular coefficients.
Moreover, if b,σ and f have bounded derivatives in x of any order, then one can show that v itself is regular enough, so that one can apply Ito formula to the process :

$$\psi_\theta = v(\theta, X_\theta)$$

Then :

$$\psi_t = \psi_s + \int_s^t (v'_\theta + Lv)(\theta, X_\theta)\,d\theta + \int_o^t \nabla v(\theta, X_\theta).\sigma(\theta, X_\theta)\,dw_\theta$$

One can show than the expectation of the above Ito integral is zero. The Lebesque integral is zero, from (2.9).
Finally :

$$E_{sx}\,\psi_s = E_{sx}\,\psi_t, \text{ which is (2.11).}$$

Remark 2.4 Using the Markov property of X_t, one can prove the "reverse" of theorem 2.3, namely that the quantity defined by (2.11) satifies equation (2.9)-see DAVIS-MARCUS [7].

Here and in the sequel, (.,.) denotes the scalar product in $L^2(R^N)$.

Proposition 2.5 (v(s), p(s)) is constant, for $s \in [o,t]$.

Proof : From the properties of the solutions of (2.9) and (2.10), on can show-see e.g. BENSOUSSAN-LIONS [1]- that (v(s),p(s)) is differentiable a.e., and :

$$\frac{d}{ds}(v(s),p(s)) = <\frac{dv}{ds}(s),p(s)> + <v(s),\frac{dp}{ds}(s)> \quad \text{a.e.}$$

$$= - <L\,v(s), p(s)> + <v(s), L^*p(s)>$$

$$= o$$

Applying proposition 2.5, and (2.11), we get :

$$(p(t),f) = (p(o), v(o))$$
$$= \int_{\mathbb{R}^N} p_o(x) E_{ox}[f(X_t)]\,dx$$

So that, from (2.7):

$$(p(t),f) = E[f(X_t)] \qquad (2.12)$$

Since (2.12) is true $\forall f \in C_o(R^N)$, and also $\forall t \geq o$, we have proved:
Theorem 2.6: $\forall t \geq o$, $p(t,.)$ is the density of the law of X_t.

2.4 The Feynman-Kac formula

Let α be a bounded measurable function defined on R^N, and consider the backward P.D.E.:

$$\left.\begin{aligned} \frac{dv}{ds}(s) + L\,v(s) + \alpha\, v(s) = o, \quad s \leq t \\ v(t) = f \end{aligned}\right\} \qquad (2.13)$$

with $f \in C_o(R^N)$.

Theorem 2.3 generalises to :

Theorem 2.7 : $\forall (s,x) \in R+ \times R^N$,

$$v(s,x) = E_{sx}[f(X_t)\exp \int_s^t \alpha(X_\theta)\,d\theta] \qquad (2.14)$$

Hint of proof : The proof is very similar to that of theorem 2.3, once one defines :

$$\psi_\theta = v(\theta, X_\theta)\text{esp}\ [\ \int_s^\theta \alpha(X_\mu)\,d\mu]$$

(2.14) is called the Feynman-Kac formula.

Define now the process $(Z_t, t \geq o)$ by :

$$\frac{dZ_t}{dt} = \alpha(X_t)Z_t,\ Z_o = 1$$

and consider the equation adjoint to (2.13):

$$\left.\begin{aligned} \frac{dp}{ds}(s) = L^*p(s) + \alpha p(s) \\ p(o) = p_o \end{aligned}\right\} \qquad (2.15)$$

It then follows by arguments similar to those in §2.3 :

$$(p(t),f) = E[\ f(X_t)Z_t],\ \forall t \geq o \quad \forall f \in C_o(R^N)$$

$$(p(t),1) = E[\ Z_t\ ]$$

The same procedure gives interesting results when Z_t is a (possibly infinite dimensional) vector-valued process, $\alpha(X_t)$ being replaced by an operator, and $\exp[\int_0^t \alpha(X_s)ds]$ by a semi-group - see BOUC-PARDOUX [2].

The relevant generalisation of this procedure to non-linear filtering will consist in replacing Z_t by an exponential martingale, E_{sx} in (2.14) by a conditional expectation, and (2.13) by a backward stochastic PDE.

3. THE GENERAL PROBLEM

3.1 Formulation of the problem

We consider the following stochastic differential system :

$$\left.\begin{aligned} dX_t &= b(t,X_t)dt + \sigma(t,X_t)\, dW_t \\ dY_t &= h(t,X_t)dt + g(t)dW_t + \tilde{g}(t)d\tilde{W}_t \end{aligned}\right\} \quad (3.1)$$

where X_t and Y_t take values in R^N and R^D respectively, $\binom{W_t}{\tilde{W}_t}$ is an R^{N+D}-valued standard Wiener process.

The hypotheses on b and σ are those of §2.1. We suppose moreover :

h is a bounded measurable function from $R_+ \times R^N$, with values in R^D. (3.2)

g and $\tilde{g}$ are bounded and measurable functions from R_+, into $R^{D\times N}$ and R^{D^2} respectively. (3.3)

$g(t)g^*(t) + \tilde{g}(t)\tilde{g}^*(t) = I$ (3.4)

$\exists\, \beta > o$ s.t. $\tilde{g}(t)\tilde{g}^*(t) \geq \beta I,\ \forall t \geq o$ (3.5)

Remark 3.1 (3.4) means that we have normalised the observation noise, making a one-to-one change of observation process. (3.5) is a non-degeneracy hypothesis, which will be crucial for the Zakaï equation to have a nice solution. It should be clear that when (3.5) is not satisfied, the conditional law of X_t, given $\{Y_s,\ s \leq t\}$ may not have a density with respect to Lebesgue measure.

Let $\Omega = C(R_+ ; R^{N+D})$,

$$\mathcal{G}_t^s = \sigma\{X_\theta, Y_\theta ; s \leq \theta \leq t\},\ \mathcal{G}^s = \mathcal{G}_\infty^s,\ \mathcal{G}_t = \mathcal{G}_t^o .$$

Define -see §2.1 - P_{sx} as the law of the process $(X_t, Y_t ; t \geq o)$, solution of (3.1), with the initial condition :

$X_\theta = x,\ Y_\theta = 0 ;\ 0 \leq \theta \leq s$;

and P as the law of the same process, with the initial condition :

X_o = given random variable, whose law admits the density p_o, satisfying (2.6)

$Y_o = o$

Define finally the observation σ-field :

$$\mathcal{F}_t^s = \sigma\{ Y_\theta - Y_s \, ; s \leq \theta \leq t \} , \quad \mathcal{F}_t = \mathcal{F}_t^o$$

We want to caracterize the conditional law of X_{t_1} -under P- given $\mathcal{F}_{t_2}$; i e . to compute quantities of the form $E[f(X_{t_1})/\mathcal{F}_{t_2}]$, say for any $f \in C_o(R^N)$; when $t_1 = t_2$ (filtering), when $t_1 > t_2$(prediction) and when $t_1 < t_2$ (smoothing).

3.2 The reference probability. Bayes formula

We now introduce new probability measures-called reference probabilities, which will serve us as tools for the derivations. Define

$$Z_t^s = \exp [\int_s^t h(X_\theta) \, dY_\theta - \frac{1}{2} \int_s^t h^2(X_\theta) d\theta] ^{(2)} , \quad Z_t = Z_t^o$$

and new measures $\overset{o}{P}_{sx}$ and $\overset{o}{P}$ on $(\Omega, \mathcal{G})$, by: $\forall t \geq s$,

$$\left.\frac{d\overset{o}{P}_{sx}}{dP_{sx}}\right|_{\mathcal{G}_t} = (Z_t^s)^{-1} , \quad \left.\frac{d\overset{o}{P}}{dP}\right|_{\mathcal{G}_t} = Z_t^{-1}$$

It follows from the Girsanov-Cameron-Martin formula - see e.g. STROOCK-VARADHAN [16]- that :

$$\left.\begin{aligned} dX_t &= [b(t,X_t) - c^*h(t,X_t)]dt + \sigma(t,X_t)dW'_t \\ dY_t &= g(t)dW'_t + \tilde{g}(t)d\tilde{W}'_t \end{aligned}\right\} \quad (3.6)$$

$\overset{o}{P}$ a.s. [and $\overset{o}{P}_{sx}$ a.s., for $t \geq s$], where $\binom{W'_t}{\tilde{W}'_t}$ is a $\overset{o}{P}$- standard Wiener process with values in R^{N+D} ; and $c = g\sigma^*$.

The following result - which is easy prove - is a version of Bayes formula :

Lemma 3.2 Let $0 < s \leq t$, and f be a real valued bounded measurable function on R^N. Then

$$E [f(X_s)/ \mathcal{F}_t] = \frac{\overset{o}{E}[f(X_s)Z_t / \mathcal{F}_t]}{\overset{o}{E}[Z_t / \mathcal{F}_t]} \text{ a.s.} \qquad (3.7)$$

If we can compute the numerator of the right-hand side of (3.7) for any bounded and continuous f, then we can compute its denominator, and hence also the left-hand side of (3.7), for the same class of functions f.

In addition to the infinitesimal operator L of X_t - defined in §2.1 - we will use the following (family indexed by t of) operators belonging to $\mathcal{L}(H^1 ; (L^2(R^N))^D)$: $\forall u \in H^1$,

$$(B_k(t)u)(x)= h_k(t,x)u(x)+ \sum_{i=1}^{N} c_{ki}(t,x)\frac{\partial u}{\partial x_i}(x) \quad k = 1,\dots,D$$

h is the observation function, and c-defined above- is related to the joint quadratic variation of X and Y :

$$< X,Y >_t = \int_0^t c(s,X_s)ds$$

Then B expresses the relation between X and Y.

We define $B^*(t)\in \mathcal{L}(L^2(R^N);[(H^1)']^D)$

by $B^*(t)u = \begin{pmatrix} B_1^*(t)u \\ \dots\dots \\ B_D^*(t)u \end{pmatrix}$ where $B_k^*(t)$ is the adjoint of $B_k(t)$, k=1...D.

It follows from (2.5) that

$B^*(t)\in \mathcal{L}(H^1 ; (L^2(R^N))^D)$.

3.3 Two associated S.P.D.Es

We consider the following backward stochastic P D E :

$$\left.\begin{array}{l} dv(s)+ L\,v(s)ds + B\,v(s)\,\overset{(3)}{dY_s} =o, s \leqslant t \\ \qquad v(t)=f \end{array}\right\} \quad (3.8)$$

and the foward S.P.D.E.:

$$\left.\begin{array}{l} dp(s)= L^*p(s)ds + B^*p(s)dY_s,\ s \geqslant o \\ p(o)= p_o \end{array}\right\} \quad (3.9)$$

Remark 3.3 We are looking for a solution to (3.9) as an $\mathcal{F}_s$-adapted process, and (3.9) is to be considered as an Ito equation.

The solution to (3.8) we are looking for will be $\mathcal{F}_t^s$-adapted. The stochastic integral in (3.8) has to be considered as a "backward Ito integral". Indeed, if

$$\tilde{Y}_s \triangleq Y_s - Y_t,$$

$$d\tilde{Y}_s = dY_s$$

But $\tilde{Y}_s$ is a "backward $\overset{o}{P} - \mathcal{F}_t^s$ wiener process",i.e. $\forall s_1 < s_2 \leq t$, $\tilde{Y}_{s_1} - \tilde{Y}_{s_2}$ is a gaussian r.v. with mean o, and covariance operator $(s_2 - s_1)I$, independent of $\mathcal{F}_t^{s_2}$. The definition of the stockastic integral in (3.8) is then the Ito definition, up to a time reversal.

The relevant coercivity condition which is needed for equations (3.8) and (3.9) is :

$$\left.\begin{aligned} &\exists \lambda \in R,\ \mu > o,\ \text{s.t.} \quad \forall u \in H^1, \\ &< -Lu,u > + \lambda \| u \|^2_{L^2(R^N)} \geq \mu \| u \|^2_{H^1} + \tfrac{1}{2} \| B u \|^2_{(L^2(R^N))^D} \end{aligned}\right\} \quad (3.10)$$

Both (2.4) and (3.5) are crucial, for (3.10) to hold. Using (3.10), one can prove - see [13] :

Theorem 3.4 Equations (3.8) and (3.9) have unique solutions :

$$v,p \in L^2(\Omega \times]o,t[, d\overset{o}{P} \times dt\,;\, H^1) \cap$$

$$\cap L^2(\Omega, d\overset{o}{P}\,;\, C([o,t];L^2(R^N)))$$

where $v(s)$ is $\mathcal{F}_t^s$-adapted, and $p(s)$ is $\mathcal{F}_s$-adapted.

3.4 The filtering problem

The main step in the derivation is the following sort of generalisation of the Feynman-Kac formula :

Theorem 3.5 : $\forall s \in [o,t]$, the following equality holds $dx \times d\overset{o}{P}_{sx}$ a.e.:

$$v(s,x) = \overset{o}{E}_{sx}[f(X_t)Z_t^s / \mathcal{F}_t^s] \qquad (3.11)$$

Hint of proof : Imitating the proof of Theorem 2.7, we define $\psi_\theta = v(\theta,X_\theta)Z_\theta^s$. The difficulty now is that ψ_θ depends on both the past and future of Y. Therefore there is no way to compute the differential $d\psi_\theta$.
Then one has to discretize the time interval [s,t], and compute small increments of the process ψ_θ. The proof, which is rather long, can be found in [13]. We will only give a formal (but we believe convincing) verification of the fact that v, defined by (3.11), must satisfy equation (3.8).

As is show in [13], it suffices to prove the result with arbitrarily smooth coefficients. Assume then f and the coefficients are smooth, and that we can conclude that v(s,x), defined by (3.11), is twice continuously differentiable in x.

Let $s = t_o < t_1 < \ldots < t_n = t$ be a mesh, with $t_{i+1} - t_i = (t-s)/n$. Admit for the moment the following equality, which follows from the Markov property of X., and will be proved later in lemma 3.9 :

$$\overset{o}{E}{}^{\mathcal{F}^{t_i}_t}_{t_i x}[f(X_t)Z_t^{t_i}] = \overset{o}{E}{}^{\mathcal{F}^{t_i}_t}_{t_i x}[Z^{t_i}_{t_{i+1}}\, \overset{o}{E}{}^{\mathcal{F}^{t_{i+1}}_t}_{t_{i+1}, X_{t_{i+1}}}(f(X_t)Z_t^{t_{i+1}})] \quad (3.12)$$

It follows from (3.12) :

$$v(t_i,x) - v(t_{i+1},x) = \overset{o}{E}{}^{\mathcal{F}^{t_i}_t}_{t_i x}[Z^{t_i}_{t_{i+1}}\, v(t_{i+1},X_{t_{i+1}}) - v(t_{i+1},x)]$$

It follows from P.Levy's theorem-see e.g. BENSOUSSAN-LIONS [1]- that under measure $\overset{o}{P}$,

$$\tilde{Y}_t = \int_0^t (I-g^*(s)g(s))^{-1/2}\,(dW'_s - g^*(s)dY_s)$$

is a standard Wiener process with values in R^N, independent of Y_t. Then :

$$dW'_\theta = g^*(\theta)dY_\theta + j(\theta)d\tilde{Y}_\theta$$

$$dX_\theta = (b(X_\theta) - c^*h(X_\theta))d\theta + c^*(X_\theta)dY_\theta + \tilde{c}(X_\theta)d\tilde{Y}_\theta$$

where $j = (I - g^*g)^{1/2}$, $\tilde{c} = \sigma j$.

It then follows from Ito formula :

$$Z^{t_i}_{t_{i+1}}\, v(t_{i+1},X_{t_{i+1}}) - v(t_{i+1},x) = \int_{t_i}^{t_{i+1}} Z_\theta^{t_i} L\, v(t_{i+1}, X_\theta)d\theta$$

$$- \int_{t_i}^{t_{i+1}} Z_\theta^{t_i} \nabla v(t_{i+1}, X_\theta)\, c^*h(X_\theta)d\theta$$

$$+ \int_{t_i}^{t_{i+1}} Z_\theta^{t_i} \nabla v(t_{i+1},X_\theta)\, c^*(X_\theta)dY_\theta + \int_{t_i}^{t_{i+1}} Z_\theta^{t_i} \nabla v(t_{i+1},X_\theta)\tilde{c}(X_\theta)d\tilde{Y}_\theta$$

$$+ \int_{t_i}^{t_{i+1}} Z_\theta^{t_i} h(X_\theta)v(t_{i+1},X_\theta)dY_\theta + \int_{t_i}^{t_{i+1}} Z_\theta^{t_i} \nabla v(t_{i+1},X_\theta)c^*\, h(X_\theta)d\theta$$

Remark that $v(t_{i+1},x)$ is $\mathcal{F}_t^{t_{i+1}}$ - adapted, then independent of $\mathcal{F}_\theta$, $\theta \leqslant t_{i+1}$, so that all terms above make sense. It follows from the independence between Y. and $\tilde{Y}$., that $\overset{o}{E}_{t_i x}(-/\mathcal{F}_t^{t_i})$ of the $d\tilde{Y}$ integral is zero. Finally :

$$v(s,x) = f(x) + \sum_{i=0}^{n-1} [v(t_i,x) - v(t_{i+1},x)]$$

$$= f(x) + \sum_{i=0}^{n-1} \overset{o}{E}{}^{\mathcal{F}_t^{t_i}}_{t_i x} \int_{t_i}^{t_{i+1}} L(s) v(t_{i+1}, X_s) Z_s^{t_i} \, ds$$

$$+ \sum_{i=0}^{n-1} \overset{o}{E}{}^{\mathcal{F}_t^{t_i}}_{t_i x} \int_{t_i}^{t_{i+1}} B(s) \, v(t_{i+1}, X_s) Z_s^{t_i} \, dY_s$$

Taking the limit as $n \to \infty$ yields :

$$v(s,x) = f(x) + \int_s^t L v(\theta,x) d\theta + \int_s^t B v(\theta,x) dY_\theta$$

The last equality being true $\forall (s,x)$, implies (3.8). Remark that the last stochastic integral is a backward one, whereas the previous ones were foward.

The second result needed is :

Theorem 3.6 The process $\{(v(s), p(s)), o \leqslant s \leqslant t\}$ is a.s. constant.

Hint of proof : Again $(v(s),p(s))$ depends both on the past and the future of Y. Therefore there is no way to compute its differential. It suffices to show that

$$\forall \theta_1, \theta_2 \text{ s.t. } \quad s < \theta_1 < \theta_2 < t,$$

$$(v(\theta_1), p(\theta_1)) = (v(\theta_2), p(\theta_2)) \quad \text{a.s.}$$

Define the mesh $\theta_1 = t_0 < t_1 < \ldots < t_n = \theta_2$, with $t_{i+1} - t_i = \frac{\theta_2 - \theta_1}{n} = \Delta t$,

and consider the following time - discretised approximation of (3.8) and (3.9) :

$$v^{i+1} - v^i + \left(\int_{t_i}^{t_{i+1}} L(s)ds\right)v^i + \left(\int_{t_i}^{t_{i+1}} B(s)dY_s\right)v^{i+1} = 0, \quad o \leqslant i < n$$

$$p^{i+1} - p^i = \left(\int_{t_{i+1}}^{t_{i+2}} L^*(s)ds\right)p^{i+1} + \left(\int_{t_i}^{t_{i+1}} B^*(s)dY_s\right)p^i, \quad o \leqslant i < n$$

$$v^n = \int_{\theta_2}^{\theta_2+\Delta t} v(s)ds \qquad p^o = \int_{\theta_2-\Delta t}^{\theta_2} p(s)ds$$

The sequence $\{v^i, p^i;\ o \leqslant i \leqslant n\}$ is well defined for small enough Δt.
Take the scalar product in $L^2(R^N)$ of the first equality with p^i, and of the second with v^{i+1}. Sum the two resulting equalities. Iterate for $i = o, 1 \ldots n-1$. We get :

$$(v^o, p^o) - \Delta t \int_{\theta_1}^{\theta_1+\Delta t} \langle L(s)v^o, p^o\rangle ds =$$

$$= (v^n, p^n) - \Delta t \int_{\theta_2}^{\theta_2+\Delta t} \langle L(s)\, v^n, p^n\rangle ds$$

The result follows by letting Δt go to zero.

Now we have the solution of the filtering problem :

Corollary 3.7 $\forall t \geqslant o$, $p(t,x)\left(\int_{R^N} p(t,x)dx\right)^{-1}$ is the density of the conditional law of X_t, given $\mathcal{F}_t$.

Proof : It $f \in C_o(R^N)$, by the two preceding theorems,

$$(p(t),f) = (p_o, v(o))$$
$$= \int_{R^N} p_o(x)\overset{o}{E}_{ox}[f(X_t)Z_t / \mathcal{F}_t]dx$$

$$(p(t),f) = \overset{o}{E}[\,f(X_t)\,Z_t\,/\,\mathcal{F}_t] \tag{3.13}$$

This is true $\forall f \in C_o(R^N)$. Then $p(t,x) \geqslant o$ a.e., a.s., and by monotone convergence :

$$(p(t),\ 1) = \overset{o}{E}[\,Z_t\ /\ \mathcal{F}_t]$$

Now it follows from lemma 3.2 :

$$E[\, f(X_t)/\mathcal{F}_t \,] = \frac{(p(t),f)}{(p(t),1)},$$

$\forall f \in C_o(R^N)$, and also $\forall t \geq o$

$p(t,x)$ is called the " unnormalised conditional density ".

3.5 The prediction problem

We want to compute the conditional law of X_t, given $\mathcal{F}_s$, $o \leq s < t$.
This problem reduces easily to a filtering problem. Define :

$$\overline{Y}_\theta = Y_{\theta \wedge s} + \tilde{W}_\theta - \tilde{W}_{\theta \wedge s}$$
$$\overline{\mathcal{F}}_t = \sigma\{\, \overline{Y}_\theta ,\ \theta \leq t \,\}$$

It follows from the fact that

$\sigma\{\, \tilde{W}_\theta - \tilde{W}_s ,\ s \leq \theta \leq t \,\}$ and $\mathcal{F}_s \vee \sigma(X_t)$ are independent, that

$$E[\, f(X_t)/\mathcal{F}_s \,] = E[\, f(X_t)/\overline{\mathcal{F}}_t \,]$$

Then, from the results of §3.4, the density of the conditional law of X_t, given $\mathcal{F}_s$, is $\overline{p}(t,x)(\overline{p}(t),1)^{-1}$, where $\overline{p}(t)$ is the value at time t of the solution of :

$$\left.\begin{aligned} d\overline{p}(\theta) &= L^*\overline{p}(\theta)\,d\theta + 1_{\{\theta \leq s\}} B^* p(\theta)\, dY_\theta \\ \overline{p}(o) &= p_o \end{aligned}\right\} \quad (3.14)$$

3.6 The smoothing problem

We want to compute the conditional law of X_s, given $\mathcal{F}_t$, $o \leq s < t$.
First consider the backward S.P.D.E :

$$\left.\begin{aligned} dv(\theta) + L\, v(\theta)\, d\theta + B\, v(\theta)\, dY_\theta &= o, \quad \theta \leq t \\ v(t) &= 1 \end{aligned}\right\} \quad (3.15)$$

Again, (3.15) has a unique solution, but in a broader class than that of (3.8), since $1 \notin L^2(R^N)$. The idea is to use the fact that $1 \in L^2(R^N, (1+|x|^2)^{-N} dx)$.
It follows from (3.11), by monotone convergence :

$$v(s,x) = \overset{o}{E}_{sx}[\, Z_t^s / \mathcal{F}_t^s \,] \quad (3.16)$$

Now consider again the Zakaï equation :

$$dp(\theta)= L^*p(\theta)d\theta + B^*p(\theta)dY_\theta \ , \ \theta \geq o$$
$$p(o)= p_o$$

Theorem 3.8 $p(s,x)v(s,x)(p(s),v(s))^{-1}$ is the density of the conditional law of X_s, given $\mathcal{F}_t$.
The proof of the Theorem makes use of the :

Lemma 3.9

(i) $\overset{o}{E}[f(X_s)Z_t / \mathcal{F}_t]= \overset{o}{E}^{\mathcal{F}_t}[f(X_s)Z_s \overset{o}{E}_{sX_s}^{\mathcal{F}_t^s}(Z_t^s)]$

(ii) If $(x,\omega) \to G(x,\omega)$ is $\mathcal{B}_N^{(4)} \otimes \mathcal{F}_t^s$ measurable with values in R_+, then :
$\overset{o}{E}[G(X_s)Z_s / \mathcal{F}_t] = \int_{R^N} p(s,x)G(x)dx$

First admit that lemma 3.9 holds true, and proced to :
Proof of theorem 3.8
Let $f \in C_o(R^N)$, and $f \geq o$. We use (i), then (3.16), then (ii) :

$$\overset{o}{E}[f(X_s)Z_t / \mathcal{F}_t]= \overset{o}{E}^{\mathcal{F}_t}[f(X_s)Z_s \overset{o}{E}_{s,X_s}^{\mathcal{F}_t^s}(Z_t^s)]$$
$$= \overset{o}{E}^{\mathcal{F}_t}[f(X_s)v(s,X_s)Z_s]$$
$$= \int_{R^N}p(s,x)v(s,x)f(x)dx$$

Then, by monotone convergence, we get :

$$\overset{o}{E}[Z_t / \mathcal{F}_t] = (p(s),v(s))$$

The result now follows from lemma 3.2

Proof of lemma 3.9

(i) It suffices to show that $\forall \varphi$ r.v. $\mathcal{F}_s$-measurable and bounded, $\forall \psi$ r.v. $\mathcal{F}_t^s$-measurable and bounded, the scalar products in $L^2(\Omega,d\overset{o}{P})$ of $\varphi\psi$ with both sides of the equality are equal. We are going to use the Markov property of the process $(X_\theta, Y_{\theta \vee s} - Y_\theta)$, whose state at time $\theta = s$ depends only on X_s.

$$\overset{o}{E}[\,f(X_s)Z_t\ \varphi\psi\,] = \overset{o}{E}[\,f(X_s)Z_s\ \varphi\ \overset{o}{E}{}^{\mathcal{F}_s}(Z_t^s\psi)\,]$$

$$= \overset{o}{E}[\,f(X_s)Z_s\ \varphi\ \overset{o}{E}_{sX_s}(\psi\overset{o}{E}{}_{sX_s}^{\mathcal{F}_t^s}(Z_t^s))\,]$$

$$= \overset{o}{E}[\,f(X_s)Z_s\ \overset{o}{E}{}_{sX_s}^{\mathcal{F}_t^s}(Z_t^s)\varphi\psi\,]$$

(ii) From the monotone class theorem, it suffices to prove the equality for G of the form :

$G(x,\omega) = g(x)\ 1_A\ (\omega)$

where $g \in C_o(R^N)$, $g \geq o$; and $A \in \mathcal{F}_t^s$.

It then suffices to show that $\forall\ \varphi,\ \psi$ as above,

$$\overset{o}{E}\ [\,g\,(X_s)1_A\ Z_s\ \varphi\psi\,] = \overset{o}{E}\ [\,\varphi\psi\ 1_A(p(s),g)\,]$$

We will make use of (3.13), and the independence under $\overset{o}{P}$ between $\mathcal{F}_s$ and $\mathcal{F}_t^s$.

$$\overset{o}{E}[\,g(X_s)1_A Z_s\,\varphi\,\psi] = \overset{o}{E}[g(X_s)\ Z_s\,\varphi\,]\,\overset{o}{E}\,[1_A\,\psi\,]$$

$$= \overset{o}{E}[\varphi\,(p(s),g)\,]\overset{o}{E}[1_A\,\psi\,]$$

$$= \overset{o}{E}[\varphi\,\psi\ 1_A(p(s),g)]$$

4. INDEPENDENT SIGNAL AND NOISE

4.1 Orientation

We want now to consider the specific case where X. and the noise in Y. are independent, i.e. g = o.
Then the system (3.1) reduces to :

$$\left.\begin{aligned} dX_t &= b(t,X_t)dt + \sigma(t,X_t)dW_t \\ dY_t &= h(t,X_t)dt + d\tilde{W}_t \end{aligned}\right\} \qquad (4.1)$$

We make all the hypotheses of §2.1, and suppose (also this is a little unnecessarily too strong, see [15]):

$$h \in C_b^{1,2}(R_+ \times R^N\ ;\ R^D)^{(5)} \qquad (4.2)$$

P, P_{sx}, $\overset{o}{P}$, $\overset{o}{P}_{sx}$, Z_t^s and $\mathcal{F}_t^s$ are defined as in §3.

The two associated S.P.D.Es are now :

$$\left.\begin{array}{l} dv(s)+ L\,v\,(s)ds + \sum_{i=1}^{D} h^i v(s)dY_s^i = o, \; s \leq t \\ \qquad\qquad v(t) = f \end{array}\right\}$$

$$\left.\begin{array}{l} dp(s) = L^* p(s)ds + \sum_{i=1}^{D} h^i p(s)dY_s^i, \; s \geq o \\ p(o)=p_o \end{array}\right\}$$

Since the operators "multiplication by h^i" and "multiplication by h^j" commute, we can certainly use the device of DOSS [8]and SUSSMANN [17], to express the solutions in terms of solutions of ordinary PDE's. This has been done indeed by LIPTSER-SHIRYAYEV [10],(6) and CLARK [4] who interpreted this result in terms of "robustness". See the discussion in DAVIS [6], this volume.

The idea is to find equations for the quantities :

$$u(t,x) = v(t,x) \exp\,[Y_t\, h\,(t,x)]$$

$$q(t,x) = p(t,x) \exp\,[-Y_t\, h\,(t,x)],$$

instead of v and p.

But instead of doing these transformations on the above equations, let us give a direct derivation of the robust equations for u and q. This derivation has two interesting features.
First, it is rather elementary : we are going to use only the theory of P.D.E., the Girsanov transformation, and the Feynman-Kac formula. Second, it does work as well when b and σ depend on the whole past of Y. The ideas in the next subsection are closely related to those in DAVIS [6], this volume. See also the bibliography therein.

4.2 The filtering problem

First note that under $\overset{o}{P}$ -or $\overset{o}{P}_{sx}$:

$$dX_t = b(t,X_t)dt + \sigma(t,X_t)dW_t \qquad (4.3)$$

and Y_t is a standard Wiener process, independent of X_t. In other words :

$$\overset{o}{P}_{sx}(dX,dY)= \overline{P}_{sx}(dX) \times W_{so}(dY)$$

where $\overline{P}_{sx}$ is the law of the solution of (4.3), with initial condition $X_\theta = x$, $\theta \leq s$; and W_{so} is Wiener measure on $C([s,+\infty[\,;\, R^D)$, starting at o. We define similarly $\overline{P}$, corresponding to the initial density p_o at time o. It then follows :

$$\overset{o}{E}_{sx}[f(X_t)Z_t^s / \mathcal{F}_t^s] = \overline{E}_{sx}[f(X_t)Z_t^s], \quad W_{so} \quad \text{a.s.} \tag{4.4}$$

Now let us integrate by parts the stochastic integral which appears in Z_t^s :

$$\int_s^t h(\theta,X_\theta)\,dY_\theta = h(t,X_t)Y_t - h(s,X_s)Y_s - \int_s^t Y_\theta(h'_\theta + Lh)(\theta,X_\theta)\,d\theta - \int_s^t Y_\theta[(\sigma\nabla h)(\theta,X_\theta),dW_\theta]$$

Define :

$$\overset{\vee}{Z}{}_t^s = \exp\{-\int_s^t Y_\theta[\sigma\nabla h(X_\theta),dW_\theta] - \frac{1}{2}\int_s^t Y_\theta^2[a\nabla h,\nabla h](X_\theta)\,d\theta\}$$

$$e(\theta,x,y) = \frac{y^2}{2}[a\nabla h, \nabla h](\theta,x) - y\,(h'_\theta + Lh)(\theta,x) - \frac{h^2(\theta,x)}{2}$$

Then :

$$\exp[Y_s h(X_s)]Z_t^s = \exp[Y_t h(X_t)].\overset{\vee}{Z}{}_t^s.\exp[\int_s^t e(\theta,X_\theta,Y_\theta)\,d\theta]$$

We now fix a Y.-trajectory, and define a new law for X (which depends on the fixed trajectory of Y.):

$$\left.\frac{d\overset{\vee}{P}_{sx}}{d\overline{P}_{sx}}\right|_{\mathcal{H}_t^s} = \overset{\vee}{Z}{}_t^s, \text{ where } \mathcal{H}_t^s = \sigma\{X_\theta, s \leqslant \theta \leqslant t\}$$

Finally define :

$$u(s,x) = \exp[Y_s h(x)]v(s,x), \text{ i.e :}$$

$$\begin{aligned} u(s,x) &= \exp[Y_s h(x)]\overset{o}{E}_{sx}[f(X_t)Z_t^s / \mathcal{F}_t^s] \\ &= \overline{E}_{sx}[\exp[Y_s h(X_s)]Z_t^s f(X_t)] \\ &= \overset{\vee}{E}_{sx}\{\exp[Y_t h(X_t)]f(X_t)\exp\int_s^t e(\theta,X_\theta,Y_\theta)\,d\theta\} \end{aligned}$$

We recognize here the Feynman-Kac formula, i.e. u must be the solution of :

$$\left.\begin{aligned} \frac{du}{ds}(s) + \overset{\vee}{L}(s,Y_s)u(s) + e(s,Y_s)u(s) &= o, \quad s \leqslant t \\ u(t) &= f\exp[Y_t h] \end{aligned}\right\} \tag{4.5}$$

where $\overset{\vee}{L}(s,Y_s)u = L(s)u - \sum_{i=1}^{N}(aY_s.\nabla h)_i \frac{\partial u}{\partial x_i}$.

Let $f \in C_o(R^N)$. From the resultsin BENSOUSSAN-LIONS [1], to each trajectory of Y. corresponds a unique solution of (4.5):

$$u \in L^2(o,t;H^1) \cap C([o,t];\, L^2(R^N))$$

Now, u being this solution of (4.5), it follows from Theorem 2.7 and the above calculations :

Theorem 4.1 $\forall\, s \in [o,t\,]$, the following equality holds $dx \times dW_{so}$ a.e.:

$$u(s,x)= \exp[Y_s h(x)]\overset{o}{E}_{sx}[f(X_t)Z_t^s \;/\; \mathcal{F}_t^s]$$

Consider now the forward equation :

$$\left.\begin{aligned} &\frac{dq}{ds}(s)= \check{L}^*(s,Y_s)q(s)+ e(s,Y_s)\, q(s) \\ &q(o)= p_o \end{aligned}\right\} \quad (4.6)$$

Again, (4.6) has a unique solution, and it follows from the argument in proposition 2.5 :

Proposition 4.2 $(q(s),u(s))$ is constant on $[o,t]$.

Finally :

Corollary 4.3 $\forall t \geqslant o$, $q(t,x)\exp[Y_t h(t,x)]$ is the "unnormalized conditional density of X_t, given $\mathcal{F}_t$".

Proof : Using proposition 4.2 and Theorem 4.1, we get :

$$\begin{aligned}(q(t)\exp[Y_t h(t)],f) &=(q(t),u(t)) \\ &=(q(o),u(o)) \\ &=\int_{R^N} p_o(x)\overset{o}{E}_{ox}[f(X_t)Z_t \;/\; \mathcal{F}_t]\,dx \\ &=\overset{o}{E}\,[f(X_t)\,Z_t \;/\; \mathcal{F}_t\,]\end{aligned}$$

This equality being true $\forall f \in C_o(R^N)$, is also true for $f=1$, by monotone convergence. The corollary then follows from lemma 3.2.

4.3 The prediction problem

Since we have assumed $\frac{\partial h}{\partial t}$ to be bounded, we cannot use the trick of §3.5. But we are going to make use of the Markov property of X_t. Let again $f \in C_o(R^N)$.

$$E^{\mathcal{F}_s}[f(X_t)]= E^{\mathcal{F}_s}E^{\mathcal{G}_s}[f(X_t)]$$

$$= E^{\mathcal{F}_s} E_{sX_s}[f(X_t)]$$

$$= E^{\mathcal{F}_s}[\tilde{u}(s,X_s)]$$

where $\tilde{u}$ is the solution of the backward Kolmogorov equation :

$$\left.\begin{array}{l} \frac{d}{d\theta}\tilde{u} + L\tilde{u} = o, \quad \theta \leqslant t \\ \tilde{u}(t)=f \end{array}\right\}$$

From corollary 4.3 :

$$E^{\mathcal{F}_s}[f(X_t)] = \frac{(q(s)\exp[Y_s h(s)], \tilde{u}(s))}{(q(s), \exp[Y_s h(s)])}$$

where q denotes the solution of (4.6).
Now let $p(\theta)$, $\theta \geqslant s$ be the solution of :

$$\left.\begin{array}{l} \frac{dp}{d\theta}(\theta) = L^* p(\theta), \quad \theta \geqslant s \\ p(s) = q(s)\exp[Y_s h(s)] \end{array}\right\} \quad (4.7)$$

From proposition 2.5,

$$(q(s)\exp[Y_s h(s)], \tilde{u}(s)) = (p(t),f)$$

But since (4.7) is a Kolmogorov forward equation,

$$(p(t),1) = (p(s),1)$$

Finally :

$$E(f(X_t)/\mathcal{F}_s) = \frac{(p(t),f)}{(p(t),1)}$$

and we have proved :

Theorem 4.4 $\forall t \geqslant s$, $p(t,x)$ is the unnormalised conditional density of law of X_t, given $\mathcal{F}_s$, where p denotes the solution of (4.7).

4.4 The smoothing problem

Consider the following equations :

$$\left.\begin{array}{l} \frac{du}{d\theta}(\theta) + \check{L}(\theta,Y_\theta)u(\theta) + e(\theta,Y_\theta)u(\theta) = o, \quad \theta \leqslant t \\ u(t)=\exp[Y_t h(t)] \end{array}\right\} \quad (4.8)$$

$$\left.\begin{aligned} &\frac{dq}{d\theta}(\theta) = \check{L}^*(\theta,Y_\theta)q(\theta) + e(\theta,Y_\theta)q(\theta) \\ &q(o) = p_o \end{aligned}\right\} \quad (4.9)$$

Again, (4.8) has a unique solution (also $\exp[Y_t h(t,x)]$ is not a $L^2(R^N)$ function - see BENSOUSSAN-LIONS [1]), and it follows from Theorem 4.1, by monotone convergence :

$$u(s,x) = \exp[Y_s h(s,x)]\overline{E}_{sx}[Z_t^s] \quad (4.10)$$

$$\overset{o}{E}[f(X_s)Z_t / \mathcal{F}_t] = \overline{E}[f(X_s)Z_t] \quad \text{a.s.}$$

From the above integration by parts, Z_t can be considered for a fixed trajectory of Y. Now fix such a trajectory, and use the Markov property of X_t, (4.10) and Corollary 4.3 :

$$\begin{aligned} \overline{E}[f(X_s)Z_t] &= \overline{E}[f(X_s)Z_s \overline{E}_{sX_s}(Z_t^s)] \\ &= \overline{E}[Z_s \exp[-Y_s h(s,X_s)]f(X_s)u(s,X_s)] \\ &= (q(s), f\,u(s)) \end{aligned}$$

Finally, $\forall f \in C_o(R^N)$, $\forall s \in [o,t]$,

$$\overset{o}{E}[f(X_s)Z_t / \mathcal{F}_t] = (q(s), f\,u(s)) \quad \text{a.s.} \quad (4.11)$$

Then (4.11) is also true for $f = 1$, and in view of Lemma 3.2, we have proved :

Theorem 4.5 $\forall s \in [o,t]$, $q(s,x)u(s,x)$ is the unnormalised conditional density of the law of X_s, given $\mathcal{F}_t$.

4.5 Unbounded coefficients

Until now, we have always assumed all coefficients b, σ and h to be bounded. This is a rather stringent hypothesis. Let us look again at the q equation, in the case where the coefficient don't depend on t, for simplicity :

$$\frac{dq}{ds} = \check{L}^*(Y_s)q(s) + e(Y_s)q(s), \text{ where :} \quad (4.12)$$

$$e(Y_s,x) = \frac{Y_s^2}{2}(a\nabla h, \nabla h)(x) - Y_s(Lh)(x) - \frac{h^2(x)}{2}$$

Suppose first that we just want the function h(x) to be unbounded. If we choose $h(x) = |x|^p$, then e is bounded above by a constant depending on Y., and (4.12) has a unique global solution. The same won't be true if for instance $h(x) = |x|^p \sin x$. On the other hand, if the second order coefficients of $\check{L}^*$ are bounded, the first order coefficients at most linear, and $e(x) \leqslant C|x|^{2-\varepsilon}$ $(\varepsilon > 0)$, then one can prove global existence and uniqueness for (4.12).

This permits to cover the case where σ is bounded, b and h grow at most linearly in x, which includes the linear case of the Kalman-Bucy theory. For details and proofs, see [15].

4.6 Numerical approximation

Computing an approximate solution of the Zakaï equation is one possible way of solving a practical non-linear filtering problem, at least when the dimension N is small.
Let us say a few words on the time-discretisation. We go back to the Zakaï equation :

$$\left.\begin{aligned} &dp(t) = L^* p(t)dt + h\, p(t)dY_t \\ &p(o) = p_o \end{aligned}\right\} \quad (4.13)$$

Following MILHSTEIN [12], a good time-discretised approximation of (4.13) is (when D = 1):

$$p_{i+1} - p_i = L^* p_{i+1}\, \Delta t + h\, p_i\, (Y_{t_{i+1}} - Y_{t_i}) + \\ + \frac{1}{2} h^2 p_i\, [(Y_{t_{i+1}} - Y_{t_i})^2 - \Delta t]$$

or even better :

$$(I - \Delta t\, L^*) p_{i+1} = p_i\, \exp[h(Y_{t_{i+1}} - Y_{t_i}) - \frac{h^2}{2}\Delta t] \qquad (4.14)$$

Then, since $p_o(x) \geqslant o$, $p_i(x) \geqslant o$, $\forall i$, $\forall x$.
In fact, the sequence p_i defined by (4.14) has an interpretation in terms of the unnormalised conditional density in a discrete time filtering problem.

Moreover, (4.14) is a time-discretised approximation of the robust equation. One then gets convergence for each sample path of Y., and even estimates of the approximation error. It remains of course to discretise the space, in order to get an algorithm that can be used on a computer. For all details, see LE GLAND[9].

5 POINT PROCESS OBSERVATION

We now consider the case where the observation is a Poisson Point Process, whose intensity is a given function of the unobserved process X_t.

5.1 Formulation of the problem

The unobserved process X_t is again the solution of :

$$dX_t = b(t,X_t)dt + \sigma(t,X_t)dW_t$$

we make all the hypothesis of §2.1.
Define $\Omega^1 = C(R_+ ; R^N)$, $\mathcal{H}_s = \sigma\{X_\theta, \theta \geq s\}$, $\mathcal{H} = \mathcal{H}_0$. Denote by P^1_{sx} the law of X. on $(\Omega^1, \mathcal{H}_s)$, with the initial condition $X_s = x$, and by P^1 the law of X. on $(\Omega^1, \mathcal{H})$, with the initial density $p_0(x)$ satisfying (2.6).

Let now N_t be a standard Poisson process defined on $(\Omega^2, \mathcal{F}, P^2)$. Define :

$$(\Omega, \mathcal{G}, \overset{o}{P}) = (\Omega^1 \times \Omega^2, \mathcal{H} \otimes \mathcal{F}, P^1 \times P^2)$$

and $\overset{o}{P}_{sx} = P^1_{sx} \times P^2$.

Given ψ, bounded and measurable application from $R_+ \times R^N$ into R_+, we define :

$$\rho(t) = \psi(t,X_t)$$

$$Z^s_t = \prod_{\left\{\substack{s < \theta \leq t \\ N_\theta \neq N_{\theta-}}\right\}} \rho(\theta) \quad \exp\left[-\int_s^t (\rho(\theta)-1)d\theta\right]$$

$$\mathcal{G}_t = \sigma\{X_\theta, N_\theta ; \theta \leq t\}, \ \mathcal{F}^s_t = \sigma\{N_\theta - N_s, \ s \leq \theta \leq t\}, \ \mathcal{F}_t = \mathcal{F}^o_t.$$

$$\left.\frac{dP}{d\overset{o}{P}}\right|_{\mathcal{G}_t} = Z^o_t$$

It can be shown- see BREMAUD [3] that under P, the law of X. is unchanged, whereas N_t is a doubly stochastic Poisson Process, with intensity $\rho(t)$, i.e:

$N_t - \int_o^t \rho(s)ds$ is a $\mathcal{F}_t$ - P martingale. Our problem consists in computing $E[f(X_{t_1}) / \mathcal{F}_{t_2}]$ for $t_1 = t_2$ (filtering), for $t_1 > t_2$ (prediction), and for $t_1 < t_2$ (smoothing).
Define :

$$h(s,x) = \psi(s,x) - 1$$

5.2 The filtering problem

Consider the following SPDEs:

$$\left.\begin{array}{l} dv(s) + Lv(s)ds + v(s)h(s)[dN_s - ds] = o,\ s \leqslant t \\ v(t) = f \end{array}\right\} \quad (5.1)$$

where $f \in C_o(R^N)$

$$\left.\begin{array}{l} dp(s) = L^*p(s)ds + p(s^-)h(s)[dN_s - ds],\ s \geqslant o \\ p(o) = p_o \end{array}\right\} \quad (5.2)$$

The dN_s integrals can be interpreted as Stieltjes integrals. We are looking for solutions whose paths are right-continuous and posses left-hand limits ; with values in $L^2(R^N)$; $p(s^-) \triangleq \lim_{\substack{\theta \uparrow s \\ \theta < s}} p(\theta)$. N_s is a pure jump process, with a.s. a finite number of jumps of size 1 on each finite time interval. The pathwise existence and uniqueness of solutions to(5.1) and (5.2) is then obvious.

Let us now prove :

Theorem 5.1 $\forall$ s $\in$ [o,t], the following holds $dx \times dP^2$ a.e.:

$$v(s,x) = \overset{o}{E}_{sx}[f(X_t)Z_t^s \,/\, \mathcal{F}_t^s] \quad (5.3)$$

Proof : Again, it suffices to prove (5.3) with smooth coefficients, in which case v is regular in s and x, between the jump times of N_t. It then follows from Ito formula that, if $V_\theta = v(\theta, X_\theta)$, between the jump times of N_θ,

$$dV_\theta = (v'_\theta + Lv)(\theta, X_\theta)d\theta + (\nabla v\,\sigma)(\theta, X_\theta)dW_\theta$$

Then at any time :

$$dV_\theta = dv(\theta, X_\theta) + Lv(\theta, X_\theta)d\theta + (\nabla v\sigma)(\theta, X_\theta)dW_\theta$$

$$dZ_\theta^s = Z_{\theta-}^s h(X_\theta)\ [dN_\theta - d\theta]$$

Since Z_θ^s is a process with bounded variation, it follows from the generalised Ito formula (see MEYER [11]):

$$d[V_\theta\ Z_\theta^s] = Z_{\theta-}^s\ dV_\theta + V_\theta\ dZ_\theta^s$$

$$= Z_{\theta-}^s[dv + Lvd\theta + vh(dN_\theta - d\theta)](\theta, X_\theta)$$

$$+ Z_{\theta-}^s(\nabla v\,\sigma)(\theta, X_\theta)dW_\theta$$

But, since W_t and N_t are independent under $\overset{o}{P}_{sx}$, $\overset{o}{E}_{sx}(-/\mathcal{F}_t^s)$ of the dW-integral is zero-Moreover, from(5.1), the first term in $d[V_\theta Z_\theta^s]$is zero. Then :

$$\overset{o}{E}_{sx}(v(t,X_t)- v(s,X_s)/ \mathcal{F}^s_t)=o,$$

which is (5.3).

Theorem 5.2 The process $\{(v(s),p(s)),o \leqslant s \leqslant t\}$ is a.s. constant.

Proof : By the argument of proposition 2.5, $(v(s),p(s))$ is constant between the jumps of N_s.
Now if $N_s \neq N_{s^-}$,

$$p(s)= p(s^-)(1 + h(s))$$

$$v(s^-)= v(s)(1 + h(s)), \text{ then}$$

$$(p(s),v(s))=(p(s^-),v(s^-))$$

We now get, by the same argument as in Corollary 3.5 :

Corollary 5.3 $\forall t \geqslant o$, $p(t,x)$ is the unnormalised conditional density of the law of X_t, given $\mathcal{F}_t$.

5.3 The prediction problem

We want to caracterise the conditional law of X_t, given $\mathcal{F}_s$, $s \leqslant t$. We can reduce this problem to the above filtering problem exactly as in §3.5. once ρ is replaced by :

$$\overline{\rho}(\theta)= 1_{\{\theta \leqslant s\}}\, \rho(\theta)+ 1_{\{\theta > s\}}$$

The unnormalised conditional density is then $p(t)$, where p is the solution of :

$$dp(\theta)= L^*p(\theta)d\theta + 1_{\{\theta \leqslant s\}}\, p(\theta^-)h(\theta)[dN_\theta - d\theta]$$

$$p(o) = p_o$$

5.4 The smoothing problem

We want to caracterize the conditional law of X_s, given $\mathcal{F}_t$, $s \leqslant t$.

Consider the following S.P.D.Es.

$$dv(\theta)+ L v (\theta)d\theta + v(\theta)h(\theta)[dN_\theta - d\theta]=o, \quad s \leqslant t$$
$$v(t)=1$$

$$dp(\theta)= L^*p(\theta)d\theta + p(\theta^-)h(\theta)[dN_\theta - d\theta]$$
$$p(o)= p_o$$

Again, these equations have unique solutions, and from (5.3), by monotone convergence :

$$v(s,x) = \overset{o}{E}_{sx}[Z_t^s / \mathcal{F}_t^s]$$

Using the same arguments as in §4.4, we get, for $f \in C_o(R^N)$:

$$\overset{o}{E}[f(X_s)Z_t / \mathcal{F}_t] = \overset{o}{E}^{\mathcal{F}_t}[Z_s f(X_s) \overset{o}{E}_{sX_s}^{\mathcal{F}_t^s}(Z_t^s)]$$

$$= \overset{o}{E}^{\mathcal{F}_t}[Z_s f(X_s) v(s,X_s)]$$

$$= (p(s), f\,v(s))$$

and finally :

Theorem 5.4 $\forall s \in [o,t]$, $p(s,x)v(s,x)$ is the unnormalised conditional density of the law of X_s, given $\mathcal{F}_t$.

Remark 5.5 N_t could be replaced by a random Poisson measure-see [14] concerning the filtering problem.
One could also combine the continuous observation, and the point process observation.

6 FOOTNOTES

(1) $C_o^2(R^N)$ denotes the space of twice continuously differentiable functions from R^N into R, with compact support.

(2) In order to simplify the notations, we write everything as if D = 1. Products should be replaced by scalar products...

(3) Here again $B\,v\ dY$ means $\sum_{k=1}^{\infty} B_k v\, dY_k$

(4) $\mathcal{B}_N$ denotes the Borel σ-field on R^N.

(5) $C_b^{1,2}(R_+ \times R^N;\ R^D)$ denotes the space of functions from $R_+ \times R^N$ into R^D, which are once continuously differentiable with respect to t, twice with respect to x, the function and its derivatives being bounded.

(6) In fact originally by B.L. ROZOVSKII, see ref [140] in [10].

7 REFERENCES

[1] A. BENSOUSSAN-J.L. LIONS *Application des inéquations variationnelles en contrôle stochastique.* Dunod (1978).

[2] R. BOUC - E. PARDOUX Moments of semi-linear random evolutions. Publication de Mathématiques Appliquées Marseille - Toulon 80-5, to appear.

[3] P. BREMAUD *Point processes and queues : Martingale dynamics,* book to appear.

[4] J.M.C. CLARK The design of robust approximations to the stochastic differential equations of non-linear filtering. in : *Communication Systems and Random Process Theory*, Ed.J. Skwirzynski, Sijthoff & Noordhoff (1978).

[5] R.F. CURTAIN This volume

[6] M.H.A. DAVIS Pathwise non-linear filtering, this volume

[7] M.H.A. DAVIS - S.I. MARCUS This volume

[8] H. DOSS Liens entre équations différentielles stochastiques et ordinaires. *Ann.Inst. H. Poincaré,* 13,pp.99-125, (1977).

[9] F. LE GLAND This volume, and thèse 3° Cycle, Paris IX, to appear

[10] R. LIPTSER - A. SHIRYAYEV *Statistics of random processes.* Springer (1978).

[11] P.A. MEYER Un cours sur les intégrales Stochastiques. in *Sém Proba X,* lecture notes in Math. 511, Springer (1976).

[12] G.N. MILSHTEIN Approximate integration of stochastic differential equations. *Siam Theory Prob. and Applic.* 19, 3, pp. 557-562, (1974).

[13] E. PARDOUX Stochastic partial differential equations and filtering of diffusion processes. *Stochastics* 3, pp. 127-167, (1979).

[14] E. PARDOUX Filtering of a diffusion process with Poisson-type observation in *Stoch. Cont. Theory and Stoch. Diff. Syst.*, Eds M. Kohlmann and W. Vogel, lecture Notes in Control and Inf. Sciences, 16. Springer(1979)

[15] E. PARDOUX Equations du filtrage non-linéaire, de la prédiction et du lissage. Publication de Mathématiques Appliquées Marseille Toulon 80 - 6, to appear

[16] D. W. STROOCK - S.R.S. VARADHAN *Multidimensional diffusion processes*. Springer (1979).

[17] H. J. SUSSMANN On the gap between deterministic and stochastic ordinary differential equations. *Ann. of Prob.* 6, pp. 19 - 41, (1978).

ESTIMATION PROBLEMS WITH LOW DIMENSIONAL FILTERS*

John Baillieul

Scientific Systems, Inc.

Recently intense interest has been aroused in the system theory community concerning the connections between well-known problems in nonlinear estimation theory and the representation theory of Lie algebras. These connections are especially interesting in problems of estimating the state of a continuous time, finite state Markov chain observed in the presence of Gaussian white noise. In this paper we delineate characteristics of problems which admit low dimensional estimation algebras, and it is shown why these constitute a class of "easy" nonlinear filtering problems.

INTRODUCTION

Among the most exciting developments in nonlinear filtering theory are those centering around the recently established connections with Lie algebra representation theory. (See the articles by Brockett, Hazewinkel and Mitter in these proceedings.) A prime example of this is the recent paper by Brockett and Clark [1], which gives a detailed account of how the representation theory of semisimple Lie algebras provides essential insight for solving finite state-continuous time (FSCT) estimation problems. The purpose of the present paper is to further develop certain of these ideas, to characterize more or less completely those filtering problems for which the dimension of the estimation algebra is low (low = 4 in this case) and to give examples to suggest how these results may be useful in analyzing some electric energy system models.

---

*This research was supported by the Dept. of Energy under Contract No. DE-AC01-79ET20361.

M. Hazewinkel and J. C. Willems (eds.), Stochastic Systems: The Mathematics of Filtering and Identification and Applications, 559–564.

Throughout we shall consider the class of irreducible finite state-continuous time (FSCT) Markov processes for which the probability distribution evolves according to

$$\dot{p} = Ap \tag{1}$$

and whose states $X = \{x_1, \ldots, x_n\}$ are observed in the presence of white noise.

$$dy = x\,dt + dw \tag{2}$$

(Note, we assume the x_i's are real numbers.) The <u>FSCT estimation problem</u> then is to compute the conditional probability $\rho(t,x|Y_t)$, where Y_t indicates the y process on the interval $[0,t]$. We shall begin with two examples which illustrate the types of applications we have in mind for this theory.

<u>Example 1</u>: Suppose an electric circuit supplies m users (all of whom we shall assume have the same electricity use characteristics) who use current only intermittently. The probability of a user initiating use of the electric current in a small unit of time Δt is $\lambda\Delta t + o(\Delta t)$ while the probability of someone already using current ceasing in Δt is $\mu\Delta t + o(\Delta t)$. The the resulting FSCT model may be written as

$$\begin{pmatrix} \dot{p}_0(t) \\ \dot{p}_1(t) \\ \cdot \\ \cdot \\ \cdot \\ \dot{p}_m(t) \end{pmatrix} = \begin{pmatrix} -m\lambda & \mu & 0 & \cdots & 0 \\ m\lambda & -[(m-1)\lambda+\mu] & 2\mu & \cdots & 0 \\ 0 & (m-1)\lambda & -[(m-2)\lambda+2\mu] & \ddots & 0 \\ \vdots & \vdots & \vdots & \ddots & \vdots \\ & & & & m\mu \\ 0 & 0 & 0 & \cdots\ \lambda & -m\mu \end{pmatrix} \begin{pmatrix} p_0(t) \\ p_1(t) \\ \cdot \\ \cdot \\ \cdot \\ p_m(t) \end{pmatrix}$$

where $p_i(t)$ = probability that i users are drawing current at time t. We assume the state of this system, $x(t) \in \{0,1,\ldots,m\}$, is observed (i.e., a meter reading is taken) in the presence of white noise; the observation equation is thus given by (2).

<u>Example 2</u>: We suppose there is an electric network of the form described in Figure 1. Here we assume that a randomly opening and reclosing switch is operating, and, as one easily verifies assuming one ohm resistances, the current i is either $3/4$ amp or $1/2$ amp depending on whether the switch is closed or open. We suppose the probability of an open switch closing in Δt units of time is $\lambda\Delta t + o(\Delta t)$ while the probability of a closed switch opening in Δt units of time is $\mu\Delta t + o(\Delta t)$. If $p_0(t)$ = the probability that the switch is open at time t while $p_1(t)$ is the probability that the

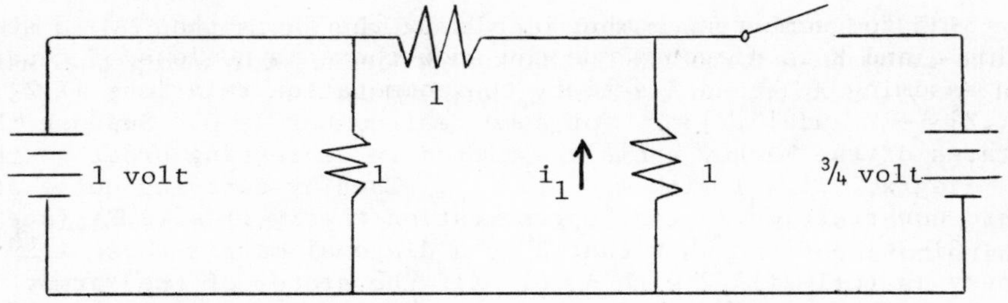

Figure 1.

switch is closed at time t then the standard FSCT model gives

$$\begin{pmatrix} \dot{p}_0(t) \\ \dot{p}_1(t) \end{pmatrix} = \begin{pmatrix} -\lambda & \mu \\ \lambda & -\mu \end{pmatrix} \begin{pmatrix} p_0(t) \\ p_1(t) \end{pmatrix}$$

Assuming we measure the current i in the presence of white noise, the observation equation is

$$dy = i\,dt + dw$$

FOUR DIMENSIONAL ESTIMATION ALGEBRAS

Given an FSCT estimation problem specified by (1) and (2) we let $B = \mathrm{diag}(x_1,\ldots,x_n)$ and form the Lie algebra $\{A - \tfrac{1}{2}B^2, B\}_{LA}$. This is known as the estimation algebra associated to (1) and (2), and in [1] it is shown that its dimension provides an upper bound on the number of sufficient statistics required to propagate the condition distribution $\rho(t,x|Y_t)$. Also in [1] it is shown that estimation algebras for FSCT estimation problems are of dimension at least four, and moreover, for each positive integer n there are n state Markov chains for which the dimension of the estimation algebra actually attains this lower bound. Thus, such problems for which $gl(2,\mathbb{R})$ is the estimation algebra are in some sense the easiest, and we shall devote the remainder of this paper to their characterization.

Theorem 1: Suppose the FSCT estimation problem associated to (1) and (2) has a four dimensional estimation algebra. Then if the states of the Markov chain are listed in decreasing or increasing order: (i) they are evenly spaced, and (ii) the matrix A is tridiagonal.

Proof: Write (as per lemma 2 in [1]) $\{A-\tfrac{1}{2}B^2,B\}_{LA} = L^1 \oplus \{\alpha I\}$ where L^1 is (an irreducible representation of) a three dimensional semisimple Lie algebra. Let X, Y and Z be a basis for L^1 such that

$B = Z+\alpha I$ for some real number α. Using the facts that L^1 is semi-simple and Z is diagonal one can show there is no loss of generality in assuming X, Y and Z satisfy the commutation relations $[X,Z] = \lambda Z$, $[Y,Z] = -\lambda Z$ and $[X,Y] = Z$, for some real number $\lambda \neq 0$. Suppose the states of the Markov chain are listed in decreasing order so that $B = \mathrm{diag}(x_1,\ldots,x_n)$ with $x_1 > \ldots > x_n$. Then by carrying out a standard construction in the representation theory of $sl(2,\mathbb{R})$ (see [2] pp. 31-34) one can show that Z is a diagonal matrix whose ii^{th} entry is $(n+1-2i)\lambda/2$ with $\lambda > 0$. (If the states of the Markov chain were listed in increasing order, Z would have the same form with $\lambda < 0$.) Since $B = Z+\alpha I$ for some real number α, this proves statement (i) in the theorem.

While ordering the states of the Markov chain uniquely determines Z in this way, there remains some ambiguity about X,Y. Let $\bar{X}$ be the n×n matrix with 1's along the subdiagonal and zeros elsewhere and let $\bar{Y}$ be the n×n matrix with the entries $-\lambda/2(n-1),\ldots,$ $-\lambda/2k(n-k),\ldots,-\lambda/2(n-1)$ descending along the superdiagonal and zeros elsewhere. It can be shown that X and Y can be any matrices related to $\bar{X}$ and $\bar{Y}$ respectively by a change of basis of the form $R = \mathrm{diag}(r_1,\ldots,r_n)$, $r_i > 0$. But it is easy to see that under all such changes of basis X remains a matrix with nonzero entries only on the subdiagonal while Y remains a matrix with nonzero entries only on the superdiagonal. Since X, Y and Z must take these respective forms and since $A = \alpha_1 I + \alpha_2 X + \alpha_3 Y + \alpha_4 Z - \frac{1}{2}B^2$ for some real α_1, α_2, α_3, α_4, we have proved statement (ii). This concludes the proof of the theorem.

The next theorem addresses the question: among FSCT estimation problems in which the states of the Markov chain are evenly spaced and for which the matrix A is tridiagonal, can we precisely characterize those for which the estimation algebra is $gl(2,\mathbb{R})$? Suppose we fix a set of real numbers $x_1 > x_2 > \ldots > x_n$ with $x_{i+1} - x_i = s$ for some positive real s. Consider the set of FSCT estimation problems specified by (1) and (2) with state set $\{x_i\}_{i=1}^n$ and for which the infinitesimally stochastic matrix A is tridiagonal. (Note that the set S of all tridiagonal infinitesimally stochastic matrices

$$A = \begin{pmatrix} -a_1 & a_2 & 0 & \cdot & \cdot & \cdot & 0 \\ a_1 & -a_2-a_3 & a_4 & \cdot & \cdot & \cdot & 0 \\ 0 & a_3 & -a_4-a_5 & \cdot & \cdot & \cdot & 0 \\ 0 & 0 & a_5 & \cdot & & & 0 \\ \vdots & \vdots & \vdots & & \cdot & & \vdots \\ & & & & & \cdot & \\ 0 & 0 & 0 & \cdots & & a_{2n-3} & -a_{2n-2} \end{pmatrix} \tag{3}$$

(where $a_i > 0$) is a $2n-2$ dimensional cone in the set of $n \times n$ real matrices.)

Theorem 2: For this class of estimation problems the set of all infinitesimally stochastic matrices A which give rise to estimation algebras isomorphic to $gl(2,\mathbb{R})$ is the intersection of S with a two dimensional subvariety of $gl(n,\mathbb{R})$ specified by the polynomial equations

$$\frac{a_1a_2}{n-1} = \frac{a_3a_4}{2(n-2)} = \dots = \frac{a_{2n-3}a_{2n-2}}{n-1}$$

and

$$\begin{aligned} 2(a_2+a_3) - a_1 - a_4 - a_5 &= s^2 \\ 2(a_4+a_5) - a_2 - a_3 - a_6 - a_7 &= s^2 \\ &\vdots \\ 2(a_{2n-4}+a_{2n-3}) - a_{2n-2} - a_{2n-5} - a_{2n-6} &= s^2 \end{aligned}$$

where $s = x_{i+1} - x_i$ is the distance between states.

Proof: In the proof of theorem 1 it was seen that if A is an infinitesimally stochastic matrix giving rise to an estimation algebra isomorphic to $gl(2,\mathbb{R})$, it may be represented by

$$A = \alpha_1 I + \alpha_2 Z + \alpha_3 X_R + \alpha_4 Y_R + \tfrac{1}{2}B^2 \tag{4}$$

where $Z = \mathrm{diag}(n-1, n-3, \dots, -n+1)$ and X_R and Y_R are related to the matrices X and Y in the proof of theorem 1 by a change of basis $R = \mathrm{diag}(r_1, \dots, r_n)$ with $r_i > 0$. Writing things out in detail one can show that X_R is a matrix with the entries $(n-1)\rho_1^{-1}$, $(n-2)\rho_2^{-1}$, $\dots, \rho_{n-1}^{-1}$ descending along the subdiagonal and zeros elsewhere while Y_R has the entries $\rho_1, 2\rho_2, \dots, (n-1)\rho_{n-1}$ descending along the superdiagonal with zeros elsewhere, where the ρ_i's are positive real numbers. It is clear from this representation that the polynomial equations written in the statement of the theorem must be satisfied. On the other hand, suppose the entries in A (given in (3)) satisfy these equations. Define $\alpha_3 = a_1/(n-1)$, $\alpha_4 = a_2$ and let $\rho_i = a_{2i}/(\alpha_4 i)$ for $i = 2,3,\dots,n-1$. It follows easily from the theorem hypothesis that $a_{2i-1}a_{2i} = i(n-i)\alpha_3\alpha_4$, so that we also have $\rho_i = (n-i)\alpha_3/a_{2i-1}$. Next, define α_1, α_2 by means of the equations $\alpha_1+(n-1)\alpha_2+\tfrac{1}{2}x_1^2 = -a_1$ and $\alpha_1+(n-3)\alpha_2+\tfrac{1}{2}x_2^2 = -a_2-a_3$. Again, from the hypothesis of the theorem, it follows that α_1, α_2 also satisfy the equations $\alpha_1+(n+1-2i)\alpha_2+\tfrac{1}{2}x_i^2 = -a_{2i-2}-a_{2i-1}$ for $i = 2,\dots,n-1$ as well as $\alpha_1 + (-n+1)\alpha_2+\tfrac{1}{2}x_n^2 = -\alpha_{2n-2}$. Thus, the matrix A in (3) is indeed representable in the form (4), and it follows that the estimation algebra is isomorphic to $gl(2,\mathbb{R})$. That the set of such matrices A is a two dimensional subvariety may be verified using techniques from elimination theory. (See [3].) The details of this calculation are omitted.

SUBOPTIMAL FILTERS

Denote the class of all matrices which are of the form (3) and which satisfy the equations in theorem 2 by L. Given an FSCT estimation problem specified by (1) and (2) where A is of the form (3), let $\hat{A} \in L$ satisfy $\|\hat{A}-A\| = \min_{M \in L} \|M-A\|$. As is shown in [1], the unnormalized conditional probability, $\rho(t,x|Y_t)$, is propagated according to the Ito equation

$$d\rho = A\rho dt + dyB\rho$$

Since the estimation algebra for this problem is typically n^2-dimensional, it seems natural to try a suboptimal filtering scheme based on propagating solutions to

$$dp = \hat{A}pdt + dyBp$$

(for which $\{\hat{A}-\frac{1}{2}B^2, B\}_{LA}$ is four dimensional). The details of this program will be analyzed in a forthcoming paper.

REFERENCES

1. Brockett, R. W. and J. M. C. Clark, "The Geometry of the Conditional Density Equations," Proc. of the Oxford Conf. on Stochastic Systems, Oxford, England, 1978.

2. Humphreys, J. E., *Introduction to Lie Algebras and Representation Theory*, New York: Springer-Verlag GTM Series, 1972.

3. Van der Waerden, B. L., *Einführung in die Algebraische Geometrie*, Springer-Verlag, Grund. der Math. Wissenschaften 51, 1973.

GROUP INVARIANCE METHODS IN NONLINEAR FILTERING OF DIFFUSION PROCESSES

John S. Baras[1]

Electrical Engineering Department
University of Maryland
College Park, Maryland 20742

ABSTRACT

Given two "nonlinear filtering problems" described by the processes

$$dx^i(t) = f^i(x^i(t))\,dt + g^i(x^i(t))\,dw^i(t)$$
$$dy^i(t) = h^i(x^i(t))\,dt + dv^i(t)\,, \quad i=1,2, \tag{1}$$

we define a notion of strong equivalence relating the solutions to the corresponding Mortensen-Zakai equations

$$du_i(t,x) = \mathcal{L}^i u_i(t,x)dt + \mathcal{L}_1^i u_i(t,x)dy_t^i\,, \quad i=1,2, \tag{2}$$

which allows solution of one problem to be obtained easily from solutions of the other. We give a geometric picture of this equivalence as a group of local transformations acting on manifolds of solutions. We then show that by knowing the full invariance group of the time invariant equations

$$du_i(t,x) = \mathcal{L}^i u_i(t,x)dt\,, \quad i=1,2, \tag{3}$$

we can analyze strong equivalence for the filtering problems. In particular if the two time invariant parabolic operators are in the same orbit of the invariance group we can show strong equivalence for the filtering problems. As a result filtering problems are separated into equivalent classes which correspond to orbits of invariance groups of parabolic operators. As specific example we treat V. Beneš's case establishing from this point of view the necessity of the Riccati equation.

M. Hazewinkel and J. C. Willems (eds.), Stochastic Systems: The Mathematics of Filtering and Identification and Applications, 565–572.

1. INTRODUCTION

Very recently new ideas and techniques have been applied to a long standing problem in stochastic systems theory: "the nonlinear filtering problem". A large portion of this new work is geometrical in nature. Thus Brockett [1]-[2] and Mitter [3]-[4] have emphasized the significance of certain Lie-algebra of partial differential operators associated with each nonlinear filtering problem, while Marcus et al [5] and Baras and Blankenship [6] have provided explicit examples where these concepts lead to significant developments in the solution of nonlinear filtering problems.

Our objective here is to describe a geometric way of characterizing computationally equivalent nonlinear filtering problems. This work is inspired by similar ideas in the theory of ordinary differential equations which go under the names "Similarity Methods" or "Group Invariance Methods"[9]-[10]. It will be apparent from the present paper that the fundamental concept in this problem of "equivalence" is that of invariance groups of (4). To make things precise consider two nonlinear filtering problems (vector) as in (1) of the abstract and the corresponding Mortensen-Zakai equations in Stratonovich form

$$\frac{\partial u_i(t,x)}{\partial t} = (\mathcal{L}^i - \frac{1}{2}\| h^i(x)\|^2)u_i(t,x)+h^{iT}(x)u_i(t,x)y^i(t), \quad i=1,2 \qquad (4)$$

Definition: The two nonlinear filtering problems above are strongly equivalent if u_2 can be computed from u_1, and vice versa, via the following types of operations:

Type 1: $(t,x^2) = \alpha(t,x^1)$, where α is a diffeomorphism.
Type 2: $u_2(t,x) = \psi(t,x)u_1(t,x)$, where $\psi(t,x)\geq 0$ and $\psi^{-1}(t,x)\geq 0$.
Type 3: Solving a set of ordinary (finite dimensional) differential equations (i.e. quadrature).

Brockett [2], has analyzed the effects of diffeomorphisms in x-space and he and Mitter [4] the effects of so called "gauge" transformations (a special case of our type 2 operations) on (4). Type 3 operations are introduced here for the first time, and will be seen to be the key in linking this problem with mathematical work on group invariance methods in o.d.e. and p.d.e.'s.

Our approach starts from the abstract version of (4):

$$\frac{\partial u_i}{\partial t} = (A^i + \sum_{j=1}^{p} B_j^i \dot{y}_j(t))u_i; \quad i=1,2 , \qquad (5)$$

where A^i, B_j^i are given by

$$\mathcal{L}^i := \text{forward gen. of } x^i$$
$$A^i := \mathcal{L}^i - \frac{1}{2} h^{iT} h^i \qquad (6)$$
$$B^i_j := \text{Mult. by } h^i_j \text{ (jth comp. of } h^i) .$$

We are thus dealing with two parabolic equations. We will first examine whether the evolutions of the time invariant parts can be computed from one another. This is a classical problem and the methods of section 2 apply. In section 3 we shall give an extension to the full equation (5) under certain conditions on B^i_j. We shall then apply this result to the examples studied by Beneš and recover the Riccati equation as a consequence of strong equivalence. Further results and details can be found in [8].

The estimation Lie algebra introduced by Brockett [2] is the Lie algebra.

$$\Lambda^i(E) = \text{Lie algebra generated by } A^i \text{ and } B^i_j, \; j=1,\ldots p, \quad i=1,2 . \qquad (7)$$

Again we shall assume that for problems considered the operators A, B_j have a common, dense invariant set of analytic vectors in X and that the mathematical relationship between $\Lambda^i(E)$ and the existence-uniqueness theory of (4) is well understood. For results of this nature we refer to [6][7].

2. USING THE INVARIANCE GROUP OF A PARABOLIC P.D.E. IN SOLVING NEW P.D.E.'S

Consider the general, linear, nondegenerate elliptic partial differential operator

$$L := \sum_{i,j=1}^{n} a_{ij}(x) \frac{\partial^2}{\partial x_i \partial x_j} + \sum_{i=1}^{n} b_i(x) \frac{\partial}{\partial x_i} + c(x)\,\text{id}. \qquad (8)$$

and assume that the coefficients a_{ij}, b_i, c are smooth enough, so that L generates an analytic semigroup, denoted by exp(tL), for at least small $t \geq 0$, on some locally convex space X of initial functions ϕ and appropriate domain Dom(L).

Let V be the set of solutions to

$$\frac{\partial u}{\partial t} = Lu$$
$$u(0,x) = \phi(x) \qquad (9)$$

in X, as we vary ϕ. The aim is to find a local Lie transformation group G which transforms every element of V into another element of V. Such a group will be called an <u>invariance group</u> of (9) or of L. This of course is a classical topic of mathematical research initiated by Sophus Lie.

Theorem 1 [10]: Every transformation g in the invariance group G of a linear parabolic equation is of the form

$$u(t,x) \to \nu(p(t,x))u(p(t,x)) + \psi(x) \tag{10}$$

where p is a transformation acting on the variables (t,x), ψ a fixed solution of the parabolic equation.

We consider now one-parameter subgroups of the invariance group G of a given parabolic partial differential equation. According to standard Lie theory the infinitesimal generators of these one-parameter subgroups form the Lie algebra $\Lambda(G)$ of the local Lie group G [9]. We shall, using standard Lie theory notation denote X_s by exp(sX) where X is the infinitesimal generator of the one parameter group $\{X_s\}$. The infinitesimal generators of G are given by

$$Z = \alpha(t,x)\frac{\partial}{\partial t} + \sum_{i=1}^{n} \beta_i(t,x)\frac{\partial}{\partial x_i} + \gamma(t,x)\,\mathrm{id}. \tag{11}$$

for some functions α, β_i, γ of t and x. If u solves (9) so does v(s)=exp(sZ)u, for small s. However v is also the solution of

$$\frac{\partial v}{\partial s} = \alpha\frac{\partial v}{\partial t} + \sum_{i=1}^{n} \beta_i \frac{\partial v}{\partial x_i} + \gamma v \ , \quad v(0) = u, \tag{12}$$

a first order hyperbolic p.d.e. (solvable by the method of characteristics). Clearly since $\partial/\partial t - L$ is linear

$$Zu \in V \text{ if } u \in V. \tag{13}$$

(13) implies

$$dZ/dt = [L,Z] \text{ on } V . \tag{14}$$

In (14) [,] denotes commutator and dZ/dt is symbolic of $\alpha_t \frac{\partial}{\partial t} + \sum_{i=1}^{n} \beta_{i,t} \frac{\partial}{\partial x_i} + \gamma_t \,\mathrm{id}$. Thus the elements of $\Lambda(G)$ in this case satisfy a Lax equation. Furthermore it can be shown [10] that α is independent of x, i.e. $\alpha(t,x) = \alpha(t)$ and that every Z satisfies an o.d.e.

$$d_\ell \frac{d^\ell Z}{dt^\ell} + d_{\ell-1}\frac{d^{\ell-1}Z}{dt^{\ell-1}} + \ldots + d_0 Z = 0$$

where $\ell \leq \dim G$.

The most widely known example, for which $\Lambda(G)$ has been computed explicitly is the heat equation. The infinitesimal generators in this case are six, as below

$$\frac{\partial}{\partial t}, 2t\frac{\partial}{\partial t} + x\frac{\partial}{\partial x}, \frac{\partial}{\partial x}, \; 1, 2t\frac{\partial}{\partial x} + x, \; 4t^2 \frac{\partial}{\partial t} + 4tx\frac{\partial}{\partial x} + x^2 \; . \tag{15}$$

Knowing the invariance group of (9) can help in solving certain "perturbations" of (9). We folow Rosencrans [10]. Thus we consider a linear parabolic equation like (9) and we assume we know the infinitesimal generators Z of the nontrivial part of G. Thus if u solves (9), so does v(s)=exp(sZ)u but with some new initial data, say, $R(s)\phi$. That is

$$e^{sZ}e^{tL} = e^{tL}R(s) \text{ on } X. \tag{16}$$

Now $R(\cdot)$ is a semigroup. Let M be its generator:

$$M\phi = \lim_{s\to 0} \frac{R(s)\phi-\phi}{s} \; , \; \phi\varepsilon\mathrm{Dom}(M) \; . \tag{17}$$

It is straightforward to compute M, given Z as in (16). Thus

$$M\phi = \alpha(0)L\phi + \sum_{i=1}^{n}\beta_i(0,x)\frac{\partial\phi}{\partial x_i} + \gamma(0,x)\phi \; . \tag{18}$$

Thus, to solve the initial value problem

$$\partial w/\partial s = Mw \; , \; w(0) = \phi \tag{19}$$

where

$$M = \alpha(0)L + \sum_{i=1}^{n}\beta_i(0,x)\frac{\partial}{\partial x_i} + \gamma(0,x) \text{ id}$$

we follow the steps given below.

Step 1: Solve $u_t = Lu$, $u(0) = \phi$.

Step 2: Find generator Z of G corresponding to M and solve

$$\frac{\partial v}{\partial s} = \alpha(t)\frac{\partial v}{\partial t} + \sum_{i=1}^{n}\beta_i(t,x)\frac{\partial v}{\partial x_i} + \gamma(t,x)v \; , \; v(0) = u \tag{20}$$

via the method of characteristics. Note this step requires the solution of ordinary differential equations only.

Step 3: Set t=0 to v(s,t,x).

This procedure allows easy computation of the solution to the "perturbed" problem (19) if we know the solution to the "unperturbed" problem (9). The "perturbation" which is of degree $\leq$ 1st, is given by the part of M:

$$P = \sum_{i=1}^{n} \beta_i(0,x)\frac{\partial}{\partial x_i} + \gamma(0,x).\,\mathrm{id}. \tag{21}$$

We shall denote by $\Lambda(P)$ the set of all perturbations like (21).

Definition: The Lie algebra $\Lambda(P)$ will be called the perturbation algebra of the elliptic operator L.

Theorem 2 [10]: The perturbation algebra $\Lambda(P)$ of an elliptic operator L, is isomorphic to a Lie subalgebra of $\Lambda(G)$ (i.e. of the Lie algebra of the invariance group of L). Moreover $\dim(\Lambda(P)) = \dim(\Lambda(G))-1$.

One significant question is: can we find the perturbation algebra $\Lambda(P)$ without first computing $\Lambda(G)$, the invariance Lie algebra? The answer is affirmative and is given by the following result [10].

Theorem 3 [10]: Assume L has analytic coefficients. An operator P_0 of order one or less (i.e. of the form (21) is in the perturbation algebra $\Lambda(P)$ of L iff there exist a sequence of scalars $\lambda_1,\lambda_2,\ldots$ and a sequence of operators $P_1,P_2,\ldots$ of order less than or equal to one such that $[L,P_n]=\lambda_n L+P_{n+1}$, $n\geq 0$ and $\Sigma\lambda_k t^k/k!$, $\Sigma P_k t^k/k!$ converge at least for small t.

It is an easy application of this result to compute the perturbation algebra of the heat equation in one dimension or equivalently of $L = \partial^2/\partial x^2$. It turns out that $\Lambda(P)$ is a 5-dimensional and spanned by

$$\Lambda(P) = \mathrm{Span}(1,x,x^2, \frac{\partial}{\partial x}, x\frac{\partial}{\partial x}). \tag{22}$$

So the general perturbation for the heat equation looks like

$$P = (ax+b)\frac{\partial}{\partial x} + (cx^2 + dx + e)\,\mathrm{id} \tag{23}$$

where a,b,c,d,e are arbitrary constants.

The implications of these results are rather significant. Indeed consider the class of linear parabolic equations $u_t=Lu$, where L is of the form (8). We can define an equivalence relationship on this class by : "L_1 is equivalent to L_2 if $L_2=L_1+P$ where P is an element of the perturbation algebra $\Lambda^1(P)$ of L_1". Thus elliptic operators or equivalently linear parabolic equations are divided into equivalent classes (orbits); within each class (orbit) $\{L(k)\}$ (k indexes elements in the class) solutions to the initial value problem $u(k)_t=L(k)u(k)$ with fixed data ϕ (independent of k) can be obtained by quadrature (i.e. an o.d.e. integration) from any one solution $u(k_0)$.

3. SUFFICIENT CONDITIONS FOR STRONG EQUIVALENCE AND APPLICATIONS.

We return now to the problem posed in section 1. Namely to discover conditions that imply strong equivalence of two nonlinear filtering problems. Our main result is:

Theorem 4: Given two nonlinear filtering problems (see (1)(2)), such that the corresponding Mortensen-Zakai equations (see (4)) have unique solutions, continuously dependent on $y(\cdot)$. Assume that using operations of type 1 and 2 (see definition in section 1) these stochastic p.d.e. can be transformed in bilinear form

$$\frac{\partial u_i}{\partial t} = (A^i + \sum_{j=1}^{p} B^i_j \xi^i_j(t))u_i, \quad i=1,2$$

such that: (i) $A^i, i=1,2$, are nondegenerate elliptic, belonging to the same equivalence class (see end of section 2) (ii) $B^i_j, j=1,\ldots p$, $i=1,2$ belong to the perturbation algebra $\Lambda(P)$ of (i). Then the two filtering problems are strongly equivalent.

Proof: See [8].

Let us apply this result to the Beneš case[11]. We consider the linear filtering problem (scalar x,y)

$$dx(t)=dw(t) \ , \ dy(t)=x(t)dt+dv(t) \tag{24}$$

and the nonlinear filtering problem (scalar x,y)

$$dx(t)=f(x(t))dt+dw(t) \ , \ dy(t)=x(t)dt+dv(t) \ . \tag{25}$$

The corresponding Mortensen-Zakai equations in Stratonovich form are: for the linear

$$\frac{\partial u_1(t,x)}{\partial t} = \frac{1}{2}\left(\frac{\partial^2}{\partial x^2} - x^2\right)u_1(t,x)+x\dot{y}(t)u_1(t,x) \ ; \tag{26}$$

for the nonlinear

$$\frac{\partial u_2}{\partial t} = \frac{1}{2}\left(\frac{\partial^2}{\partial x^2} - x^2\right)u_2(t,x) - \frac{\partial}{\partial x}(fu_2)+x\dot{y}(t)u_2(t,x). \tag{27}$$

We wish to show that (24)(25) are strongly equivalent only if f (the drift) is a global solution of the Riccati equation

$$f_x + f^2 = ax^2 + bx + c \ . \tag{28}$$

First let us apply to (25)(27) an operation of type 2. That is let (defines v_2)

$$u_2(t,x) = v_2(t,x)\exp\left(\int_0^x f(u)du\right). \tag{29}$$

Then the new function v_2 satisfies

$$\frac{\partial v_2(t,x)}{\partial t} = \frac{1}{2}\left(\frac{\partial^2}{\partial x^2} - x^2 - V(x)\right)v_2(t,x) + x\dot{y}(t)v_2(t,x) , \tag{30}$$

where

$$V(x) = f_x + f^2 . \tag{31}$$

We apply the theorem to (26)(30). So

$$A^1 = \frac{1}{2}\left(\frac{\partial^2}{\partial x^2} - x^2\right) , \quad A^2 = \frac{1}{2}\left(\frac{\partial^2}{\partial x^2} - x^2 - V\right)$$

while $B^1 = B^2 = $ Mult. by x. From the results of section 2, the only possible equivalence class is that of the heat equation. Clearly from (23) $A^1, B^1, B^2 \varepsilon \Lambda(P)$ for this class. For $A^2 \varepsilon \Lambda(P)$ it is necessary that V be quadratic, i.e. f satisfies the Riccati equation (28).

REFERENCES

[1] R.W. Brockett and J.M.C. Clark, "Geometry of the Conditional Density Equation", Proc.Int.Conf. on An. and Opt. of Stoch. Syst., Oxford, England, 1978.
[2] R.W. Brockett, "Classification and Equivalence in Estimation Theory", Proc.18th IEEE Conf.on Dec. and Control, 1979, pp.172-175.
[3] S.K. Mitter, "Filtering Theory and Quantum Fields" to appear in Asterisque, 1980.
[4] S.K. Mitter, "On the Analogy Between Mathematical Problems of Non-linear Filtering and Quantum Physics", Rich.di Automatica, 1980.
[5] S.I. Marcus, M. Hazewinkel and Chugn' huan Lie, "Algebraic Structures in Nonlinear Estimation" Proc. of 1980 Joint Aut.Cont.Conf.
[6] J.S. Baras and G.L. Blankenship, "Nonlinear Filtering of Diffusion Processes: A Generic Example" to appear.
[7] J.S. Baras, S.K. Mitter and D. Ocone, "Nonlinear Filtering of Diffusion Processes in Unbounded Domains", to appear.
[8] J.S. Baras, "Group Invariance Methods in Nonlinear Filtering of Diffusion Processes", Proc. of 1980 IEEE Conf. on Dec. and Cont.
[9] G.W. Bluman and J.C. Cole, Similarity Methods for Differential Equations, Springer-Verlag, 1974.
[10] S.I. Rosencrans, "Perturbation Algebra of an Elliptic Operator", J. of Math. Anal. and Appl., 56, 1976, pp. 317-329.
[11] V. Beneš, "Exact Finite Dimensional Filters for Certain Diffusions with Nonlinear Drift", to appear in Stochastics, 1980.

[1] Work partially supported by ONR contract N00014-79-C-0808, Univ. of Maryland, AFOSR grant AFOSR-77-3261B at LIDS of MIT, Joint Services Electronics Program grant N00014-75-C-0648 at Harvard University.

ASYMPTOTIC BOUNDS ON THE MINIMAL ERROR OF NON-LINEAR FILTERING

B.Z. Bobrovsky
School of Engineering
Tel-Aviv University
Tel-Aviv, Israel

M. Zakai
Dept. of Electrical Engineering
Technion
Haifa, Israel

The minimal mean square error for the scalar, standard non-linear filtering problem is considered. Asymptotic lower and upper bounds on the error are derived for the case where the intensity of the observation noise tends to zero.

1. INTRODUCTION

The research in non-linear filtering theory was strongly influenced by the explicit results of linear filtering and was accompanied with the hope, or ambition, for similar achievements. The difficulties involved can, however, be better appreciated by comparing the non-linear filtering problem with that of non-linear parameter estimation problems of communication theory on which theorists worked extensively for a long time. In particular, consider the following pulse position modulation (or radar range estimation) problem, let $s(t)$ be a known function of finite energy and consider the received signal

$$r(t) = s(t-t_o) + \sqrt{N_o}\, n(t) \quad ; \quad -\frac{T}{2} \leq t \leq \frac{T}{2}$$

where $n(t)$ is a white Gaussian noise of unit spectral density. The problem being that of estimating the location parameter t_o. The main results for this problem can be summarized as follows (cf. [8], the survey paper [9], and [5] for recent work):
(a) The solution for the minimal mean square error estimator is known, however in most applications its implementation is not practical (b) Explicit expressions for the minimal mean square error are generally unknown however, bounds on the minimal error

M. Hazewinkel and J. C. Willems (eds.), Stochastic Systems: The Mathematics of Filtering and Identification and Applications, 573–581.

are available. The estimation problem is of practical interest for the case where the noise is very small, more specifically; practical systems deal with the case where the standard deviation of the estimation error is of the order of $10^{-2}T$ or less while the a-priori uncertainty is of the order of T. Consequently, asymptotic results for $N_0 \to 0$ are particularly useful and it turns out that the available performance bounds are tight in this region. (c) The design philosophy of practical estimators is as follows : The error associated with certain suboptimal estimators is shown to approach the lower bound as $N_0 \to 0$ therefore these suboptimal estimators are asymptotically optimal. Such estimators are widely used in practice. In many practical non-linear filtering problems the ratio of the a postriori to a-priori error variance is also very small. It is therefore believed that a similar approach of error bounds and asymptotically optimal approximate filters will be an effective tool for designing non-linear filters. Some recent results in this direction are summarized in the following sections.

2. A LOWER BOUND

Consider the scalar filtering problem where the signal x_t and the observation y_t satisfy

$$\left.\begin{aligned} dx_t &= m(x_t)dt + \sigma(x)dw_t \\ dy_t &= h(x_t)d_t + \sqrt{N_o}\, d\nu_t \end{aligned}\right\} \qquad (1)$$

where ν and w are independent Brownian motions. Assume that $m(\cdot)$, $h(\cdot)$ are differentiable and the derivatives $m'(x)$, $h'(x)$ are bounded and continuous. In this section it is assumed that σ is a constant, σ_0. Let A_t and B_t be defined by

$$A_t = Em'(x_t)$$

$$\frac{\sigma_o^2}{N_o} B_t^2 = E(m'(x_t))^2 - A_t^2 + \frac{\sigma_o^2}{N_o} E(h'(x_t))^2$$

Consider the Gaussian linear filtering problem

$$\left.\begin{aligned} du_t &= A_t u_t d_t + \sigma_o dw_t \\ dv_t &= B_t u_t d_t + \sqrt{N_o}\, d\ \ t \end{aligned}\right\} \qquad (2)$$

Then the mean square filtering error of (1) is lower bounded by the corresponding error for the linear system (2) [1] :

$$E(x_t - \hat{x}_t)^2 \geq E(u_t - \hat{u}_t)^2 \qquad (3)$$

where $(\hat{\cdot})_t$ denotes the conditional expectation conditioned on the path $\{y_\theta, 0 \leq \theta \leq t\}$. A recent generalization to (3) is as follows [2] : let $Z(\cdot)$ be a real valued function, assume that there exists a function $Z_0(\cdot)$ such that $(Z(\alpha+\delta)-Z(\alpha))/\delta \leq Z_0(\alpha)$ for all α and all $\delta \leq \delta_0$, $\delta_0 > 0$, further assume that $E|Z_0(x_t)|^2 < \infty$ then

$$E(\hat{Z}(x_t) - Z_t(x_t))^2 \geq E(Z'(x_t))^2 E(\hat{u}_t - u_t)^2 \qquad (4)$$

In the original system (1), σ is assumed to be a constant, (4) enables the derivation of bounds for the case where $\sigma = \sigma(x)$ by a suitable transformation from x_t to another variable.

3. ASYMPTOTIC BOUNDS ($N_0 \to 0$)

Consider the filtering problem of equation (1), $m(\cdot)$ and $\sigma(\cdot)$ are assumed to satisfy a global Lipschitz condition and $h(\cdot)$ is assumed to be twice continuously differentiable and of polynomial growth. Let $\Pi_t(\alpha)$ denote the conditional probability distribution of x_t given $\{y_\theta, 0 \leq \theta \leq t\}$, then the pair $(x_t, \Pi_t(\cdot))$ will be assumed to posses a stationary invariant probability measure. e_t will denote the error $(x_t - \hat{x}_t)$ and $e_{z,t}$ will denote $(Z(x_t) - \hat{Z}(x_t))$.

Definition. If for any $\varepsilon > 0$ there exists a $N_{00} > 0$ such that for all N_0, $N_0 \leq N_{00}$

$$E\, e_t^2 \leq (1+\varepsilon) f_1, \qquad (E\, e_t^2 \geq (1-\varepsilon) f_2)$$

where $f_1(f_2)$ depends only on the a-priori statistics of the x_t process then $f_1(f_2)$ will be said to be an asymptotic upper (lower) bound on the filtering error. We will write in this case

$$E\, e_t^2 \lesssim f_1 \quad (E\, e_t^2 \gtrsim f_2) \quad \text{and} \quad E\, e_t^2 \approx f \quad \text{if} \quad f_1 = f_2 = f.$$

Proposition 1. Let $e_{h,t} = (h(x_t) - \hat{h}(x_t))$, then

$$E^2(e_{h,t}^2) \lesssim N_0\, E\{\sigma^2(x_t)(h'(x_t))^2\}\,.$$

Outline of proof. Consider the Kushner equation for the conditional moments (cf. chapter 8 of [7])

$$d\hat{\phi}(x_t)dt = (m(x_t)\phi'(x_t))^{\wedge}\, dt + \frac{1}{2}(\sigma^2(x_t)\phi''(x_t))^{\wedge}\, dt +$$

$$+ N_o^{-1/2}((h(x_t) - \hat{h}(x_t))\phi(x_t))^{\wedge}\, d\bar{w}_t$$

where $\bar{w}_t$ is the standerized innovation process. Set $\phi(x) = h(x)$, $h(x_t)$ will be denoted by h_t, calculate $d\hat{h}_t$ and then by Ito's formula calculate $d(\hat{h}_t)^2$. Set $\phi(x) = h^2(x)$ and calculate $d(h_t^2)^{\wedge}$. Note that by the stationarity assumption $d\, E(h_t^2 - (\hat{h}_t)^2 = 0$. Therefore, by the above calculations

$$E(\sigma_t^2(h_t')^2) - N_o^{-1}E((e_{h,t}^2)^{\wedge})^2 = 2E[m_t h_t'(h_t-\hat{h}_t)]+E[\sigma_t^2 h_t''(h_t-\hat{h}_t)]$$

The right hand side of this equation is upper bounded by $K\, E\, e_{h,t}^2$ where K is some constant, since $E\, e_{h,t}^2 \to 0$ as $N_o \to 0$ it follows that

$$E((e_{h,t}^2)^{\wedge})^2 \approx N_o E(\sigma_t^2(h_t')^2)$$

and the result of proposition 1 follows by Jensen's inequality.

As an application of proposition 1, consider the case where $\sigma(x) = \sigma_o$; $h(x) = h_o \cdot x$ (with $m(x)$ remaining non-linear), it follows from proposition 1 that $E\, e_t^2 \lesssim N_o^{1/2}\sigma h_o^{-1}$ and it follows from (3) that in this case $E\, e_t^2 \gtrsim N_o^{1/2}\sigma h_o^{-1}$ which shows the asymtotic tightness of proposition 1. In fact, it was shown in [10] that this upper bound can be achieved by linear filtering which means that even for non-linear $m(x)$, optimal linear filtering is asymptotically optimal.

Proposition 2. If, in addition to the previous assumptions

$$(h'(x))^{-2} = \left(\frac{dh(x)}{dx}\right)^{-2}$$

is continuous and of polynomial growth and if

$$N_o^{-q/2} \cdot E|e_t|^{2n} \xrightarrow[N_o \to 0]{} 0 \qquad (5)$$

for all n and all $q < n$, then

$$(E\, e_t^2)^2 \lesssim N_o \cdot E(\sigma^2(x_t)(h'(x_t))^{-2}) \qquad (6)$$

The proof is similar to that of proposition 1, details and related results are given in [3].

As an example consider the saturated sensor problem :

$$dx_t = -\alpha x_t dt + \sigma_o dw_t, \ \alpha>0; \ dy_t = (\lambda \text{arc tg } x_t/\lambda)dt + N_o^{1/2} d\nu_t \ ,$$

assumption (5) is assumed to be satisfied. The asymptotic upper bound of proposition 2 will now be compared with an asymptotic lower bound derived from (3). From proposition 2 we have :

$$E\,(e_t^2) \lesssim \sigma_o \rho E^{1/2}\left\{\frac{(x_t^2+\lambda^2)^2}{\lambda^4}\right\} \qquad (7)$$

$$\lesssim \rho_o \sigma_o (3(\lambda/V)^{-4} + 2(\lambda/V)^{-2} + 1)^{1/2}$$

where

$$V = E^{1/2} x_t^2 = (\sigma_o^2/2\alpha)^{1/2} \quad \text{and} \quad Ex^4 = 3(Ex^2)^2 \quad \text{since} \quad x_t$$

is Gaussian. Turning now to a lower bound, it follows from (3) that

$$E\, e_t^2 \gtrsim \sigma_o \rho / (E(h'(x_t))^2)^{1/2} \ . \qquad (8)$$

Let $B^2 = E(h'(x_t))^2$, since x_t is Gaussian,

$$B^2 = \frac{\lambda}{V}\frac{1}{\sqrt{2\pi}}\int_{-\infty}^{\infty}\left(\frac{1}{1+\eta^2}\right)^2 \text{Exp} - \frac{\lambda^2\eta^2}{2V^2}\, d\eta \ .$$

In order to evaluate this expression, let

$$I_1(\beta) = \frac{1}{\sqrt{2\pi\sigma^2}}\int_{-\infty}^{\infty}\frac{\text{Exp} - \frac{x^2}{2\sigma^2}}{x^2+\beta^2}\, dx \ ,$$

then from p. 338 of [6] we have

$$I_1(\beta) = \frac{1}{\beta\sigma}\,\text{Exp}\,\frac{\beta^2}{2\sigma^2}\cdot\int_{\beta/\sigma}^{\infty}\text{Exp} - \frac{\eta^2}{2}\, d\eta \ .$$

Let

$$I_2(\beta) = \frac{1}{\sqrt{2\pi\sigma^2}} \int_{-\infty}^{\infty} \frac{\text{Exp}\ -\frac{x^2}{2\sigma^2}}{(x^2+\beta^2)^2}\, dx \quad .$$

Then, it follows by differentiating and interchanging the order of differentiation and integration that

$$I_2(\beta) = -\frac{1}{2\beta}\frac{\partial}{\partial\beta} I_1(\beta) .$$

Therefore

$$I_2(\beta) = \frac{1}{2\beta\sigma}\left\{\frac{1}{\sigma\beta} + \left(\frac{1}{\beta^2} - \frac{1}{\sigma^2}\right)\text{Exp}\ \frac{\beta^2}{2\sigma^2} \cdot \int_{\beta/\sigma}^{\infty} \text{Exp}\ -\frac{\eta^2}{2}\, d\eta\right\}$$

and (8) becomes

$$B^2 = \frac{\lambda}{2V}\left(1 - \left(\frac{\lambda}{V}\right)^2\right)\left(\text{Exp}\ \frac{\lambda^2}{2V^2}\right)\int_{\lambda/V}^{\infty} \text{Exp}\ -\frac{\eta^2}{2}\, d\eta + \frac{1}{2}\left(\frac{\lambda}{V}\right)^2 \quad . \tag{9}$$

Equations (7), (8) and (9) yield the asymptotic upper and lower bound. In particular, for $(\lambda/V) = 1$, which represents substantial saturation but not hard limiting, the ratio of the upper to the lower bound is 1.73.

Remark. As mentioned earlier, for the case where $h(x) = h_0 \cdot x$ and $\sigma(x) = \sigma_0$, the upper bound of proposition 2 is achievable asymptotically with a linear filter. The following heuristic arguments indicate that for nonlinear invertible $h(x)$ and $\sigma(x)=\sigma_0$, (5) is achievable by a cascade of a linear filter followed by the instantaneous non-linearity $h^{-1}(x)$ followed again by a linear filter. Let $dz_t = -\alpha z_t dt + \alpha dy_t$ where y_t is as given by the second line of (1) and $\alpha>0$. Assuming that α is chosen large enough so as to avoid appreciable distortion on $h(x_t)$, we can write in the steady state, $z_t \approx h(x_t) + n_t$ where n_t is a wide band noise, the bandwidth of which is much larger than the bandwidth of $h(x_t)$. Since N_0 is assumed small so is En_t^2. Set now $v_t = h^{-1}(z_t)$, then

$$v_t \approx x_t + n_t (dh^{-1}(\theta)/d\theta)\Big|_{\theta=h(x_t)} = x_t + n_t (h'(x_t))^{-1} \tag{10}$$

Assuming that the bandwidth of the spectral density of n_t is much wider than that of the $(h'(x_t))^{-1}$ process, $n_t(h'(x_t))^{-1}$

can be considered as a white noise of spectral density $N_0 E(h'(x_t))^{-2}$ which is uncorrelated to x_t. Apply now an optimal linear filter to (10), it follows from equation (28) of [10] that the asymptotic filtering error will be as given by (6).

4. BOUNDS DERIVED FROM A RELATION BETWEEN SHANNON AND FISHER INFORMATION

Consider equation (1) under the assumptions of proposition 1, but without the assumption of stationarity. Assume that x_t of (1) possesses a probability density $p_t(\beta) = d\,\mathrm{Prob}\{x_t \leqslant \beta\}/d\beta$. Let $\Pi_t(\beta) = \mathrm{Prob}\{x_t \leqslant \beta \mid y_\theta, 0 \leqslant \theta \leqslant t$ and assume that for almost all ω, $d\Pi_t(\beta)/d\beta = \pi_t(\beta)$ exists for all β and t. The mutual information between the random path $(y_\theta, 0 \leqslant \theta \leqslant t)$ and the random variable x_t will be denoted $I(x_t, y_0^t)$ and is given by

$$I(x_t, y_0^t) = E\,\ell n\,\frac{\pi_t(x_t)}{p_t(x_t)} = E\int_{-\infty}^{\infty}\left(\ell n\,\frac{\pi_t(\beta)}{p_t(\beta)}\right)\pi_t(\beta)d\beta\ ,$$

where the expectation in the second expression is with respect to x_t and the paths $(y_\theta, 0 \leqslant \theta \leqslant t)$ and the expectation in the last expression is with respect to $(y_\theta, 0 \leqslant \theta \leqslant t)$.

It has recently been shown by Bucy [4] that under certain regularity conditions (cf. also [3] and p. 188 Vol. II of [7]) the following result holds :

$$\frac{1}{2}N_0^{-1}E(h(x_t) - \hat{h}(x_t))^2 - \frac{d}{dt}I(y_0^t, x_t) =$$

$$= \frac{1}{2}E\int_{-\infty}^{\infty}\pi_t(\beta)\sigma^2(\beta)\left(\frac{\partial \ell n \pi_t(\beta)}{\partial\beta}\right)^2 d\beta - \frac{1}{2}\int_{-\infty}^{\infty}p_t(\beta)\sigma^2(\beta)\left(\frac{\partial \ell n p_t(\beta)}{\partial\beta}\right)^2 d\beta\ . \tag{11}$$

<u>Proposition 3</u>. Under the assumption for the validity of (11) and under the stationarity assumption

$$N_0^{-1}Ee_h^2 \geqslant \frac{(E|Z'(x_t)\sigma(x_t)|)^2}{E\,e_z^2} - E\left(\frac{2m(x_t)}{\sigma(x_t)} - \sigma'(x_t)\right)^2 \tag{12}$$

where $Z(x)$ is assumed to be differentiable, $E|Z'(x_t)\sigma(x_t)| \leqslant \infty$ and $\sigma(\beta)\Pi_t(\beta)(|Z(\beta)|+1) \to 0$ a.s. as $(\beta) \to \infty$.

<u>Remark</u>. The special case of (8) where $m(x) = -\alpha x$, $\sigma(x) = \sigma_0$ and $Z(x) = h(x)$ was derived in [4].

<u>Outline of proof</u>. Consider (11) and note that : (a) by

stationarity $dI(y_o^t, x_t)/dt = 0$, (b) the second term in the right hand side of (11) can be evaluated from the explicit solution of the stationary Fokker Plank equation, (c) the first term in the right hand side of (11) can be upper bounded by the same arguments as in the standard proof of the Cramer-Rao bound. For details of the proof and the following two applications, cf. [3]. Let $m(x) = -\alpha x$, $\sigma(x) = \sigma_o$ and $h(x) = x^2$, proposition 1 yields the upper bound

$$E\, e_h^2 \lesssim N_o^{1/2} \sigma_o^2 (2/\alpha)^{1/2} \quad \text{and (12) yields} \quad E\, e_h^2 \gtrsim 2\pi^{-1/2} N_o^{1/2} \alpha^{-1/2}$$

therefore the ratio between the upper and lower bounds is $\sqrt{\pi/2} \approx 1.25$.

Turning to the second application, let $\sigma(x) = \sigma_o$ and $h(x) = x$; as is well known, for the linear case $m(x) = -\alpha x$, $(\alpha>0)$, as $N_o \to 0$, $(e_t^2)^\wedge \sigma_o^{-1} N_o^{-1/2} \approx 1$ in the steady state. By the results of propositions 1 and 3 it can be shown that even without the linearity assumption on $m(x)$, it still holds that $(e_t^2)^\wedge \sigma_o^{-1} N_o^{-1/2}$ converges in quadratic mean to 1 as $N_o \to 0$ and

$$(e_t^3)^\wedge N_o^{-3/4}$$

converges in quadratic mean to zero. These results are related to the problem of deriving finite dimensional approximations to equations for conditional moments.

REFERENCES

(1) Bobrovsky, B.Z., and Zakai, M. : 1976, "A Lower Bound on the Estimation Error for Certain Diffusion Processes", IEEE Trans. Inform. Theory, IT-22, pp.45-52.
(2) Bobrovsky, B.Z., and Zakai, M. : Jan. 1981, "On Lower Bounds for the Nonlinear Filtering Problem". To be published in the IEEE Trans. on Inf. Th.
(3) Bobrovsky, B.Z., and Zakai, M. : March 1980, "Asymptotic a-priori Estimates for the Error in the Nonlinear Filtering Problem", E.E. Publication 371.
(4) Bucy, R.S. : 1979, Information Sciences, 18, pp.179-187.
(5) Golubev, G.K. : 1979, "Computation of Efficiency of Maximum-likelihood Estimate when Observing a Discontinuous Signal in White Noise", Problems of Information Transmission, (English translation of Problemi Peredachi Informatsii), 15, pp.206-211.
(6) Gradshteyn, I.S., and Rizhik, I.M. : 1965, Tables of Integrals, Series and Products, Academic Press, New York.
(7) Lipster, R.S., and Shiryayev, A.N. : 1977, Statistics of Random Processes, Springer Verlag, New York.
(8) Sakrison, D.J. : 1970, "Notes on Analog Communication", Van Nostrand, New York.

(9) Seidman, L.P. : May 1970, "Performance Limitations and Error Calculations for Parameter Estimation", Proceedings of the IEEE, Vol. 58, pp.644-652.

(10) Zakai, M., and Ziv, J. : 1972, "Lower and Upper Bounds on Optimal Filtering Error of Certain Diffusion Processes", IEEE Trans. Inf. Th., IT-18, pp.325-331.

AN APPROXIMATION TO OPTIMAL NONLINEAR FILTERING WITH DISCONTINUOUS OBSERVATIONS

G.B. Di Masi and W.J. Runggaldier*

LADSEB-CNR and Istituto di Elettrotecnica, Università di Padova, Italy
*Seminario Matematico, Università di Padova, Italy

The paper deals with a possible approach to the problem of finite-dimensional recursive filtering in the nonlinear case, when the signal is a diffusion process and the observations have both continuous and discontinuous components. The approach used consists in the approximation of the optimal filter by means of a converging sequence of finite-dimensional filters. These are given in terms of funtionals of continuous-time Markov chains and can be recursively computed via a finite-dimensional Zakai equation.

1. INTRODUCTION

Consider a diffusion process x_t described by the Itô equation

$$dx_t = a(x_t)dt + b(x_t)dw_t \ , \quad 0 \leqslant t \leqslant T \ , \ x_o = 0 \tag{1}$$

where w_t is a standard Wiener process. Suppose that x_t is only partially observed through the process y_t $(0 \leqslant t \leqslant T)$ given by

$$y_t = \int_0^t c(x_s)ds + v_t + N_t \tag{2}$$

where v_t is a standard Wiener process and N_t is doubly stochastic Poisson with rate λ_t that may depend on x_t, namely

$$\lambda_t = \lambda(x_t) \tag{3}$$

For simplicity all quantities will be considered as scalars. Denoting by $\mathcal{F}_t^y$ the σ-algebra generated by $\{y_s, 0 \leqslant s \leqslant t\}$, the nonlinear filtering problem for the present model consists in the evaluation of the estimate

M. Hazewinkel and J. C. Willems (eds.), Stochastic Systems: The Mathematics of Filtering and Identification and Applications, 583–590.

$$\hat{f}(x_t) = E\{f(x_t) \mid \mathscr{F}_t^y\} \quad , \tag{4}$$

where f is a Borel function on R.

The aim of the present paper is to provide an approximation to the estimate (4) along the lines of what has been done by the authors in [3] (see also [4, Ch.7]) for the case of a signal corrupted by additive white noise. The result will be a sequence of finite-dimensional recursive filters that converge to the optimal filter in (4). As in [3], in order to construct the approximating filters, the process x_t will be approximated (in the sense of weak convergence) by a sequence of continuous-time Markov chains. For computational purposes it will be convenient to have finite-state chains. In what follows we will therefore consider a slightly modified model. More precisely, given an open and bounded set $G \subset R$, we will stop the original diffusion (1) on its first exit form G and use the same symbols x_t and y_t to denote the stopped diffusion and the corresponding observation according to (2). It is possible to obtain a sequence of finite-state approximating chains also without necessarily stopping the original diffusion; however we will use here the above stopped model for the sake of simplicity.

Our approximation will be developed under the following assumptions on (1)-(4):

A.1. The functions a,b,c,f and λ are continuous and bounded and λ is strictly positive.

A.2. The processes v_t and x_t are independent and for each $t \in [0,T]$, v_t is independent of $\{N_s, 0 \leq s \leq t\}$.

A.3. Equation (1) has a unique solution in the weak sense.

A.4. $P\{\tau = \bar{\tau}\} = 1$, where τ and $\bar{\tau}$ are the exit times of the diffusion (1) from G and its closure $\bar{G}$ respectively.

2. ABSOLUTELY CONTINUOUS CHANGE OF PROBABILITY MEASURE

For our purposes the more convenient approach to nonlinear filtering is the so-called measure transformation approach or reference-probability method, rather than the innovations-approach (see the tutorial by S. Marcus). Assuming that x_t and y_t are defined on a probability space $\{\Omega, \mathscr{F}, P\}$, the aim of the present section is therefore to construct a new probability measure on $\{\Omega, \mathscr{F}\}$ that is mutually absolutely continuous with respect to P and such that

i) under P_1 the processes x_t and y_t are independent

ii) the restrictions of P_1 and P to the σ-algebras generated by x_t are the same.

To this end define on $\{\Omega, \mathscr{F}\}$ a measure P_1 by means of its Radon-Nikodym derivative with respect to P, namely

$$\frac{dP_1}{dP} = \exp\left[-\int_0^T c(x_s)dv_s - \tfrac{1}{2}\int_0^T c^2(x_s)ds - \right.$$

$$- \int_0^T \log \lambda(x_s)dN_s - \int_0^T [1- \lambda(x_s)] \, ds \Big] \tag{5}$$

Denoting by E_1 the expectation with respect to the measure P_1 and defining the increasing families of σ-subalgebras

$$\mathscr{F}_t = \mathscr{F}^x_T \vee \mathscr{F}^y_t \quad \text{and} \quad \mathscr{G}_t = \mathscr{F}^x_t \vee \mathscr{F}^y_t , \tag{6}$$

where $\mathscr{F}^x_t = \sigma\{x_s, 0 \leqslant s \leqslant t\}$ and $\mathscr{F}^y_t = \sigma\{y_s, 0 \leqslant s \leqslant t\}$, each containing all P-null subsets in $\mathscr{F}$, let

$$L_t = E_1 \{ \frac{dP}{dP_1} \mid \mathscr{G}_t \} \tag{7}$$

Then L_t represents a version of the Radon-Nikodym derivative of the restriction of P to $\mathscr{G}_t$ with respect to the corresponding restriction of P_1 and we have

Theorem 1: If A.1 and A.2 are satisfied then the following statements hold

S.1) P_1 is a probability measure and $P_1 \sim P$.

S.2) $z_t := y_t - N_t = \int_0^t c(x_s)ds + v_t$ is an $(\mathscr{F}_t, P_1)$ - standard Wiener process and N_t is an $(\mathscr{F}_t, P_1)$ - standard Poisson process.

S.3) $E_1 \{ e^{iv(y_t - y_s)} \mid \mathscr{F}_s \} = \exp [(e^{iv} - 1 - \frac{1}{2}v^2)(t-s)] := \psi(v, t-s)$.

S.4) L_t is the unique solution to the integral equation

$$L_t = 1 + \int_0^t L_s \, c(x_s) d(y_s - N_s) + \int_0^t L_{s-} [\lambda(x_s) - 1] d(N_s - s), \; P_1\text{-a.s.}$$

Proof (outline only):
S.1), S.2) and S.4) follow from [6, Thms.1,2] while S.3) is obtained using the differentiation formula [1,p.1026] .

The following corollaries provide the desired properties i) and ii) for the measure P_1.

Corollary 1. If A.1 and A.2 are satisfied, then under P_1 the process y_t has independent increments and x_t and y_t are independent.

Proof. From S.3) we have

$$E_1 \{ e^{iv(y_t - y_s)} | \mathscr{F}^y_s \} = E_1 \{ E_1 \{ e^{iv(y_t - y_s)} \mid \mathscr{F}_s \} | \mathscr{F}^y_s \} = \psi(v, t-s)$$

which shows that y_t is an independent-increment process.

Analogously

$$E_1\{e^{iv(y_t-y_s)}|\mathcal{F}_T^x\} = E_1\{E_1\{e^{iv(y_t-y_s)}|\mathcal{F}_s\}|\mathcal{F}_T^x\} = \psi(v,t-s)$$

which shows that the increment distributions, and therefore all joint distributions of y_t, do not depend on $\{x_t, 0\leqslant t\leqslant T\}$.

<u>Corollary 2</u>. Under A.1 and A.2, L_t in (7) is such that $E_1\{L_t|\mathcal{F}_t^x\}=1$. Furthermore, the restrictions of P_1 and P to $\mathcal{F}_t^x(t\in[0,T])$ are the same.

<u>Proof</u>. The first result follows from S.4), noticing that $\mathcal{F}_t^x$ and $\mathcal{F}_t^y$ are independent under P_1 and therefore L_t is an $(\mathcal{F}_t^y, P_1)$ - martingale for every fixed trajectory of the process x_t. The second result follows immediately from the first, observing that for every $\mathcal{F}_t^x$ - adapted process φ_t we have

$$E\,\varphi_t = E_1 L_t\,\varphi_t = E_1\{E_1\{L_t\,\varphi_t|\mathcal{F}_t^x\}\} = E_1\{\varphi_t E_1\{L_t|\mathcal{F}_t^x\}\} = E_1\,\varphi_t.$$

3. THE APPROXIMATING FILTER

With the measure P_1 introduced in the previous section it is well known (see f.e. [1] , [2]) that the optimal filter in (4) can be given the following representation

$$E\{f(x_t)|\mathcal{F}_t^y\} = \frac{E_1\{f(x_t)L_t|\mathcal{F}_t^y\}}{E_1\{L_t|\mathcal{F}_t^y\}} \tag{8}$$

Since under P_1 the processes x_t and y_t are independent, the conditioning in the right hand side of (8) only fixes the trajectory of the process y_t, while the expectation is an ordinary expectation over the process x_t. This fact, combined with Corollary 2, then leads to an equivalent filter representation, which will prove particularly useful in the sequel. In fact, letting $\bar{x}_t$ be a version of x_t independent of y_t, denoting by $\bar{L}_t$ the process L_t in (7) when x_t is replaced by $\bar{x}_t$, and defining

$$V_t(f) := E\{f(\bar{x}_t)\bar{L}_t|\mathcal{F}_t^y\} \quad , \tag{9}$$

then the following representation holds

$$E\{f(x_t)|\mathcal{F}_t^y\} = \frac{V_t(f)}{V_t(1)} \tag{10}$$

The original problem of approximating $\hat{f}(x_t)$ by a sequence of finite-dimensional recursive filters reduces now to the problem of approximating the functional $V_t(f)$ by a sequence of functionals $V_t^h(f)$ that can be computed-recursively by a finite-dimensional procedure. This can be done as in [3] along the following lines:

- Construct a sequence of finite-state continuous-time Markov chains x_t^h that converge weakly to the process x_t
- Let $\bar{x}_t^h$ be a version of x_t^h, independent of y_t
- Let $\bar{L}_t^h$ be the process L_t in (7) when x_t is replaced by $\bar{x}_t^h$
- Define $V_t^h(f) = E\{ f(\bar{x}_t^h)\bar{L}_t^h | \mathscr{F}_t^y \}$

Then we have (see [3])

Theorem 2: Under the assumptions A.1-A.4, $V_t^h(f)$ converge to $V_t(f)$ P-a.s. for each $t \in [0,T]$ as h goes to zero.

4. RECURSIVE EVALUATION OF THE APPROXIMATING FUNCTIONALS

The aim of the present section is to show that the functionals $V_t^h(f)$ are completely determined by quantities that can be computed recursively.

Defining

$$U_t^h(\bar{x}_t^h) = E\{ \bar{L}_t^h | \mathscr{F}_t^y \vee \sigma\{\bar{x}_t^h\}\} \quad ,$$

$$p_t^h(i) = P\{\bar{x}_t^h = i\} \quad ,$$

$$p_{st}^h(i,j) = P\{\bar{x}_t^h = j \mid \bar{x}_s^h = i\} \, , \; s \leqslant t \; ,$$

$$q_t^h(i) = U_t^h(i) p_t^h(i) \quad ,$$

where we have denoted by i and j the generic states of the finite chain $\bar{x}_t^h$, we have

$$V_t^h(f) = E\{ f(\bar{x}_t^h)\bar{L}_t^h | \mathscr{F}_t^y \} = E\left\{ f(\bar{x}_t^h) E\{ \bar{L}_t^h | \mathscr{F}_t^y \vee \sigma\{x_t^h\}\} \mid \mathscr{F}_t^y \right\}$$

$$= E\{ f(\bar{x}_t^h)\, U_t^h(\bar{x}_t^h) \} = \sum_i f(i) q_t^h(i) \qquad (11)$$

We now show that the $q_t^h(i)$'s can be recursively computed. To this end first notice that an integral equation analogous to the one in S.4) can be derived also for $\bar{L}_t^h$, namely

$$\bar{L}_t^h=1+\int_0^t \bar{L}_s^h\, c(\bar{x}_s^h)d(y_s-N_s)+\int_0^t \bar{L}_{s-}^h\,[\boldsymbol{\lambda}(\bar{x}_s^h)-1]\,d(N_s-s). \tag{12}$$

From this, after some manipulations, we get

$$V_t^h(f)=E\,\{\,f(\bar{x}_t^h)\,\}+\int_0^T E\{\,\bar{L}_s^h\,c(\bar{x}_s^h)f(\bar{x}_t^h)\,|\,\mathscr{F}_s^y\,\}\,d(y_s-N_s)$$

$$+\int_0^t E\{\,\bar{L}_{s-}^h\,[\boldsymbol{\lambda}(\bar{x}_s^h)-1]\,f(\bar{x}_t^h)\,|\,\mathscr{F}_{s-}^y\,\}\,d(N_s-s). \tag{13}$$

Observe that, by the Markovianity of $\bar{x}_t^h$, we have for the second integrand in (13)

$$E\{\,\bar{L}_{s-}^h\,[\boldsymbol{\lambda}(\bar{x}_s^h)-1]\,f(\bar{x}_t^h)\,|\,\mathscr{F}_{s-}^y\,\}$$

$$=E\Big\{\,[\boldsymbol{\lambda}(\bar{x}_s^h)-1]\,E\{\,\bar{L}_{s-}^h f(\bar{x}_t^h)\,|\,\mathscr{F}_{s-}^y\vee\,\boldsymbol{\sigma}\{\bar{x}_s^h\}\}\,|\,\mathscr{F}_{s-}^y\,\Big\}$$

$$=E\Big\{\,[\boldsymbol{\lambda}(\bar{x}_s^h)-1]\,U_{s-}^h(\bar{x}_s^h)E\,\{\,f(\bar{x}_t^h)\,|\,\boldsymbol{\sigma}\{\bar{x}_s^h\}\}\,\Big\}$$

$$=\sum_j[\,\boldsymbol{\lambda}(j)-1]\,U_{s-}^h(j)\,\sum_i f(i)p_{st}^h(j,i)p_s^h(j)$$

$$=\sum_j[\,\boldsymbol{\lambda}(j)-1]\,q_{s-}^h(j)\,\sum_i f(i)p_{st}^h(j,i) \tag{14}$$

and analogously for the first integrand, where we can drop the minus sign in L_{s-}^h, $\mathscr{F}_{s-}^y$, U_{s-}^h and q_{s-}^h.
Substituting the thus obtained expressions and (11) into (13), which holds for any f, we get

$$q_t^h(i)=p_t^h(i)+\int_0^t \sum_j p_{st}^h(j,i)c(j)q_s^h(j)d(y_s-N_s)$$

$$+\int_0^t \sum_j p_{st}^h(j,i)\ [\lambda(j)-1]\ q_{s-}^h(j)d(N_s-s) \qquad (15)$$

or, in matrix form

$$q_t^h=p_t^h+\int_0^t \Pi_{st}^h C^h q_s^h\ d(y_s-N_s)+\int_0^t \Pi_{st}^h\ [\Lambda^h-I^h]\ q_{s-}^h d(N_s-s), \qquad (16)$$

where, denoting by N^h the number of states of x_t^h, p_t^h and q_t^h are N^h-column vectors with i-th elements $p_t^h(i)$ and $q_t^h(i)$, C^h and Λ^h are N^h-diagonal matrices with c(i) and λ(i) respectively in the i-th position, Π_{st}^h is the transpose of the state transition matrix of $\bar{x}_t^h$ and I^h is the N^h-identity.

Equation (15) provides the desired recursive relation, which in differential form becomes

$$dq_t^h = A_t^h q_t^h\ dt + C^h\ q_t^h\ d(y_t-N_t)+[\Lambda^h-I^h]\ q_{t-}^h\ d(N_t-t), \qquad (17)$$

where A_t^h is the transpose of the intensity matrix of $\bar{x}_t^h$. This equation can be shown to be the Zakai equation for the filtering problem (1)-(4) when x_t is replaced by x_t^h (see f.e. also [5]). Such equations have been the starting point of most of the contributions to nonlinear filtering in this volume. The reader interested in efficient solutions to (17) is therefore referred to those papers.

REFERENCES

1 R.Boel, P.Varaiya and E.Wong, "Martingales on jump processes". SIAM J. Control 13, pp.999-1061 (1975).

2 P.Brémaud and J.Jacod, "Processus ponctuels et martingales: résultats récents sur la modélisation et le filtrage". Adv. Appl. Prob. 9, pp.362-416 (1977).

3 G.B.Di Masi and W.J.Runggaldier, "Continuous-time approximations for the nonlinear filtering problem". To appear on Appl. Math. and Opt.

4 H.J.Kushner, "Probability methods for approximations in stochastic control and for elliptic equations". Academic Press (1977).

5 E.Pardoux, "Filtering of a diffusion process with Poisson-type observation", in "Stochastic Control Theory and Stochastic differential Systems". (Kohlmann and Vogel eds.). Lecture Notes in Contr. and Inf. Sci., 16, Springer-Verlag (1979).

6 M.V.Vaca and D.L.Snyder, "Estimation and decision for observations derived from martingales: Part I, Representations". IEEE Trans. IT 22, pp.691-707 (1976).

7 M.V.Vaca and D.L.Snyder, "Estimation and decision for observations derived from martingales: Part II". IEEE Trans. IT 24, pp.32-45 (1978).

SOME RESULTS AND SPECULATIONS ON THE ROLE OF LIE ALGEBRAS IN FILTERING.

Michiel Hazewinkel
Dept. Math., Erasmus Univ. Rotterdam
P.O. Box 1738
3000DR ROTTERDAM
The Netherlands

Steven I. Marcus
Dept. Electrical Engineering
Univ. of Texas at Austin
AUSTIN, Texas 78712 USA

1. INTRODUCTION. SETTING THE STAGE.

Consider a stochastic dynamical system of the type

$$(1.1) \qquad dx_t = f(x_t)dt + G(x_t)dw_t, \quad dy_t = h(x_t)dt + dv_t$$

where f,G,h are (sufficiently regular) vector and matrix valued functions, and w and v are unit variance Wiener processes independent of the initial state x(0) and independent of each other. We are interested in ways of calculating the conditional expectation $\hat{\phi}(x_t)$ (best least squares estimates) of functions $\phi(x_t)$ given the observations $y^t = \{y_s: 0 \leq s \leq t\}$ through time t. In particular we are interested in finite dimensional recursive filters for $\hat{\phi}(x_t)$. By definition this means a machine driven by the observations:

$$(1.2) \qquad d\eta_t = \alpha(\eta_t)dt + \beta(\eta_t)dy_t$$

defined on a finite dimensional manifold M (so that $\eta_t \in M$ and

M. Hazewinkel and J. C. Willems (eds.), Stochastic Systems: The Mathematics of Filtering and Identification and Applications, 591–604.

$\alpha(\eta_t)$, $\beta(\eta_t)$ are vectorfields on M), such that for a suitable output function

$$(1.3) \qquad \gamma(\eta_t) = \hat{\phi}(x_t)$$

(Equations (1.2), (1.3) together form a finite dimensional recursive filter for the statistic $\hat{\phi}(x_t)$.

Now a certain unnormalized version $\rho(x,t)$ of the conditional density for x_t given y^t satisfies the Duncan-Mortensen-Zakai equation. Written in Fisk-Stratonovic form this equation is

$$(1.4) \qquad d\rho(x,t) = (\mathfrak{L} - \tfrac{1}{2}\sum_{i=1}^{p} h^i(x)^2)\rho(t,x)dt + \sum_{i=1}^{p} h^i(x)\rho(t,x)dy_t^i$$

where $\mathfrak{L}$ is the Fokker-Planck operator

$$(1.5) \qquad \mathfrak{L}(.) = \frac{1}{2}\sum_{i,j} \frac{\partial^2}{\partial x_i \partial x_j}((GG^T)^{i,j}.) - \sum_{i=1}^{n} \frac{\partial}{\partial x_i}(f^i.)$$

(Here f^i, h^i is the i-th component of f,h and $(GG^T)^{i,j}$ is the (i,j)-th entry of the product of the matrix G with its transpose); cf. [7] for a derivation of the Duncan-Mortensen-Zakai equation. The Lie algebra of differential operators generated by $\mathfrak{L} - \frac{1}{2}\Sigma h^i(x)^2$ and $h^1(x), \ldots, h^p(x)$ is called the estimation Lie algebra. (Here $h^i(x)$ is the multiplication operator $\rho(x) \mapsto h^i(x)\rho(x)$). We refer to the two appendices on "manifolds and vectorfields" and on "Lie algebras" in this volume for basic background information on these topics.

Both Brockett and Mitter have independently proposed the study of this estimation Lie algebra as an approach to the filtering properties of (1.1). This idea has been quite remarkably successful. Some evidence for this lies in the following. First equation (1.4) is bilinear (albeit infinite dimensional) and the Lie

algebra generated by the matrices A,B in a control system $\dot{x} = Ax + (Bx)u$ is known to be influential ([5]).
Second in the case of a linear system

$$(1.6) \quad dx_t = Axdt + Bdw_t, \; dy_t = Cx_t dt + dv_t$$

the Lie algebra of equation (1.5) and the Lie algebra of the Kalman filter of (1.6) are closely related [2]. The third point requires more explanation. Suppose that a finite dimensional filter (1.2), (1.3) existed. The equations are supposed to be in Fisk-Stratonovic form so that they make sense on a manifold [6]). Then we have two ways for calculating $\hat{\phi}(x_t)$: once via (1.2), (1.3) and once via (1.4) followed by normalization and integration. We can assume (1.2), (1.3) to be minimal and by a conjectured generalization of Sussmann's minimal realization result [20] we would have a homomorphism of the estimation Lie algebra onto the Lie algebra generated by the vectorfields $a(\eta_t)$ and $b(\eta_t)$ in (1.2). This is precisely what happens in the case of linear systems [2]. And inversely given such a homomorphism of Lie algebras satisfying an additional isotropy subalgebra condition a suitable generalization of the results of [13] or [23] would give a filter. Thus we would have a correspondence between statistics which are finite dimensionally recursively computable and certain homomorphisms of Lie algebras of the estimation algebra into Lie algebras of vectorfields on manifolds. Most of what follows makes little sense unless this is more or less true. There is, fortunately, a fair amount of positive evidence (linear case [2,4], finite state case [4,5] certain bilinear systems [15,26], cubic sensor [21,11]).

There are still more reasons for the importance of the estimation algebra involving representation theory, functional integration and deep analogies with quantum physics [17,18,19].

2. EXAMPLES OF ESTIMATION ALGEBRAS.

2.1. <u>The simplest nonzero linear system</u>, [2]. The stochastic dynamical system is $dx_t = dw_t$, with observations $dy_t = x_t dt + dv_t$. The estimation algebra is four dimensional with basis

$\frac{1}{2}\frac{d^2}{dx^2} - \frac{1}{2}x^2, x, \frac{d}{dx}, 1$. It is a well-known Lie algebra (especially in physics). It is called the oscillator algebra.

2.2. <u>Heisenberg-Weyl</u> algebras. Let W_n denote the associative algebra $\mathbb{R}\langle x_1,\dots,x_n; \frac{d}{dx_1},\dots,\frac{d}{dx_n}\rangle$ of all (partial) differential operators in $\frac{\partial}{\partial x_1}, \dots, \frac{\partial}{\partial x_n}$ (of any order) with polynomial coefficients. As an associative algebra it is generated by the symbols $x_1, \dots, x_n, \frac{\partial}{\partial x_1}, \dots, \frac{\partial}{\partial x_n}$ subject to the relations suggested by the notations used, i.e. $x_i \frac{\partial}{\partial x_i} x_i - x_i \frac{\partial}{\partial x_i} = 1$, $x_i x_j = x_j x_i$, $\frac{\partial}{\partial x_i}\frac{\partial}{\partial x_j} = \frac{\partial}{\partial x_j}\frac{\partial}{\partial x_i}$, and $x_i \frac{\partial}{\partial x_j} = \frac{\partial}{\partial x_j} x_i$ if $i \neq j$. A basis for W_n (as a vectorspace over $\mathbb{R}$) consists of the monomials $x^\alpha \frac{\partial^\beta}{\partial x^\beta} = x_1^{\alpha_1} \dots x_n^{\alpha_n} \frac{\partial^{\beta_1}}{\partial x_1^{\beta_1}} \dots \frac{\partial^{\beta_n}}{\partial x_n^{\beta_n}}$, $\alpha_i, \beta_j \in \mathbb{N} \cup \{0\}$. In this paper W_n is always considered as a Lie algebra (with the bracket operation $[D,D'] = DD'-D'D$). The Lie algebra W_n has a one dimensional centre $\mathbb{R}.1$ (consisting of scalar multiples of the identity operator) and $W_n/\mathbb{R}.1$ is simple.

2.3. <u>The cubic sensor</u>. The system is $dx_t = dw_t$ with observations, $dy_t = x_t^3 dt + dv_t$. In this case the estimation algebra is equal to all of W_1. For a proof cf. [10].

2.4. <u>Quadratic observations</u>. Now consider $dx_t = dw_t$, $dy_t = x_t^2 dt + dv_t$. Then the estimation algebra is $W_1^{(2)}$ which is the subalgebra of W_1 spanned by all monomials of the form $x^i \frac{d^j}{dx^j}$ with $i - j$ even.

2.5. Example of mixed linear bilinear type. The system is $dx_{1t} = dw_{1t}$, $dx_{2t} = x_{1t}dt + dx_{1t}dw_{2t}$ with observations $dy_t = x_{2t}dt + dv_t$. Here the estimation algebra turns out to be equal to W_2,[10].

2.6. Example. The system is $dx_{1t} = dw_t$, $dx_{2t} = x_{1t}^2 dt$ with observations $dy_{1t} = x_{1t}dt + dv_{1t}$, $dy_{2t} = x_{2t}dt + dv_{2t}$. Here again the estimation algebra is W_2, [10].

2.7. Example,[15]. The system is $dx_{1t} = dw_t$, $dx_{2t} = x_{1t}^2 dt$ with observations $dy_{1t} = x_{1t}dt + dv_t$. In this case the estimation Lie algebra has as a basis the operators

$$A = - x_1^2 \frac{\partial}{\partial x_2} + \frac{1}{2}\frac{\partial^2}{\partial x_1^2} - \frac{1}{2} x_1^2,\quad B_i = x_1 \frac{\partial^i}{\partial x_2^i},\quad C_i = \frac{\partial}{\partial x_1}\frac{\partial^i}{\partial x_2^i},$$

$D_i = \frac{\partial^i}{\partial x_2^i}$ $i = 0,1,2, \ldots$ with the bracket relations

$[A,B_i] = C_i$, $[A,C_i] = B_i + 2B_{i+1}$, $[B_i,C_j] = - D_{i+j}$ and all other brackets between basis elements equal to zero.

2.8. Example. The system is $dx_t = dw_t$ with observations $dy_t = (x_t+\varepsilon x_t^3)dt + dv_t$. Here ε is a (small) parameter. In this case one finds that the estimation algebra is equal to W_1 for all $\varepsilon \neq 0$ (and of course equal to the oscillator algebra if $\varepsilon=0$).

2.9. Example. The system is $dx_t = dw_{1t} + \varepsilon x_t dw_{2t}$ with observations $dy_t = x_t dt + dv_t$. In this also one finds that the estimation algebra is equal to W_1 for all $\varepsilon \neq 0$.

2.10. Degree increasing estimation algebras. Consider systems of the form $dx_t = f(x_t)dt + G(x_t)dw_t$, $dy_t = h(x_t)dt + dv_t$ and assume that f, G and h are smooth and that all components of f and G are zero for x = 0. Consider the Lie algebra of all differential operators of the form $\Sigma f_\alpha(x) \frac{\partial^\alpha}{\partial x^\alpha}$, α a multiindex, $f_\alpha(x)$ smooth (finite sums). This algebra acts on the space $F(\mathbb{R}^n)$ of all smooth functions in $x_1, \ldots, x_n$. Let $F_i(\mathbb{R}^n)$ denote the subspace of all functions $\phi \in F(\mathbb{R})$ such that

$$\frac{\partial^\alpha \phi}{\partial x^\alpha}(0) = 0 \text{ for all } \alpha \text{ with } |\alpha| = \alpha_1 + \ldots + \alpha_n \leq i. \text{ Then}$$

$F(\mathbb{R}^n)/F_i(\mathbb{R}^n)$ is a finite dimensional vectorspace (isomorphic to the vectorspace of all polynomials in $x_1, \ldots, x_n$ of total degree $\leq i$). Now under the assumptions on f and G stated, the Fokker-Planck operator $\mathfrak{L}$ maps $F_i(\mathbb{R}^n)$ into itself and multiplication with h(x) always does so. Hence for these systems the estimation algebra L maps $F_i(\mathbb{R}^n)$ into itself. Let $L_i = \{D \in L \mid DF(\mathbb{R}^n) \subset F_i(\mathbb{R}^n)\} = \mathrm{Ker}(L \to \mathrm{End}(F(\mathbb{R}^n)/F_i(\mathbb{R}^n)))$. Then L_i is an ideal of L, L/L_i is finite dimensional, $L \supset L_1 \supset \ldots$ and if f, G, h are all three analytic then $\cap L_i = \{0\}$.

2.11. <u>Pro-finite dimensional algebras</u>. An infinite dimensional Lie algebra L will be called profinite dimensional if there exists a sequence of ideals $L_1 \supset L_2 \supset \ldots$ such that L/L_i is finite dimensional for all i and $\cap L_i = \{0\}$. Thus the degree increasing estimation algebras of 2.10 above are examples if f, G, h are analytic (or at least not flat at 0). Another example of a profinite dimensional Lie algebra is 2.7. The relevance of this property for the existence of (approximate) filters will be discussed in 6.1 below.

2.12. <u>Identification of linear systems with noise corrupted coefficients</u>.

The system is $dx_t = a_t x_t dt + dw_{1t}$, $da_t = dw_{2t}$ with observations $dy_t = x_t dt + dv_t$. The estimation algebra is again W_2.

3. WEYL ALGEBRAS.

As we saw in section 2 above the (Heisenberg-)Weyl algebras W_n often occur as estimation algebras. Thus, according to the introduction, it becomes important to study the homomorphisms of W_n into the Lie algebras V(M) of vectorfields on finite dimensional manifolds M.

3.1. Nonimbedding theorem. Let M be a finite dimensional smooth manifold. Then for all $n \geq 1$ there are no nonzero homomorphisms of Lie algebras $W_n \to V(M)$ or $W_n/\mathbb{R}.1 \to V(M)$.

3.2. The cubic sensor. For the cubic sensor the conjectured generalization of Sussmann's minimal realization result has been proved (during this conference in fact) [21,11] and as a consequence of this and 3.1, 2.3 we have

3.3. Theorem. For the cubic sensor 2.3 there exist no nonzero statistics which can be computed by finite dimensional filters (1.2) - (1.3).

Of course this theorem says nothing about approximate methods. The reader is also invited in this connection to look at the contribution by M. Zakai in this volume [22].

It seems most likely that the proof of theorem 3.1 can be adapted easily to yield a similar result for $W_1^{(2)}$ which would give an analogue of theorem 3.3 for example 2.4.

4. A NUMBER OF OPEN PROBLEMS.

The results of sections 2 and 3 above suggest a large number of open problems.

4.1. Problem. First and foremost there is the question of the appropriate generalizations of the results of Krener and Sussmann discussed in section 1.

4.2. Problem. Determine (up to isomorphism) all finite dimensional Lie subalgebras of W_1 and more generally W_n. An obvious example is Q_n which as a vector space is spanned by the monomials $x^\alpha \frac{d^\beta}{dx^\beta}$ with $|\alpha| + |\beta| \leq 2$. Thus Q_1 is 6 dimensional. Another example is the subalgebra spanned as a vector space by $\frac{\partial}{\partial x}$, $x\frac{\partial}{\partial x}$, $1, x, \ldots, x^m$ for some m. Conjecturally all finite dimensional subalgebras of W_1 are isomorphic to subalgebras of one of these. Thus the algebra spanned by $x\frac{\partial}{\partial x}$, $x^2\frac{\partial}{\partial x}$, $\frac{\partial}{\partial x}$, 1 which is isomorphic to $gl_2(\mathbb{R})$ is also isomorphic to the subalgebra of Q_1 spanned by $\frac{d^2}{dx^2}$, x^2, 1, $x\frac{d}{dx}$. Another example of a finite subalgebra of W_1 is the linear span of

$$1,\ x,\ \frac{d}{dx},\ x^2,\ x\frac{d}{dx}+x^3,\ \frac{d^2}{dx^2}+2x^2\frac{d}{dx}+x^4$$ which is isomorphic to Q_1.

4.3. Problem. Are there finite dimensional estimation algebras (in W_n) which are not isomorphic to the estimation algebra of a linear system? In particular can the classical finite dimensional Lie algebras arise as estimation Lie algebras?

4.4. Problem. Consider the Lie algebra of all expressions $\Sigma f_i(x)\frac{\partial}{\partial x_i}+g(x)$, $f_i(x)$, $g(x)$ smooth functions on $\mathbb{R}^n$. Can this Lie algebra arise as an estimation Lie algebra (up to isomorphism)?

4.5. Problem. The classical simple infinite dimensional (filtered) Lie algebras of Lie and Cartan are all subalgebras of the algebra $\hat{V}_n$ of formal vectorfields in n-variables. Can one of these algebras arise as an estimation Lie algebra? There are many infinite dimensional Lie algebras contained in the V(M). One example of an infinite dimensional estimation algebra which can be embedded in a V(M) occurs in [14]. More are needed.

4.6. Problem. If there is no noise in the state equations the Fokker-Planck operator degenerates to a first order differential operator and the resulting estimation algebra is always naturally an algebra of vectorfields. What does this imply for filtering, and what happens if the noise term in the state equations is given a coefficient ε and we let ε go to zero?

4.7. Problem. Develop tests for the finite dimensionality of the Lie algebra generated by a finite set of elements of W_n.

5. MORPHISMS BETWEEN SYSTEMS, COMPATIBLE REPRESENTATIONS AND ISOTROPY SUBALGEBRAS.

5.1. Isotropy subalgebras. Let $L \subset V(M)$ be a Lie algebra of vectorfields on M. Let $x \in M$. Then the isotropy subalgebra L_x of L at x consists of all $X \in L$ such that the tangent vector at x of X is zero. Equivalently if X is seen as a derivation on the algebra F(M) of smooth functions on M, cf. [12] on the appendix on manifolds and vectorfields in this volume, then $X \in L_x$ iff $(Xf)(x) = 0$ for all $f \in F(M)$. Now let $\phi : M \to N$ be a morphism of smooth manifolds and suppose that ϕ is compatible with a homomorphism of Lie algebras $\alpha : L \to V(N)$. This means that $(\alpha X)_{\phi(m)} = d\phi(X_m)$ for all $m \in M$. In terms of derivations it means that

$$X(\phi^*(g)) = \phi^*(\alpha(X)(g)), \quad g \in F(N) \tag{5.2}$$

where $\phi^*(g)$ is the function on M defined by $\phi^*(g)(m) = g(\phi(m))$. Another way of stating (5.2) is that ϕ^* is a homomorphism of L-modules where V(N) acquires its L-module structure via α. It immediately follows from (5.2) that if $\phi : M \to N$ and $\alpha : L \to V(N)$ are compatible, then for all $x \in M$, $\alpha(L_x) \subset V(N)_{\phi(x)}$. This is the extra condition on homomorphisms of Lie algebras involved in Krener's theorem [13]; cf. also Sussmann's paper [23].

5.3. Estimation algebras with representations. Thus to construct finite dimensional filters we need not just any homomorphism of Lie algebras from the estimation Lie algebra into a V(M), we need one which is compatible with the natural representation of the estimation algebra acting on (unnormalized) densities $\rho(x)$ and V(M) acting on F(M). That is we need a homomorphism of Lie algebras α: $L \to V(M)$ together with a linear map ψ: {functions on densities} $\to$ F(M) which is a homomorphism of L-modules (where V(M) acquires its L-module structure via α). It is easy to find homomorphisms of Lie algebras $L \to V(M)$ which do not satisfy this extra condition. Thus for example in [14] there occurs an estimation Lie algebra with basis a, b_1, b_2,... and brackets $[a,b_i] = b_{i+1}$, $[b_i,b_j] = 0$. An ad hoc representation of this Lie algebra by means of vectorfields is $a \mapsto e^y \frac{\partial}{\partial y}$, $b_i \mapsto (i-1)! e^{iy} \frac{\partial}{\partial x}$, and this realization of L does not correspond to a filter for the conditional density.

6. APPROXIMATE AND SUBOPTIMAL FILTERS.

6.1. Power series expansions. Let us consider again the case of the degree increasing estimation algebras of section 2.10 above. In this case we had a homomorphism of Lie algebras $L \to L/L_i \to \mathrm{End}(F/F_i)$ (where F is the space of smooth functions on $\mathbb{R}^n$). Now F/F_i is a finite dimensional vectorspace, say $F/F_i \simeq \mathbb{R}^r$. Choose coordinates $\eta_1, \ldots, \eta_r$ in $\mathbb{R}^r$ and map $A \in \mathrm{End}(\mathbb{R}^r)$ to the vectorfield $\Sigma\, a_{ij}\eta_i \frac{\partial}{\partial \eta}$. This gives us a homomorphism of Lie algebras $L \to V(\mathbb{R}^r)$ and this homomorphism comes together with a natural map {space of smooth densities} $\to \mathbb{R}^r$, viz. $\rho \mapsto (\frac{\partial \rho^\alpha}{\partial x^\alpha}(0))_\alpha$ where α runs through all multiindices such that $|\alpha| \leq i$, and, virtually by the definition of the various maps, $L \to V(\mathbb{R}^r)$ is compatible with {space of smooth densities} $\to \mathbb{R}^r$. Thus the isotropy subalgebra condition is automatically fulfilled in this case. So that (modulo the appropriate generalizations of [13], [23]) we should obtain a sequence of filters for various statistics $\psi_1, \psi_2, \psi_3, \ldots$. The fact that $\cap L_i = \{0\}$ if f,G,h are analytic should correspond to a statement that the statistics $\psi_1, \psi_2, \ldots$ determine $\rho(x,t)$ uniquely.

In fact if $\rho(x,t)$ admits a power series expansion $\rho(x,t) = \Sigma x^\alpha \rho_\alpha(t)$, then these various statistics ought to be the $\sum_{|\alpha|\leq i} x^\alpha \rho_\alpha(t)$. Quite possibly these filters exist even when $\rho(x,t)$ cannot be shown to admit a power series expansion and then converge to $\rho(x,t)$ in some singular way. More generally one may hope for generalized power series expansions when the estimation algebra is profinite dimensional (in an isotropy subalgebra respecting way).

6.2. Perturbation and deformation techniques. As we have seen the estimation Lie algebras of examples 2.8 and 2.9 are both equal to W_1 for all $\varepsilon \neq 0$. Yet the associated "Lie algebras mod ε^n" are finite dimensional for all n [8]. There should be approximate filters corresponding to these Lie algebras corresponding (more or less) to the calculation of the first n terms in a power series development (if it exists) of $\rho(t,x)$ in powers of ε, $\rho(t,x) = \rho_0(t,x) + \varepsilon\rho_1(t,x) + \varepsilon^2\rho_2(t,x) + \ldots$ Similar ideas seem to be involved in [1].

6.3. Suboptimal filters. If one throws away the second observation in example 2.6 one finds example 2.7 which has an estimation algebra of profinite dimensional type. Moreover for this particular example the various ideals do correspond to filters for various moments [15]. These are suboptimal filters in the case of the original system. The question arises whether quite generally a quotient of a sub-Lie-algebra of the estimation algebra corresponds (under suitable compatibility, i.e. isotropy subalgebra, conditions) to a suboptimal filter for some statistic. We are also curious to know whether there exists an estimation Lie algebra L which is not itself realizable in a V(M) but which is a union of subalgebras $L = \sum_{i=1}^{\infty} L_i, L_1 \subset L_2 \subset \ldots$ such that each L_i is realizable in some V(M).

6.4. Changes in output structure. Quite generally the following question seems to merit investigation: What happens to the estimation algebra when the output structure is changed, e.g. when an output is added, when the output is processed through another system before being observed, when a component of the state is made observable, ... etc.

ACKNOWLEDGEMENT.

The work of S.I. Marcus was supported in part by the U.S. National Science Foundation under grant ENG 76-11106.

REFERENCES.

1. G.M. Blankenship, Some approximation Methods in Nonlinear Filtering, In: Proc. IEEE CDC 1980 (Dec., Albuquerque).
2. R.W. Brockett, Remarks on Finite Dimensional Nonlinear Estimation, In: C. Lobry (ed), Analyse des systèmes (Bordeaux 1978), 47-56, Astérisque 75-76, Soc. Math. de France, 1980.
3. R.W. Brockett, Classification and Equivalence in Estimation Theory, Proc. 1979 IEEE CDC (Ft Lauderdale, Dec. 1979).
4. R.W. Brockett, J.M.C. Clark, The Geometry of the Conditional Density Equation, Proc. Int. Conf. on Analysis and Opt. of Stoch. Systems, Oxford 1978.
5. R.W. Brockett, Lectures on Lie algebras in systems and Filtering, In: M. Hazewinkel, J.C. Willems (eds), Stochastic Systems: The Mathematics of Filtering and Identification and Applications, D. Reidel Publ. Co., 1981, this volume.
6. J.M.C. Clark, An Introduction to Stochastic Differential Equations on Manifolds, In: D.Q. Mayne, R.W. Brockett (eds), Geometric Methods in System Theory, Reidel, 1973, 131-149.
7. M.H.A. Davis, S.I. Marcus, An Introduction to Nonlinear Filtering, In: M. Hazewinkel, J.C. Willems (eds), Stochastic Systems: the Mathematics of Filtering and Identification and Applications, D. Reidel Publ. Co., 1981, this volume.
8. M. Hazewinkel, On Deformations, Approximations and Nonlinear Filtering, Submitted IFAC, Kyoto, 1981.
9. M. Hazewinkel, S.I. Marcus, unpublished
10. M. Hazewinkel, S. Marcus, On Lie Algebras and Finite Dimensional Filtering, submitted to Stochastics.

11. M. Hazewinkel, S.I. Marcus, H.J. Sussmann, Nonexistence of Exact Finite Dimensional Filters for the Cubic Sensor Problem. In Preparation.

12. S. Helgason, Differential Geometry, Lie groups and Symmetric Spaces, Acad. Press, 1978.

13. A.J. Krener, On the Equivalence of Control System and the Linearization of Nonlinear Systems, SIAM J. Control 11(1973), 670-676.

14. P.S. Krishnaprasad, S.I. Marcus, Some Nonlinear Filtering Problems arising in Recursive Identification, In: M. Hazewinkel, J.C. Willems (eds) Stochastic Systems: The Mathematics of Filtering and Identification and Applications, D. Reidel Publ. Co., 1981, this volume.

15. C.-H. Liu, S.I. Marcus, The Lie Algebraic Structure of a Class of Finite Dimensional Nonlinear Filters, In: "Filterdag Rotterdam 1980", M. Hazewinkel (ed), Report 8011, Econometric Institute, Erasmus Univ., Rotterdam, 1980.

16. S.I. Marcus, S.K. Mitter, D. Ocone, Finite Dimensional Nonlinear Estimation for a Class of Systems in Continuous and Discrete Time, Proc. Int. Conf. on Analysis and Optimization of Stochastic Systems, Oxford 1978.

17. S.K. Mitter, On the Analogy between the Mathematical Problems of Nonlinear Filtering and Quantum Physics, Richerche di Automatica, to appear.

18. S.K. Mitter, Filtering Theory and Quantum Fields, In: C. Lobry (ed), Analyse des systèmes (Bordeaux 1978), 199-206, Astérisque 75-76, Soc. Math. de France, 1980.

19. S.K. Mitter, Lectures on Filtering and Quantum Theory, In: M. Hazewinkel, J.C. Willems (eds), Stochastic Systems: The Mathematics of Filtering and Identification and Applications, D. Reidel Publ. Co., 1981, this volume.

20. H.J. Sussmann, Existence and Uniqueness of Minimal Realizations of Nonlinear Systems, Math. Syst. Theory 10 (1977), 263-284.

21. H.J. Sussmann, Rigorous results on the cubic sensor problem, In: M. Hazewinkel, J.C. Willems (eds), Stochastic Systems: The Mathematics of Filtering and Identification and Applications, D. Reidel Publ. Co., 1981, this volume.
22. M. Zakai, A footnote to the papers which prove the nonexistence of finite dimensional filters, In: M. Hazewinkel, J.C. Willems (eds), Stochastic systems: the mathematics of filtering and identification and applications, D. Reidel Publ. Co., 1981, this volume.
23. H.J. Sussmann, An extension of a theorem of Nagano on transitive Lie algebras, Proc. Amer. Math. Soc. 45(1974), 349-356.

DETERMINISTIC ESTIMATION AND ASYMPTOTIC STOCHASTIC ESTIMATION[1]

Omar Hijab

Division of Applied Sciences
Pierce Hall
Harvard University
Cambridge, Mass. 02138 USA

Let $t \to x(t)$ be a state process and consider observations $t \to y(t)$ of a signal $t \to h(x(t))$ in the presence of additive white noise $\dot{y} = h(x) + v$. The stochastic filter is the map that associates to each observation record $y(\tau)$, $0 \leq \tau \leq t$, the conditional mean $E(\ \phi(x(t))\ |\ y(\tau),\ 0 \leq \tau \leq t\ )$. In this paper it is shown that the output of the stochastic filter converges to the output of what is known as the deterministic filter as the variances of the impinging noises go to zero, exactly in analogy with the physical fact that quantum mechanics converges to classical mechanics as Plank's constant goes to zero.

0. INTRODUCTION

Consider a state process $x(\cdot)$ evolving according to the following dynamics

$$\dot{x} = f(x) + g(x)u \qquad x \in R^n \tag{1}$$

where $u(\cdot)$ is "white noise" and $x(0) = x_0$ is a random initial state independent of $u(\cdot)$. Consider observations given by

$$\dot{y} = h(x) + v \qquad y(0) = 0 \tag{2}$$

where $v(\cdot)$ is another "white noise" independent of the state process $x(\cdot)$. In stochastic estimation one is interested in the time evolution of the conditional mean

$$\hat{\phi}(t) \equiv E(\ \phi(x(t))\ |\ y(\tau),\ 0 \leq \tau \leq t\ )$$

M. Hazewinkel and J. C. Willems (eds.), Stochastic Systems: The Mathematics of Filtering and Identification and Applications, 605–612.

of a scalar function $\phi(x(t))$ of the state at time t given the observation record $y(\tau)$, $0 \le \tau \le t$, for $t \ge 0$.

The equations (1),(2) represent a state-space model that gives rise to observations $y(\cdot)$ that are continuous and satisfy $y(0) = 0$. Let $C(0,t)$ denote the space of all continuous observation records $y(\tau)$, $0 \le \tau \le t$. The <u>stochastic estimator</u> or <u>stochastic filter</u> corresponding to the model (1),(2) is the causal map

$$\begin{array}{lcl} y(\tau),\ 0 \le \tau \le t, & \to & \hat{\phi}(t) \\ C(0,t) & \to & R \end{array} \tag{3}$$

that assigns to each observation record the corresponding value of the conditional mean. The filter then depends on f,g,h and ϕ and the probability distribution μ_0 of x_0 in R^n.

For each $\lambda > 0$ consider the following alternate state-space model

$$\dot{x} = f(x) + \sqrt{\lambda} g(x) u \qquad x \in R^n \tag{1_λ}$$

$$\dot{y} = h(x) + \sqrt{\lambda} v \qquad y(0) = 0. \tag{2_λ}$$

Here $u(\cdot)$ and $v(\cdot)$ are standard "white noises" as before and $x(0) = x_0^\lambda$ is a random initial state in R^n having a probability distribution μ_0^λ of a certain form (see section 1).

For each $\lambda > 0$ the model (1_λ),(2_λ) gives rise to a filter

$$\begin{array}{lcl} y(\tau),\ 0 \le \tau \le t, & \to & \hat{\phi}_\lambda(t) \\ C(0,t) & \to & R. \end{array} \tag{3_λ}$$

We thus have a family of filters parametrized by $\lambda > 0$. In the linear Kalman-Bucy case (f, h, ϕ linear, g constant and $x_0^\lambda \sim N(0,\lambda)$) it is easily checked that $\hat{\phi}_\lambda(t)$ is independent of λ, equal to $\hat{\phi}_1(t)$ for all $\lambda > 0$.

In the general nonlinear case we have that

$$\hat{\phi}_0(t) = \lim_{\lambda \downarrow 0} \hat{\phi}_\lambda(t) \tag{4}$$

exists and the map

$$\begin{array}{lcl} y(\tau),\ 0 \le \tau \le t, & \to & \hat{\phi}_0(t) \\ C(0,t) & \to & R \end{array} \tag{5}$$

is what is known as the <u>deterministic estimator</u> or <u>deterministic</u>

filter[2] corresponding to the state - space model (1), (2).

Although we do not state the analogy explicitly, the above result mirrors the physical fact that quantum mechanics converges to classical mechanics as Planck's constant (λ here) goes to zero. In various forms this fact is well-known to physicists; see the lectures of Mitter in these proceedings.

The plan of the paper is as follows. We first go over the "robust" version of the stochastic filter as described in the lectures of Davis and state without proof the results that we need. We then go over deterministic estimation. Finally we insert the parameter λ and indicate the derivation of equation (4), the asymptotic WKB expansion and the asymptotic formula for the conditional variance.

We shall use (1) as a blanket reference since the contents there include a detailed exposition of the results announced here, their proofs, and relevant bibliographical references.

1. STOCHASTIC ESTIMATION

In what follows the vector fields $f, g : R^n \to R^n$ will have compact support; the function $h:R^n \to R$ will have compact support; and all maps will be smooth. The vector fields f, g can be thought of as first order differential operators as follows. If $\phi:R^n \to R$ is smooth set

$$g(\phi)(x) \equiv \frac{\partial\phi}{\partial x_1}(x)g_1(x) + \cdots + \frac{\partial\phi}{\partial x_n}(x)g_n(x).$$

The function $f(\phi)$ is defined similarly. One can repeatedly apply f, g to ϕ forming, for example,

$$A_\lambda(\phi) \equiv (f + \frac{\lambda}{2} g^2)(\phi) \equiv f(\phi) + \frac{\lambda}{2} g(g(\phi)).$$

If $u(\cdot)$ and $v(\cdot)$ are standard scalar "white noises" then (1_λ) and (2_λ) are well-defined for $\lambda > 0$. The solution $x_\lambda(\cdot)$ of (1_λ) is then a Markov diffusion governed by A_λ :

$$E(\ \phi(x_\lambda(t)) \mid x_\lambda(0) = x_0^\lambda) = \phi(x_0^\lambda) + tA_\lambda(\phi)(x_0^\lambda) + o(t) \qquad (6)$$

as $t \downarrow 0$.

The usual theory of conditional expectations then yields the fact that $\phi(t)$ and $\phi_\lambda(t)$ are well-defined almost surely on the space $C(0,t)$ of all continuous observation records. Endow $C(0,t)$ with the sup norm topology. Let $\lambda > 0$.

Theorem. For each $t \geq 0$ and bounded smooth ϕ there is a continuous map

$$\begin{array}{ll} y(\tau),\ 0 \leq \tau \leq t, & \rightarrow \mu_t^\lambda(\phi) \\ C(0,t) & \rightarrow R \end{array}$$

such that

(i) $\hat{\phi}_\lambda(t) = \mu_t^\lambda(\phi)/\mu_t^\lambda(1)$ almost surely on $C(0,t)$;

(ii) if $y(\tau)$, $0 \leq \tau \leq t$, is continuously differentiable then

$$\mu_t^\lambda(\phi) = E\left(\exp\left(-\frac{1}{\lambda}\int_0^t V(\tau,x_\lambda(\tau))d\tau\right)\phi(x_\lambda(t))\right)$$

where $\quad V(t,x) = \frac{1}{2}h(x)^2 - h(x)\dot{y}(t)$;

(iii) if $y(\tau)$, $0 \leq \tau \leq t$, is continuously differentiable and ϕ has compact support then

$$\frac{d\mu_t^\lambda}{dt}(\phi) = \mu_t^\lambda(\{A_\lambda - \frac{1}{2\lambda}h^2\}\phi) + \mu_t^\lambda(\{\frac{1}{\lambda}h\}\phi)\frac{dy}{dt};$$

(iv) μ_0^λ is the probability distribution of x_0^λ in R^n.

For the proof see (1). From now on we take (i) as the definition of the map (3_λ). One may think of the map (3) as the input-output map of a control system, the Zakai system, having:

- state variable μ
- initial state μ_0
- dynamics given by $\frac{d\mu}{dt} = \{A^* - \frac{1}{2}h^2\}\mu + \{h\}\mu\,\frac{dy}{dt}$
- inputs $y(\cdot)$
- output map $\mu \rightarrow \mu(\phi)/\mu(1)$.

In what follows we assume that there are measures $\nu_0, \nu_1, \cdots$ and a smooth function S_0 such that

$$\mu_0^\lambda = e^{-\frac{1}{\lambda}S_0}\{\nu_0 + \lambda\nu_1 + \cdots + \lambda^\ell\nu_\ell + o(\lambda^\ell)\} \qquad (7)$$

as $\lambda\downarrow 0$, for all $\ell \geq 0$.

2. DETERMINISTIC ESTIMATION

Consider the state-space model (1),(2). In this section we go over the construction of the deterministic filter (5). This is a map that assigns to each continuous observation record $y(\tau)$, $0 \leq \tau \leq t$, a state $\hat{x}(t)$ in R^n as follows.

For each x_0 in R^n and continuous control $u(\tau)$, $0 \leq \tau \leq t$, the corresponding state trajectory $x(\tau)$, $0 \leq \tau \leq t$, satisfying equation (1) and $x(0) = x_0$ is well-defined. Given a state trajectory and a continuous function of time $v(\tau)$, $0 \leq \tau \leq t$, equation (2) defines a continuously differentiable observation record $y(\tau)$, $0 \leq \tau \leq t$, satisfying $y(0) = 0$.

Fix a smooth nonnegative function $S_0 : R^n \to R$. To each input triple $(x_0, u(\cdot), v(\cdot))$ and $t \geq 0$ assign the energy

$$\tilde{J}_t(x_0, u(\cdot), v(\cdot)) \equiv S_0(x_0) + \frac{1}{2}\int_0^t u(\tau)^2 + v(\tau)^2 d\tau, \tag{8}$$

at time t. A <u>minimum energy input triple given</u> $y(\tau)$, $0 \leq \tau \leq t$, is a triple $(\hat{x}_0, \hat{u}(\cdot), \hat{v}(\cdot))$ that minimizes (8) over all $(x_0, u(\cdot), v(\cdot))$ satisfying (1) and (2) subject to the constraint of fixed observations $y(\tau)$, $0 \leq \tau \leq t$. After replacing $v(\cdot)$ by $\dot{y}(\cdot) - h(x(\cdot))$ this is equivalent to an unconstrained minimization of (8). A <u>deterministic</u> (or <u>minimum energy</u>) <u>estimate of the state at time t given</u> $y(\tau)$, $0 \leq \tau \leq t$, is the endpoint $\hat{x}(t)$ of the state trajectory corresponding to a minimum energy input triple. For example at $t = 0$ a deterministic estimate of the state is a global minimum $\hat{x}_0$ of S_0. Set

$$J_t(x_0, u(\cdot), v(\cdot)) \equiv \tilde{J}_t(x_0, u(\cdot), v(\cdot)) - \frac{1}{2}\int_0^t \dot{y}(\tau)^2 d\tau,$$

the "unnormalized" energy at time t. For each $y(\tau)$, $0 \leq \tau \leq t$, set

$$S(t,x) \equiv \min\{ J_t(x_0, u(\cdot), v(\cdot)) \mid (1),(2) \text{ hold and } x(t) = x\}.$$

$S(t,x)$ is then the minimum (unnormalized) energy required to get to the state x in time t conditional on observing $y(\tau)$, $0 \leq \tau \leq t$.

It turns out that, for $t \geq 0$ sufficiently small, S is smooth and is the unique solution of the (unnormalized) conditional energy equation

$$\dot{S} + f(S) + \frac{1}{2}(g(S))^2 = \frac{1}{2}h^2 - h\dot{y} \tag{HJB}$$

satisfying $S(0,x) = S_0(x)$ for all x in R^n. Moreover if S_0 has a unique nondegenerate global minimum $\hat{x}_0$ then for $t \geq 0$ sufficiently small $S_t = S(t,\cdot)$ has a unique nondegerate global minimum $\hat{x}(t)$, <u>the</u> deterministic estimate of the state at time t.

Although the above discussion defines $\hat{x}(t)$ for continuously differentiable $y(\tau)$, $0 \leq \tau \leq t$, one can extend the definition to arbitrary observation records in $C(0,t)$ by effecting a gauge transformation

$$S \leftarrow S + hy$$

in (HJB). If we set $\hat{\phi}_0(t) \equiv \phi(\hat{x}(t))$ then the map (5)[4] is well-defined for small time $t \geq 0$.

In what follows we assume that S_0 is a positive definite function having a global minimum at $\hat{x}_0$ (for the exact definition see (1)). To such an S we associate the quadratic form

$$|\xi|_S^2 = \Sigma\, a_{ij}\xi_i\xi_j$$

where $A = (a_{ij})$ is the inverse of the Hessian of S evaluated at the global minimum $\hat{x}$. We also assume that the measures ν_0, ν_1, ··· appearing in the expansion (7) above have smooth densities a_0, a_1,··· of compact support (with respect to Lebesgue measure) and that $a_0(\hat{x}_0) > 0$. Finally, if P is a second order constant coefficient differential operator

$$P = \Sigma\, a_{ij}\partial^2/\partial x_i\partial x_j + \Sigma\, b_i\partial/\partial x_i + c$$

then set $\sigma_P(\xi) \equiv \Sigma\, a_{ij}\xi_i\xi_j$.[5]

3. ASYMPTOTIC ESTIMATION

We first indicate the results of this paper at time t = 0. This means we have to describe the limiting behaviour of the mean

$$\hat{\phi}_\lambda(0) = E(\phi(x_0^\lambda))$$

and the variance

$$\widehat{\phi^2}_\lambda(0) - \hat{\phi}_\lambda(0)^2$$

as λ goes to zero. The following proposition is the method of steepest descent.

Proposition. Let ϕ have compact support and let S be as above. Then there is a constant[6] H > 0 and a second order constant coefficient differential operator P such that

$$\sigma_P(\xi) = |\xi|_S^2 \quad \text{and}$$

$$\int_{R^n} e^{-\frac{1}{\lambda}S(x)}\phi(x)dx = (2\pi\lambda)^{n/2}\cdot\frac{e^{-\frac{1}{\lambda}S(\hat{x})}}{H}\{\ \phi(\hat{x}) + \frac{\lambda}{2}(P\phi)(\hat{x}) + o(\lambda)\ \}$$

as $\lambda \downarrow 0$.

If in the above proposition we replace S by S_0, ϕ by ϕa_i, and use expansion (7) then as a corollary we have the following.

Theorem. There is a second order constant coefficient differential operator D_0 such that

(i) $D_0(1) = 0$

(ii) $\sigma_{D_0}(\xi) = |\xi|^2_{S_0}$

(iii) $\hat{\phi}_\lambda(0) = \phi(\hat{x}_0) + \frac{\lambda}{2}(D_0\phi)(\hat{x}_0) + o(\lambda)$

as $\lambda \downarrow 0$.

As a corollary to the above theorem we have the following.

<u>Theorem</u>. Using the above notation

$$\widehat{\phi^2}_\lambda(0) - \hat{\phi}_\lambda(0)^2 = \frac{\lambda}{2}\left|\frac{\partial\phi}{\partial x}(\hat{x}_0)\right|^2_{S_0} + o(\lambda)$$

as $\lambda \downarrow 0$.

To prove the analogous results for positive time, we need the following theorem, sometimes known as <u>the WKB expansion</u>.

<u>Theorem</u>. For $t \geq 0$ sufficiently small there are measures $\nu^0_t, \nu^1_t, \cdots$ such that for ϕ having compact support and for all $\ell \geq 0$,

$$\mu^\lambda_t(\phi) = e^{-\frac{1}{\lambda}S_t}\{\nu^0_t(\phi) + \lambda\nu^1_t(\phi) + \cdots + \lambda^\ell\nu^\ell_t(\phi) + o(\lambda^\ell)\}$$

as $\lambda \downarrow 0$, where S_t is the solution of (HJB) satisfying $S_t\big|_{t=0} = S_0$.

Using the WKB expansion one shows, exactly as in the case $t = 0$, the asymptotic formula for the conditional mean

$$\hat{\phi}_\lambda(t) = \phi(\hat{x}(t)) + o(1)$$

and the asymptotic formula for the conditional variance

$$\widehat{\phi^2}_\lambda(t) - \hat{\phi}_\lambda(t)^2 = \frac{\lambda}{2}\left|\frac{\partial\phi}{\partial x}(\hat{x}(t))\right|^2_{S_t} + o(\lambda).$$

The corresponding differential operator D_t then satisfies

(i) $D_t(1) = 0$

(ii) $\sigma_{D_t}(\xi) = |\xi|^2_{S_t}$

(iii) $\hat{\phi}_\lambda(t) = \phi(\hat{x}(t)) + \frac{\lambda}{2}(D_t\phi)(\hat{x}(t)) + o(\lambda)$

as $\lambda \downarrow 0$.[7]

4. NOTES

[1]The contents of this paper appear in the author's Ph.D. thesis written under the direction of A. J. Krener.

[2]The deterministic filter was derived in the 1960's by R. E. Mortensen and others.

[3]A real function α of a real variable $\lambda > 0$ is said to be $o(\lambda^{\ell})$ as $\lambda \downarrow 0$ iff $\lambda^{-\ell}|\alpha(\lambda)| \downarrow 0$ as $\lambda \downarrow 0$.

[4]One may also think of the map (5) as the input-output map of a control system having dynamics given by (HJB).

[5]σ_P is usually known as the symbol of P.

[6]H^2 is the determinant of the Hessian of S evaluated at the global minimum $\hat{x}$.

[7]Compare this last equation with equation (6).

5. REFERENCE

(1) Hijab,Omar, "Minimum Energy Estimation", Ph.D. dissertation, University of California, Berkeley, December 1980.

NON-LINEAR FILTERING EQUATION AND A PROBLEM OF PARAMETRIC ESTIMATION

François LE GLAND

INRIA and Paris-IX (France)

A MLE is designed for the problem presented in the introduction : first we use the NLF equation as a way to compute the likelihood ratio $J(\theta)$, then a discretization (in time) of the NLF equation is introduced, which leads to an approximated likelihood ratio $J^k(\theta)$. A weak convergence result shows the pointwise convergence of J^k to J. In the last section we present the main result of this paper and, as a corollary, we deduce the convergence of the approximated MLE's to the MLE of the initial problem.

1. INTRODUCTION

We are interested in the following problem : using a radio-astronomical device, we are able to observe a process η_t which can be described in the following way :

$$\eta_t = h(\theta, X_t) + \xi_t \tag{1}$$

where : ξ_t is a gaussian white noise modelling measurement errors
h is a regular function of the couple (θ, x) (this will be made more precise in section 3)
θ is a parameter to be estimated
X_t is a random perturbation, with values in $\mathbb{R}^n$, related to the atmospheric turbulence.

In an integrated form, (1) becomes :

M. Hazewinkel and J. C. Willems (eds.), Stochastic Systems: The Mathematics of Filtering and Identification and Applications, 613–620.

$$Y_t = \int_o^t h(\theta, X_s)\,ds + W_t \tag{2}$$

We want to estimate the parameter θ (a feature of the star we are observing) from the only data available : the sample path $t \to Y_t$.

2. THE MAXIMUM LIKELIHOOD PROBLEM

First we need a model for the random perturbation X_t. For mathematical reasons we choose a (weak sense) diffusion process, associated with the infinitesimal generator :

$$L = a_{ij}(.)\, D^2_{ij} + b_i(.)\, D_i \tag{3}$$

We set : $G_t = \sigma(X_s\ ;\ s \leq t)$, $F_t = \sigma(Y_s\ ;\ s \leq t)$ and $B_t = G_t \vee F_t$.

Assume that $b(.)$ and $a(.)$ are bounded continuous fonctions on $\mathbb{R}^n$, and consider the unique solution P of the "martingale problem" [8] associated with (3), then set $\tilde{P} = P \times Q$ where Q is the Wiener measure, so that under $\tilde{P}$: X_t is the diffusion process associated with (3) and Y_t is a brownian motion independant of X_t.

We then define another law P_θ, by its RN derivative w.r.t. $\tilde{P}$, restricted to B_T :

$$(dP_\theta/d\tilde{P})_{B_T} = \exp\left(\int_o^T h(\theta, X_s)\,dY_s - \frac{1}{2} \int_o^T h^2(\theta, X_s)\,ds\right) =$$

$$= Z_T(\theta).$$

By the Girsanov-Cameron -Martin theorem, we can see that P_θ is the probability measure under which our previous description (2) of the observation process holds. Then we introduce the likelihood ratio of P_θ w.r.t. $\tilde{P}$, given F_T as the conditional expectation :

$$J(\theta) = \tilde{E}(Z_T(\theta) \mid F_T)$$

and our problem becomes :

(P) $\max_{\theta \in \Theta} J(\theta)$, where Θ is a compact subset of some euclidean space.

3. THE NLF EQUATION AS A WAY TO COMPUTE A LIKELIHOOD RATIO

It is well known (see e.g. lectures by S. Marcus and E. Pardoux, and [7]) that the unique solution (under weak technical hypotheses) of the backward stochastic PDE :

$$\begin{aligned} & dv_t(\theta) + Lv_t(\theta)dt + h(\theta,.)v_t(\theta)dY_t = 0 \\ & v_T(\theta) = f \end{aligned} \tag{4}$$

satisfies :

$$v_o(\theta,x) = \tilde{E}_x(f(X_T)Z_T(\theta) \mid F_T) \tag{5}$$

Now if we introduce the forward (Zakaï) equation associated with (4) :

$$\begin{aligned} & du_t(\theta) + Au_t(\theta)dt = h(\theta,.)\, u_t(\theta)dY_t \\ & u_o(\theta) = \bar{u} \end{aligned} \tag{6}$$

where $A = - L^*$ and $\bar{u}$ is the density (w.r.t. Lebesgue measure on $\mathbb{R}^n$) of the r.v. X_o, it is well known that the two processes $u_t(\theta)$ and $v_t(\theta)$ are dual, and then, that $u_t(\theta)$ is the unnormalized conditional density of the r.v. X_T given F_T. As a consequence we have :

$$J(\theta) = \tilde{E}(Z_T(\theta) \mid F_T) = (u_T(\theta),\ 1).$$

From this point we will make the following regularity assumptions on the function h :

A1. for every θ , $h(\theta,.)$ belongs to $S = C_b^2(\mathbb{R}^n)$

A2. as a function of θ , h is infinitely differentiable with values in S.

If A1 holds, we are able to introduce the so called "robust" version of equation (5). We set :

$$\begin{aligned} & u_t(\theta) = p(t,\theta) \exp h(\theta,.)Y_t \\ & A^y_\theta(t) = \exp(- h(\theta,.)Y_t) A(. \exp h(\theta,.)Y_t) \end{aligned} \tag{7}$$

and the function $p(\theta)$ can be seen to satisfy the following ordinary (non stochastic) PDE :

$$\begin{aligned} & p'_t(\theta) + (A^y_\theta(t) + \frac{1}{2} h^2(\theta,.))\, p_t(\theta) = 0 \\ & p_o(\theta) = \bar{u} \end{aligned} \tag{8}$$

If moreover A2 holds, it is possible according to [1] to express the gradient of $p(\theta)$ w.r.t. the parameter θ as the solution of an ordinary PDE similar to (8) and to compute $J'(\theta)$, the derivative of the likelihood ratio. This approach will be developped in my thesis, and allows to achieve the maximization problem (P) using a gradient algorithm.

4. DISCRETIZATION (in time) OF THE NLF EQUATION

For simplicity, we will drop in this section, any reference to the parameter θ. We present an implicit scheme that will appear as a discretization of both (6) and (8). Following an idea of Milshtein [5] for approximations to Ito equations, we will approximate such quantities as :

$$\int_{kn}^{k(n+1)} h\, u(s)\, dY_s$$

by : $(h\, \Delta Y_n + h^2((\Delta Y_n)^2 - \frac{k}{2}))\, u_n^k$.

Then a natural way to discretize (6) is to consider the sequence u_n^k defined by :

$$u_{n+1}^k + k\, A\, u_{n+1}^k = \psi_n\, u_n^k \text{ and } u_o^k = \bar{u} \tag{9}$$

where

$$\psi_n = \exp(h\, \Delta Y_n - \frac{1}{2} h^2 k) \tag{10}$$

and

$$\Delta Y_n = Y_{k(n+1)} - Y_{kn}$$

We then define : $u_k(t) = u_n^k \exp\,(h(Y_t - Y_{kn}) - \frac{1}{2} h^2 (t - kn))$ if $kn \leq t < k(n+1)$, so that u_k is adapted w.r.t. $(F_t\ ;\ t \geq o)$. In order to preserve the duality (4)-(6) we must set :

$$(I - kL)(\psi_n^{-1} v_n^k) = v_{n+1}^k$$

and $\quad v_N^k = f$ where $kN \leq T < k(N+1)$

(Proof : $(u_{n+1}^k, v_{n+1}^k) = (u_{n+1}^k, (I-kL)(\psi_n^{-1} v_n^k)) = ((I+kA)u_{n+1}^k, \psi_n^{-1} v_n^k)$

$$= (u_n^k, v_n^k)).$$

Going back to the "robust" equation (8) we can define another sequence p_n^k according to the following discretization of (8) :

$$p_{n+1}^k + k\, A^y(k(n+1))\, p_{n+1}^k = p_n^k \exp - \frac{1}{2} h^2 k \tag{11}$$

and set : $u_n^k = p_n^k \exp h\, Y_{kn}$. If we multiply (11) by

$\exp h Y_{k(n+1)}$ and use (7), we get :

$$\bar{u}^k_{n+1} + k A \bar{u}^k_{n+1} = \bar{u}^k_n \exp(h \Delta Y_n - \frac{1}{2} h^2 k)$$

Comparing with (9),(10) we have : $\bar{u}^k_n = u^k_n$

An interesting consequence of this identification result, it that $u_k(t)$ has a probabilistic interpretation, as will be seen in the next section, while the "robust" framework makes possible to evaluate the approximation error (i.e. the difference between p(t) and $p_k(t)$) for every sample path of the observation process Y_t, as will appear in my thesis.

5. STOCHASTIC INTERPRETATION OF THE DISCRETIZED EQUATION : AN APPROXIMATED PROBLEM

It can be seen that, similarily to (5), the following relation holds :

$$v^k_o(\theta,x) = \tilde{E}_x(f(X^k_T)\ Z^k_T(\theta) \mid F_T)$$

where the process X^k_t is deduced from X_t by a change of time : we consider a sequence if i.i.d. (under the measure $\tilde{P}$) r.v'.s τ_n, exponential with mean 1, and independant of both X_t and Y_t ; we set :

$$\sigma_n = \sum_{i=1}^{n} \tau_i \text{ and } X^k_t = X_{k\sigma_n} \text{ if } kn \leq t < k(n+1) \qquad (12)$$

We then introduce the σ-fields : $G^k_t = \sigma(X^k_s ; s \leq t)$ and $B^k_t = G^k_t \vee F_t$ and :

$$Z^k_T(\theta) = \exp(\int_o^T h(\theta,X^k_s)dY_s - \frac{1}{2}\int_o^T h^2(\theta,X^k_s)ds)$$

Now consider the probability measure P^k_θ defined by its R N derivative w.r.t. $\tilde{P}$, restricted to B^k_T $(dP^k_\theta/d\tilde{P})_{B^k_T} = Z^k_T(\theta)$

so that under P^k_θ we have :

X^k_t is defined by (12)

$$dY_t = h(\theta,X^k_t)dt + dW_t$$

That is to say P^k_θ is the probability measure under which relation (2) holds, assuming the random perturbation is now modelled by the Markov process X^k_t. Moreover, the relation :

$$(u_k(T,\theta),\ 1) = \tilde{E}(Z^k_T(\theta) \mid F_T) = J^k(\theta)$$

defines the likelihood ratio of P^k_θ w.r.t. $\tilde{P}$, given F_T, and our problem becomes :

$$(P_k) \quad \max_{\theta\in\Theta} J^k(\theta)$$

Remark it is easy to show that $X^k \xrightarrow{W} X$ so that for every θ,

$$J^k(\theta) \to J(\theta) \text{ as } k \to o.$$

6. THE MAIN THEOREM

We begin with a martingale description of the Markov process X^k and then deduce the main result of this paper.

The sequence η_n^k defined by $\eta_n^k = X_{k\sigma_n}$ is a Markov chain whose transition function is pU_p where $p = 1/k$ and U_p is the resolvent of the semi-group P_t associated with the diffusion process X_t (i.e. $U_p = \int_o^\infty e^{-ps} P_s ds$). Consider now g in $S = C_b^2(\mathbb{R}^n)$ so that Lg belongs to $C_b(\mathbb{R}^n)$ according to the boundedness of a(.) and b(.) in (3). Then set : $\Lambda(g,g) = Lg^2 - 2gLg = 2a_{ij}(.)\, D_i g\, D_j g$

Proposition : under the above assumptions, the sequence $\mu_n^k(g)$ defined by $\mu_n^k(g) = g(\eta_n^k) - g(\eta_o^k) - k \sum_{i=o}^{n-1} pUp\, Lg(\eta_i^k)$ is a square integrable martingale, and its increasing process $\alpha_n^k(g)$ associated with Doob's decomposition of the submartingale $(\mu_n^k(g))^2$, satisfies :

$$\alpha_n^k(g) \le k \sum_{i=o}^{n-1} pUp\, \Lambda(g,g\, ;\, \eta_i^k).$$

We can then derive an useful expression for the likelihood ratio $J^k(\theta)$.

$$\int_o^T g(X_s^k)\,dY_s = \sum_{i=o}^{N-1} g(\eta_i^k)(Y_{k(i+1)} - Y_{ki}) + g(\eta_N^k)(Y_T - Y_{kN}) =$$

$$= g(X_T^k)Y_T - \sum_{i=o}^{N-1} Y_{k(i+1)}(g(\eta_{i+1}^k) - g(\eta_i^k)) = g(X_T^k)Y_T -$$

$$- k \sum_{i=o}^{N-1} Y_{k(i+1)}\, pUp\, Lg(\eta_i^k) - \nu_N^k(g)$$

where the sequence $\nu_n^k(g)$, defined by $\nu_n^k(g) = \sum_{i=o}^{n-1} Y_{k(i+1)}(\mu_{i+1}^k(g) - \mu_i^k(g))$ is a square integrable martingale, whose increasing process $\beta_n^k(g)$ satisfies :

$$\beta_n^k(g) \leq k \sum_{i=o}^{n-1} Y_{k(i+1}^2 \, pUp \wedge (g,g \; ; \eta_i^k) \tag{13}$$

But, according to A1 (section3), $h(\theta,.)$ fulfilles the assumptions of the proposition, so :

$$J^k(\theta) = \tilde{E} \exp (R^k(h(\theta,.)) - \nu_N^k(h(\theta,.))) \tag{14}$$

where : $R^k(g) = g(X_T^k)Y_T - k \sum_{i=o}^{N-1} Y_{k(i+1)} pUpLg(\eta_i^k) - \frac{1}{2} \int_o^T g^2(X_s^k)ds$ (15)

<u>Theorem</u> :as functions defined on $S = C_b^2(\mathbb{R}^n)$ with values in $\mathbb{R}$, $(J^k, k \in \mathbb{R}_+^*)$ are locally lipschitzian, uniformly w.r.t. k, i.e. for every bounded set B in $C_b^2(\mathbb{R}^n)$, there is a constant K(B) independant of k, such that : $\forall\, g,g_o \in B; |J^k(g)-J^k(g_o)| \leq K(B)||g-g_o||_S$.

<u>Proof</u> : using (14),(15) and the relation $|e^x-e^y| \leq |x-y|(e^x+e^y)$ we have, for arbitrary g, g_o in B :

$$|J^k(g)-J^k(g_o)|^2 \leq 4\, \tilde{E}((R^k(g)-R^k(g_o))^2 + (\nu_N^k(g)-\nu_N^k(g_o))^2) \times$$
$$\times\, \tilde{E}(\exp 2(R^k(g)-\nu_N^k(g)) + \exp 2(R^k(g_o)-\nu_N^k(g_o))).$$

We just need to remark that : $\tilde{E}(\nu_N^k(g)-\nu_N^k(g_o))^2 = \tilde{E}\, \beta_N^k(g-g_o)$ and then apply (13), and to write :

$$\exp 2(R^k(g)-\nu_N^k(g)) = \exp(2R^k(g)+\phi_c(2)\beta_N^k(g)) \exp(-2\nu_N^k(g) -$$
$$-\phi_c(2)\beta_N^k(g)) \tag{16}$$

where $\phi_c(\lambda) = (\exp(\lambda c)-1-\lambda c)/c^2$ and c is a constant to be chosen later.

The first exponential in (16) is easily bounded by a constant c(B) which depends on T, $\sup_{t \leq T} |Y_t|$ and on c, so that:

$$\tilde{E} \exp 2(R^k(g)-\nu_n^k(g)) \leq c(B)\, \tilde{E} \exp(-2\nu_N^k(g)-\phi_c(2)\beta_N^k(g)).$$

The remaining expectation is bounded by 1, according to a martingale lemma in [6], provided the constant c is chosen so that :

$$\sup_{n \geq o} (\nu_n^k(g) - \nu_{n+1}^k(g)) \leq c \tag{17}$$

so it is then enough to choose $c \geq \bar{c}$ in (16), where $\bar{c}$ satisfies (17) for arbitrary g in B. (it can easily be proved that such a $\bar{c}$ exists). The proof is now complete. Moreover, according to A1-A2 (section 3) $h(\Theta)$ is bounded in $S = C_b^2(\mathbb{R}^n)$, since Θ is compact, hence : $|J^k(\theta)-J^k(\theta_o)|=|J^k(h(\theta,.))-J^k(h(\theta_o,.))|\leq K(h(\Theta))||h(\theta,.))-$

$-h(\theta_o,.)||_S\ K'(\Theta)d(\theta,\theta_o)$.

Using both this result and the remark in end of section 5, it is easy to prove :
Corollary : consider $\hat{\theta}_k$ achieving the problem (P_k) (section 5) (such an element exists as J^k is continuous on a compact) ; then any accumulation point of the sequence $\hat{\theta}_k$ achieves the problem (P).

Remark : all this derivation can be done when a discretization in both time and space (see e.g. Kushner [3] , [4] and Di Masi -Runggaldier [2]) is designed in a similar way, and one can get the same result of convergence for the approximated MLE's.

REFERENCES

[1] G. CHAVENT : "Analyse fonctionnelle et identification de coefficients répartis dans les équations aux dérivées partielles" (Thèse d'Etat) 1971.

[2] G. DI MASI and W. RUNGGALDIER : "Continuous-time approximations for the nonlinear filtering problem", submitted to Applied Math. and Optimization.

[3] H.J. KUSHNER : "Probability methods for approximations in stochastic control and for elliptic equations", 1977. Academic Press.

[4] H.J. KUSHNER : "A robust discrete space approximation to the optimal nonlinear filter for a diffusion", 1979. Stochastics (3), 2, pp. 75-83.

[5] G.N. MILSHTEIN : "A method of second order accuracy integration of stochastic differential equations", 1978. Th. Proba. Appl. (23), 2, pp. 396-401.

[6] J. NEVEU : "Martingales à temps discret", 1972. Masson.

[7] E. PARDOUX : "Stochastic PDE's and filtering for diffusion processes", 1979. Stochastics (3), 2, PP. 127-167.

[8] D.W. STROOCK and S.R.S. VARADHAN : "Multidimensional diffusion processes", 1979. Springer Verlag.

REGULARITY OF CONDITIONAL LAWS IN NON-LINEAR FILTERING THEORY AND STOCHASTIC CALCULUS OF VARIATIONS

Dominique MICHEL

Attachée de recherche au C.N.R.S. France

Using the theory of stochastic partial differential equations, Krylov-Rozovskii and Pardoux obtained regularity results for conditional laws arising in non-linear filtering theory. One gives here, in a more general setting, a new method for getting that sort of results, based on the "stochastic calculus of variation" of Paul Malliavin.

INTRODUCTION.

One gives, here, (cf.(6)), a new method for showing the existence and regularity of conditional laws in non-linear filtering theory. This method is quite different from those of Rozovskii (1) and Pardoux (7), (8), as it does not use the filtering equation and the theory of stochastic partial differential equations. The processes that are considered have a rather general form : they have stochastic differentials whose coefficients can depend on all the past of the observed process. No restrictive condition is needed for the initial conditional law. This shows that filtering is regularizing.
The main tool of this work is the "stochastic calculus of variation" that Paul Malliavin developped in 1976 in order to show the absolute continuity of the laws of a diffusion process.

0. NOTATIONS.

0.1. $\mathcal{C}(\mathbb{R}^n)$ is the space of continuous maps from $\mathbb{R}^+$ to $\mathbb{R}^m$. $\mathcal{B}_n$ is the borelian σ-algebra on it, $\mathcal{B}_{n,t}$ the sub-algebra

M. Hazewinkel and J. C. Willems (eds.), Stochastic Systems: The Mathematics of Filtering and Identification and Applications, 621–628.

generated by $\{c(s)\ ,\ s \leq t\ ,\ c \in \mathcal{C}(\mathbb{R}^n)\}$.

0.2. Θ_n is the sub-space of $\mathcal{C}(\mathbb{R}^n)$, formed by maps which take values 0 at time 0 . $\mathcal{B}_n$ and $\mathcal{B}_{n,t}$ are the traces of $\mathcal{A}_n$ and $\mathcal{A}_{n,t}$ on Θ_n . μ_m is the Wiener-measure on Θ_n , whose generic element is denoted by w .

0.3. $\mathcal{C}^k_b(\mathbb{R}^n)$ is the space of $\mathcal{C}^k$-functions from $\mathbb{R}$ to $\mathbb{R}^n$ whose derivatives are bounded until order k .

0.4. $\mathcal{C}^k_o(\mathbb{R}^n)$ is the subspace of $\mathcal{C}^k_b(\mathbb{R}^n)$ whose elements vanish at infinity.

1. THE MAIN RESULT.

Let $(\Omega, \mathcal{F}, P)$ be a probability space with an increasing family of algebras $\{\mathcal{F}_t\}$, contained in $\mathcal{F}$, and $(W_1,\ldots,W_n)$, $n = m + p$ independant, $\mathcal{F}_t$-adapted Wiener processes on this space. Let a, b, A, B be $m \times 1$, $m \times n$, $p \times 1$ and $p \times n$ dimensional matrices whose entries (generically noted f) are functions on $\mathbb{R}^+ \times \mathbb{R}^m \times \mathcal{C}(\mathbb{R}^p)$ satisfying :

(H_o) for each (t,x) , $c \rightarrow f(t,x,c)$ is $\mathcal{A}_{m,t}$-mesurable ,

(H_1) there exist an increasing function K, $0 < K < 1$ and a constant L , such that :

$$|f(t,x,c) - f(t,x,c')| \leq L \left[\int_o^t |c_s - c'_s|^2 dK(s) + |x - x'|^2 + |c_t - c'_t|^2\right]$$

$$f^2(t,x,c) \leq L \left[\int_o^1 (1 + |c_s|^2) dK(s) + |x|^2 + |c_t|^2\right] .$$

(H_2) BB^* is invertible and there exists a constant C such that :

a. $(BB^* Z, Z) \geq C|Z|^2 \quad \forall Z \in \mathbb{R}^p$

b. $(b(I - B^*(BB^*)^{-1} B)b^* Z, Z) \geq C|Z|^2 , \quad \forall Z \in \mathbb{R}^n$.

Let (Θ_t, ξ_t) be the solution of the stochastic system, for $0 \leq t \leq T$, T fixed :

$$d\Theta_t = a(t, \Theta_t, \xi)dt + b(t, \Theta_t, \xi)dW_t .$$

$$d\xi_t = A(t, \Theta_t, \xi)dt + B(t, \xi)dW_t .$$

Such a strong solution exists because of (H_1) .

1.1. Lemma.

There exist independant Wiener processes $\tilde{W}^1$ and $\tilde{W}^2$, with values in $\mathbb{R}^m$ and $\mathbb{R}^p$ such that :

$$d\Theta_t = \tilde{a}(t, \Theta_t, \xi)dt + \tilde{b}_1(t, \Theta_t, \xi)d\tilde{W}_t^1 + \tilde{b}_2(t, \Theta_t, \xi)d\xi_t$$

$$d\xi_t = (BB^*)^{1/2} d\tilde{W}_t^2 + A(t, \Theta_t, \xi)dt \quad \text{where :}$$

$$\tilde{b}_1 = [b(I - B^*(BB^*)^{-1} B)b^*]^{1/2} ;$$

$$\tilde{b}_2 = bB^*(BB^*)^{-1} \ ; \ \tilde{a} = a - b\, B^*(BB^*)^{-1} A .$$

1.2. Theorem.

Let (Θ, ξ) be a process as before, satisfying, in addition, the hypotheses :

(H_3) Θ_o and ξ_o are in $L^p(\Omega , P)$ for all p .

(H_4) the entries of the matrices a, b, A, B are of class $\mathcal{C}^{k+4}$ w.r. to Θ and continuous w.r. to all variables as well as their Θ-derivatives, which are supposed to be bounded .

Then almost surely in ξ the conditional measure $P(\Theta_t \in dx \,|\, \mathcal{F}_t^\xi)$ is absolutely continuous w.r. to Lebesgue measure for all t, $0 < t \leqslant T$; the density is of class $\mathcal{C}^k$ and satisfies :

$$\sup_{\varepsilon \leqslant t \leqslant T} \|f_t\|_{\mathcal{C}_b^k(\mathbb{R}^m)} < +\infty \quad \text{for all } \varepsilon > 0 .$$

The bounds on the right can be explicitly calculated from the bounds of the entries of the matrices a, b, A, B and their Θ-derivatives.

2. THE FIRST STEP : A LEMMA OF FUNCTIONAL ANALYSIS.

Lemma. Let μ be a bounded Radon measure on $\mathbb{R}^d$ such that there exist $C_o > 0$ and $N \in \mathbb{N}$:

$$\left| \int D^\alpha \varphi(x)\, \mu(dx) \right| \leqslant C_o \|\varphi\|_\infty \quad \forall \varphi \in \mathcal{C}_b^\infty(\mathbb{R}^d) \ , \ \forall \alpha \ , \ |\alpha| \leqslant N .$$

Then μ is absolutely continuous with respect to Lebesgue measure. If $N > d$, the density f belongs to $\mathcal{C}_b^k(\mathbb{R}^d)$ with $k = N - d - 1$ and : $\|f\|_{\mathcal{C}_b^k(\mathbb{R}^d)} \leqslant C_o\, A(d,N)$.

3. SECOND STEP : THE GENERALIZED BAYES FORMULA.

3.1. Lemma. Let $(\Omega', \mathcal{F}', P')$ a probability space with an increasing family of σ-algebras $\{\mathcal{F}_t'\}$ contained in $\mathcal{F}'$, $(\beta_1, \ldots, \beta_n)$ n independant, $\mathcal{F}'_t$-adapted Wiener processes, on this space. Let (α_1, α_2), $(\gamma_1, \gamma_2) : \Omega' \times [0,T] \to (\mathbb{R}^m \otimes \mathbb{R}^m) \times (\mathbb{R}^p \otimes \mathbb{R}^m) \times \mathbb{R}^m \times \mathbb{R}^p$ bounded and $\mathcal{F}'_t$-adapted v.a. Consider :

$$\zeta_i(t) = \int_0^t \alpha_i(s) d\beta^i(s) + \int_0^t \gamma_i(s) ds + \zeta_i(0) \quad \text{where}$$

$\beta^1 = (\beta_1, \ldots, \beta_m)$, $\beta^2 = (\beta_{m+1}, \ldots, \beta_n)$ and $E'(\zeta_i^2(0)) < +\infty$. Then, the stochastic system :

$$dX_t = b_1(t, X_t, \zeta_2) d\zeta_1(t) + b_2(t, X_t, \zeta_2) d\zeta_2(t) + a(t, X_t, \zeta_2) dt,$$

$X_0 \in L^2(\Omega', P')$ admits a unique solution and there exists a borelian functional Q on $[0,T] \times \mathbb{R}^m \times \mathcal{C}_0(\mathbb{R}^m \times \mathbb{R}^p)$, such that : $\forall (\alpha, \beta, \gamma), \forall X_0$, $X_t = Q_t(X_0, \zeta_1, \zeta_2)$, $P_0 \otimes \nu_{\alpha\beta\gamma}$ almost surely where :

(i) P_0 is the law of X_0

(ii) $\nu_{\alpha\beta\gamma}$ is the image of the Wiener measure under the map : $\beta \to (\zeta_1, \zeta_2)$, (α, γ) being fixed.

3.2. Proposition.

Let φ be a continuous map from $\mathbb{R}^m$ to $\mathbb{R}$ such that : $E(|\varphi(\Theta_t)|) < +\infty$. Then :

$$E(\varphi(\Theta_t) \mid \mathcal{F}_t^\xi) = \int_{\mathbb{R}^m \times \mathcal{C}_0(\mathbb{R}^m)} \varphi(Q_t(x,w,\xi)) \rho_t(x,w,\xi) dF_{\xi_0}(x) d\mu_m(w)$$

where $dF_{\xi_0}(x) = P(\Theta_0 \in dx \mid \xi_0)$, and

$$\rho_t(x,w,\xi) = \exp\Big(\int_0^t (A(s, Q_s(x,w,\xi), \xi) - \bar{A}(s,\xi))^* [BB^*(s,\xi)]^{-1/2} d\bar{w}_s$$

$$- \frac{1}{2} \int_0^t (A(s, Q_s(x,w,\xi), \xi) - \bar{A}(s,\xi))^* (BB^*(s,\xi))^{-1} (A(s, Q_s(x,w,\xi), \xi) - \bar{A}(s,\xi)) ds \Big)$$

with $\bar{A}(s,\xi) = E(A(s, Q_s(x,w,\xi), \xi) \mid \mathcal{F}_s^\xi)$ and

$$d\bar{w}_s = (BB^*)^{-1/2}(s,\xi) [d\xi_s - \bar{A}(s,\xi) ds] .$$

Proof : Use lemmas 1.1., 3.1. together with generalized Bayes formula (cf. (2), p. 279).

3.3. One has then reduced the initial problem to that of performing an integration by parts in the integral

$$I_t(x,\xi) = \int D^i \varphi(Q_t(x,w,\xi)) \rho_t(x,w,\xi) d\mu_m(w) .$$

An analogous problem in finite dimension: Let μ_o be the measure on $\mathbb{R}$ with density w.r.t. to Lebesgue measure $e^{-x^2/2}/2\pi$.

$L = \partial^2/\partial x^2 - x\, \partial/\partial x$ is a self adjoint operator w.r.t. to μ_o on a suitable space of functions and :

$\partial f/\partial \mathbf{x} . \partial g/\partial x = 1/2[L(fg) - f\, L\, g - g\, L\, f]$. So :

$$\int_{\mathbb{R}} \frac{\partial f}{\partial x}(g(x))\, h(x) d\mu_o(x) = \int_{\mathbb{R}} \left(\frac{\partial (f \circ g)}{\partial x} . \frac{\partial g}{\partial x} \right) \left(\frac{\partial g}{\partial x}\right)^{-2} h\, d\, \mu_o$$

So, if the term between the brackets is in $L^1(\mu_o)$, we can bound the first integral by $K\|f\|_\infty$.

In 1976, Paul Malliavin has constructed on the Wiener space an operator, called the Ornstein-Uhlenbeck operator, generalizing the operator L just above ((3),(4),(5)) . Daniel Stroock reformulated his work in 1979 ((9)) .

4. THIRD STEP : THE ORNSTEIN-UHLENBECK OPERATOR.

4.1. The O.U. process on Θ_m .

It is a Markov process on Θ_m, which has μ_m as in variant measure and is invariant under reversing of the time when one takes μ_m as initial measure. It can be defined as the unique family of measures $\{P_\Theta, \Theta \in \Theta_m)\}$ on $([0,\infty[, \Theta_m)$ such that :

(i) $P_\Theta(\Theta_o = \Theta) = 1$

(ii) $(\varphi(e^{t/2}\Theta_t(f)) - \frac{\|f\|}{2} \int_o^t e^s \varphi''(e^{s/2}\Theta_s(f))ds, \mathcal{B}^2_{m,t}, P_\Theta)$ is a martingale, where :

(a) $\varphi \in \mathcal{C}^\infty_o(\mathbb{R})$;

(b) $f \in \mathcal{C}_o(\mathbb{R}^m), \|f\| = \|f\|_{L^2}$

(c) $\Theta_t(f) = - \int_o^\infty f'(s)\Theta_t(s)ds$

(d) $\mathcal{B}^2_{m,t}$ is the σ-algebra on $\mathcal{C}([0,\infty[, \Theta_m)$ generated by $\{\Theta_s, s \leq t , \Theta \in \mathcal{C}([0, \infty[, \Theta_m)\}$.

4.2. The O.U. operator.

(i) This is the infinitesimal generator of the self adjoint extension of the semi-group generated by the O.U. process on $L^2(\mu_m)$. We denote it by A, with domain $D(A)$.
(ii) Define $D_p = \{\Phi \in L^p(\mu_m), \exists \Psi \in L^p(\mu_m), (\Phi(\Theta_t) - \int_0^t \Psi(\Theta_s)ds, \mathcal{B}^2_{m,t}, \tilde{P})$ is a martingale$\}$ where $\tilde{P} = \int P_\Theta d\mu_m(\Theta)$. Such a Ψ is unique and denoted by $\mathcal{A}_p \Phi$. Then : $D_2 = D(A)$; $A = D_2$.
(iii) $L^2(\mu_m)$ can be realized as $\oplus \mathcal{H}_n$ where $\mathcal{H}_n$ is the space of n-iterated Wiener integrals. Define : $D(\mathcal{L}) = \{\Phi \in L^2(\mu_m)$, $\Phi = \sum_n \Phi_n$, $\Phi_n \in \mathcal{H}_n$, $\sum_n n^2 E(|\Phi_n|^2) < +\infty\}$ and $\mathcal{L}\Phi = -1/2 \sum_n n \Phi_n$, if $\Phi \in D(\mathcal{L})$. Then : $D(\mathcal{L}) = D(A)$ and $A = \mathcal{L}$.

<u>Proposition</u>. - If Φ, Ψ belong to $D(A)$, $\Phi\Psi$ belongs to D_1. Define then : $\nabla\Phi . \nabla\Psi = 1/2[\mathcal{A}_1(\Phi\Psi) - \Phi A \Psi - \Psi A \Phi]$.

4.3. The action of A on solution of stochastic equations.

4.3.1. **<u>Definition</u>. Define :** $K_p = \{\Phi \in D_{2p} / \nabla\Phi . \nabla\Phi \in L^{2p}(\mu_m \otimes P)\}$; $K = \bigcap_{p \geq 2} K_p$.

4.3.2. <u>Proposition</u>. Let $\eta : \Omega \times \Theta_m \times \mathbb{R}^+ \to \mathbb{R}^n$ be a v.a. such that : $E(\sup_{t \leq T}(|\eta_t|^{2p} + |L\eta_t|^{2p} + |\nabla\eta_t . \nabla\eta_t|^{2p})) < +\infty \quad \forall p$.
Let X_t be the solution of :
$dX_t = C(t,\eta_t,X_t,\xi)dW^1_t + D(t,\eta_t,X_t,\xi)d\xi_t + E(t,\eta_t,X_t,\xi)dt$
where C,D,E satisfie Lipschitz conditions.
Then if C,D,E are of $\mathcal{C}^2$ class w.r. to η and x and have bounded derivatives w.r. to η and x, X_t belongs to K. In addition, AX_t and $((\nabla X^i_t . \nabla X^j_t))_{i,j}$ satisfie linear stochastic differential equations whose coefficients are $\mathcal{C}^\infty$ functions of first and second derivaties w.r. to η and x of C,D,E.

5. FINAL STEP.

Let C be the matrix with entries $\nabla Q^i_t . \nabla Q^j_t$, Δ_t her determinant, D the matrix of her cofactors. Then :

$$\nabla(\varphi \circ Q_t) . \nabla Q^j_t = \partial\varphi / \partial\Theta_i \, C^{ij} .$$

So : $\partial\varphi / \partial\Theta_i = \Delta_t^{-1} d_{ij} \; (\nabla(\varphi \circ Q_t).\nabla Q_t^j)$ and :

$$I_t(x,\xi) = \int_{\Theta_m} \nabla(\varphi \circ Q_t).\nabla Q_t^j d_{ij} \Delta_t^{-1} \rho_t(x,w,\xi) d\mu_m(w).$$

By 4.3.2. , Q_t is in K . By limit theorems , one can show that ρ_t and Δ^{-1} are in K , using condition $(H_2.b)$. It is then possible to perform an integration by parts and one gets :

$$I_t(x,\xi) = \int_{\Theta_m} (\varphi \circ Q_t)\tilde{Q}_t(x,w,\xi)\rho_t(x,w,\xi)d\mu_m(w)$$

where $\tilde{Q}_t = -AQ_t^j d_{ij} \Delta_t^{-1} - (\nabla Q_t^j . \nabla(d_{ij}\Delta^{-1}\rho_t)) \rho_t^{-1}$.

Iterating this method for higher derivatives of φ , one gets the desired estimates by applying lemma 1.1. All the estimates obtained are uniform on every interval $[\varepsilon,T]$, and this allows to get the result almost surely for all t, $0 < t \leqslant T$.

REFERENCES

(1) Krylov, N.V., Rozovskii, B.L. : "On conditional distributions of diffusion processes", 1978, vol. 12, n° 2, pp. 336-356.

(2) Liptser, R.S., Shiryayev, A.N. : "Statistics of random processes", 1979, Appl. of Math., n° 5, Springer-Verlag.

(3) Malliavin, P. : "Stochastic calculus of variation and hypoelliptic operators", 1978, Proc. Int. Symposium on stochastic diff. equations, (Kyoto, 1976), Tokyo.

(4) Malliavin, P. : "C^k-hypoellipticity with degeneracy" (parts 1 and 2), 1978, Stochastic analysis, edited by Friedman and Pinsky, Academic Press.

(5) Malliavin, P. : "Regularité de lois conditionnelles et calcul des variations stochastique ", 1979, C.R.A.S., 289, série A, pp. 357-360.

(6) Michel, D. : "Régularité des lois conditionnelles en théorie du filtrage non linéaire et calcul des variations stochastique ", 1980, C.R.A.S., 290, série A, p. 387, to appear in J. of Funct. analysis.

(7) Pardoux, E. : "Stochastic partial differential equations and filtering of diffusion processes", 1979, Stochastics, vol. 3, pp. 127-167.

(8) Pardoux, E. : "Equation du filtrage non linéaire, de la prédiction et du lissage".1980, Preprint de l'Université de Provence.

(9) Stroock, D. : "The Malliavin calculus and its applications to second order parabolic equations", 1979, Preprint.

FINITE DIMENSIONAL ESTIMATION ALGEBRAS IN NONLINEAR FILTERING

Daniel Ocone
Department of Mathematics
University of Wisconsin - Madison

1. INTRODUCTION

The lectures by Brockett, Mitter, and others at this conference outline a novel approach, inspired by geometric control and quantum field ideas, to nonlinear filtering. Among the intriguing possibilities of the new theory are methods for finding filtering problems with finite-dimensionally computable solutions. Our talk reports preliminary work done on this issue.

We consider the usual, signal-observation model

$$\begin{aligned} dx(t) &= f(x(t))\,dt + g(x(t))\,db(t) \qquad & x(0) &= x_0 \\ dy(t) &= h(x(t))\,dt + dw(t) \qquad & y(0) &= 0 \end{aligned} \tag{1}$$

in which x, b, and y, are respectively, $\mathbb{R}^n$-, $\mathbb{R}^m$-, and $\mathbb{R}^p$-valued processes, and b and w are independent, standard Brownian motions. As is well-known, the conditional density of $x(t)$ given $y(s)$, $0 \leq s \leq t$, has an unnormalized version $p(x, t| y(s), 0 \leq s \leq t)$ governed by Zakai's [11] equation

$$\begin{aligned} \text{đ}p(x,t) &= \{\mathcal{L}^* - \tfrac{1}{2}\sum_{i=1}^{p} h_i^2(x)\}\, p(x,t)\,dt + \sum_{i=1}^{p} h_i(x)p(x,t)\,dy_i(t) \\ \mathcal{L}^*\phi &= \tfrac{1}{2}\sum_{i,j=1}^{n} \frac{d^2}{dx_i dx_j}(gg')_{ij}\,\phi - \sum_{i=1}^{n}\frac{d}{dx_i} f_i(x)\,\phi \end{aligned} \tag{2}$$

((2) is written in Stratonovich form, and in it the $y(s)$ dependence of $p(x, t)$ is omitted for notational simplicity.) The

M. Hazewinkel and J. C. Willems (eds.), Stochastic Systems: The Mathematics of Filtering and Identification and Applications, 629–636.

algebro-geometric approach proposes to derive information about $p(x,t)$ by studying the estimation algebra ([4]), $\Lambda := \{\mathcal{L}_0, h_i, i = 1, \ldots, p\}_{LA}$,

$$\mathcal{L}_0 = \mathcal{L}^* - \tfrac{1}{2} \sum_{i=1}^{p} h_i^2 .$$

This is the Lie algebra generated by the operators $\mathcal{L}_0$ and h_i, $i = 1, \ldots, p$ using the bracket $[A, B] = A \circ B - B \circ A$. (When we say a function of h belongs to Λ, we mean the operator 'multiplication by $h(x)$'.) Professor Brockett describes the usefulness and meaning of the estimation algebra in his lectures. There he develops the so-called Wei-Norman method, which solves,

$$\dot{z} = f(z) + u(t)\, g(z),$$

when $f(z)$ and $g(z)$ generate a finite dimensional Lie algebra of vector fields, by parameterizing the solution manifold independently of u and then expressing the parameters as functionals of u. An extension to Zakai's equation (replace dy/dt by $u(t)$) plausibly suggests itself; if $\dim \Lambda < \infty$, we expect that $p(x,t)$ can be parameterized by a finite set of statistics explicitly calculated by a Wei-Norman technique.

Happily, our expectations are borne out in all currently known instances of problems with explicit solutions for $p(x,t)$. The class of such problems was recently extended beyond the Kalman-Bucy case by V. Beneš, and is given, up to diffeomorphisms of the signal space, in the following result (Beneš, [1]): if a) $g(x) = I$, $h(x) = Hx$ and b) $f(x) = \nabla V(x)$ where $V(x)$ is a function on all of $\mathbb{R}^n$ satisfying $\Delta V(x) + \|\nabla V(x)\|^2 + \|Hx\|^2 = x^T Ax + B^T x + C$, $A \geq 0$, then for initial densities $p(x, 0)$ of any form $p(x,t)$ can be expressed in terms of a finite, recursively generated set of statistics. Indeed, for problems (1) satisfying a) and b), the estimation algebra has finite dimension and the solutions may be obtained by Wei-Norman calculations (see [9]). We remark that Beneš's original proof was probabilistic and not Lie algebraic; for yet another demonstration, see the lectures by Professor Mitter at this conference.

Our work addresses two questions that now arise. 1) Are there other examples for which $\dim \Lambda < \infty$? 2) If so, can explicit solutions to the corresponding filtering problems be found? In section 2, we address 1), and, in the case of scalar x and y, find new examples. In sections 3 and 4, we develop the new examples and show that the Lie algebraic method does <u>not</u> lead to solutions. Some resulting theoretical issues are raised in conclusion. More detailed accounts of this work may be found in [8] and [9]. In particular, some errors of misstatement and omission in [9] stand corrected here and in [8].

2. ESTIMATION ALGEBRA FINITE DIMENSIONALITY

Finite dimensionality conditions are stated here under the assumptions: 1) $f, h \in C^{\infty}(U)$ where U is an open, connected domain of $\mathbb{R}^n$, and 2) $m \geq n$, $g(x) = G$ = constant $n \times m$ matrix of rank n.

Theorem 1. Suppose $\dim \Lambda < \infty$. If a function $h \in \Lambda$ then h is a polynomial of degree ≤ 2.

Sketch of proof (see [8]). Λ must contain both of the following sequences:

a) $\{[\mathrm{ad}\, \mathcal{L}_0]^k h\}_{k=0}^{\infty}$, $\quad [\mathrm{ad}\, A]\, B := [A, B]$

and

b) $\{a_k\}_{k=0}^{\infty}$ $\quad a_0 = h, \quad a_k = [[\mathcal{L}_0, a_{k-1}], a_{k-1}]$.

Now a calculation will show

$$[\mathrm{ad}\, \mathcal{L}_0]^k h = \sum_{\ell_1, \ldots \ell_k = 1}^{m} D^k h(A_{\ell_1}, \ldots, A_{\ell_k}) \frac{d^k}{dx_{\ell_1} \cdots dx_{\ell_k}}$$

+ terms with lower order differential operators

where $D^k h$ is the $k^{\underline{th}}$ differential of h as a symmetric k-linear function, and $A_\ell = \ell^{th}$ column of $A = GG^T$. From this, unless h is polynomial a) will include operators of arbitrarily high order. If h(x) is polynomial, b) will be a sequence of polynomials and $\deg a_k$ will grow without bound unless $\deg h(x) \leq 2$.

Theorem 1 provides a quite restrictive necessary condition. In the scalar case with g = constant, we can use it find all examples with $\dim \Lambda < \infty$.

Theorem 2. Let $n = m = p = 1$, $g = 1$. Then $\dim \Lambda < \infty$ only if

i) $h(x) = \alpha x$

$$f' + f^2 = ax^2 + bx + c \qquad (3)$$

or ii) $h(x) = \alpha x^2 + \beta x, \quad \alpha \neq 0$ and

$$f' + f^2 = -h^2 + a(2\alpha x + \beta)^2 + b + c(2\alpha x + \beta)^{-2} \tag{4}$$

or $$f' + f^2 = -h^2 + ax^2 + bx + c\,. \tag{5}$$

Remarks: In theorem 2, a constant may be added to h ($h(x) = \alpha x + c$), but this only adds a deterministic component to $y(t)$, and is thus irrelevant to filtering. Theorem 2 can also be used to find all finite dim. Λ when $g(x)$ is nonconstant and sgn $g(x)$ is constant, since in this case a state space diffeomorphism $\tilde{x} = \alpha(x)$ can be found for which $\tilde{g} = 1$ and estimation algebras are isomorphic under state space diffeomorphism (Brockett, [3]). Finally, condition i) is the same as that of Beneš except that globality of $f(x)$ is not required; a version of ii) is also noted in Brockett (see [4]).

3. BEHAVIOR OF f(x)

We want next to specify filtering problems and Zakai equations corresponding to the various h and f of theorem 2 and then to determine when explicit solutions are possible. In this regard, we cannot simply use (1) and (2) without further qualification, because the functions f will in general become singular at finite values of x. The next result summarizes the possible behaviors of $f(x)$.

<u>Theorem 3.</u> If f satisfies (3) with $a < 0$, (4) or (5), $f(x)$ must have a singularity in any unbounded interval (i.e. f is defined and finite only on bounded intervals.). If f satisfies (3), (4), or (5) and f is singular at $x = r$, then

$$f(x) \sim \frac{1}{x - r} \qquad x \to r \tag{6}$$

unless, possibly, $f(x)$ satisfies (4) and $r = -(2\alpha)^{-1}\beta$. These results follow easily from letting $f = v'/v$ and analyzing the resulting equation for v by the Sturm-Louiville comparison theorem. Note that, for various choices of the coefficients a, b, and c, solutions to (3) may have global solutions (no singularities) or solutions on semi-infinite intervals.

Now let f and h satisfy one of the conditions of theorem 2 on a domain $U = (r_0, r_1)$ such that f is singular at any bounded endpoint of U. Suppose, moreover that the singularity is as in (6). It can then be shown from stochastic differential equation theory [6] that

$$dx(t) = f(x(t)) + db(t) \qquad x(0) \in U \quad \text{a.s.} \tag{7}$$

has a unique solution that remains in U for all time; essentially the singularities of $f(x)$ will prevent $x(t)$ from reaching the boundaries. In diffusion theory lingo, the finite boundaries are entrance boundaries of $x(t)$. If we let

$$dy(t) = h(x(t))\,dt + dw(t) \tag{8}$$

the Zakai equation, written formally with $\dot{y}(t)$, becomes

$$\frac{\partial p}{\partial t}(x,t) = \mathcal{L}_0 p(x,t) + h(x)\,p(x,t)\,\dot{y}(t) \tag{9}$$

$$\lim_{x \to r}{}' \left\{ \frac{\partial}{\partial x} - 2f(x) \right\} p(x,t) = 0 \qquad r \text{ a finite boundary of } U.$$

Note that a boundary condition appears even though $x(t)$ never reaches the boundary.

In our Wei-Norman analysis of problems with the finite estimation algebras given in theorem 2, we will restrict ourselves to (7) - (9) above with its special assumption that at finite boundaries of U, $f(x)$ has a $1/x\text{-}r$ singularity. Since $x(t)$ will remain between singularities of $f(x)$ this is a natural choice. In any excluded cases for which a Zakai equation of form similar to (9) can be defined, the analysis extends, at least formally, and the conclusions are the same. (See [8], [9]).

4. WEI-NORMAN ANALYSIS

We next wish to describe the following result: if U is bounded or semi-infinite the Wei-Norman method will not produce an explicit, finite dimensional solution to the problem given in (7) - (9). When $U = \mathbb{R}$, the Wei-Norman method will work, but theorem 3 implies that U may equal $\mathbb{R}$ only if $h(x) = \alpha x$ and $f' + f^2 = ax^2 + bx + c$ and this situation is covered by the aforementioned theorem of Beneš. Thus we do not succeed in finding new, explicitly solvable problems.

To indicate why the Wei-Norman technique fails, we consider the prototypical example

$$\begin{aligned} dx(t) &= \frac{1}{x(t)}\,dt + db(t) \qquad x(0) \in (0,\infty) \\ dy(t) &= x(t)\,dt + dw(t) \end{aligned} \tag{10}$$

for which $U = (0,\infty)$ and $f(x) = \frac{1}{x}$ satisfies (3) for $a = b = 0$. The general analysis is similar to the treatment of (10) in all essential aspects (see [8]). The signal $x(t)$ in in (10) is often referred to as the Bessel process.

It is first necessary to review the Wei-Norman technique for Zakai's equation. We treat the simplest problem, $dx(t) = db(t)$, $x(0) = 0$, $dy(t) = x(t)dt + dw(t)$, for which the Zakai equation is

$$\frac{\partial p}{\partial t}(x,t) = Ap(x,t) + xp(x,t)\,y(t) \tag{11}$$

where $A = \frac{1}{2}\,\partial^2/\partial x^2 - \frac{1}{2}x^2$. Calculation shows that $[A, x] = \partial/\partial x$, $[A, \partial/\partial x] = x$, and $[\partial/\partial x, x] = I$, and hence that the estimation algebra Λ is finite dimensional with basis $\{A, x, \partial/\partial x, I\}$. Now suppose to each $X \in \{A, x, \partial/\partial x, I\}$ we can associate a group on semigroup of operators e^{tX}. The idea is to try a solution of (11) of the form

$$p(x,t) = [e^{g_1(t)A}\, e^{g_2(t)x}\, e^{g_3(t)\,\partial/\partial x}\, e^{g_4(t)}\, p_0](x) \tag{12}$$

where p_0 = initial density and the $g_i(t)$ are functions to be determined. The feasibility of (12) depends on the existence of relations

$$e^{tX_i}X_j = \sum_{k=1}^{d} c_k^{ij}(t)\, X_k\, e^{tX_i} \qquad 1 \le i,\ j \le d \tag{13}$$

for the basis elements X_i, $1 \le i \le d$, of Λ. From matrix Lie algebra theory, we expect, in fact, that

$$e^{tX_i}X_j = \sum_{k=0}^{\infty} \frac{t^k}{t!}\,[\mathrm{ad}X_i]^k X_j\, e^{tX_i} \qquad 1 \le i,\ j \le d \tag{14}$$

If (14) holds, then by differentiating (12) w.r.t. t, repeatedly employing (14) and substituting in (11), we find that p will be a solution if the $g_i(t)$'s satisfy a certain system of differential equations. To make the scheme concrete, define $e^{tA}\varphi$ so that $u(x,t) = e^{tA}\varphi(x)$ satisfies $\partial u/\partial t = Au$, $u(x,0) = \varphi(x)$, $(e^{tx}\varphi)(x) = e^{tx}\varphi(x)$, $(e^{t\,\partial/\partial x}\varphi)(x) = \varphi(x+t)$. Relations (14) can then be verified directly, and so $g_1, \ldots, g_4$ can be found. These calculations will indeed give the right answer (see [8]).

Now for problem (11), the Zakai equation is (8):

$$\begin{aligned} \frac{\partial p(x,t)}{\partial t} &= \{\tfrac{1}{2}\,\partial^2/\partial x^2 - \partial/\partial x\,\tfrac{1}{x} - \tfrac{1}{2}x^2\}\, p(x,t) + xp(x,t)\,\dot{y}(t) \\ \lim_{x \downarrow 0}\, [\partial/\partial x - 2/x]\, p(x,t) &= 0 \end{aligned} \tag{15}$$

If we let $p(x,t) = xq(x,t)$, $q(x,t)$ must satisfy

$$\frac{\partial}{\partial t} q(x,t) = \{\tfrac{1}{2}\partial^2/\partial x^2 - \tfrac{1}{2}x^2\}\, q(x,t) + xq(x,t)\dot{y}(t)$$
$$q(0,t) = 0 \qquad t > 0 . \tag{16}$$

The Lie algebra and Wei-Norman analyses on (15) and (16) are equivalent, but (16) is easier to handle. In fact, the algebra $\tilde{\Lambda} := \{\frac{1}{2}\partial^2/\partial x^2 - \frac{1}{2}x^2, x\}_{LA}$ associated to (16) is the same as the Λ defined for (11), except that the operators of $\tilde{\Lambda}$ act on functions defined only on $\mathbb{R}^+$. We are led again to try a solution to (16) of form (12). However, now the operator e^{tA} must be defined so that $u(x,t) = (e^{tA}\varphi)(x)$ solves $\partial u/\partial t = Au$, $u(x,0) = \varphi(x)$ $u(0,t) = 0$, in order to satisfy the boundary condition $q(0,t) = 0$. When this definition is made, (14) no longer holds. For instance (14) requires that $e^{tA}x = (\cosh t)xe^{tA} + (\sinh t)\,\partial/\partial x\, e^{tA}$. For sufficiently well-behaved φ $u(z,t) = (e^{tA}x\varphi)(z)$ and $v(z,t) = [(\cosh t)\, e^{tA}\varphi](z) + [\sinh t\, \partial/\partial x\, e^{tA}\varphi](z)$ will both satisfy $\partial\psi/\partial t = A\psi$, but $u(0,t) = 0$ for any φ and $v(0,t) \neq 0$ for most φ because $(\partial/\partial x\, e^{tA}\varphi)(0)$ need not be zero. Hence u and v are not the same. Likewise, any formula of type (13) will fail, and, hence, the Wei-Norman method cannot proceed.

Our calculations show that the condition $\dim \Lambda < \infty$ may not guarantee a finite dimensionally computable solution. What are additional conditions on the operators that would determine when Lie algebraic calculations work? This problem is closely related to that of associating Lie group representations to representations of Lie algebras by unbounded operators on a Hilbert space. In fact, the failure of the Wei-Norman method on (16) is intimately related to the fact that the Lie algebra Span $\{d/dx, -ix, -i\}$ on $L^2(\mathbb{R}^+)$ does not correspond to a Lie group representation on $L^2(\mathbb{R}^+)$ ([8]). The representation problem has been considered extensively by Nelson ([7]) and Flato, et al ([5]), who study the issue of operator domains. Flato, et al ([5]) show that a n.a.s.c. for finding a Lie group representation is the existence of a common, dense, invariant domain of analytic vectors for the Lie algebra. Similarly, using results of this theory, it can be shown that $\tilde{\Lambda}$ of (16) does not have a common, dense set of analytic vectors that satisfy the boundary condition $\varphi(0) = 0$. Domain considerations and analytic vectors should play a role in estimation algebra theory. (see also [3], [10]).

Thus what we need is an estimation algebra integration theory that specifies domain conditions for obtaining finite dimensional solutions from finite dimensional algebras. However, we should find other explicitly solvable examples or we might produce a theory with no applications!

Acknowledgements. I thank Prof. S. K. Mitter, Prof. John Baras, and V. E. Beneš for discussions and advice on this topic. The research was supported by the Air Force Office of Scientific Research under grant no. AFOSR-77-3281B.

REFERENCES

1. Beneš, V., "Exact Finite Dimensional Filters for Certain Diffusions with Nonlinear Drift," to appear in Stochastics.

2. Brockett, R. W., "Remarks on Finite Dimensional Nonlinear Estimation," to appear in Asterisque.

3. Brockett, R. W., "Classification and Equivalence in Estimation Theory," Proceedings, 18th IEEE Conference on Decision and Control, December 1979 Ft. Lauderdale, Fla.

4. Brockett, R.W., and Clark, J.M.C., "On the Geometry of the Conditional Density Equation," in Analysis and Optimization of Stochastic Systems, ed. D.L.R. Jacobs, Academic Press, 1980.

5. Flato, Simon, Snellman, and Sternheimer, "Simple Facts about Analytic Vectors and Integrability," Ann. Scient. Ec. Norm. Sup., 4e serie, t. 5, 1972, pp. 423-434.

6. Gihman, Skorohod, Stochastic Differential Equations, Springer-Verlag, New York, 1972.

7. Nelson, E., "Analytic Vectors," Ann. Math. Soc., 75, 1959, pp. 572-615.

8. Ocone, D., Topics in Nonlinear Filtering Theory, Ph.D. Thesis, M.I.T., June 1980.

9. Ocone, D., "Nonlinear Filtering Problems with Finite Dimensional Estimation Algebras," Proceedings of the 1980 JACC Conference, San Francisco, CA, Aug. 13-15, 1980.

10. Mitter, S.K., "Filtering Theory and Quantum Fields," to appear in Asterisque.

11. Zakai, M., "On the Optimal Filtering of Diffusion Processes," Z. Wahr. verw. Geb., 11, 1969, pp. 230-243.

RIGOROUS RESULTS ON THE CUBIC SENSOR PROBLEM

Hector J. Sussmann

Mathematics Department
Rutgers University
New Brunswick, N.J. 08903, U.S.A.

ABSTRACT

We sketch some rigorous results on the cubic sensor problem, without detailed proofs. These results, together with the work of Hazewinkel and Marcus in this volume, imply in a rigorous way the nonexistence of smooth finite dimensional filters for any nontrivial statistic of the problem.

We consider the problem of finding filters for conditional statistics of X given the observations Y, where $X = \{X_t : t \geq 0\}$ and $Y = \{Y_t : t \geq 0\}$ are real-valued stochastic processes defined on a probability space (Ω, A, P), progressively measurable with respect to an increasing family of σ-algebras $\{A_t : t \geq 0\}$, and satisfying

$$\begin{aligned} X_t &= X_0 + W_t \\ Y_t &= \int_0^t X_s^3 ds + V_t, \end{aligned} \tag{1}$$

where $\{W_t\}$, $\{V_t\}$ are standard Wiener processes with respect to the A_t, and where A_0, the W_t and the V_t are independent. We assume that (Ω, A, P) is so chosen that the X, Y, W and V processes have continuous sample paths.

We let $\mathcal{Y}_t$ denote the σ-algebra generated by the observa-

M. Hazewinkel and J. C. Willems (eds.), Stochastic Systems: The Mathematics of Filtering and Identification and Applications, 637–648.

tions Y_s, $0 \leq s \leq t$. If $\phi:\mathbb{R} \to \mathbb{R}$ is any Borel function such that $\phi(X_t)$ is integrable, we use $\widehat{\phi(X_t)}$ to denote the conditional expectation of $\phi(X_t)$ given $\mathcal{Y}_t$. A <u>family of conditional measures</u> (FCPM) for (1) is a map $m:\mathbb{R}_+\times\Omega \to \mathbb{P}(\mathbb{R})$, where $\mathbb{P}(\mathbb{R})$ is the set of Borel probability measures on the real line, such that

(i) $\omega \to m(t,\omega)$ is $\mathcal{Y}_t$-measurable (as a map with values in the Banach space $\mathrm{Bor}(\mathbb{R})$ of finite Borel measures on $\mathbb{R}$) for each $t \geq 0$, and

$$\text{(ii)} \int_{\mathbb{R}} \phi(x)dm(t,\omega)(x) = \widehat{\phi(X_t)}(\omega) \quad \text{for a.e.}\,\omega \tag{2}$$

for all bounded Borel functions $\phi:\mathbb{R} \to \mathbb{R}$.

It is easy to show that an FCPM for (1) exists, and that it is unique in the sense that, if m_1 and m_2 are FCPM, then $m_1(t,\cdot) = m_2(t,\cdot)$ a.s. for every t.

Now let Φ be any real-valued function whose domain is a subset S of $\mathbb{P}(\mathbb{R})$. Assume that

$$m(t,\omega) \in S \quad \text{for a.e.}\,\omega, \text{ for all } t > 0, \tag{3}$$

if m is an FCPM for (1). (In view of the uniqueness, whether or not (3) holds does not depend on the choice of m.) Then we can define a process $\tilde{\Phi} = \{\tilde{\Phi}_t : t \geq 0\}$ by letting

$$\tilde{\Phi}_t(\omega) = \Phi(m(t,\omega)). \tag{4}$$

Any process $\tilde{\Phi}$ obtained in this fashion will be called a <u>conditional statistic</u> for (1).

We now construct a "canonical" FCPM. Let $\mathcal{W}$ denote the space of continuous functions $w:[0,\infty) \to \mathbb{R}$ such that $w(0) = 0$, and let P_0 be Wiener measure. Let $y(\cdot) \in \mathcal{W}$. Define, formally,

$$V(x,y,t) = \frac{1}{2}\int_0^t (x+w_s)^6 ds - \int_0^t (x+w_s)^3 dy(s), \tag{5}$$

where w_s is the evaluation map $w \to w(s)$. We would like to define a measure $\mu(m_0,y,t) \in \mathrm{Bor}(\mathbb{R})$ by the condition that

$$\int_{\mathbb{R}} \phi(x)d\mu(m_0,y,t)(x) = \int_{\mathbb{R}}\int_{\mathcal{W}} \phi(x+w_t)e^{-V(x,y,t)}dP_0(w)dm_0(x). \tag{6}$$

Here m_0 is some fixed initial Borel measure on $\mathbb{R}$.

However, we cannot use (6) directly to define a measure, because it is not completely clear that V is well defined, in view of the fact that the second integral in (5) is not well defined if y is only continuous. To get around this difficulty we use the transformation (widely employed elsewhere in this volume) arising from formally evaluating the integral by parts. We get

$$V(x,y,t) = V^*(x,y,t) - y(t)(x+w_t)^3, \tag{7}$$

where

$$V^*(x,y,t) = \frac{1}{2}\int_0^t A^*(x,y,s)\,ds + \int_0^t B^*(x,y,s)\,dw_s \tag{8}$$

$$A^*(x,y,t) = (x+w_t)^6 + 6y(t)(x+w_t) \tag{9}$$

$$B^*(x,y,t) = 3(x+w_t)^2 y(t) \tag{10}$$

Therefore, if we define a measure $\mu^*(m_0,y,t)$ that satisfies

$$\int_{\mathbb{R}} \phi(x)\,d\mu^*(m_0,y,t)(x) = \int_{\mathbb{R}}\int_{W} \phi(x+w_t)e^{-V^*(x,y,t)}\,dP_0(w)\,dm_0(x), \tag{11}$$

we find, formally, that $\mu(m_0,y,t)$ is the product of $\mu^*(m_0,y,t)$ times the function

$$x \to e^{y(t)x^3}. \tag{12}$$

To prove that (11) indeed does define a measure, we let

$$U^*(x,y,t) = e^{-V^*(x,y,t)}. \tag{13}$$

Then we find

$$U^*(x,y,t) = 1 + \frac{1}{2}\int_0^t [(-A^*+B^{*2})U^*](x,y,s)\,ds - \int_0^t (B^*U^*)(x,y,s)\,dw_s. \tag{14}$$

The second integral is a local martingale, and $-A^*+B^{*2}$ is bounded above by a fixed constant K. Therefore

$$\int U^*(x,y,t)\,dP_0 \le e^{Kt}. \tag{15}$$

From now on, the letter K denotes a constant, which may vary from one formula to the next, but which can be chosen independently of x, t and y, as long as

$$||y||_t = \sup\{|y(s)| : 0 \le s \le t\} \tag{16}$$

remains bounded. It is clear that the K of (15) satisfies these conditions.

Replacing V^* by pV^* (if $1 \leq p < \infty$) yields a similar estimate

$$\int U^*(x,y,t)^p dP_0 \leq e^{Kt} \tag{17}$$

with a p-dependent K. So $U^*(x,y,t)$ is in $L^p(W,P_0)$ for $1 \leq p < \infty$, and its L^p norm is bounded by e^{Kt}, with $K = K(p)$. If q is the conjugate exponent of p, and $\phi \in L^q(\mathbb{R})$, then

$$\int |\phi(x+w_t)|^q dP_0 = (2\pi t)^{-1/2} \int_{\mathbb{R}} |\phi(x+z)|^q e^{-z^2/2t}\, dz$$

$$\leq Kt^{-1/2} \int_{\mathbb{R}} |\phi(z)|^q\, dz$$

and so the L^q norm of $\phi(x+w_t)$ is bounded by $C||\phi||_{L^q}$, where C denotes, here and in subsequent formulas, a constant that can be chosen to be independent of x, t and y, as long as $||y||_t$ remains bounded, and t is bounded away from 0. (So that C is allowed to "blow up" as $t \to 0$, whereas K is not.)

Hence

$$\left| \int_W \phi(x+w_t) e^{-V^*(x,y,t)} dP_0(w) \right| \leq Ce^{Kt} ||\phi||_{L^q} \tag{18}$$

so that the right hand side of (11) is well defined, and bounded by $Ce^{Kt}||\phi||_{L^q}|m_0|$, where $|m_0|$ is the total variation of m_0. This shows that $\mu^*(m_0,y,t)$ actually has a density $x \to u^*(m_0,y,t,x)$, with the property that, for $1 < p < \infty$

$$u^*(m_0,y,t,\cdot) \in L^p(\mathbb{R}) \tag{19}$$

and

$$||u^*(m_0,y,t,\cdot)||_{L^p} \leq Ce^{Kt}|m_0|. \tag{20}$$

Now let $0 \leq a < \frac{1}{4}$, and $t > 0$. For $0 \leq s \leq t$, let $b = \frac{a}{t}$ and

$$\tilde{U}^*(x,y,s) = e^{bs(x+w_s)^4} U^*(x,y,s). \tag{21}$$

Then

$$\tilde{U}^*(x,y,s) = 1+\frac{1}{2}\int_0^s (\tilde{A}^*\tilde{U}^*)(x,y,\sigma)d\sigma+\int_0^s (\tilde{B}^*\tilde{U}^*)(x,y,\sigma)dw_\sigma, \quad (22)$$

where

$$\tilde{A}^*=-A^*+B^{*2}+b(x+w_s)^4+12bs(x+w_s)^2+16b^2s^2(x+w_s)^6-12bs(x+w_s)^5 \quad (23)$$

and

$$\tilde{B}^* = 4bs(x+w_s)^3-B^*. \quad (24)$$

Now, $\tilde{A}^*$ is a polynomial in $x+w_s$, with coefficients that remain uniformly bounded as long as $||y||_t$, b, and bs remain bounded. The leading term is

$$(16b^2s^2-1)(x+w_s)^6,$$

which is less than $(16a^2-1)(x+w_s)^6$ (since $0 \leq s \leq t$). So $\tilde{A}^*$ is bounded above by $C = C(a)$, and then we get

$$\int_W \tilde{U}^*(x,y,t)dP_0 \leq e^{Ct}. \quad (25)$$

From this it follows that

$$\int_{\mathbb{R}} e^{ax^4}|u^*(m_0,y,t,x)|dx \leq e^{Ct}|m_0| \quad (26)$$

whenever $a < \frac{1}{4}$. (Here $C = C(a)$.)

If m_0 itself has a density u_0 which satisfies

$$|||u_0|||_a = \int_{\mathbb{R}} e^{ax^4}|u_0(x)|dx < \infty$$

for some $a < \frac{1}{4}$, then we can use the same argument with $\tilde{U}^*$ replaced by $\overline{U}^*$, where

$$\overline{U}^*(x,y,t) = e^{a(x+w_t)^4}U^*(x,y,t).$$

In this case a formula similar to (22) holds, except that now $\tilde{A}^*$ is replaced by $\overline{A}^*$, which is a polynomial like $\tilde{A}^*$, but does not contain any term with a coefficient a/t. Hence we now obtain a bound

$$\int_{\mathcal{W}} \overline{U}^*(x,y,t)dP_0 \leq e^{Kt}, \tag{27}$$

i.e. a bound such as (25), except that now the bound remains valid all the way up to $t = 0$. Then

$$|||u^*(m_0,y,t,\cdot)|||_a \leq e^{Kt}|||u_0|||_a. \tag{28}$$

One can prove much more, namely, that u^* is infinitely differentiable in x, and that all the derivatives satisfy similar bounds. The idea of the proof is as follows: to show $\frac{\partial^n u^*}{\partial x^n}$ exists, and to get an estimate for it, we must integrate $u^*(m_0,y,t,x)$ times $\phi^{(n)}(x)$, and estimate the result in terms of $||\phi||_{L^q}$. This amounts to computing the integral in the right side of (11), but with $\phi^{(n)}$ instead of ϕ. Now $\phi^{(n)}(x+w_t)$ is also "derivative of $\phi(x+w_t)$ with respect to the variable w_t", which is one of the integrated variables in the integral with respect to P_0. So it is clear, at least at the formal level, that the derivative can be shifted to the other factor $e^{-V^*}dP_0$, via "integration by parts". This can actually be done rigorously, and the result is a sequence of formulas

$$\int_{\mathbb{R}}\int_{\mathcal{W}} \phi^{(n)}(x+w_t)e^{-V^*(x,y,t)}dP_0dm_0$$
$$= \int_{\mathbb{R}}\int_{\mathcal{W}} \phi(x+w_t)D_n e^{-V^*(x,y,t)}dP_0dm_0,$$

where the D_n turn out to be stochastic integrals, which belong to L^p for $1 \leq p < \infty$, and have L^p norms bounded by $CP(x)(1+t^m)$, where $C = C_{n,p}$ and $P(x) = P_{n,p}(x)$, and the $P_{n,p}$ are polynomials in x.

From this it follows that, for $a < \frac{1}{4}$,

$$\int_{\mathbb{R}} e^{ax^4}\left|\frac{\partial^n u^*}{\partial x^n}(m_0,y,t,x)\right|dx \leq Ce^{Ct}|m_0|.$$

If m_0 has a density u_0 which has n derivatives, and satisfies

$$|||u_0|||_{a,n} = \max_{0\le k\le n} |||u_0^{(k)}|||_a < \infty,$$

then a similar reasoning yields bounds

$$|||u^*(m_0,y,t,\cdot)|||_{a',n} \le Ke^{Kt}|||u_0|||_{a,n}$$

whenever $a' < a < \frac{1}{4}$.

Let V be the space of all infinitely differentiable functions $\phi:\mathbb{R}\to\mathbb{R}$ that satisfy

$$|\phi|_{a,n} = \sup_x |\phi^{(n)}(x)e^{ax^4}| < \infty$$

for all $n \ge 0$ and all $a < \frac{1}{4}$. Then the $|\ldots|_{a,n}$ are norms, which turn V into a complete locally convex topological vector space. Then the bounds we have just obtained imply that, for $t > 0$,

$$u^*(m_0,y,t,\cdot) \in V \quad \text{for all } m_0, \tag{29}$$

$$|u^*(m_0,y,t,\cdot)|_{a,n} \le Ce^{Ct}|m_0| \tag{30}$$

(with $C = C_{a,n}$).

For $u_0 \in V$, let $\tilde{u}^*(u_0,y,t,\cdot)$ be $u^*(m_0,y,t,\cdot)$, where m_0 is the measure with density u_0. Then the map $u_0 \to \tilde{u}^*(u_0,y,t,\cdot)$ is continuous in V, uniformly in y,t as long as t, $||y||_t$ remain bounded.

If we now define

$$u(m_0,y,t,x) = e^{y(t)x^3}u^*(m_0,y,t,x),$$

$$\tilde{u}(u_0,y,t,x) = u(m_0,y,t,x) \quad \text{if} \quad dm_0 = u_0dx,$$

we find that $u(m_0,y,t,\cdot) \in V$ and that the same bounds as above hold for u.

One can write $U^*(x,y,t)-U^*(x,\tilde{y},t)$ as a stochastic integral as was done for $U^*(x,y,t)$ in (11), and use this to estimate its expectaction in terms of $||y-\tilde{y}||_t$. Finally, one obtains

THEOREM I. Let W be topologized by the family of seminorms $||\ldots||_t$. Then

(i) the map

$$(m_0,y,t) \to u(m_0,y,t,\cdot)$$

<u>from</u> $Bor(\mathbb{R})\times\mathcal{W}\times(0,\infty)$ <u>to</u> $\mathcal{V}$ <u>is continuous, and</u>

(ii) <u>the map</u>

$$(u_0,y,t) \to \tilde{u}(u_0,y,t,\cdot)$$

<u>from</u> $\mathcal{V}\times\mathcal{W}\times[0,\infty)$ <u>to</u> $\mathcal{V}$ <u>is continuous</u>.

Now return to the problem (1). Let m_0 be the probability distribution of X_0. Define $\nu(y,t)$ to be the probability measure with density

$$d\nu(y,t)(x) = \frac{u(m_0,y,t,x)dx}{\int_{\mathbb{R}} u(m_0,y,t,\xi)d\xi} . \tag{31}$$

Let $m(t,\omega) = \nu(y_\omega,t)$, where y_ω is the sample path $s \to Y_s(\omega)$. Then $m(t,\omega)$ is a probability measure for each $t > 0$, $\omega \in \Omega$. The Kallianpur-Striebel formula implies

THEOREM II. m <u>is an</u> FCPM <u>for</u> (1).

Now let $\Phi:S \to \mathbb{R}$, where $S \subseteq \mathbb{P}(\mathbb{R})$. If $\mathbb{P}(\mathbb{R}) \cap \mathcal{V} \subseteq S$ then (3) holds, and so Φ defines a conditional statistic $\tilde{\Phi}$ for (1). Consider a stochastic differential equation

$$dz = f(z)dt + g(z)dY, \; z \in M \tag{32}$$

with C^∞ coefficients f,g, driven by the semimartingale Y, together with an initial state $z_0 \in M$, and a smooth output map $h:M \to \mathbb{R}$. The state space M is a smooth manifold, and the equation is understood in the sense of Stratonovich. Eq. (32), with the initial condition $z(0) = z_0$, has a solution $t \to z(t)$, defined up to a stopping time τ. We say that (32), z_0 and h define a <u>smooth finite-dimensional filter</u> (SFDF) for $\tilde{\Phi}$ if, for each t,

$$\tilde{\Phi}_t(\omega) = h(z(t)(\omega))$$

for almost all ω such that $\tau(\omega) > t$.

Now, it is well known that the solutions of (32) can be

defined "at the level of the individual sample paths", by defining what it means for a curve $t \to \xi(t)$ to be a solution of

$$dz = f(z)dt + g(z)dy,\ z(0) = z_0 \tag{33}$$

where now y is just a continuous function with $y(0) = 0$, i.e. $y \in \mathcal{W}$. It turns out that, with this definition of "solution", the usual existence and uniqueness properties hold and, moreover, the solution depends continuously on y (with the topology given by the seminorms $||\ldots||_t$). Moreover, if we now define a stochastic process $t \to z(t)$, whose sample path for a given $\omega \in \Omega$ is the solution of (33) with $y = y_\omega$, then z is (a version of) the solution of (32) with initial condition $z(0) = z_0$. We call this z the canonical solution of (32) with i.c. $z(0) = z_0$.

Now suppose that (32), with z_0 and h, is an SFDF for $\tilde{\Phi}$. Then, for each $t > 0$

$$h(z(t)(\omega)) = \tilde{\Phi}_t'(\omega) \tag{34}$$

for almost all ω such that $\tau(\omega) > t$, where z is the canonical solution. The left side of (34) is of the form $\psi_t(y_\omega)$, where ψ is a continuous function of the path y. The right side is of the form $\Phi(\nu(y_\omega,t))$ where, again, $\nu(\cdot,t)$ is a continuous function of y. If we assume that Φ is continuous on V, then the right side of (34) is a continuous function of the path y, for each $t > 0$. Since any set of full measure in the space of y-paths is dense, the fact that (34) holds almost surely implies that it holds surely.

Now consider the control system

$$\mathring{z} = f(z) + \alpha g(z),\ z \in M \tag{35}$$

with initial condition z_0, and output h. The controls are piecewise constant functions α defined on closed intervals $[0,T]$. For any control α, and any $z' \in M$, we let $t \to \pi(\alpha,z',t)$ denote the solution of (35) with initial condition z' at $t = 0$.

If α is a control, we let $\eta(\alpha)$ be the indefinite integral of α, for which $\eta(\alpha)(0) = 0$. Let $Q(\alpha):\mathrm{Bor}(\mathbb{R}) \to V$ be the map which assigns to each $m_0 \in \mathrm{Bor}(\mathbb{R})$ the function $u(m_0,\eta(\alpha),T,\cdot)$, and let $\Psi: V_+ \to \mathbb{R}$ be given by

$$\Psi(\phi(\cdot)) = \Phi(\phi(\cdot)/\int_{\mathbb{R}} \phi(x)dx) \tag{36}$$

where V_+ is the set of all $\phi \in V$ that are nonnegative but not identically zero. The preceding considerations imply:

THEOREM III. Suppose that (32), with z_0 and h define a SFDF for $\tilde{\Phi}$, where $\Phi: S \to \mathbb{R}$, $V \cap \mathbb{P}(\mathbb{R}) \subseteq S$, and Φ is continuous on $V \cap \mathbb{P}(\mathbb{R})$. Then the equality

$$\Psi(M(\alpha)(m_0)) = h(\pi(\alpha, z_0, T)) \tag{37}$$

holds for all $T > 0$, and all piecewise constant $\alpha: [0,T] \to \mathbb{R}$ such that $\pi(\alpha, z_0, T)$ is defined.

Let $L(\sigma)$ denote, for $\sigma \in \mathbb{R}$, the operator

$$L(\sigma) = \frac{1}{2}\frac{d^2}{dx^2} - \frac{x^6}{2} + \sigma x^3. \tag{38}$$

Then each $L(\sigma)$ generates a continuous semigroup in $L^2(R)$, which we denote by $t \to e^{tL(\sigma)}$. The transformations $e^{tL(\sigma)}$ actually extend to $\mathrm{Bor}(\mathbb{R})$, and have the property that $e^{tL(\sigma)}m_0 \in V$ for every $m_0 \in V$, $t > 0$. Moreover, the semigroups $\{e^{tL(\sigma)}\}$ of transformations of V are jointly smooth, in the following sense: if $p \to v(p)$ is a family of elements of V, which depend smoothly on the scalar or vector parameter p, and if $\sigma_1, \ldots, \sigma_k \in \mathbb{R}$, then

$$(p, (t_1, \ldots, t_k)) \to e^{t_1 L(\sigma_1)} e^{t_2 L(\sigma_2)} \cdots e^{t_k L(\sigma_k)} v(p)$$

is smooth.

Now, if α is a control, let us write $\alpha = [\sigma_1, t_1; \sigma_2, t_2, \ldots, \sigma_k, t_k]$ to indicate that α equals σ_k during time t_k, then σ_{k-1} during time t_{k-1}, and so on. For a control α, the construction of $u(m_0, \eta(\alpha), t, \cdot)$ can be carried out directly by means of formula (6), which in this case makes sense rigorously. Moreover, one proves easily that

$$\begin{aligned} &u(m_0, \eta([\sigma_1, t_1, \ldots, \sigma_k, t_k]), t_1 + \ldots + t_w, \cdot) \\ &\quad = e^{t_1 L(\sigma_1)} \cdots e^{t_k L(\sigma_k)} m_0. \end{aligned} \tag{39}$$

On the other hand, we can express $\pi(\alpha, z_0, t)$ in a similar fashion, namely

$$\pi([\sigma_1,t_1;\ldots;\sigma_k,t_k],z_0,t_1+\ldots+t_k) = F(t_1,\sigma_1)\ldots F(t_k,\sigma_k)(z_0) \tag{40}$$

where $t \to F(t,\sigma)$ denotes the flow of the vector field $f+\sigma g$.

If equality (37) holds, then we can take some control α, and let $u_0 = M(\alpha)(m_0)$, $z_0' = \pi(\alpha,z_0,T)$ (if α is defined on $[0,T]$). We can then apply (37) to all controls that are the concatenation of α with some other control. This gives

$$\Psi(e^{t_1L(\sigma_1)}\ldots e^{t_kL(\sigma_k)}(u_0)) = h(F(t_1,\sigma_1)\ldots F(t_k,\sigma_k)(z_0')). \tag{41}$$

Now let Λ^* be the Lie algebra of vector fields generated by f and g, and let I be the ideal of Λ that consists of all the vector fields a such that

$$[b_1,[b_2,[\ldots[b_n,a]\ldots]]]h(z) = 0$$

for all z in $R(z_0')$, the set of points reachable from z_0', and all $b_1,\ldots,b_n$ in Λ^*. Let $\Lambda = \Lambda^*/I$. Also, let Λ_0 be the Lie algebra of differential operators on $\mathbb{R}$ generated by the $L(\sigma)$. Assume that Ψ is smooth, in the sense that $\Psi(v(p))$ is smooth in p whenever $v(\cdot)$ is a family of elements of V depending smoothly on p. Then we can define a surjective homomorphism $\lambda:\Lambda_0 \to \Lambda$ as follows. First map the free Lie algebra Γ in two generators γ,δ onto Λ, by mapping γ,δ to the classes of f, $f+g$ modulo I. Call this map ε. Similarly, define $\varepsilon_0:\Gamma \to \Lambda_0$ by $\varepsilon_0(\gamma) = L(0)$, $\varepsilon_0(\delta) = L(1)$. Under the stated assumptions, it can be proved that ε vanishes on the kernel of ε_0, and so λ exists.

Finally, it can also be proved that Λ necessarily has the following property:

(P) There is a nontrivial homomorphism from Λ into some Lie algebra of vector fields on some smooth manifolds.

If Λ has (P), then Λ_0 must have (P) as well. But Hazewinkel and Marcus prove that Λ does not have (P). So we conclude:

THEOREM. _Let_ $\Phi:S \to \mathbb{R}$, _where_ $V \cap \mathbb{P}(\mathbb{R}) \subseteq S$, _be such that the_ *corresponding* $\Psi : V_+ \to \mathbb{R}$ *is continuous and smooth, and that the conditional statistic* $\tilde{\Phi}$ *is not a constant. Then* $\tilde{\Phi}$ *cannot be represented as the output of a SFDF.*

The details of the proofs will appear elsewhere.

A FOOTNOTE TO THE PAPERS WHICH PROVE THE NONEXISTENCE OF FINITE DIMENSIONAL FILTERS

Moshe Zakai

Technion, Israel Institute of Technology
Haifa, Israel.

Let the signal be defined by $dx_t = m(x_t)dt + \sigma(x_t)dw_t$ and observed via $dy_t = h(x_t)dt + d\nu_t$ where w and ν are independent Brownian motions. The filtering problem is the problem of determining the conditional distribution of x_t conditioned on $\{y_\theta, 0\leqslant\theta\leqslant t\}$ and a basic question is that of existence of finite dimensional filters, i.e., when can the filtering problem be solved by a set of finite dimensional stochastic differential equations driven by the observation y_t. Following the results on the nonexistence of finite dimensional filters, the question arises whether there exists a set of finite dimensional equations driven by y_t which determines the conditional moment $\hat{x}_t = E(x_t|y_\theta, 0\leqslant\theta\leqslant t)$ or $\hat{h}_t = E(h(x_t)|y_\theta, 0\leqslant\theta\leqslant t)$. The purpose of this note is to point out a related result of J. Hammer (to be published) which states that if $\hat{h}_\theta$ is known in the interval $(t-\varepsilon)<\theta\leqslant t$ for some arbitrary $\varepsilon<0$, then it determines $(h^n_\alpha)\hat{} = E\{h^n(x_\alpha)|y_\theta, 0\leqslant\theta\leqslant\alpha\}$ for all $n = 2,3,\ldots$ and all α in $(t-\varepsilon,t]$; hence under some assumptions on $h(x)$, $\{\hat{h}_\theta,(t-\varepsilon)\leqslant\theta\leqslant t\}$ determines the conditional distribution of x_t. In other words, if a finite dimensional set of equations for $\hat{h}_t$ exists, it can be used to derive the conditional probability. It should be pointed out, however, that the possibility of the existence of finite dimensional equations for the conditional moments <u>does not</u> contradict the results of the non-existence of finite dimensional filters since the procedure for determining the conditional distribution of x_t from $\{\hat{h}_\theta, t-\varepsilon\leqslant\theta\leqslant t\}$ is not finite.

A simplified version of the result of J. Hammer will now be stated and proved for more detailed results we refer the reader to Hammer's forthcoming paper. <u>Assume that</u> h_θ <u>is known for all</u> θ <u>in the interval</u> $(t-\varepsilon,t]$, $\varepsilon>0$ <u>and that</u> $(h^n_t)\hat{}$, $n = 1,2,\ldots$

M. Hazewinkel and J. C. Willems (eds.), Stochastic Systems: The Mathematics of Filtering and Identification and Applications, 649–650.

<u>satisfies the Kushner equation for the conditional moments for all n, then</u> $\{\hat{h}_\theta, (t-\varepsilon) \leqslant \theta \leqslant t\}$ <u>determines</u> $(h_t^n)^\wedge$ <u>for all</u> $n \geqslant 0$.

<u>Proof:</u>
By the moment equation $\hat{h}_t$ has the representation

$$d(h_t^n)^\wedge = a_\theta^{(n)} d\theta + ((h_\theta^{n+1})^\wedge - (h_\theta^n)^\wedge \hat{h}_\theta) d\bar{\nu}_\theta$$

where $\bar{\nu}_t$ is the innovation process which is a standard Brownian motion and the explicit form of $a_\theta^{(n)}$ is of no interest to us here. The quadratic variation of $\hat{h}_\theta$ over $((t-\varepsilon),t]$ is therefore given by

$$\int_{t-\varepsilon}^{\theta} ((h_\alpha^2)^\wedge - (\hat{h}_\alpha)^2)^2 d\alpha$$

and this determines $(h_\theta^2)^\wedge$ without ambiguity since, by Jensen's inequality $(h^2)^\wedge \geqslant (\hat{h})^2$. Assume now that $\hat{h}_\theta$, $(h_\theta^2)^\wedge$, $(h_\theta^n)^\wedge$ are known for all θ in $((t-\varepsilon),t]$ and for some $n \geqslant 2$, then it follows from the equation for conditional moments that the cross-quadratic variation of $(h_\theta^n)^\wedge$ with $\hat{h}_\theta$ is given by

$$\int_{t-\varepsilon}^{\theta} ((h_\alpha^{(n+1)})^\wedge - (h_\alpha^n)^\wedge \cdot \hat{h}_\alpha)((h_\alpha^2)^\wedge - (\hat{h}_\alpha)^2) d\alpha$$

and since $(h^n)^\wedge$, $(h^2)^\wedge$, $\hat{h}$ are assumed to be known, this cross-quadratic variation determines $(h^{(n+1)})^\wedge$. This proves that $(h_t^n)^\wedge$ can be evaluated recursively for all integers.

<u>Remark</u>
Instead of requiring that $\{\hat{h}_\theta, (t-\varepsilon) \leqslant \theta \leqslant t\}$ be known we may require that for some <u>odd</u> $n_o > 1$, $(h_\theta^{n_o})^\wedge$ and $(h_\theta^{(n_o+1)})^\wedge$ are known in the same interval $((t-\varepsilon),t]$. The proof is as follows, the quadratic variation of $(h^{n_o})^\wedge$ is

$$\int_{t-\varepsilon}^{\theta} \left(\left(h_\alpha^{(n_o+1)}\right)^\wedge - (h_\alpha^n)^\wedge \cdot \hat{h}_\alpha\right)^2 d\alpha,$$

since $(h^{n_o+1})^\wedge$ and $(h^{n_o})^\wedge$ are known, this quadratic variation determines $\hat{h}_\alpha$ and the determination of $\hat{h}_\theta$ is unambiguous since for n_o odd,

$$\left(h_\theta^{(n_o+1)}\right)^\wedge \geqslant \left[\left(h_\theta^{n_o}\right)^\wedge\right]^{(n_o+1)/n_o} \geqslant \left(h_\theta^{n_o}\right)^\wedge \cdot \hat{h}_\theta \;.$$

List of other contributed papers

Besides the tutorials, the invited lecture series and the contributed papers which are included in this volume there were quite a number of other meritorious contributed papers. We would have liked to include virtually all of them but considerations of space and immediate relevance to the main themes of the conference precluded this.

Below there is a list of these contributions. Most exist at least in preprint form. There are especially a number of most interesting papers on applications in geology and biology. We have added the addresses of the authors for the convenience of the reader who wishes to secure these preprints.

D. Aeyels (Lab. Theor. Electriciteit, Rijks Univ. Gent, Sint-Pietersnieuwstr. 41, 9000 Gent, Belgium), The genericity problem of observed vector fields.

A.C. Antoulas (Math. Syst. Th., ETH Zürich, 8092 Zürich, Switzerland), Factorization theory for nonsingular polynomial matrices.

J. Baillieul (Scientific Systems Inc., Suite 309-310, 186, Alewife Brook Parkway, Cambridge, Mass 02138, USA), Chaotic motion in nonlinear feedback systems.

J. Baras (Dept. EE, Univ. of Maryland, College Park, MD 20742, USA), Nonlinear filtering of diffusion processes: a generic example.

C. Byrnes (Dept. Appl. Math., Harvard Univ., Pierce Hall, Cambridge, Mass 02138, USA), Oscillations in nonlinear closed loop feedback systems.

N. Christopeit, K. Helmes (Inst. f. angew. Math., Univ. Bonn, Wegelerstrasse 10, 5300 Bonn, BRD), Discrete dynamic programming approach to the Benes problem.

P. Collingwood (Dept. Engr, Univ. of Warwick, Coventry CV4 7AL, England), Some remarks on the Lie algebra generated by the conditional density equation associated with the nonlinear filter problem.

M. Delfour (Centre de Recherche Appl., Univ. de Montréal, Case Postale 6128, Succ. A, Montréal H3C3J7, Canada), Abstract Riccati equations arising from filter problems for PDE's and infinite dimensional systems.

H. Dym (Weizman Inst. of Science, Rehovot, Israel), On an extension problem, generalized Fourier analysis and an entropy formula.

Sj.D. Flåm (Chr. Michelsen Inst., Fantoftvegen 38, N 5036 Fantoft, Bergen, Norway), On steady states in stochastic systems: stochastic models of management of renewable resources.

J. Franke (Fachbereich Math., Univ. Frankfurt, Robert-Mayer-Str. 6-10, Frankfuurt D-6000, BRD), A general approach to robust prediction of discrete time series.

A. Gerardi, F. Marchetti (Ist. Mat. "G. Castelnuovo", Univ. di Roma, Cittá Univ., 00100 Roma, Italy), Lateral diffusion on cells and eigenvalue estimates.

B. Hanzon (Dept. Math., Erasmus Univ. Rotterdam, P.O.Box 1738, 3000 DR Rotterdam, The Netherlands), The number of consecutive observations needed for ARMA model identification.

S. Haykin (Comm. Research Lab., McMaster Univ., 1280 Main Str. West, Hamilton, Ontario L8S5L7, Canada), The adaptive lattice filter.

O. Hijab (Dept. Appl. Math., Harvard Univ., Pierce Hall, Cambridge Mass 02138, USA), Nonlinear filtering and the WKB approximation.

M. Kohlmann (Inst. f. angew. Math., Univ. Bonn, Wegelerstrasse 10, D-5300 Bonn, BRD), On the existence of optimal partially observed controls.

A.J. Krener (Dept. Math., Univ. of California at Davis, Calif 95616, USA), The deterministic and stochastic realization problem for acausal linear systems.

M. Lamnabhi (Lab. Signaux et systèmes, E.S.E., Plateau du Molon, 91190, Gif-sur-Yvette, France), Noncommuting variables and symbolic computation in statistical physics.

U.E. Makov (Dept. Math., Chelsea College, Univ. of London, Manresa Rd, London, SW 36LX England), Approximated filters for multiple tracking.

S.I. Marcus (Dept. Electrical Engr., Univ. of Texas at Austin, Texas 78712, USA), Modeling and approximation of stochastic differential equations by semimartingales.

A. Van der Schaft (Math. Inst., Univ. of Groningen, P.O.Box 800, 9700 AV Groningen, The Netherlands), Hamiltonian dynamics with external forces and observations.

Bj. Ursin (Div. Petroleum engineering and applied geophysics, Norwegian Inst. of Techn., Trondheim, Norway), Inverse methods for wave propagation in layered media.

Bj. Ursin, Wave propagation in layered media.

M. West (Dept, Math., Univ. of Nottingham, Univ. Park, Nottingham NG7 2RD England), A Bayesian approach to nonlinear filtering.

D.E. Williams (Royal Aircraft establishment, Space Dept., Farnborough, Hants, GUI 46TD, England), A geometric derivation of the Kalman Filter equations.

A. Winter (Geological survey of Denmark, Thoravej 31, 2400 Kobenhavn, Denmark), Estimation of subsurface properties of the earth with applications to oil recovery.

List of participants (including invited speakers)

D. Aeyels (Gent)
A.C. Antoulas (ETH, Zürich)
L. Arnold (Bremen)
R. Astier(Palaiseau)
J. Baillieul (S^2I, Cambridge)
J. Baras (College Park)
S.G. Bengtsson (ASEA, Västerås)
K. Bennaceur (Valbonne)
S. Bittante (Milano)
B. Bobrovsky (Tel Aviv)
R. Boel (Gent)
A. Bressan (Padova)
R.W. Brockett(Harvard)
C.I. Byrnes (Harvard)
P.E. Caines (Harvard)
P. Collingwood (Warwick)
C. Costantini (Rome)
N. Cristopeit (Bonn)
G. Crosta (Varese)
R.F. Curtain (Groningen)
R.W.R. Darling (Warwick)
M.H.A. Davis (Imperial College, London)
M. Deistler (TU, Vienna)
W. De Koning (Delft)
M. Delfour (Montréal)
P. Dewilde (Delft)
G.B. Di Masi (CNR, Padova)
C. Duhamel (Paris-Sud)
H. Dym (Weizman Inst.)
L. Finesso (Padova)
Sj.D. Flåm (C. Michelsen Inst, Fantoft)
D. Florens (Montpellier)
J. Franke (Frankfurt)
J.-J.J. Fuchs (IRISA, Rennes)
M. Fliess (Gif-sur-Yvette)
M. Fochler (Nürnberg)
A. Gerardi (Rome)
A. Ginensky (Chicago)
D. Guegan (Paris-Sud)
H.Ö. Gülcür (Ankara)
E.J. Hannan (Canberra)
B. Hanzon (Rotterdam)
H. Hauptmann (Hamburg)
S. Haykin (Hamilton, Ontario)
M. Hazewinkel (Rotterdam)
K. Helmes (Bonn)
O. Hijab (Berkeley)
C. Jacob (Jouy-en-Josas)
H. Jespersen (Lyngby)
A. Johnson (Delft)
H.D. Joos (DFVLR, Wessling)
T. Kailath (Stanford)
M. Kohlmann (Bonn)
A.J. Krener (Davis)
P.S. Krishnaprasad (College Park, MD)
M. Lamnabhi (Gif-sur-Yvette)
Y.D. Landau (Grenoble)
A.A. Lazar (Princeton)
F. Le Gland (INRIA, Le Chesnay)
A. Lindquist (Lexington, KY)
L. Ljung (Linköping)
U.E. Makov (Chelsea College, London)
F. Marchetti (Rome)
S.I. Marcus (Austin, TX)
C. Martin (Cleveland)
D. Michel (Paris VI)
S.K. Mitter (MIT)
P. Murino (Napoli)
M.M. Newman (Belfast)
D. Ocone (MIT)
E. Özhan (Ankara)
E. Pardoux (Marseille)
G. Picci (CNR, Padova)
M. Piccione (Rome)
H.N.J. Poulisse (Nijmegen)
F. Perrot (Gif-sur-Yvette)
N.S. Papageorgiou (Harvard)
A.P. Roberts (Belfast)
W. Roth (ETH, Zürich)
W. Runggaldier (CNR, Padova)
B. Rickmann (MSDS, Stanmore)
J.M.G. Sa da Costa (Manchester)
M. Scheutzow (Kaiserslautern)
H.J. Sussmann (Rutgers Univ.)
Bj. Ursin (Trondheim)
R.V. Valqui Vidal (Lyngby)
A.J. Van der Schaft (Groningen)
J.A. Van Gelderen (Delft)

J. Van Schuppen (MC, Amsterdam)
A.C.M. Van Swieten (Hilversum)
A.C.F. Vorst (Rotterdam)
M. West (Nottingham)
V. Wihstutz (Bremen)
J.C. Willems (Groningen)
D.E. Williams (Farnborough)
A. Winter (Geological Survey, Kopenhagen)
M. Zakai (Technion, Haifa)

SUBJECT INDEX